# TABLE OF CONTENTS (Con't.)

## CHAPTER 14. PRODUCTION TECHNIQUES

## CHAPTER 15. PACKING, STORING, AND SHIPPING PROCEDURES

# TABLE OF CONTENTS (Con't.)

## CHAPTER 16. COMPONENT DESIGN

# TABLE OF CONTENTS (Con't.)

## TABLE OF CONTENTS (Con't.)

# TABLE OF CONTENTS (Con't.)

# TABLE OF CONTENTS (Con't.)

# TABLE OF CONTENTS (Con't.)

# TABLE OF CONTENTS (Con't.)

# TABLE OF CONTENTS (Con't.)

## CHAPTER 27. NUCLEAR DECAY TIMERS

# LIST OF ILLUSTRATIONS

## LIST OF ILLUSTRATIONS (Con't.)

## LIST OF ILLUSTRATIONS (Con't.)

## LIST OF ILLUSTRATIONS (Con't.)

## LIST OF ILLUSTRATIONS (Con't.)

## LIST OF ILLUSTRATIONS (Con't.)

## LIST OF ILLUSTRATIONS (Con't.)

# LIST OF ILLUSTRATIONS (Con't.)

## LIST OF ILLUSTRATIONS (Con't.)

## LIST OF ILLUSTRATIONS (Con't.)

## LIST OF TABLES

# LIST OF TABLES (Con't.)

## LIST OF TABLES (Con't.)

## LIST OF SYMBOLS*

*Units and subscripts are defined where used

| | | |
|---|---|---|
| $A$ | = | area |
| $a$ | = | acceleration |
| AC | = | alternating current |
| $BR$ | = | beat rate |
| $b$ | = | width |
| $C$ | = | capacitance |
| $C$ | = | Coulomb capacity |
| $C$ | = | viscous damping coefficient |
| $c$ | = | velocity of light |
| $d$ | = | diameter |
| DC | = | direct current |
| $E$ | = | equivalent weight |
| $E$ | = | rebound coefficient |
| $E$ | = | Young's modulus of elasticity |
| $F$ | = | Faraday's constant |
| $F$ | = | force |
| $f$ | = | frequency |
| $G$ | = | gain |
| $G$ | = | shear modulus of elasticity |
| $G$ | = | torque |

| | | |
|---|---|---|
| $g$ | = | acceleration due to gravity |
| $H$ | = | energy per cycle |
| $H$ | = | power |
| $h$ | = | thickness |
| $I$ | = | current |
| $I$ | = | moment of inertia |
| $i$ | = | gear velocity ratio |
| $J$ | = | ratio of moments of inertia |
| $K$ | = | time magnification |
| $k$ | = | proportionality constant |
| $k$ | = | spring constant |
| $L$ | = | inductance |
| $L$ | = | latitude |
| $Lo$ | = | longitude |
| $\ell$ | = | length |
| $M$ | = | constant |
| $m$ | = | mass |
| $N$ | = | number of gear teeth |
| $N$ | = | number of turns in a spring |
| $n$ | = | factor |
| $n$ | = | number of cycles |
| $n$ | = | span of pallet |
| $P$ | = | pressure |
| $p$ | = | number of poles |

$Q$ = flow

Q = quality factor

$R$ = recovery

$R$ = resistance

$R$ = sliding friction coefficient

$R$ = universal gas constant

$R_e$ = Reynolds number

$r$ = burning rate

$r$ = radius

$S$ = length-to-diameter ratio

$s$ = distance

$T$ = period

$T$ = temperature

$T$ = tension

$t$ = time

$V$ = voltage

$V$ = volume

$v$ = velocity

$W$ = weight

$w$ = flow rate

$X$ = forcing function

$Z$ = impedance

$\alpha$ = escapement angle

$\alpha$ = temperature coefficient of the modulus of elasticity

| | | |
|---|---|---|
| $\beta$ | = | current gain |
| $\beta$ | = | temperature coefficient of frequency |
| $\gamma$ | = | temperature coefficient of linear expansion |
| $\Delta$ | = | deflection |
| $\delta$ | = | change in voltage ratio |
| $\epsilon$ | = | gear ratio parameter |
| $\eta$ | = | gear train efficiency |
| $\eta$ | = | viscosity |
| $\theta$ | = | angle |
| $\theta$ | = | angular displacement |
| $\lambda$ | = | length ratio |
| $\mu$ | = | coefficient of friction |
| $\nu$ | = | kinematic viscosity |
| $\rho$ | = | density |
| $\sigma$ | = | stress |
| $\sigma$ | = | total pallet travel angle |
| $\tau$ | = | pitch |
| $\phi$ | = | angle through which gear turns |
| $\phi$ | = | escape wheel drop |
| $\phi$ | = | flux |
| $\chi$ | = | gear train ratio |
| $\chi$ | = | pallet and tooth lead |
| $\Omega$ | = | precessional angular velocity |
| $\omega$ | = | angular velocity |

## PREFACE

The Engineering Design Handbooks of the US Army Materiel Command have evolved over a number of years for the purpose of making readily available basic information, technical data, and practical guides for the development of military equipment. While aimed primarily at US Army materiel, the handbooks serve as authoritative references for needs of other branches of the Armed Services as well. The present handbook covers timing systems and components.

This handbook presents both theoretical and practical data pertaining to design methods and procedures for timing systems and devices. The subjects covered are precision reference timers, electronic timers, mechanical timers, pyrotechnic timers, flueric timers, and a few others.

Prepared as an aid to military designers, this handbook should also be of benefit to scientists and engineers engaged in other related research and development programs or who have the responsibility for the planning and interpretation of experiments and tests concerning the performance of materiel related to timers.

The handbook was prepared by The Franklin Institute Research Laboratories, Philadelphia, Pennsylvania. It was written for the Engineering Handbook Office of Duke University, prime contractor to the US Army Materiel Command. Its preparation was under the technical guidance and coordination of a special committee with representation from various agencies of the US Army Materiel Command.

TIMING SYSTEMS AND COMPONENTS

INTRODUCTION

CHAPTER 1

THE NATURE OF TIMING SYSTEMS*

A timer is a programming device; its purpose is to control the time interval between an input signal and an output event or events†. There are four essential components in all timers: (1) a start system that initiates the programming action, (2) a power supply that sustains the timing action, (3) a time base or regulator, and (4) an output system that performs the required operation at the end of the desired time interval.

## 1-1 PURPOSES AND FACTORS AFFECTING USE

In selecting the components of timing systems, the designer must first determine the purpose for which the system is to be used and the factors influencing the selection of components. Some of the factors to be considered in the choice of the basic mechanism are:

(1) Time Range. What timing intervals, time delays, sequencing, and programming are to be provided?

(2) Time Variation. Is the system to be designed to provide a fixed time interval, or is it to be variable? Is the time period to be adjustable locally or remotely; manually or automatically? What is the range of adjustment which will be required?

(3) Reliability. What reliability level is required?

(4) Accuracy. What timing accuracy or what repeatability is required?

(5) Safety. Must the timer be fail safe? Must it reset after interruption?

(6) Power Source. Will the timer be powered by a spring, g-weights, battery, AC mains, barometric pressure change, or other means?

(7) Input Signal. What is the input signal?

(8) Output Signal. What is the output signal? It is a mechanical motion; electronic, digital, or analog?

(9) Environment. What are the environmental extremes to which the timer will be exposed, and in which it must operate?

(10) Cost. Will the cost of the timer be compatible with the cost of the total system?

(11) Maintenance. Is maintenance or repair required? If so, how can it be facilitated?

## 1-2 TIMER TYPES

The types of timer discussed in this handbook are listed in Table 1-1. When discussing the various designs, timers are classified into one of the types shown depending upon the method used to generate the time base.

---

*Principal contributors to this handbook were Gunther Cohn, Joseph F. Heffron, Paul F. Mohrbach, Daniel J. Mullen, Raymond R. Raksnis, Melvin R. Smith, Ramie H. Thompson, and Robert F. Wood.

†Distinct terms for timing systems and components are defined in the Glossary.

**TABLE 1-1**

**TIMERS DISCUSSED IN HANDBOOK**

| Timer Type | Handbook Organization | | |
|---|---|---|---|
| | Part | Chapters | Introductory Chapter |
| Precision Reference Timers | One | 2–5 | 2 |
| Electronic Timers | Two | 6–11 | 6 |
| Mechanical Timers | Three | 12–16 | 12 |
| Pyrotechnic Timers | Four | 17–21 | 17 |
| Flueric Timers | Five | 22–25 | 22 |
| Miscellaneous Timing Devices (Electro-mechanical Timers, Nuclear Time Base Generators) | Six | 26, 27 | |

Each part (One to Five) in this handbook begins with an introductory chapter that contains detailed definitions pertaining to the particular timer type, lists of advantages and disadvantages, and discussions of specific military applications, requirements, and auxiliary

equipment used. This same information is contained in the two chapters comprising Part Six. Therefore, this general introductory material is not included in this chapter. Succeeding chapters in each Part discuss system design considerations, production techniques, packaging, storing, and shipping procedures, and component design details to the extent that those topics apply to the particular timer type.

## 1-3 RELATION OF ACCURACY, OUTPUT, POWER, AND COST

General timer characteristics are listed in Table 1-2. As a general rule, there is a direct relationship between the accuracy of a timing device and its output power and cost. Those timing devices that are most accurate, such as the quartz crystal controlled units and the cesium beam standards, are likely to have the least output power and to be the highest in cost. Those timers that have a lower order of accuracy, such as the pyrotechnic delays and the untuned-escapement mechanical timers, are likely to provide more output power and to be lower in cost.

**TABLE 1-2**

**GENERAL CHARACTERISTICS OF TIMERS**

| Features | Precision Reference | Electronic | Mechanical | Pyrotechnic | Flueric | Electro-chemical |
|---|---|---|---|---|---|---|
| Input to start | Voltage pulse | Voltage | Voltage or mechanical | Voltage, flame, or firing pin | Fluid pressure | Voltage, chemical release |
| Time base | Crystal or atomic | Oscillator | Escapement, motor, tuning fork | Pyrotechnic burning rate | Oscillator | Rate of chemical reaction |
| Time range | $10^{-9}$ sec to years | $10^{-3}$ to $10^3$ sec | Seconds to days | $10^{-3}$ to $10^3$ sec | 1 to $10^3$ sec | Minutes to days |
| Accuracy | 1 part in $10^6$ to 1 part in $10^{12}$ | ±0.1% | ±5% to 1 part in $10^6$ | ±10% | ±1% | ±4-10% |
| Output | Voltage pulse or time interval | Voltage | Mechanical | Flame | Fluid pressure, voltage | Chemical reaction, voltage, chemical |

## 1-4 MILITARY APPLICATIONS

All of the timing systems and components discussed in this handbook are of interest to the military. Many of the timers are components of ammunition, fuzing, or control devices. Particularly in fuzes, where delays are crucial to safe and effective performance, timers are almost always present. Timers and time delay devices for various fuze types are covered in different chapters, depending on the type of timer used. Timers for most safing and arming devices, being mechanical, are discussed in Chapter 13. If information is sought about a particular fuze in this handbook, it is best located through the Index.

Since this is a timer handbook, fuzes and fuze explosive components are discussed herein only to the extent that knowledge of their operation aids the design of timing systems and components. For design details of fuzes and their components, see the following references:

(1) AMCP 706-179, Engineering Design Handbook, *Explosive Trains*.

(2) AMCP 706-210, Engineering Design Handbook, *Fuzes*.

(3) MIL-HDBK-137, *Fuze Catalog*, Department of Defense, 20 February 1970.
  Vol. 1, *Current Fuzes (U)* (Confidential report).
  Vol. 2, *Obsolete and Terminated Fuzes*.
  Vol. 3, *Fuze Explosive Components (U)* (Confidential report).

PART ONE – PRECISION REFERENCE TIMERS

CHAPTER 2

INTRODUCTION TO PRECISION REFERENCE TIMERS

## 2-1 TIME STANDARDS

Timekeeping has two distinct aspects: determination of date (or epoch) and determination of interval. Date is concerned with when an event occurred whereas interval is some fixed multiple or fraction of the unit of time and is independent of a starting point[1]*. For measurement purposes, an accurate time scale is set up so that, from a chosen origin point, a constant unit is laid off until the resulting scale extends over the interval of interest.

A system of time measurement requires regularly occurring, uniform, periodic phenomena as a reference base. Since man's earliest concern with time was linked with the passing of the days and nights, a system of time measurement evolved that was based on rotation of the earth. The rate of the rotation is determined by measuring the motion of some point on the earth's surface with respect to some celestial object or position. It was established that all time measurements based on the rotation of the earth are subject to nonuniformity due to periodic and irregular variations in the speed of rotation. These variations are detectable and compensatable, but the corrections, of necessity, must be made after the occurrence of the event. Time systems of this type are now known as rotational or nonuniform time whereas systems of measurement that are independent of the earthly day are called uniform time. A number of the time systems or standards of

*Superscript numbers refer to References listed at the end of each chapter.

time which have been developed are described in the paragraphs that follow.

## 2-1.1 APPARENT SOLAR TIME

The apparent solar day is the interval of time between two successive lower transits of the sun's center over the same meridian, where a meridian is defined as a great circle passing through a given point and the two poles. Hence, a meridian is the intersection of a plane, through the poles and the given point, with the surface of the earth[2]. A system has been established of 24 standard meridians originating in Greenwich, England, and spaced every 15 deg about the surface of the earth. This system is used for navigational purposes and for time standardization. A lower transit occurs at apparent midnight. However, due in part to the fact that the earth's orbit is elliptical rather than circular and the orbital plane does not coincide with the plane of the equator, the apparent solar days vary in length.

## 2-1.2 MEAN SOLAR TIME

The system of mean solar time was devised to overcome the problem of variable day length. It is, however, actually based on sidereal time. Each day is of the same length, this length being equal to the average length of all the days in a solar year. The effect is the same as if the earth's orbit were circular and in the same plane as the equator. There is no means of observing mean solar time directly; nor can the apparent solar time be determined directly with adequate accuracy. In practice sidereal time is observed, and mean and apparent solar times are calculated. The difference between apparent solar time, and mean solar time varies approximately ± 16 min in the course of a year. This difference is known as the equation of time. It should be noted that, while mean solar days are of uniform length, they are approximately 4 min longer than the period of the earth's rotation[3] with respect to a fixed star.

## 2-1.3 SIDEREAL TIME

The system of sidereal time avoids some of the problems caused by the earth revolving around the sun, seasonal and other changes, and provides a convenient method of locating celestial bodies. A sidereal day is defined as the interval between two successive upper transits of the vernal equinox over the same meridian[4]. The vernal equinox is formed by the intersection of the plane of the ecliptic with the plane of the earth's equator. This intersection is also called the "First Point of Aries" and is a fundamental reference point for locating celestial bodies[1]. However, due to the complex motion of the earth's poles, the vernal equinox does not remain fixed and consequently the sidereal day is not only somewhat shorter than the period of the earth's rotation but is also of variable length. These effects are small, and suitable corrections are possible.

## 2-1.4 UNIVERSAL TIME

Universal Time (UT) closely approximates mean solar time but in practice is derived from sidereal time which, in turn, is determined from the meridian transits of selected stars. Universal time is, therefore, a form of rotational time and is subject to the attendant irregularities. Over the years there have been improvements in detecting and predicting the variations in the earth's rotation, and universal time has been revised accordingly. Consequently there are now several subdivisions of this system[3]:

(1) $UT_0$ – Universal time calculated directly from observed sidereal time. This measure of universal time contains irregularities due to polar motions as well as rotational variations.

(2) $UT_1$ – Universal time derived by corrections of $UT_0$ for observed polar motion. However, irregularities due to variations in rotation remain.

(3) $UT_2$ – Universal time derived by correcting $UT_0$ for both observed polar motions and extrapolated seasonal variation in the rate of the earth's rotation. This measure of universal time is virtually free of nonuniformity due to periodic changes in rotation but irregularities caused by irregular variation in rotation are not wholly corrected.

(4) UTC – Internationally Coordinated Universal Time (UTC) approximates $UT_2$ and is used as the basis for all civil time keeping. As instituted on 1 Jan 1972, coordinated universal time proceeds at the same rate as atomic time (see par. 2-1.6) and differs from the latter by an exact multiple of one atomic second. Step changes in UTC of precisely one second (leap second) are introduced on 1 Jan and/or 1 July when required to keep the difference between UTC and $UT_2$ to less than ± 0.7 sec. The time signal emitted by the HF standard time and frequency transmitters (such as WWV) are coded to permit the users of these signals to determine the difference to UTC-UT, to within 0.1 sec.

## 2-1.5 EPHEMERIS TIME

Ephemeris time is based on the revolution of the earth around the sun. It is obtained in practice from observations of the motion of the moon about the earth. In October 1956, the International Committee of Weights and Measures defined the second of ephemeris time as the fraction 1/31,556,925.9747 of the tropical year for January 0, 1900 at 12 hours ephemeris time (January 0, 1900 = December 31, 1899)[1]. Since the unit of ephemeris time is thus of constant length by definition, ephemeris time is a uniform time scale.

## 2-1.6 ATOMIC TIME

When an electron makes a transition from one energy state to another in an atom, it absorbs or emits energy. The amount of energy absorbed or emitted is equal to the difference in the energy of the two states. If the absorption or emission is in the form of electromagnetic energy (photons), its frequency is proportional to the change in electron energy. This means that an electron making a given transition in a given atom emits or absorbs a definite amount of energy at a specific frequency. It is this particular characteristic of atoms that serves as the basis for atomic clocks[4]. The most extensively developed atomic oscillator currently used as a time standard is the cesium beam resonator. This device utilizes a specific transition of the cesium atom. The frequency of oscillation has been *established*, in 1956, as 9,192,631,770 cycles per ephemeris second. Atomic time is the time based on this transition. The internationally coordinated atomic time scale is designated IAT. One second in the international systems of units has been *defined*, in 1967, as the duration of 9,192,631,770 cycles of this specific transition of the cesium atom. The IAT second is thus equal to the ephemeris second to within the errors in the determination of the latter. In the course of observations extending over many years, no difference in the rates of the atomic vs the ephemeris time scales has yet been found. Except for a (constant) difference in the origins of these two time scales, atomic time and ephemeris time are thus synonymous. From a practical standpoint, however, atomic time has all but replaced ephemeris time because of its vastly superior accessibility.

## 2-1.7 U S STANDARD TIME

U S Standard Time differs from the internationally coordinated universal time by an

integral number of hours. The Master clock of the U S Naval Observatory, Washington, D.C., determines Standard Time for the United States. The master clock consists of an atomic resonator, a quartz crystal oscillator, and a clock movement. The Naval Observatory determines universal time and ephemeris time from astronomical observations, and publishes data that enable one to obtain the different kinds of time used in geodesy, navigation, and scientific work. The various clocks and frequency standards used in generation of signals for precision timers are discussed in par. 3-1.

## 2-2 FREQUENCY STANDARDS

Time standards and frequency standards are based upon dual aspects of the same phenomenon[5]. The reciprocal of time interval is frequency. As a practical matter, a standard of frequency can serve as the basis for time measurement. To avoid errors, when a frequency standard is used to maintain time (either interval or date) care must be taken to reference the frequency to the time scale of interest. If frequency is quoted in the units "hertz", the corresponding time interval will be in seconds, the unit of time in the international system of measure. The present international standard is based on the transition of the cesium (Cs) atom as discussed in par. 2-1.6. This Cs transition is also the fundamental reference for a frequency measurement. A Cs frequency standard consists basically of a quartz crystal oscillator, a synthesizer that translates the crystal frequency to the Cs frequency, a Cs beam tube, and a servo feedback loop that adjusts the crystal frequency so that the synthesizer output is always at the resonance frequency of the Cs atoms. If the standard is operating on the atomic time scale, one million cycles of the 1 MHz signal describe one second.

It is not necessary to have physical access to an atomic frequency standard to obtain a reference for accurate frequency measurements. A local crystal oscillator can be controlled or monitored by suitable standard frequency and time signal emissions, provided these emissions are controlled by atomic clocks at the transmitter.

A local frequency standard can be maintained to within one part in $10^{10}$ or better by comparison of its relative phase or time difference to that of a received very low frequency (VLF) or high frequency (HF) carrier[4]. Any one of a number of monitoring systems may be chosen to make this comparison possible, depending on the degree of precision required for the relative phase measurement. For the greatest precision, the local standard must have a low drift that is predictable to within a few parts in $10^{10}$ over several days[1]. Averaging the time from days to months may be required to obtain very high accuracies.

Exceptions are the transmissions from the Loran-C system (see par. 2-4.1(2)) if received within about 1500 miles from the transmitter. It is possible to transfer the stability of the master clock that controls the transmission to the local oscillator to within $1 \times 10^{-11}$ on a continuous basis, and averaging times are in the order of 100 sec. Anyone receiving these signals, which are controlled by the U S Naval Observatory frequency and time standards, has access to one of the most precise frequency standards available. (Previous frequency offsets — 3 parts in $10^8$ from a nominal value of 100 kHz — in all standard transmissions were abolished on 1 Jan. 1972.)

## 2-3 DISCUSSION OF REQUIRED PARAMETERS

Several parameters are used to compare the performance qualities of precision reference timers. Any discussion of this subject requires a basic understanding of the following terms[6]:

(1) Accuracy. The degree to which the frequency of an oscillator is the same as the frequency of an accepted primary standard or the degree to which the frequency of an oscillator corresponds to the accepted definition of frequency. The particular reference standard or the particular definition of frequency must be included in the statement of accuracy.

(2) Reproducibility. The degree to which an oscillator of a given type will produce the same frequency from unit to unit and from one occasion of operation to another. In general, the degree to which the frequency of an oscillator may be set by a calibration procedure is included within this definition of reproducibility.

(3) Intrinsic Reproducibility. The degree to which an oscillator will reproduce a given frequency without the need for calibrating adjustments either during manufacture or afterward. This quality is a characteristic of an apparatus design and not a characteristic of a resonance.

(4) Stability. The degree to which an oscillator will produce the same frequency over a period of time once continuous operation has been established. The specification of a stability value requires a statement of the time interval involved in the measurement, and the complete specification constitutes a functional relationship for which the measuring interval is the independent variable.

A frequency standard must provide a stable, spectrally pure signal if it is to yield a narrow spectrum after multiplication to the microwave region. The high signal-to-noise ratio requirements for quality communication systems have been met by the specification of narrow band widths, which in turn require stable narrow-band signals.

The quality of a precision quartz oscillator usually is expressed in terms of long-term and short-term stability. Long-term stability is sometimes called aging rate and usually is expressed in fractional parts per unit time, as

"3 parts in $10^9$ per day". It refers to slow changes with time in average frequency, arising usually from secular changes in the resonator or other elements of the oscillator. Short-term stability refers to changes in average frequency over a time span sufficiently short that long-term effects may be neglected.

(5) Spectral Purity. The spectral purity expresses the same information in the frequency domain as that expressed by stability in the time domain. A very crude oscillator will have a reasonably good spectrum at the frequency of oscillation; but, with frequency multiplication, the spectrum rapidly degrades. It is, therefore, necessary to have an extremely well-defined spectrum at the outset so that further multiplication into the microwave region will maintain acceptable quality. A simple way to improve spectral purity is to put the oscillator signal through a narrow band-pass filter before multiplication, or to phase lock a low noise oscillator to the multiplied signal.

## 2-4 MILITARY APPLICATIONS

### 2-4.1 NAVIGATION, POSITION FINDING

During peacetime the navigation of a military aircraft on a routine flight differs little from that of a similar nonmilitary aircraft. However, some military operations require high selective accuracy. Examples are a tanker seeking to rendezvous with an aircraft it is to replenish, a reconnaissance aircraft returning to its carrier, a fighter providing close support at front lines, an emergency supply aircraft, or one seeking survivors in a lifeboat. In addition many military missions require a passive system to avoid betraying one's position, and vulnerability to manmade disturbances (jamming) must be considered. Similar conditions exist for a ship at sea.

Position finding or fixing is the determination of the position of the craft (a fix) without reference to any former position. To accomplish this with sufficient accuracy, one must depend on available precision reference signals:

(1) **Loran**. Loran (*LO*ng *RA*nge *N*avigation) was developed during World War II, at the MIT Radiation Laboratory, to provide ships and aircraft with a means of precise navigation. Basically, a Loran chain consists of a master and two or more slave stations. A pulse transmitted from the master station is received via ground wave by a slave station, which, in turn, transmits its pulse at a fixed time later. This fixed time, known as the coding delay, is kept constant by monitor stations that steer the chain. The time difference between reception of a master-slave pulse pair determines a hyperbolic line; the intersection of two such lines gives position.

Standard Loran, also known as Loran-A, operates in the 2-MHz band[7]. Each station transmits pulses of 45-$\mu$sec duration, which occupy a band width of 75 kHz.

(2) **Loran-C**. Loran-C was developed to extend Loran coverage with fewer stations. The Loran-C system, a pulsed radio navigation system is operated on a 100-kHz carrier by the U S Coast Guard, and offers a means for precision transfer of time. Clock synchronization to $\pm 1$ $\mu$sec is possible within range of the east coast Loran-C chain. All Loran-C stations are now controlled by atomic clock standards and the emissions are monitored by the U S Naval Observatory[1]. Table 2-1[7] lists the locations of 30 Loran-C stations presently in operation. Loran-C signals are usable for ground wave reception to about 1500 km landward, 3000 km seaward, and about 10,000 km skyward at reduced accuracy[7].

Changes in Loran-C station emissions are published in "Time Service Announcements" published by the U S Naval Observatory. Ref. 8 is a discussion of the means of using Loran-C and includes the possible sources of error.

## TABLE 2-1. LORAN-C STATION DATA

| Chain | Period, Rate μsec | Master[a] | Slaves[a,b] W | X | Y | Z |
|---|---|---|---|---|---|---|
| East Coast | SS-0 100 000 | Cape Fear, N.C. ("T" slave sometimes broadcasts from Wildwood, N.J. at ED 84028.7 μsec) | Jupiter, Fla 13 695.5 | Cape Race, Newfoundland (4 MW) 36 389.6 | Nantucket Island, Mass. 52 542.5 | Dana, Ind. 68 564.2 |
| Central Pacific (Hawaiian) | SH-4 59 600 | Johnston Island | | Upolo Point, Hawaii 15 971.8 | Kure Island, Hawaii 35 252.4 | |
| Mediterranean | SL-4 79 600 | Simeri Chichi (Catanzaro), Italy | | Matratin, Libya 14 107.6 | Targabarun, Turkey 32 273.3 | Estartit, Spain 50 999.7 |
| North Atlantic | SL-7 79 300 | Angissoq, Greenland (500 kW) | Sandur, Iceland (4 MW) 15 068.2 | Ejde, Faeroe Islands 27 803.9 | | Cape Race, Newfoundland (4 MW) 48 212.3 |
| Northern Pacific (Alaskan) | SL-2 79 800 | St. Paul, Pribiloff Islands | | Sitkinak, Alaska 14 284.4 | Attu, Aleutian Islands 31 875.3 | Port Clarence, Alaska (1850 kW) 53 069.1 |
| Norwegian Sea | SL-3 79 700 | Ejde, Faeroe Islands | Sylt, Germany 30 065.6 | Bo, Norway 15 048.1 | Sandur, Iceland 48 944.5 | Jan Mayen, Norway 63 216.3 |
| Northwest Pacific | SS-3 99 700 | Iwo Jima (4 MW) | Marcus Island (4 MW) 15 283.3 | Hokkaido, Japan 36 685.2 | Gesashi, Okinawa 59 463.0 | Yap Island (4 MW) 80 746.5 |
| Southeast Asia | S-3 49 700 | Sattahip, Thailand | | Lampang, Thailand 13 182.8 (computed) | Con Son Island, South Vietnam 29 522.1 (computed) | |

[a] Power 250-400 kW, except where noted.
[b] Emission delay given with reference to master.

Information current but subject to change.

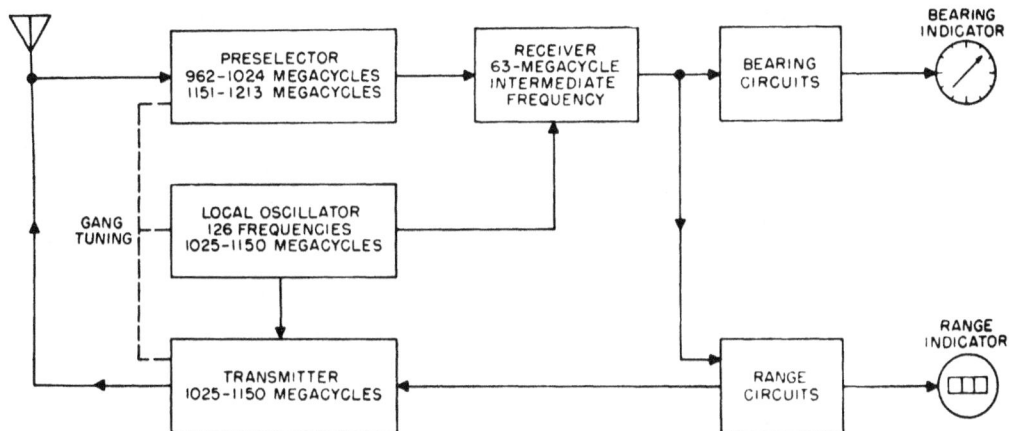

*Figure 2-1. Components of Airborne Air Traffic Control Equipment[9],[10]*

## 2-4.2 AIR TRAFFIC CONTROL

In air traffic control, military aircraft may use TACAN (*Tactical Aircraft Navigation* System) navigation equipment. TACAN is a military omnibearing and distance measurement system using the same pulses and frequencies for the distance measurement function as the Standard DME (Distance Measuring Equipment) system[9]. It is of the polar-coordinate type, i.e., there is a bearing facility that provides on the aircraft a meter indication of its direction in degrees of bearing from the ground beacon chosen by the pilot. There is also a distance facility that provides on the aircraft a meter indication in nautical miles of its distance from the ground beacon. Fig. 2-1[10] shows the components of airborne TACAN equipment such as the AN/ARN-21. A crystal-controlled 4044-Hz oscillator is used as a yardstick or time reference. In essence, the time interval between interrogation and reply is measured in terms of the corresponding number of cycles and fractions of a cycle of the 4044-Hz reference wave. The crystal controlled oscillator generates range markers that serve to measure microsecond intervals for the accurate measurement of distance[10].

## 2-4.3 COMMUNICATIONS

Modern military communication systems in which frequency stability of the transmitting and receiving equipment are supplied by a stable clock or precise interval control include synchronous digital transmission, switching, terminal, and security equipment[11].

Military applications require operation under severe environmental conditions, and usually impose restrictions on size and weight as well, particularly in airborne equipment. In mobile systems, Doppler shifts and time variations due to changing transmission path lengths must be compensated for by either automatic correction circuits or by manual readjustment of local equipment[12].

As an example, Table 2-2[13] summarizes the present and future frequency control requirements for SSB equipment. To obtain these requirements, a high degree of frequency-temperature stability must be obtained for the crystal oscillator along with low aging for the crystals and improvements in the frequency synthesizers.

The TRI-TAC communication system is planned to provide the Armed Forces with

### TABLE 2-2
### REFERENCE OSCILLATOR REQUIREMENTS[13]

|  | Present SSB Voice | Future SSB Secure Voice |
|---|---|---|
| $\Delta f$ (Total) | ± 40 Hz | ± 5 Hz |
| $\Delta f$/Equipment | ± 20 Hz | ± 2.5 Hz |
| $\dfrac{\Delta f}{f}$ (HF, 30 MHz) | $6.7 \times 10^{-7}$ | $8.3 \times 10^{-8}$ |
| $\dfrac{\Delta f}{f}$ (VHF, 76 MHz) | $2.6 \times 10^{-7}$ | $3.3 \times 10^{-8}$ |
| Temperature Range | −40° to +75°C | −40° to +75°C |
| Time (for frequency re-calibration) | 26 wk | 120 wk |

the capability of a fully tactical, automatically switched, digital, secure communications network. Three methods for system synchronization other than the master-slave relationship are frequency averaging, independent atomic clocks, and bit stuffing[14].

A new multifunction avionic system based on the use of accurately synchronized clocks at each terminal of the system has been proposed[15]. The system must be synchronized to the order of 10 to 100 nsec if range accuracy of ±10 to 100 ft is required. These accuracies are currently achievable, both by atomic frequency standards and by less accurate crystal standards in systems where it is practical to resynchronize the remote station clocks to a common system time.

## 2-4.4 IFF (IDENTIFICATION, FRIEND OR FOE)

Secondary radar, originally used during World War II to identify friendly aircraft and ships, was known as IFF. Each vehicle was equipped with a pulse transponder that, upon receiving a radar pulse, replied with a specially coded group of pulses. These pulses, displayed on a radar screen next to the radar reply, indicated that the reply was from a friend. The pulse codes frequently were changed to guard against enemy use of captured equipment.

Secondary surveillance radar, or radar beacon, is an updated version of this system. Secondary surveillance radar uses equipment on the ground to transmit an interrogation signal to the aircraft that transmits a response back to the ground station. If the transponder in the aircraft is set to respond, it transmits a signal to the ground equipment that is processed and displayed on a radar screen. The interrogation pulse groups consist of several modes (Fig. 2-2[9]), the selection of which is either controlled by the ground operator or automatically selected in a mixed sequence. Before the transmitter of the airborne transponder will reply, the proper pulse pairs must be received and processed within the transponder. The transponder reply codes are shown in Fig. 2-3[9]. Two framing pulses spaced 20.3 $\mu$sec apart with 12 information pulses between them provide the basic reply code. Thus, the code system is capable of producing 4096 different coded identification replies. In addition to the information pulses, a special position identification (SPI) pulse may be used with any of the 4096 codes upon request[9].

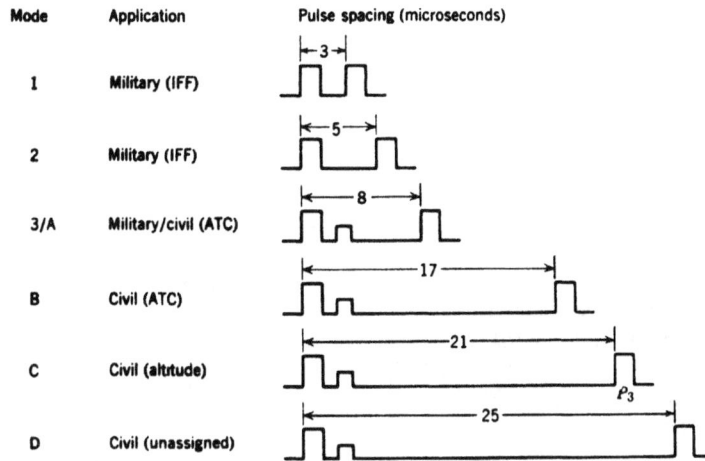

Figure 2-2. Interrogation Pulse Modes Used on Secondary Surveillance Radar[9,10]

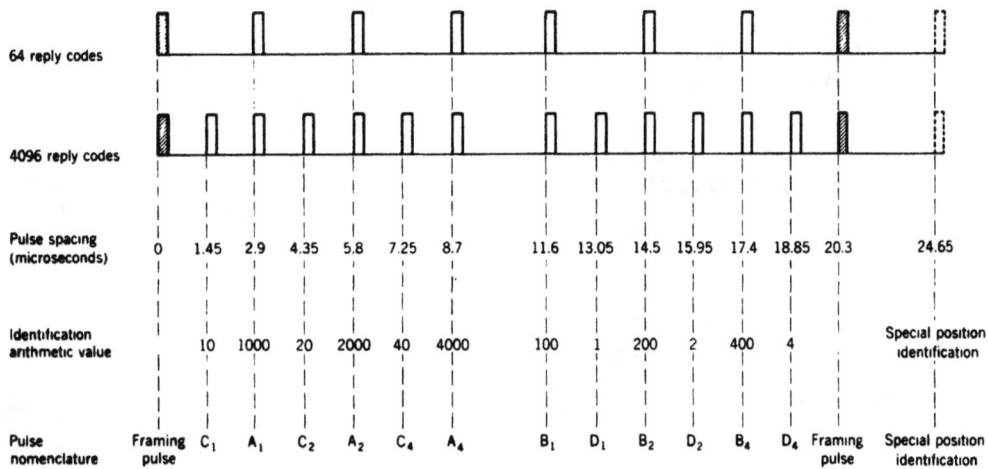

Figure 2-3. Transponder Reply Codes Used on Secondary Surveillance Radar[9]

# REFERENCES

1. *Frequency and Time Standards*, Hewlett-Packard Co., Palo Alto, Calif., Application Note 52, 1965.

2. P. B. Gove, Ed., *Webster's Third New International Dictionary*, G. and C. Merriam Company, Springfield, Mass., 1967.

3. *Encyclopedia Brittanica*, Chicago, Illinois, 1964, Vol. 22, Time, pp. 224-9.

4. J. J. Nassau, *A Textbook of Practical Astronomy*, McGraw-Hill Book Co., Inc., N.Y., 1932.

5. E. Hafner, "Frequency and Time," *New Horizons in Measurement*, 58-68 (1969).

6. A. O. McCoubrey, "A Survey of Atomic Frequency Standards", IEEE Transactions, **54**, 117-35 (February 1966).

7. L. D. Shapiro, "Time Synchronization From Loran-C", IEEE Spectrum **46**, 46-55 (August 1968).

8. P. F. Pakos, "Use of the LORAN-C System for Time and Frequency Dissemination", *Proceedings of the 23rd Annual Symposium on Frequency Control*, U S Army Electronics Command, Ft. Monmouth, N.J., May 1969.

9. M. Kayton and W. Fried, Eds., *Avionics Navigation Systems*, John Wiley and Sons, Inc., N.Y., 1969.

10. S. H. Dodington, "Airborne Tacan Equipment AN/ARN-21", Electrical Communication 33, 60-4 (March 1956).

11. MIL-STD-188C, *Military Communication System*, Dept. of Defense, 24 November 1969.

12. D. E. Johnson and J. P. Friedrichs, "Frequency Standards for Military Applications", *Proceedings of the 14th Annual Symposium on Frequency Control*, U S Army Signal Research and Development Laboratory, Ft. Monmouth, N.J., June 1960.

13. O. P. Layden, "Frequency Control for Tactical Net SSB Equipment", *Proceedings of the 23rd Annual Symposium on Frequency Control*, U S Army Electronics Command, Ft. Monmouth, N.J., May 1969.

14. J. M. Dresser, "Frequency Control Requirements for the Mallard Communication System", *Proceedings of the 23rd Annual Symposium on Frequency Control*, U S Army Electronics Command, Ft. Monmouth, N.J., May 1969.

15. T. C. Viars, "Application of Precise Time-Frequency Technology in Multifunction Systems", *Proceedings of the 23rd Annual Symposium on Frequency Control*, U S Army Electronics Command, Ft. Monmouth, N.J., May 1969.

# CHAPTER 3

## SIGNAL GENERATION SYSTEMS

### 3-1 INTRODUCTION

Recent advances in the accuracy of time keeping have been due largely to the outstanding developments in precision clocks. The modern *quartz crystals* and *atomic clocks* attain a standard of precision considerably in advance of the best pendulum clocks, and their development has led to the elimination of the pendulum clock in major time-keeping observatories and in time service throughout the world. Two types of motion have been used to produce these clocks, viz., the mechanical vibrations of quartz and the oscillations of individual atoms[1].

The most stable frequency standards are the atomic clocks, specifically cesium beam, rubidium gas cell, or hydrogen maser. Their frequency stability is compared in Table 3-1. Since the comparison is at constant ambient conditions, it does not indicate at least one other important factor, namely that the stability of rubidium cells is affected much more by the temperature than is cesium. The cesium frequency standard has the highest intrinsic reproducibility and frequently serves as a primary standard.

The family of quartz crystal oscillators forms the secondary standards. The frequency stability of these oscillators varies, depending on temperature compensation or control features (see Table 3-1 state of the art as of 1971).

In addition to these frequency standards, the tuning fork is discussed in par. 3-3. Tuning forks can be made in a package of 0.7 in.$^3$ with a frequency stability of ± 0.001% when controlled in an oven.

### 3-1.1 DEFINITIONS

#### 3-1.1.1 Clocks

The quartz-crystal clock and atomic clock are described:

(1) Quartz-crystal Clock. A standard quartz clock is a precision time piece that provides a series of electrical timing pulses at intervals of seconds or tenths of seconds; some applications require minutes to be marked, or to indicate the particular time of day. Most quartz clocks employ a crystal designed to operate at a nominal frequency of 5 MHz although frequencies of 1 MHz, 2.5 MHz, and 100 kHz are also much in use. Frequency division may be entirely electronic, or electronic in the first stages and then electromechanical[1].

(2) Atomic Clock: An atomic clock is a precision time piece controlled by an atomic or molecular spectral line[3]. These small electrical and magnetic vibrations are almost wholly independent of normal external conditions; and, because their frequencies are a property of the atoms themselves, they are identical for all atoms of the same kind.

In one of the most commonly used types of atomic clocks, the atoms used are those of cesium. However, many atomic clocks of several different kinds now have been made and are available commercially[3]. They enable time intervals to be measured with an accuracy approach 1 part in $10^{12}$ time units or 0.1 $\mu$sec per day. Clocks throughout the world can thus be synchronized to 1 $\mu$sec.

## TABLE 3-1

## CHARACTERISTICS OF ATOMIC AND QUARTZ CRYSTAL FREQUENCY STANDARDS

| Atomic Clocks | Frequency Stability (one day, constant ambient conditions) | Volume |
|---|---|---|
| Cesium beam | $2 \times 10^{-13}$ | 800-2000 in.$^3$ |
| Rubidium gas cell | $10^{-12}$ | 80-1500 in.$^3$ |
| Hydrogen maser | $2 \times 10^{-14}$ | 27 ft$^3$ |

| Quartz Crystals[a] | Frequency Stability Over Temperature Range $\pm 10^{-6}$ | Power Consumption, mW | Volume, in.$^3$ |
|---|---|---|---|
| Crystal Oscillator Temperature Range | 4 - 25 | 10—50[b] | 0.25—3 |
| $-55°$ to $+105°$ C | 25 | | |
| $-40°$ to $+90°$ C | 15 | | |
| $0°$ to $+50°$ C | 4 | | |
| Temperature Compensated Crystal Oscillator (TCXO) | 0.1—10 | 35—100[b] | 1—2.5 |
| $-55°$ to $+105°$C | 0.5—10 | | |
| $-40°$ to $+75°$C | 0.3—10 | | |
| $0°$ to $+50°$C | 0.1—1 | | |
| Temperature Controlled Crystal Oscillator (Single Oven) | 0.005—0.1 | 1—10 W[c] | 7—35 |
| $-55°$ to $+75°$C | 0.01—0.1 | | |
| $-40°$ to $+75°$C | 0.01—0.1 | | |
| $0°$ to $+50°$C | 0.005 | | |
| Temperature Controlled Crystal Oscillator (Double Oven) $0°$ to $+50°$C | 0.001—0.0001 | 5—15 W[c] | 1000-2000 |

[a]from Ref. 2
[b]Output Power = 1 mW
[c]At lowest ambient temperature

### 3-1.1.2 Frequency Standards

A frequency standard is an atomic clock that is used in frequency control because of its greater precision and accuracy. Basic to the choice of these atomic standards is their unprecedented frequency stability. For instance, the cesium and hydrogen atomic standards require no other reference for calibration. Three other devices that could be used as frequency standards are (1) thallium beam, (2) ammonia maser, and (3) rubidium maser. Atomic frequency standards operate by one of two means: (1) by determining the frequency corresponding to dipole inversion in a beam of atoms, or (2) by determining the frequency corresponding to a transition between the energy levels of atoms in a fixed sample.

### 3-1.2 PRIMARY STANDARDS

A primary frequency standard is one that provides a frequency that is well-defined without reference to any external standard. In all atomic frequency standards there are means for (1) selecting atoms in a certain energy state, (2) enabling long life times in that state, (3) exposing these atoms to microwave energy, and (4) detecting the results[4]. Two primary standards that have reached a high state of development are the hydrogen maser and the cesium beam. The cesium beam device uses passive atomic resonators to steer high quality quartz oscillators via feedback circuits. The hydrogen maser, an active device, derives its signal from stimulated emission of microwave energy amplified by electronic means to a useful power level.

### 3-1.3 SECONDARY STANDARDS

Secondary frequency standards are those that must be referenced to an accepted source such as a primary standard. Quartz crystal oscillators are used widely as high-quality secondary standards. The cesium beam device makes use of slaved quartz crystal oscillators.

Another secondary standard is the rubidium vapor standard. It is a secondary standard because it must be calibrated against a primary standard during construction; it is not self-calibrating[4].

### 3-1.4 BASIC OPERATING PRINCIPLES

#### 3-1.4.1 Quartz Crystal Clock

The resonant element of a quartz clock, usually called the crystal, is a bar, plate, or ring of quartz. To have it vibrate at a desired frequency in a particular mode, it is cut to certain dimensions and in a selected orientation to the various axes of the natural quartz crystal. The resonator assembly includes supports for the crystal and electrodes through which energy is supplied to the crystal to maintain the vibration using the piezoelectric property of crystalline quartz. The crystal is kept in oscillation by a maintaining amplifier connected to the electrodes of the resonator. The stability of the frequency of oscillation and the uniformity of the rate of the clock are primarily related to the quality of the quartz resonator and to characteristics of the maintaining amplifier, its power supply, and associated environmental control components. For more information on quartz crystal oscillators, see par. 3-2.

#### 3-1.4.2 Atomic Clock

The cesium atomic beam uses the hyperfine transition of the cesium atom. The atoms pass through a system of magnets and are deflected away from a detector unless transitions are induced by an applied field derived from a quartz clock and alternating at the spectral line frequency. Any deviation from the spectral line frequency produces an error signal that is applied to the quartz oscillator to bring it back to its correct value. The nominal value of the quartz oscillator is usually 5 MHz and a fairly complex frequency converter is required between the clock and the spectral line frequency[3]. Atomic resonance devices are discussed in more detail in par. 3-4.

## 3-2 QUARTZ CRYSTAL OSCILLATORS

Crystalline quartz has great mechanical and chemical stability, and a high mechanical $Q$ (which means that a small amount of energy is needed to sustain oscillation). The piezoelectric properties of quartz make it convenient for use in an oscillator circuit. Stressing of quartz and certain other crystals produces an electric potential at the electrodes. Conversely, placing each crystal in an electric field deforms them a small amount proportional to field strength and polarity. This property is known as the piezoelectric effect.

In practice, a quartz resonator is mounted between two electrodes, presently thin metallic coatings deposited directly on the crystal by evaporation. Mechanical support is provided at places on the crystal so chosen as to avoid, to the greatest extent possible, inhibition of the desired vibration while suppressing any unwanted vibrations. A correct alternating voltage applied across the crystal causes it to vibrate at a frequency such that mechanical resonance exists within the crystal. Details on the various shapes and their modes of vibration are discussed in par. 4-2. The various types of quartz crystal controlled oscillators are discussed in the paragraphs that follow. Oscillators for military equipment are covered by a military specification[5]. For frequency stabilities, see Table 3-1.

### 3-2.1 GENERAL PURPOSE OSCILLATOR

The simplest crystal oscillator uses neither temperature control nor temperature compensation techniques (see Fig. 3-1[6]). An amplifier, either vacuum tube or transistor type, is used with some degree of selectivity depending upon the type of crystal unit used. Degree of isolation from the load is dependent upon the output requirements and amount of

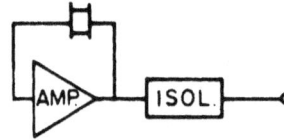

Figure 3-1. General Purpose Crystal Controlled Oscillator[6]

stability required. In this case, the main cause of instability is temperature.

Crystal aging or drift is only a problem with this form of oscillator because it is small compared with changes due to temperature and can be absorbed by periodic adjustment of the circuit phase. For stability data, see Table 3-1. More information is contained in Ref. 7 and in a design handbook, Ref. 8.

### 3-2.2 TEMPERATURE COMPENSATED CRYSTAL OSCILLATOR (TCXO)

To obtain better stability than that possible with the basic crystal oscillator, TCXO can be used[2]. This principle is shown in Fig. 3-2[6]. The increased stability, however, is at the expense of added circuit complexity more volume, and higher power consumption. Fig. 3-3[6] shows a simple compensation circuit. Frequency deviation of 1 part in $10^6$ over a temperature range of $-30°$ to $60°C$ has been shown possible with a 30 MHz overtone

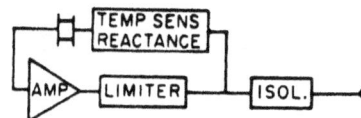

Figure 3-2. Temperature Compensated Crystal Oscillator (TCXO)[6]

AT-type unit. With more sophisticated design of the sensing network, 1 part in $10^7$ in the temperature range $-30°$ to $50°$ at 3 MHz has been reported[9]. Fig. 3-4[6] shows the frequency-temperature curve for an uncompensated and, for comparison, a compensated 25 MHz crystal unit. A bibliography on TCXO's follows the references at the end of this chapter.

Added circuit complexity for temperature compensation consists of a varactor, a thermistor-resistor network, and a voltage regulator. The latter is added because regulation of 100 or 1000 to 1 is needed, depending on how good the voltage source is and the degree of frequency stability needed (see Fig. 4-1(B)).

The resistance of the thermistor-resistor network is temperature sensitive so that a voltage which is a function of temperature is supplied to the varactor. This voltage variation changes the capacitance value of the varactor and thus the load capacitance of the crystal. This load capacitance change varies the crystal oscillator frequency in a predetermined manner to compensate for the crystal frequency-temperature variation. Depending on the stability desired, the reactance change of the varactor also must compensate for temperature-induced changes of the voltage regulator, active device, circuit elements, and buffer amplifiers. The choice of values in the thermistor-resistor network is quite complicated and usually requires computer optimization techniques to determine the network parameters.

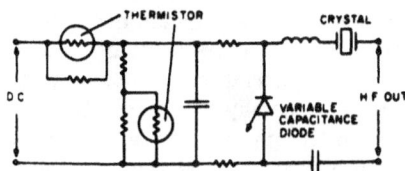

Figure 3-3. *Temperature Compensation Circuit*[6]

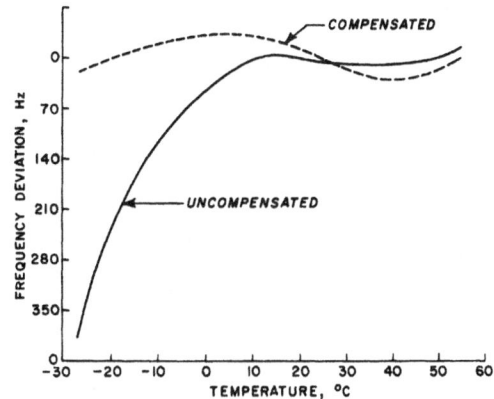

Figure 3-4. *Effect of Compensation on the Frequency of a 25-MHz Crystal Unit*[6]

The best angle of cut for the crystal depends on the stability needed and the temperature range. A compromise must be reached between the crystal with a more linear frequency-temperature slope between turning points with its greater frequency change and the crystal with a smaller frequency change but a more nonlinear slope. The more linear slope is easier to compensate but it must be compensated more precisely because of the larger frequency change. Another consideration is the location of the turnover points because synthesis of the compensating reactance is simplified when only one turn-over occurs in the temperature range of interest.

To obtain good frequency stability as a function of temperature, especially under temperature transient conditions, it is necessary that good thermal tracking exist among the crystal, varactor, and thermistors. If thermal gradients exist among these components, a large frequency change — on the order of a few parts in $10^6$ — can take place under transient conditions. Also, the crystal unit exhibits transient behavior during temperature change, due to thermal gradients within the quartz plate that can be as large as a few parts in $10^6$ for rapid temperature changes.

The TCXO exhibits good frequency accuracy from turn-on because all components are at the same temperature whether power is being supplied or not and the slight temperature rise due to turn-on has only minimal effect on frequency. In certain applications this feature is of major importance.

The aging of a TCXO is dependent primarily on the crystal and presently is between $5 \times 10^{-8}$/wk to $5 \times 10^{-10}$/wk for the fundamental mode units used. When dealing with the TCXO, as with the oven crystal oscillator discussed in par. 3-2.3, the overall frequency tolerance that can be expected is arrived at by adding the aging, over whatever time period is applicable, to the frequency-temperature stability. For instance, the $5 \times 10^{-7}$ frequency-temperature stability coupled with an aging of $1 \times 10^{-8}$/wk gives an overall frequency tolerance of $1 \times 10^{-6}$ for 1 yr.

## 3-2.3 TEMPERATURE CONTROLLED OSCILLATOR

For applications calling for a stability better than that obtainable with a TCXO, a temperature controlled crystal oscillator (oven oscillator) must be used[2]. The ovens used are of two general types — thermostatically controlled and proportionally controlled. The thermostatically controlled oven uses a bimetallic or mercury thermostat to sense the temperature and supplies heat to the oven on an "on-off" basis. This gives simple control but normally is restricted to less precise frequency control and therefore is not covered here. The proportionally controlled type uses a resistance temperature sensor and by bridge circuitry supplies heat on a continuous basis. This results in better temperature control, thus higher precision frequency control.

The improved stability over the TCXO requires more space and considerably higher power consumption. In addition, there is added circuit complexity over the basic crystal oscillator (see Fig. 4-1(C)). The oven circuitry is composed of a resistance bridge, amplifier, transistor controller, and heater winding. A voltage regulator and multistage buffer amplifier are used and an automatic gain control circuit is needed to keep the crystal drive constant. One arm of the resistance bridge is the sensor, a specially wound resistor or a thermistor whose resistance value is a function of temperature. Any temperature change in the oven is detected by this sensor resulting in bridge unbalance. The bridge voltage output is amplified and operates on the transistor controller, causing more or less current to flow through the heater winding, thus regulating the temperature. There should be good thermal coupling between the sensor and the heater winding on the oven shell so that high loop gain may be employed. The oven structure should be well insulated, and possess large heat capacity, to keep the rate of change of temperature at a minimum. Without good thermal coupling the temperature at the crystal oscillator will vary considerably, due to heating of the control circuit, resulting in poor frequency stability. The operating temperature of the oven can be changed by varying the value of one resistor in the bridge circuit. This changes the balance point and thus the operating temperature. A well-designed single stage oven will give 0.1 deg to 1.0 deg C temperature stability over a wide ambient range.

Insulation is used in the oven around the components to keep heat loss low and to minimize temperature gradients within the chamber. Generally, the gradients are proportional to the power used in the heater. The best insulation is a vacuum and the normal way to achieve this is through use of a double wall dewar flask. A less expensive and more rugged method is the use of foam material;

however, heat loss is greater than in the evacuated case.

The operating temperature of the oven must be higher, usually by 10 deg to 15 deg C, than the highest ambient temperature expected. The actual operating temperature of the oven must be adjusted to correspond to the turn-over temperature of the crystal being used in that oven.

In order to realize the best frequency-stability performance possible with crystal-controlled oscillators, double-oven temperature stabilizing techniques are employed essentially to eliminate the effects of ambient temperature changes on the frequency of oscillation. A single, carefully-designed proportionally controlled oven can maintain an ambient ratio of the order of 500:1 (i.e., internal temperature change of 0.1 deg C for an ambient range of 50 deg C). Consequently, placing one such oven completely within another of similar capability will result in an overall ambient ratio of the order of $2.5 \times 10^5$:1 (about 0.0002 deg C variation for a 50 deg C change in ambient). As a result, frequency changes caused by ambient temperature changes are reduced to be of the order of $1 \times 10^{-12}$/deg C, or even less.

If the best available overtone-mode crystal units (i.e., fifth overtone, 5 MHz) are used in oscillators of this type, and corresponding care is observed in the design of a buffer amplifier, voltage regulator, etc., the effects of other ambient influences such as supply voltage and load impedance can be made to be of the same order as the effects of temperature, with the result that crystal-controlled oscillators are available whose output frequency will remain stable within $1 \times 10^{-11}$ due to all causes — except shock vibration and static acceleration — for periods of a day or longer.

In order to achieve stability of this order, the temperature control and voltage regulator

circuits usually are placed inside the first, or outer, temperature-stabilized enclosure along with the RF buffer amplifiers. The crystal unit and oscillator circuit are placed within the second, or inner, oven enclosure.

For proper operation of a two-stage oven system, the outer enclosure must be maintained at a temperature somewhat above the highest ambient temperature to be encountered, and then the inner enclosure must be held several degrees above the temperature of the outer oven. This necessarily requires that the turn-over temperature of the crystal unit used in such an oscillator be controlled carefully in manufacture since the operating point of the inner oven must be adjusted very accurately (within about 0.1 deg C) to the turn-over temperature of the crystal. For example, if an ambient range of 0° to 50°C is specified, the operating temperature of the outer oven would be chosen in the range of 60° to 65°C, and consequently the crystal units should be specified to have turn-over temperatures in the range of 70° to 75°C.

An oven stabilized oscillator is not on frequency immediately after turn-on because the oven must warm up to its operating temperature before the crystal can be stabilized at the turnover point. To obtain a fast warmup characteristic, a separate warmup heater and associated on-off thermostat are used. A large amount of power is used for a short time in the warmup heater, thus heating the chamber quickly. The thermostat will cut off a few degrees below the desired operating temperature where the proportional control takes over. Warmup times of 5 to 30 min, from lowest ambient, are normal with a power dissipation of 4-40 W during this warmup time.

## 3-2.4 OSCILLATOR FOR SEVERE ENVIRONMENT

Phase coherence and spectral purity are very important for stable frequency sources subject to high shock and vibration in missile and space applications. Fig. 3-5[6] illustrates the use of an oven to obtain the thermal stability required. Particular attention is paid to the mounting material which is often foam. Since the environment requires a rather rigid support, the g forces are transmitted to the crystal vibrator which results in a frequency shift. The rigidity of the support relaxes with time and contributes to a high aging rate. For frequencies of 10 MHz and up, stability of the order of 1 part in $10^{10}$ per g is possible if the shock and vibration frequencies are less than the resonant resonance of the crystal mounting system. Static acceleration causes, as a rule of thumb, a relative frequency change of roughly $2 \times 10^{-9}$ per g when the acceleration vector is in the worst case direction. Less than $1 \times 10^{-10}$ per g is possible if the crystal unit can be oriented precisely in the acceleration field. It has been shown that the triple-ribbon supported vibrator in the TO-5 transistor enclosure has been found suitable for this class of service[6].

## 3-2.5 PRECISION OSCILLATORS

For the greatest stability and lowest drift rate, careful attention must be given to the oscillator circuit and the highest precision crystal unit must be used. Fig. 3-6[6] outlines the necessary controls that must be used and is typical of most precision oscillators on the market today. A double oven with proportional control is used to hold the temperature constant and to ensure that no gradients exist in the quartz plate (see par. 3-2.3). One factor governing the stability of a crystal unit is that

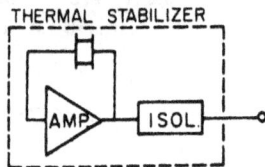

Figure 3-5. Oscillator for Severe Environments[6]

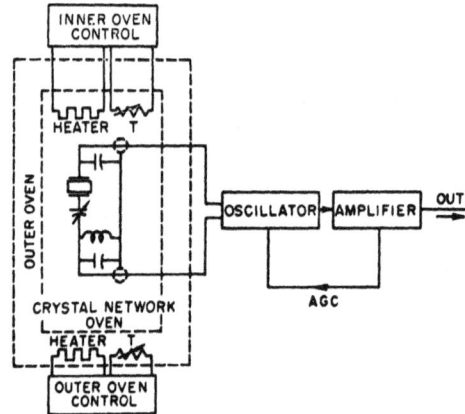

Figure 3-6. Precision Crystal Controlled Oscillator[6]

of amplitude of oscillation; therefore, a large amount of feedback is used in the amplifier and in the automatic gain control to maintain a constant level of the oscillator signal. An oscillator using the 2.5 MHz fifth overtone crystal unit design has an average daily drift rate as low as a few parts in $10^{10}$, and after three weeks of operation, the aging is less than $10^{-9}$ Hz per wk[6]. Design criteria for the various crystal oscillators are discussed in par. 4-2.

## 3-3 TUNING FORK OSCILLATORS

### 3-3.1 APPLICATION

The tuning fork can be said to be one of the oldest vibrating devices used for frequency control. Tuning forks are not used in new military equipment because crystals are more stable. The only application is in situations where cost is of prime importance. Forks are much cheaper. They have been used in such devices as precision clocks, reference clocks in computers, navigation systems, satellite timers, and central timekeeping systems. One of the problems in its early application was keeping the fork vibrating. Ideally, the pickup and drive device for maintaining vibrations should be loosely coupled to the fork. Present practice uses a magnetic transducer for both pickup and drive functions.

The transducer consists of a magnet inside a coil located close to the tine of the fork. Changes in flux, due to the vibration, produce a signal in the pickup coil. Conversely, a signal applied to another transducer, also located near the tine, will maintain motion. An adequate amplifier connected between the two will cause the fork to vibrate at the proper frequency. Typical gains required of the amplifier are 10 to 50 dB depending on fork design[10].

### 3-3.2 NONTEMPERATURE CONTROLLED OSCILLATOR

As a standard oscillator, a fork vibrates at its fundamental, or lowest, resonant frequency. By the positioning of the drive and pickup coils, and by control of the oscillator amplifier phasing, a tuning fork can be made to oscillate at its second partial resonant frequency, about six times the fundamental frequency. At the higher mode, the tuning fork exhibits a $Q$ and a frequency stability similar to those at its fundamental frequency. Present forks have $Q$'s of 15,000 to 23,000 in vacuum, and a temperature coefficient of frequency in the range of 1/5 to 1/3 ppm/deg C from $-20°$ to $50°C$. Fig. 3-7[11] shows the modes of the fork motion when oscillating both in its fundamental and in its first partial mode.

### 3-3.3 TEMPERATURE CONTROLLED/ COMPENSATED OSCILLATORS

To help alleviate the effect of temperature on frequency, alloys such as Ni-Span, Vibraloy, Nivarox are used to produce low temperature coefficients. These alloys are heat-treated to adjust the coefficient. For close coefficients over wide temperature ranges, heat treating is not consistent enough

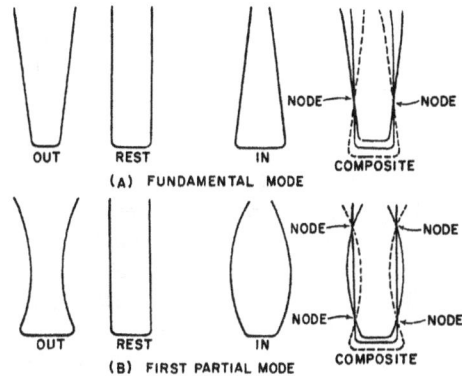

Figure 3-7. Tuning Fork Vibrations

to allow the use of alloy forks for general applications.

The most practical present method of compensation is the use of a bimetallic fork made of a laminate of a positive coefficient Elinvar alloy and a negative coefficient steel alloy. The coefficient then can be adjusted by grinding one laminate or the other.

Another method to correct the temperature coefficient problem is to use an oven. In this method, the fork is enclosed in a proportional oven controlled by vacuum tubes or transistors. At the present time, temperature versus frequency tolerances of $\pm 0.02\%$ over the military temperature range and $\pm 0.001\%$ in the room temperature range are readily achieved[10].

### 3-3.4 EFFECTS OF OTHER ENVIRONMENTAL FACTORS ON STABILITY

One of the environmental factors is fork attitude. A fork with tines down runs faster than one with tines up due to the effect of gravity. Of all frequency instruments, forks are the most susceptible to shock and vibration. Various methods to compensate for this include driving both tines with the coils between the tines, having an internal stud mounting in place of an external stud, and using foamed materials or damped springs.

Forks are subject to aging just as crystals are. Aged forks usually increase in frequency. Presoaking fork assemblies at moderately high temperatures provides a quick "aging" process so that a unit can be made compatible with its end use. In instances where precise adjustments are required, the frequency can be adjusted with a phase control. Variations up to 200 parts in $10^6$ are practical and more than enough to compensate for future aging. For design details of tuning fork oscillators, see par. 4-3.

## 3-4 ATOMIC RESONANCE DEVICES

### 3-4.1 ATOMIC ACTION

Atoms are capable of existing in several different energy states that are the result of different energy relationships between the various components of the atom. The atom can switch about among the different energy states permitted to it, but to do so it must absorb or emit energy. The relationship of major interest, as far as atomic resonators are concerned, is that between the nucleus of the atom and its outermost or valence electrons. The magnetic moment of these electrons may be aligned either with or against the magnetic field of the nucleus depending upon the direction of spin of the electrons. It is these two possible alignments that constitute the hyperfine energy differences in the ground state of the atom of concern here.

If a magnetic field of proper frequency is applied to the atom, the magnetic moment of the electron can be induced to switch its alignment relative to the nuclear magnetic field. Electrons in the lower of the two energy states absorb energy and switch to the higher state; electrons in the higher states are stimulated to switch to the lower state by the applied field, but in doing so they emit rather than absorb energy. The emitted energy is of the same frequency as the stimulating field. The frequency of the magnetic field required to cause alignment transitions is related to the magnetic moment of the electron, the spin of the electron, and the magnetic field of the nucleus.

It should be noted that any quantity of atoms will have approximately equal numbers in each of the possible hyperfine levels of the ground state. If an RF magnetic field is applied, transitions from the lower level to the higher level will approximately equal those in the opposite direction. The RF energy absorbed will equal the energy emitted and there will be no sensible indication that the transitions have taken place. It is necessary, therefore, to work with a group of atoms that are preponderantly in only one of the two possible states. There are various means of achieving this condition; the means being suited to the particular type of resonator in question (see par. 3-4.2.1).

Transition frequencies are the same for all unperturbed atoms of a given substance, and the band width over which the transitions occur is quite narrow. It is this reproducibility of frequency from atom to atom, along with the extremely sharp resonance ($Q = 100$ million) that has led to the development of atomic resonators as precision time keeping devices.

There are basically two methods of using atomic resonators for time or frequency control; viz., as passive resonators or as active resonators. The passive resonator is used to determine and control the frequency of an applied RF magnetic field. Active resonators, on the other hand, actually generate an RF signal from the energy emitted by the atoms undergoing transition.

## 3-4.2 PASSIVE RESONATORS

### 3-4.2.1 Atomic Beam Resonators

Atomic beam resonators, as the name implies, make use of atoms or molecules of various substances in the form of a beam. This beam, under proper magnetic manipulation, is employed as an extremely precise frequency determining element. The most extensively developed form of beam resonator is the cesium beam, which is the international primary time standard. Considerable work also has been done on the thallium beam. The operation of the various atomic beams is quite similar and, therefore, only the cesium beam is discussed.

The cesium beam resonator consists essentially of a source of cesium[133] atoms, focusing magnets, a transition chamber, and a detector. These components are enclosed in a magnetically-shielded vacuum envelope and are arranged as shown in Fig. 3-8[12]. There are two fundamental functions to be performed within the resonator: (1) isolation of the particular atoms that undergo the desired transition from one energy level to another, and (2) effecting and detecting the transition itself.

When the RF generator that supplies the energy to the transition chamber is tuned for maximum detector output, its frequency will be precisely the transition frequency. For a cesium beam the exact frequency is 9,192,631,770 Hz or, as the second is defined in terms of the cesium beam, it may be more correct to say that 9,192,631,770 cycles of the cesium beam oscillator equals one second of ephemeris time[13]. The band width of such a device is approximately 250 Hz.

### 3-4.2.2 Gas Cell Resonators

The operation of gas cell resonators, like that of atomic beam resonators, is based upon a hyperfine transition of atoms of certain alkali metals. The most extensively developed type of gas cell employs rubidium[87]. However, gas cells have been constructed which use atomic hydrogen, sodium[23] or cesium[133]. Operation of these cells is similar to that of the rubidium cell. The means used for changing the energy states of the atoms in gas cells is called optical pumping[14].

A passive resonator generally employs a relatively low-frequency crystal oscillator as its primary source of RF energy. This oscillator is variable over a limited range, and the center frequency is usually a integral number of kilohertz or megahertz. Frequency multipliers and synthesizers are used to raise the oscillator frequency to that required to effect the desired transition. An alternate method is to operate the crystal oscillator at a submultiple of the transition frequency and use the synthesizer to obtain the desired standard frequency from the crystal oscillator. In either case, the signal applied to the atomic resonator usually is phase modulated, and a synchronous detection network is incorporated in the output circuit of the resonator. In this manner an error signal is generated which is applied as a correction to the crystal oscillator. The entire system is, in effect, a servo controlled oscillator with a highly specialized frequency determining element. Circuit parameters are such that for very short times the frequency stability is that of the crystal oscillator and for long times that of the atomic resonator. Time, in this case, is relative to the response time of the servo loop.

## 3-4.3 ACTIVE RESONATORS

### 3-4.3.1 Masers

A maser is an atomic or molecular device that is capable of coherent amplification or generation of electromagnetic waves. The masers employed in precision timing applications are generally of the gaseous type and are operated as oscillators. These oscillators depend for their operation upon the energy

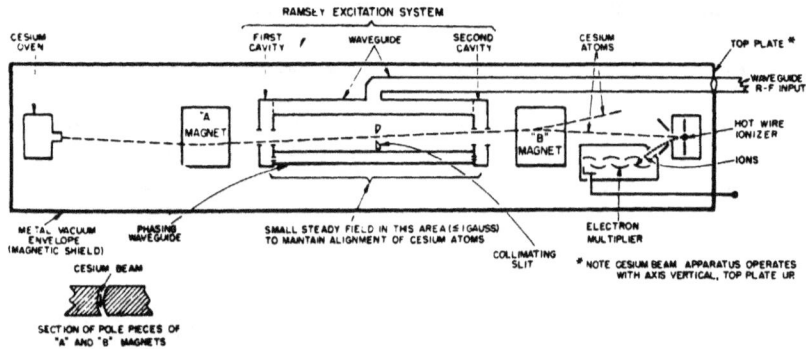

Figure 3-8. Cesium Atomic-beam Frequency-standard Apparatus

emitted by atoms or molecules undergoing transitions from a higher to a lower energy state. They, therefore, require a continuous supply of atoms or molecules in the desired energy state which may be provided either as a constant stream of selected atoms as in a beam resonator or by optical pumping as in a vapor cell.

The most extensively developed maser oscillator for timing or frequency control application is the atomic hydrogen maser. It consists essentially of a source of atomic hydrogen, a selecting magnet, a storage chamber, a resonant cavity, a vacuum envelope, and magnetic shields. The components are arranged as shown in Fig. 3-9[15].

The hydrogen maser requires a continuous stream of hydrogen atoms in the higher energy level of the hyperfine state. These are provided by passing molecular hydrogen

through a dissociator that separates the molecules into atoms. The atoms then are passed through a focusing magnet that directs atoms in the desired state into a storage bulb while rejecting atoms in unwanted states.

Atoms entering the storage bulb are confined inside for a brief period before they can escape through the entrance. During this period some of them will emit energy spontaneously and make the transition to the lower energy state. The energy emitted will stimulate still more atoms to undergo transitions and these, in turn, will release more energy. When the resonant cavity is tuned to the frequency of the emitted energy, the occurrence of the transitions will be pulled into phase with the electromagnetic wave in the cavity. If enough transitions take place to provide sufficient energy to overcome the losses in the system, oscillation will occur at the transition frequency. A coupling loop in the cavity allows a small amount of power to be drawn from this oscillation without seriously disturbing the system. The resonant frequency of the cavity is adjustable over a small range so that it can be tuned precisely to the transition frequency.

### 3-4.3.2 RF Output of Masers

Masers of the type used for frequency

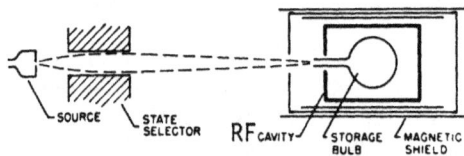

Figure 3-9. Schematic Diagram of the Hydrogen Maser[15]

control do not generate very large amounts of power. Also, to avoid shifting the frequency of the maser by variable loading, the output coupling must be kept low and isolation of the output circuitry must be provided. Useful output signal is thus reduced to a very low level. Higher level signals can be produced by phase locking the output of the maser to the signal from a precision crystal oscillator. Heterodyne circuits and frequency synthesizers are used to convert the maser output to the crystal frequency. In this manner an integral frequency at a power of several milliwatts is obtained.

## REFERENCES

1. Arthur Beer, *Vistas in Astronomy*, Pergamon Press, N.Y., 1955, pp. 438-46.

2. O. P. Layden, et al., "Crystal-Controlled Oscillators", IEEE Transactions on Inst. and Meas., **IM**-21, 277-86 (August 1972).

3. "Atomic Clocks", Engineering, **205**, 459-62 (March 1968).

4. *Frequency and Time Standards*, Application Note 52, Hewlett-Packard Co., Palo Alto, Calif., 1965.

5. MIL-O-55310, *Oscillators, Crystal, General Specification For*, Dept. of Defense, 4 December 1970.

6. E. A. Gerber and R. A. Sykes, "State-of-the-Art — Quartz Crystal Units and Oscillators", Proceedings of IEEE, **54**, 103-16 (Feb. 1966).

7. S. Schodowski, *Proceedings of Interlaboratory Seminars on Component Technology: Part I — Crystal Oscillators and Frequency Synthesizers*, Report ECOM-2865, U S Army Electronics Command Fort Monmouth, N.J., August 1967.

8. *Quartz Crystal Oscillator Circuits, Design Handbook*, Serial TP64-1072, The Magnavox Co., Fort Wayne, Ind., Contract DA36-039-AMC-00043(E), 15 March 1965.

9. P. Duckett, R. Peduto, and G. Chizak, "Temperature Compensated Crystal Oscillators", *Proceedings, 24th Annual Symposium on Frequency Control*, U S Army Electronics Command, Fort Monmouth, N.J., 27-29 April 1970, pp. 191-9.

10. F. Dostal, "The Increasing Applications of Tuning Forks and Other Vibrating Metal Resonators in Frequency Control Systems", *Proceedings, 19th Annual Symposium on Frequency Control*, U S Army Electronics Command, Fort Monmouth, N.J., April 1965, pp. 59-78.

11. M. Pleasure, "Tuning Forks as Circuit Elements", *Proceedings, 14th Annual Symposium on Frequency Control*, U S Army Signal Research and Development Laboratory, Fort Monmouth, N.J., May 1960, pp. 397-404.

12. F. D. Lewis, "Frequency and Time Standards", Proceedings of the IRE **43**, 1046-68 (Sept. 1955).

13. A. O. McCoubrey, "A Survey of Atomic Frequency Standards", Proceedings of the IEE, **54**, 116-35 (Feb. 1966).

14. M. E. Packard and B. E. Swartz. "The Optically Pumped Rubidium Vapor Frequency Standard", IRE Transactions on Instrumentation I-11, 215-23 (Dec. 1962).

15. D. Kleppner, et al., "Hydrogen-Maser Principles and Techniques", Physical Review, 138, A972-83 (May 1965).

# BIBLIOGRAPHY

## TEMPERATURE COMPENSATED QUARTZ OSCILLATORS (Par. 3-2.2)

L. F. Koerner, "Methods of Reducing Variations in Crystals Over a Wide Temperature Range", *IRE National Convention Record*, 1956, pp. 48-54.

R. A. Spears, "Thermally Compensated Crystal Oscillators", Journal Brit. IRE, **105**, 613-20 (October 1958).

L. F. Koerner, "A Portable Frequency Standard", Electronics, 32, 173-6, (May 1959).

E. A. Gerber, "Reduction of Frequency-Temperature Shift of Piezoelectric Crystals by Application of Temperature-Dependent Pressure", Proc. of the IRE, 48, 224-45 (February 1960).

O. J. Baltzer, "Temperature Compensation of Transistorized Crystal Oscillators", *Proc. 12th Southwestern IRE Conference*, Houston, Texas, 22 April 1960.

N. R. Malik, "Linearizing Frequency Temperature Characteristics of Quartz Crystals by Network Synthesis", Thesis, Graduate Department of Electrical Engineering, State University of Iowa, August 1960.

R. A. Spears, "Precision Crystal Chronometers", Electronic Technology, 37, 368-73 (October 1960).

E. A. Gerber and M. H. Miles, "Temperature Compensation of Piezoelectric Resonators by Mechanical Stress", *Proceedings, 15th Annual Symposium on Frequency Control*, U S Army Electronics Command, Fort Monmouth, N.J., 1961, pp. 49-65.

G. R. Hykes and D. E. Newell, "A Temperature Compensated Frequency Standard", *Proceedings 15th Annual Symposium on Frequency Control*, U S

Army Electronics Command, Fort Monmouth, N.J., 1961, pp. 297-303.

E. A. Gerber and M. H. Miles, "Reduction of the Frequency Temperature Shift of Piezoelectric Resonators by Mechanical Stress", Proc. of the IRE, **49**, pp. 1650-4 (November 1961).

R. J. Munn, "Temperature Compensated Quartz Crystal Units", *Proceedings, 16th Annual Symposium on Frequency Control*, U S Army Electronics Command, Fort Monmouth, N.J., 1962, pp. 169-86.

R. J Munn and K. Krieger, *Crystal Units With Improved Temperature Characteristics*, Final Report, Motorola Inc., Chicago, Ill. and USAECOM, Fort Monmouth, N.J., Contract DA-36-039-SC-87271, 30 June 1962.

W. L. Smith and W. J. Spencer, *Quartz Crystal Controlled Oscillators*, Final Report, Bell Telephone Laboratories and USAECOM, Fort Monmouth, N.J., Contract DA-36-039-SC-85373, January 31, 1963, pp. 68-70.

O. P. Layden, "Progress and Problems in Quartz Crystal Circuitry and Measurements", *Proceedings, 17th Annual Symposium on Frequency Control*, U S Army Electronics Command, Fort Monmouth, N.J., 1963, pp. 464-81.

D. E. Newell and R. H. Banghert, "Temperature Compensation of Quartz Crystal Oscillators", *Proceedings, 17th Annual Symposium on Frequency Control*, U S Army Electronics Command, Fort Monmouth, N.J., 1963, pp. 491-507.

G. R. Bart, *Improved Frequency Temperature Characteristics of Quartz Crystal Units*,

Final Report, Victor Electronic Systems Co., Chicago, Ill. and USAECOM, Fort Monmouth, N.J., Contract DA-36-039-SC-90792, 31 October 1963.

D. E. Newell and R. H. Bangert, *Frequency Temperature Compensation Techniques for Quartz Crystal Oscillators*, Final Report, The Bendix Corp., Davenport, Iowa, and USAECOM, Fort Monmouth, N.J., Contract DA-36-039-SC-90782, June 30, 1963; Interim Report, 6th Quarterly, Contract DA-36-039-AMC-02282(E), December 31, 1964.

D. E. Newell, N. Hinnah, G. K. Bistline, Jr., and J. K. Reighter, *Crystal Units With Improved Temperature Characteristics*, Final Report, McCoy Electronics Co., Mt. Holly Springs, Pa., The Bendix Corp., Davenport, Iowa, and USAECOM, Fort Monmouth, N.J., Contract DA-36-039-SC-90820, March 31, 1964.

D. E. Newell, H. Hinnah, and R. H. Bangert, "Advances in Crystal Oscillator and Resonator Temperature Compensation", *Proceedings 18th Annual Symposium on Frequency Control*, U S Army Electronics Command, Fort Monmouth, N.J., 1964, pp. 487-534.

G. Bistline, Jr., and D. B. Jacoby, "Glass Enclosed Crystal Units for Temperature Compensated Oscillators", *Proceedings, 18th Annual Symposium on Frequency Control*, U S Army Electronics Command, Fort Monmouth, N.J., 1964, pp. 193-203.

R. C. Rennick, *Temperature Compensated Crystal Oscillators for Production*, Bell Telephone Laboratories Record, October 1964.

O. T. Sokolov, "On the Compensation of the Temperature-Frequency Coefficient of Quartz in Transistor Oscillators", Radiotekhnika (Izvest. Voz), 7, (3), 1964 (in Russian).

W. H. Horton, S. B. Boor, and R. B. Angove, *Tighter Tolerance Crystal Units*, Final Report, Systems Inc., Orlando, Florida and USAECOM, Fort Monmouth, N.J., Contract DA-36-039-AMC-02335(E), November 1964.

R. H. Bangert and D. Thomann, *Design and Development of Frequency Temperature Compensated Quartz Crystal Oscillator, 0-1227 ( )/U*, Phase I – Interim Report, The Bendix Corp., Davenport, Iowa and USAECOM, Fort Monmouth, N.J., Contract DA-28-043-AMC-00042(E), December 31, 1964.

C. Hardingham, "A Temperature-Compensated Crystal Oscillator", *IREE (Aust.) R. & EE Convention*, Canberra, Australia, March 1965, Paper 12.

A. E. Anderson, M. E. Frerking, and G. R. Hykes, *Temperature Compensated Quartz Crystal Units*, Interim Report, Collins Radio Co., Cedar Rapids, Iowa and USAECOM, Fort Monmouth, N. J., Contract DA-28-043-AMC-00210(E), April 1, 1965.

D. E. Newell, H. Hinnah, and R. H. Bangert, "Recent Developments in Crystal Oscillator Temperature Compensation", *Proceedings, 19th Annual Symposium on Frequency Control*, U S Army Electronics Command, Fort Monmouth, N.J., 1965, pp. 617-41.

S. B. Boor, W. H. Horton, and R. B. Angrove, "Passive Temperature Compensation of Quartz Crystals for Oscillator Applications", *Proceedings, 19th Annual Symposium on Frequency Control*, U S Army Electronics Command, Fort Monmouth, N.J., 1965, pp. 105-24.

C. D. Dominguez, *Design and Development of Frequency Temperature Compensated Quartz Crystal Oscillator 0-1227( )/U*, Final Report, The Bendix Corp., Davenport, Iowa, and USAECOM, Fort Monmouth, N.J., Contract DA-28-043-AMC-00042(E), September 30, 1965 (AD-634 520).

S. Schodowski, "Aging of Temperature Compensated Crystal Oscillators", Proc. of the IEEE, **54**, 808-9 (May 1966).

*Temperature Compensated Crystal Oscillator*, 1st and 2nd Quarterly Reports, Standard Telephones and Cables Limited, Harlow, Essex, Ministry of Aviation, Contract KJ/X/426/C.B. 55(c)-1, February 1, 1966 and June 7, 1966.

C. D. Dominguez and I. E. Hardt, *Frequency Temperature Compensation Techniques for Quartz Crystal Oscillators*, Final Report, The Bendix Corp., Davenport, Iowa, and USAECOM, Fort Monmouth, N.J., Contract DA-36-039-AMC-02282(E), May 1, 1967 (AD-821 964).

S. Schodowski, "Crystal Oscillators and Frequency Synthesizers, Part I: Moderate Precision Quartz Oscillators", *Proceedings of Interlaboratory Seminars on Component Technology*, Part I, Report 2865, USAECOM, Fort Monmouth, N.J., August 1967, pp. 53-74.

S. Schodowski, *Temperature Performance Measurement Methods for Temperature Compensated Quartz Oscillators*, Report 2896, USAECOM, Fort Monmouth, N.J., October 1967 (AD-664 157).

C. Hardingham, "Temperature Compensated Crystal Oscillators – A Review of Recent Australian Developments", *Proceedings I.R.E.E.*, Australia, November 1967, pp. 424-34.

P. C. Vovelle "Recent Improvements to TCXO's", *Proceedings, 22nd Annual Symposium on Frequency Control*, U S Army Electronics Command, Fort Monmouth, N.J., 1968, pp. 311-24.

D. E. Newell and H. Hinnah, "Automatic Compensation Equipment for TCXO's", *Proceedings, 22nd Annual Symposium on Frequency Control*, U S Army Electronics Command Fort Monmouth, N.J., 1968, pp. 298-310.

D. E. Newell and H. D. Hinnah, *Frequency Temperature Compensation Techniques for Quartz Crystal Oscillators*, Semiannual Reports 1 and 2, CTS Knights, Inc., Sandwich, Ill. and USAECOM, Fort Monmouth, N.J., Contract DAAB07-67-C-0433, June 1968 (AD-837 531) and June 1969 (AD-859 010), respectively.

O. P. Layden, "Frequency Control for Tactical Net SSB Equipment", *Proceedings, 23rd Annual Symposium on Frequency Control*, U S Army Electronics Command, Fort Monmouth, N.J., 1969, pp. 14-7.

D. E. Newell and H. Hinnah, "A Report on Segmented Compensation and Special TCXO's", *Proceedings, 23rd Annual Symposium on Frequency Control*, U S Army Electronics Command, Fort Monmouth, N.J., 1969, pp. 187-91.

H. A. Batdorf, "Temperature Compensated Crystal Oscillators Operating from 800 kHz to 1500 kHz", *Proceedings, 23rd Annual Symposium on Frequency Control*, U S Army Electronics Command, Fort Monmouth, N.J., 1969, pp. 192-7.

P. Duckett, R. Peduto, and G. Chizak, "Temperature Compensated Crystal Oscillators", *Proceedings, 24th Annual Symposium on Frequency Control*, U S Army Electronics Command, Fort Monmouth, N.J., 1970, pp. 191-9.

S. Schodowski, "A New Approach to a High Stability Temperature Compensated Crystal Oscillator", *Proceedings, 24th Annual Symposium on Frequency Control*, U S Army Electronics Command Fort Monmouth, N.J., 1970, pp. 200-8.

S. Schodowski, *A New Approach to a High Stability Temperature Compensated Crystal Oscillator*, Report 3359, USAECOM, Fort Monmouth, N.J., November 1970.

D. E. Newell and H. Hinnah, *Frequency Temperature Compensation Techniques for Quartz Crystal Oscillators*, Final Report, CTS Knights, Inc., Sandwich, Ill. and USAECOM, Fort Monmouth, N.J., Contract DAAB07-67-C-0433, February 1971.

S. V. Whitten, Jr., "A Practical Approach to the Design of a Temperature Compensated Crystal Oscillator", *Proceedings, 1971 IEEE National Telemetering Conference*, Washington, D. C., 12-15 April 1971.

# CHAPTER 4

## BASIC DESIGN CRITERIA FOR SIGNAL GENERATION

### 4-1 INTRODUCTION

There is a continuous demand for increased precision and accuracy in frequency control. Today fast time pulses are used in radar, precision navigation systems, velocity measurement, guidance of fast-moving aircraft and missiles, and rating of ship and shore frequency standards. Microsecond synchronization of clocks for periods of 24 hr or more at a number of range stations is necessary for accurate determination of missile velocity. Standard oscillators that are used to control clocks at remote points must not vary in frequency more than 1 part in $10^{11}$ per day. Similarly, accurate synchronization of precision frequency standards ashore and afloat is essential to operation of single side band communication systems. The general theory of signal generation applicable to basic reference timers is described in the paragraphs that follow.

### 4-1.1 GENERAL THEORY

Three oscillator types used for the generation of frequency signals in basic reference timers are (1) quartz crystal oscillators, (2) tuning fork oscillators, and (3) atomic frequency standards. Each one has a special characteristic that is used for generating a constant frequency signal. In the quartz crystal oscillator, it is the mechanical vibration of the crystal; in the tuning fork oscillator, it is the mechanical vibration of the fork. Atomic frequency standards use transitions between states separated by energies corresponding to microwave frequencies (see pars. 3-2 to 3-4).

For a device to be used by a basic reference timer, its frequency must possess great stability and low drift rate. For a precision quartz crystal oscillator to possess these characteristics, detailed attention must be given to the design of the oscillator circuit and to the selection of a quartz resonator[1]. Tuning fork oscillators that are hermetically sealed and temperature compensated can be produced to operate in the frequency range of 400 to 1000 Hz and also at 50-60 Hz. With respect to atomic frequency standards, the controlled rubidium gas cell oscillator is the most compact, has a high degree of short-term stability, and a long-term stability sufficiently adequate for a wide range of applications. By virtue of the fact that cesium beam oscillators have a very high degree of long-term stability and the important property of intrinsic reproducibility, they qualify as primary frequency standards. Even though the size and weight of the cesium beam oscillators are greater than a rubidium gas cell, these instruments are portable. The atomic hydrogen maser has the highest degrees of short-term and long-term stability as well as intrinsic reproducibility when compared with available atomic oscillators. Its size, weight, and cost, however, are also greater[2].

Time standards and frequency standards are related. A standard of frequency can serve as the basis for time measurement, and vice versa, with certain restrictions. To avoid errors when a frequency standard is used to maintain time, either interval or epoch, care must be taken to identify the time scale of interest; i.e., atomic, UTC sidereal, etc.[3].

If a consistent local system of time and

frequency standards is to be maintained, the standards must be intercompared. Further, if a local system is to be kept in correspondence with national standards, a reference must be established and maintained. Radio broadcasts from frequency and time standard stations are used most often as the link to keep this reference. The various systems that may be used for synchronization and the test methods used in these systems are discussed in par. 5-3.

## 4-1.2 COMPONENTS AND CIRCUITS

Most crystal-controlled frequency standards are composed of the following components[1]:

(1) A control element, i.e., the quartz crystal unit

(2) A negative resistance element, i.e., the oscillator circuit using vacuum tubes or transistors

(3) A thermostat or temperature-control device to keep the frequency control element and other circuit elements at constant temperature

(4) Suitable frequency dividers or other means for producing lower output frequencies

(5) Integrating devices, such as clock indicators

(6) A suitable power supply.

The circuits used in quartz crystal oscillators are shown in Fig. 4-1[4]. The basic circuit of Fig. 4-1(A) is relatively simple. For temperature compensation (Fig. 4-1(B)), the added circuit complexity is shown in heavy lines. Additions consist of a varactor, thermistor-resistor network, and voltage regulator (see par. 3-2.2). For temperature control, shown in Fig. 4-1(C), oven circuitry is added, again shown in heavy lines (see par. 3-2.3).

The atomic resonance of cesium and rubidium is used in a passive sense to control the frequency of quartz crystal oscillators through the action of electronic feedback circuits. Fig. 4-2[2] shows a basic block diagram of an atomic resonator-controlled oscillator while Fig. 4-3[2] shows a complete hydrogen maser standard. Design details of the components and circuits are covered in pars. 4-2 to 4-4.

## 4-2 QUARTZ CRYSTAL OSCILLATORS

### 4-2.1 QUARTZ CRYSTAL UNITS

#### 4.2.1.1 Vibration Types

It has been stated that the type of motion used in signal generation in the quartz crystal oscillator is the mechanical vibration of the quartz crystal itself. Fig. 4-4[5] illustrates the various modes of motion that are used in practically all crystal units on the market today. Quartz bars or plates are used in the flexural, extensional, and shear modes of motion to cover the frequency range from a few hundred hertz to over 200 MHz. In order to obtain the best characteristics, particularly as a function of temperature, certain orientations have been developed with respect to the crystallographic axes as shown in Table 4-1[5].

All of the high frequency units listed are excited by an electrical field that is perpendicular to the major surfaces of the crystal. In A and B vibrators, an electric field perpendicular to the major surface usually is used to couple to the thickness shear mode. To do so, the electrodes are deposited on the central portion that is the principal frequency determining part of the quartz plate. When the resulting two-terminal resonator is connected into a circuit, it behaves as though it were an electrical network. It is so located in the oscillator circuit that its equivalent electrical network becomes a major part of the resonant circuit that controls oscillator frequency. At present, AT-cut plates vibrating either in their fundamental mode or in one of

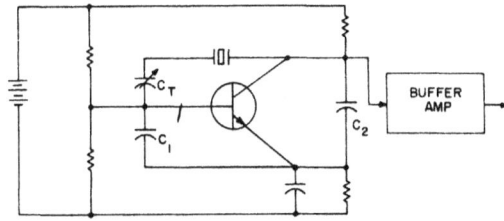

(A) Basic Circuit

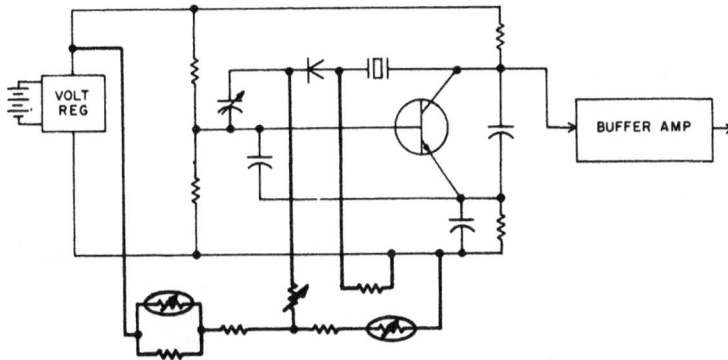

(B) Temperature Compensated Circuit

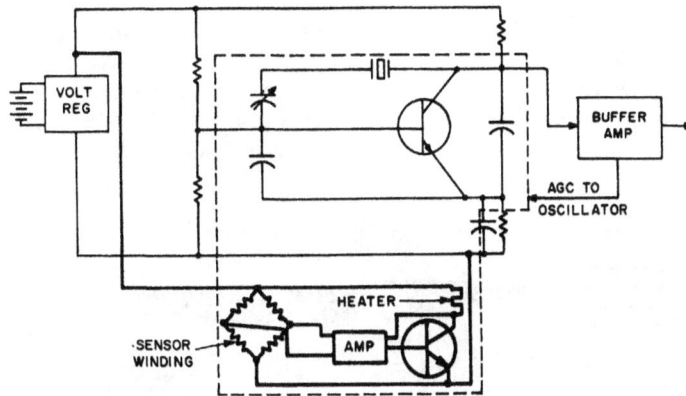

(C) Temperature Controlled Circuit

Figure 4-1. Crystal Oscillator Circuits for Frequency Standards[1]

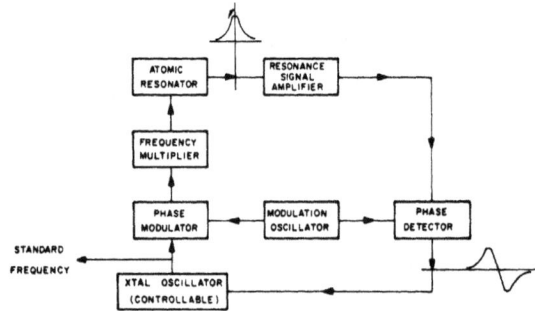

*Figure 4-2. Basic Diagram of Atomic Resonator Controlled Oscillator[2]*

their mechanical overtones are being used almost exclusively for frequency standards[5]. The quartz plates have a plano-convex shape, are polished, and carry electrodes that are applied by vapor deposition.

The attainment of a high $Q$ and low aging is possible by use of meticulous care in the preparation of the crystal unit[6]. Aging rates of $2 \times 10^{-9}$ per wk and less, and $Q$ values of $1.5 \times 10^{6}$ to $2 \times 10^{6}$ for 5-MHz fifth overtone plano-convex units are attained commercially in reasonable quantities for oven controlled use. (For a definition of $Q$, see par. 4-3.1.2.)

### 4-2.1.2 Enclosures

The performance of crystal vibrators is controlled largely by the type of mounting system used and its enclosure. Types of holders in current use are shown in Fig. 4-5[5]. Holders HC-6, HC-13, and HC-18, with crystal vibrators listed in Table 4-1, can be used to cover the frequency range from a few kHz to 200 MHz. This type of holder is sealed to the base by soldering. To eliminate the resulting contamination, all-glass holders of the HC-6 and HC-18 dimensions were developed. Designated as HC-26 and HC-27, they are used principally in the high-frequency, thick-

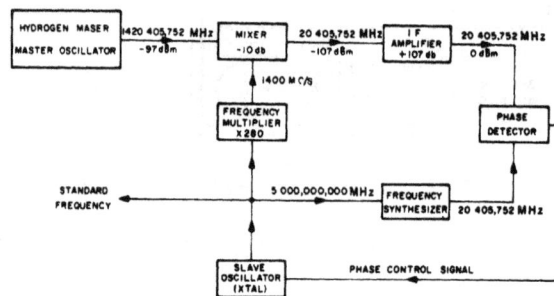

*Figure 4-3. Complete Hydrogen Maser Frequency Standard[2]*

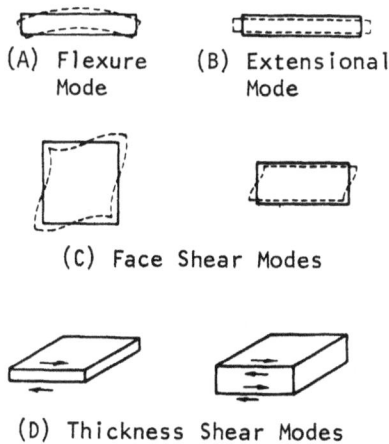

(A) Flexure Mode    (B) Extensional Mode

(C) Face Shear Modes

(D) Thickness Shear Modes

*Figure 4-4. Basic Modes in Quartz Crystal Vibrators[5]*

ness shear crystal units[7]. For high frequency overtone precision crystal units, HC-30 and T-11 all-glass holders are used[8]. They are of the drop seal type and eutectic solder may be used in the mounting system. However, in the HC-26 and HC-27 enclosures, high temperature bonding agents such as silver paste and cements must be used. The principal advantage of glass holders is that they are easier to clean, and with normal processing, there should be less contamination inside the enclosures.

*Figure 4-5. Quartz Crystal Unit Holders[5]*

A new sealing machine has been reported for use with HC-26 and HC-27 glass holders[9]. This machine has a production capability of 10,000 units per month with high reliability. A 5-MHz fundamental resonator in evacuated coldweld HC-6/U and round transistor configurations has been developed for use in temperature compensated oscillators. Specific goals were long-term aging of no more than 2 parts in $10^9$ per wk, high $Q$ (500,000 min), good frequency stability of $35 \pm 2$ parts per $10^6$ between turning points, and ability to withstand bakeout temperatures of 450°C. Figs. 4-6[10] and 4-7[10] depict the HC-6 type coldweld unit and the round transistor type coldweld holder. Materials used for the enclosures included copper-clad kovar bases, compatible glass-to-metal seals, and covers of high purity nickel. In attempts to miniaturize the enclosure as much as possible, there have been recent developments to produce holders similar to the TO-5 transistor enclosure, but having a height of 0.070 in. and a diameter of 0.250 in. The seal is made by an electron beam welding process[11].

### 4-2.1.3 Quartz Material

Synthetic quartz is now available from commercial sources, and recently attempts have been made to improve the $Q$-value. Internal friction limits $Q$ as shown in Fig. 4-8[12]. This is a plot of the internal friction of natural quartz over a wide temperature range

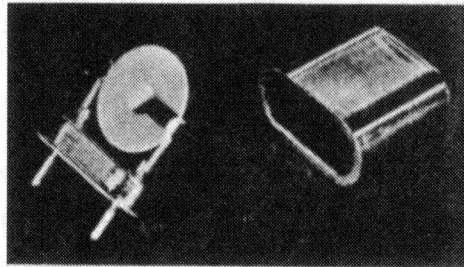

*Figure 4-6. HC-6 Type Coldweld Unit*

**TABLE 4-1**

**DESIGNATION OF QUARTZ VIBRATORS[5]**

| Vibrator Designation | Usual Reference | Mode of Vibration | Frequency Range |
|---|---|---|---|
| A | AT Cut | Thickness Shear | 0.5 to 250 MHz |
| B | BT | Thickness Shear | 1 to 30 MHz |
| C | CT | Face Shear | 300 to 1000 kHz |
| D | DT | Face or Width Shear | 200 to 750 kHz |
| E | +5°X | Extentional | 60 to 300 kHz |
| F | −18°X | Extensional | 60 to 300 kHz |
| G | GT | Extensional | 100 to 500 kHz |
| H | +5°X | Length-Width Flexure | 10 to 100 kHz |
| J | +5°X (2 plates) | Duplex Length-Thickness Flexure | 1 to 10 kHz |
| M | MT | Extensional | 60 to 300 kHz |
| N | NT | Length-Width Flexure | 10 to 100 kHz |
| K | X-Y bar | Length-Width Flexure or Length-Thickness Flexure | 2 to 20 kHz |

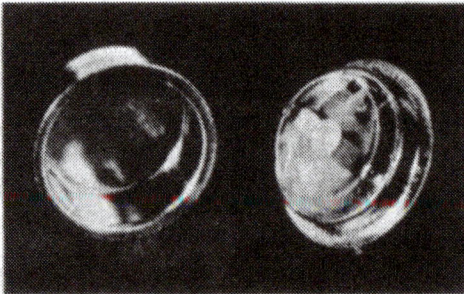

*Figure 4-7. Round Transistor Type Coldweld Holder*

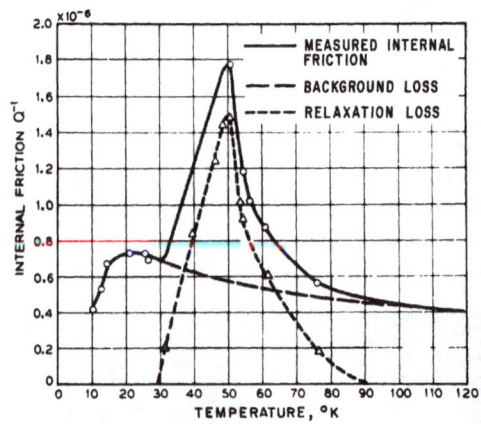

*Figure 4-8. Friction Losses in Quartz[12]*

obtained by determining the $Q$ of a 5-MHz fifth-overtone glass-enclosed AT-cut vibrator[13,14].

This crystal unit construction has been used for measuring the $Q$ of both synthetic and natural quartz because it is felt that most of the measurable loss is in the vibrating material and not in the mounting system. It has been demonstrated that the $Q$ of synthetic quartz can be improved to a value equal to natural quartz by appropriate control of the growth solution and growth rate[15]. The background relaxation shown in Fig. 4-8 has been explained as a direct conversion of acoustic waves into thermal energy[14]. The sodium carbonate method of quartz growth leads to crystals of very small hydrogen content with a $Q$ in the order of $2 \times 10^6$ at the 5-MHz level[13]. A BT-cut vibrator whose elastic constant is twice as high as an AT-cut vibrator offers higher $Q$ values[16].

### 4-2.1.4 Design Considerations

Oscillator crystal design primarily is concerned with obtaining high $Q$ units that have low resistance and exhibit good temperature behavior. The presence of unwanted modes at certain frequency ranges of conventionally used crystal designs can cause serious application difficulties. A study has been made in which the techniques of energy trapping used to design filter crystals can control the mode spectrum[17]. The latest issue of the *Military Crystal Specifications* states that the resistance of the unwanted modes should be at least double the resistance of the main mode[18]. Table 4-2[19] summarizes the characteristics of new crystal unit design of the 5-MHz fifth overtone type. Techniques for processing crystals to meet requirements of military and satellite applications have been developed[20]. Stabilities in the $10^{-11}$ range per month and recoverability to within 2 parts per $10^{10}$ after temperature interruption were realized in isolated cases.

Aging rates of thickness-shear AT-cut vibrators have been reduced in the past few years by the use of glass holders or cold-welded metal holders that allow high temperature bakeout prior to sealing. Fig. 4-9[12] illustrates how the change in frequency has been reduced by these means.

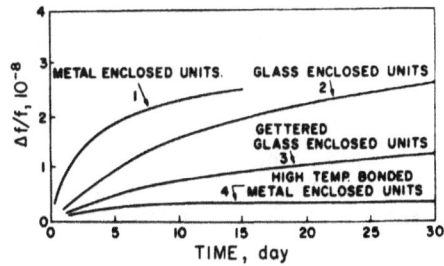

*Figure 4-9. Aging of Metal-enclosed and Glass-enclosed 5-MHz Crystal Units*[12]

### 4-2.2 OSCILLATOR CIRCUITRY

In very general terms, a crystal-controlled oscillator may be described as consisting of an amplifier, or gain circuit, together with a feedback network that contains a piezoelectric crystal unit[4]. A typical oscillator circuit is

**TABLE 4-2**

**SUMMARY OF CHARACTERISTICS OF 5-MHz CRYSTAL UNIT**

Frequency: 5000.000 kHz (5th overtone mode)
Tolerance: $\pm$ 1 part per $10^6$ at optimum operating temp. (in range 70° to 80°C)
Average $Q$:     2,500,000
Maximum effective resistance:
    Series =     140 ohm; parallel = 185 ohm
Offset and retrace characteristics:
    Freq. offset: $< 1 \times 10^{-9}$ 24 hr after restart
    Aging retrace: $< 5 \times 10^{-10}$ 72 hr after restart
Temperature coefficient:
    0.1 part per $10^6$/deg C within $\pm$ 0.5 deg C of operating temp.
Aging:     $< 5 \times 10^{-10}$/day after 3 days
           $< 7 \times 10^{-10}$/wk after 30 days
Level of drive: 70 $\mu$A $\pm$ 20%
Typical parameters:    $L$ = 9.2 H
                       $C_m$ = 0.0001 pF
                       $C_o$ = 4.3 pF

shown in Fig. 4-1(A). Self-oscillation of such a circuit will occur provided that the loop gain exceeds unity at some frequency for which the total loop phase is $2n\pi(n = 0, 1, 2...)$. The stable level of oscillating signal will be determined either by the self-limiting characteristics of the loop, or by the action of an external automatic-gain-control loop acting on the circuit.

When the desired frequency of oscillation corresponds to the lowest resistance vibrational mode of the controlling crystal unit (i.e., fundamental-mode operation of a thickness-shear type resonator), additional band-limiting networks generally are not required in the oscillator loop. However, such is not always the case. For example, a low-frequency oscillator may be designed to use a flexure-mode crystal that could exhibit lower resistance at a higher frequency corresponding to an extensional or width-shear mode of vibration. In this instance, a simple low-pass network would be required in the oscillator loop to insure operation at the desired flexure-mode frequency. Similarly, a band-pass network would be needed in order to assure operation of an oscillator circuit at one of the overtone-modes of a thickness-shear-resonator (i.e., the resistance of the desired fifth overtone mode of a VHF crystal unit could be greater than that of either the fundamental, third, or seventh-overtone modes).

The output signal spectrum of a crystal-controlled oscillator will depend upon the oscillating level, the electrical noise introduced by circuit elements, and by the bandwidths of the circuits. Generally, the spectral width is extremely small; and, for a majority of applications, only the center or average frequency behavior need be considered. Disregarding for the moment the corruption of signal purity by electrical noise, the average frequency of oscillation always will be determined by the loop phase requirement. Deviations in frequency from a

constant value will result whenever any perturbation of loop phase occurs. These perturbations generally may be considered to be of two classes, namely:

(1) Deviations in electrical circuit parameters

(2) Deviations in the crystal unit characteristics.

In class (1) are included such factors as the temperature coefficients of capacitors, inductors, resistors, and transistors; the aging of these devices; their voltage and current characteristics; and their susceptibility to mechanical disturbance. Whatever the cause, a phase perturbation in the circuit transfer phase will cause a change in the average frequency of oscillation, assuming that the crystal unit characteristics remain unperturbed.

Class (2) deviations include changes in crystal unit characteristics caused by temperature, temperature gradient, drive level, "frequency aging", and mechanical disturbances (shock, acceleration, and vibration).

The crystal oscillator circuit generally is designed to use a gain circuit having as broad a band as practicable consistent with operation of the desired crystal modes in order to reduce the network effect on frequency stability. The crystal feedback network, on the other hand, usually is designed to have as narrow a transmission band as can be obtained so as to make the frequency of oscillation depend essentially only on the crystal unit characteristics. When frequency adjustment is required, it is accomplished preferably by introducing a variable reactance that changes the transmission frequency of the crystal network without widening the transmission band. When these general practices are followed, the oscillator frequency stability will depend primarily upon the characteristics of the crystal unit, which should be chosen to have (1) a high $Q$-factor; (2) a low temperature coefficient of frequency over the intended operating temperature range; (3) a low drive-level coefficient of frequency; (4) low sensitivity to mechanical shock, acceleration, and vibration; and (5) low frequency aging. More information on oscillator circuitry is contained in Refs. 21-23.

### 4-2.3 OVENS

The design of ovens for crystal oscillators should consider requirements of the specific application. Two examples of prototype ovens developed for portable communications and satellite instrumentation — ovens A and B, Fig. 4-10, respectively — are presented. Fig. 4-10[24] shows an exploded view of the ovens while Fig. 4-11[24] presents the basic circuit diagram of the control system. Oven A was designed to maintain a single HC-6/U crystal at a preset temperature between 80° and 90°C over an ambient temperature range of −50° to 75°C. The entire package measured 3.2 in.[3]. Oven B was designed to maintain a compensated crystal oscillator at a temperature of 0° to 5°C to a stability of 50 millidegrees short term and an objective of 100 millidegrees over the variation in ambient temperature of −55° to −15°C. This package measured 5.2 in.[3]. See also par. 3-2.3.

One major factor in design is to control the heat produced at a "hot spot" such as a resistor or transistor. Sources of such heat (see Fig. 4-11) are the integrated circuit, the oven control transistors, and the bridge resistors. The transistor was sunk into the oven shell to distribute its heat to the shell and aid the heater in its functions. Fig. 4-12[24] is a plot of the typical control system with a maximum power of 3 W for quick warm-up.

The circular printed circuit board in the oven (Fig. 4-10) contains the bridge resistors, temperature adjusting potentiometer, and the differential amplifier housed in a standard TO-5 transistor case. Insulation of the entire assembly was achieved by a microfiber type of glass wool.

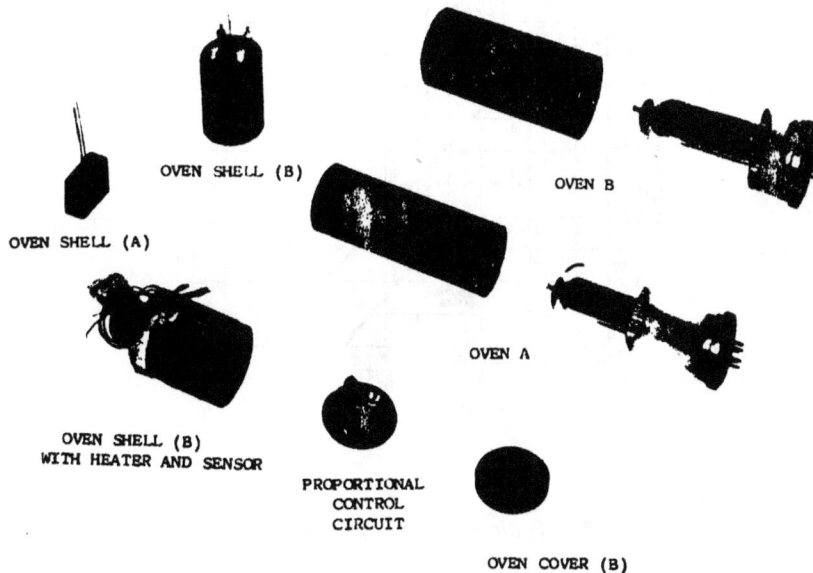

*Figure 4-10. Exploded View of Crystal Ovens*

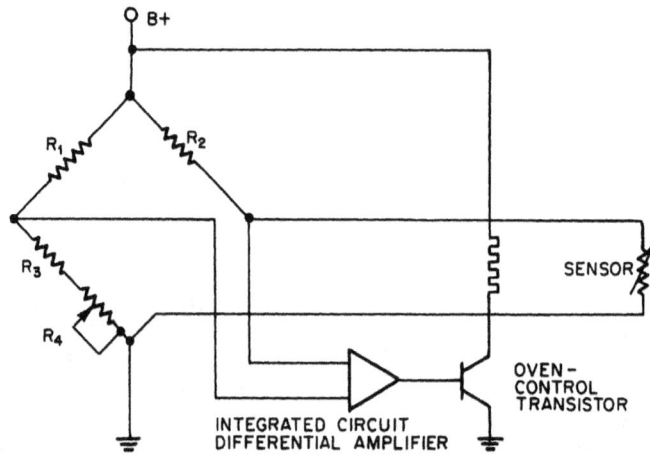

Figure 4-11. Basic Proportional Control System

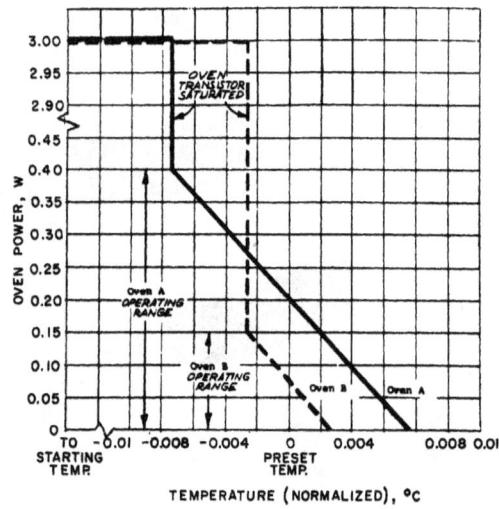

Figure 4-12. Typical Control System Characteristics

## 4-3 TUNING FORK OSCILLATORS

### 4-3.1 TUNING FORKS

#### 4-3.1.1 Dimensional Considerations

The frequency $f$ of a fork is determined by the tine dimensions. Frequency is directly proportional to tine thickness and inversely proportional to the square of tine length

$$f = k \left(\frac{h}{\ell^2}\right), \text{Hz} \qquad (4\text{-}1)$$

where

$k$ = proportionality constant, dimensionless

$h$ = tine thickness, in.

$\ell$ = tine length, in.

The remaining dimension ($b$) has no relation to frequency (Fig. 4-13[25]). Usually a low-frequency fork is long and a high-frequency fork is short, but this is not mandatory. Although most forks are driven so that they vibrate in a fundamental mode, higher frequencies sometimes are produced by driving a fork in an overtone mode.

#### 4-3.1.2 Quality Factor

The quality factor $Q$ of the fork is related to its material, dimensional relationships, and the method of driving the fork. $Q$ also can be determined from the spacing of the half power points or by the vibration decay characteristics. Forks made of certain aluminum alloys can attain twice the $Q$ of nickel alloy forks but have very poor frequency versus temperature coefficients. The $Q$ of a resonator (electrical, mechanical, or any other) is[26]

$$Q = 2\pi \left(\frac{H_s}{H_d}\right), \text{dimensionless} \qquad (4\text{-}2)$$

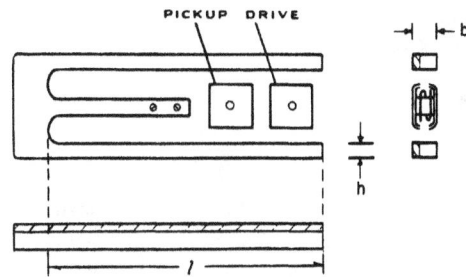

Figure 4-13. Typical Tuning Fork

where

$H_s$ = energy stored per cycle, ft-lb

$H_d$ = energy dissipated per cycle, ft-lb

#### 4-3.1.3 Temperature Coefficient of Frequency

Temperature variation is the largest single factor affecting frequency stability. The temperature coefficient of frequency $\beta$ is[27]

$$\beta = (\alpha + \gamma)/2, \text{(deg C)}^{-1} \qquad (4\text{-}3)$$

where

$\alpha$ = temperature coefficient of the modulus of elasticity, (deg C)$^{-1}$

$\gamma$ = temperature coefficient of linear expansion, (deg C)$^{-1}$

By suitable selection of materials, $\beta$ can be made to approach zero. This was the basis for development of the constant modulus alloys. The metal used for mono-metallic types has a composition as shown in Table 4-3[27]. Fork manufacturers may vary the composition to suit certain requirements of stability, temperature limits, etc.

**TABLE 4-3**

**ANALYSIS OF MONOMETALLIC TYPE FORK**

| Metal | Composition, % |
|---|---|
| Nickel | 30-38 |
| Chromium | 5-13 |
| Iron | 48-61 |
| Manganese | 0.5-2 |
| Silicon | 0.5-1 |
| Cobalt | 0.5-1 |
| Tungsten | 1-3 |

Bimetallic construction uses a block of constant modulus alloy and a strip of carbon steel, silver soldered together. The ratio of nickel alloy to carbon strip thickness determines the temperature coefficient of the tuning fork and will be in the range of 4-6 to 1 with the nickel alloy being the largest quantity[27].

**4-3.1.4 Shock and Vibration**

Both tines of a fork can be forced to vibrate parasitically in phase, in addition to the normal out-of-phase vibration under disturbance of shock or vibration. However, if the coils are placed between the tines, see Fig. 4-13, the pick-up coil will tend to cancel pick-up voltage due to in-phase tine motions and thus reduce this component in the output. Use of an internal stud to mount the fork at its center of moment reduces the shock and vibration effects by approximately 5 times in the internal stud configuration[25].

**4-3.2 OSCILLATOR CIRCUITRY**

A simplified equivalent circuit of the fork assembly in the form of a two-terminal pair network is shown in Fig. 4-14(A)[27]. As with the crystal, the fork parameter values are dependent on amplitude of vibration. In the figure, the representation is that of an electromagnetic transducer or resonator in which an output coupling is placed to produce an output that is proportional to the

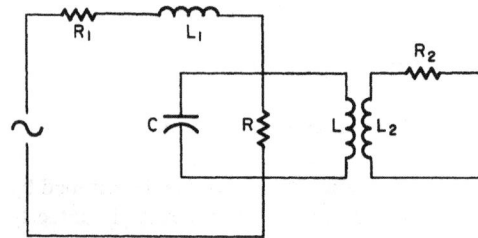

(A) Simplified Equivalent Circuit

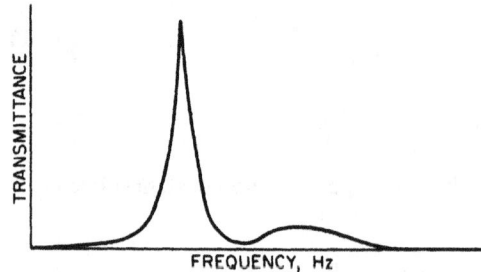

(B) Transmittance vs Frequency

*Figure 4-14. Oscillator Circuit*

amplitude of oscillation. Transmittance (ratio of generalized output to input) is shown as a function of frequency in Fig. 4-14(B). Peak response corresponds to antiresonance of the motional parameters, and the secondary response corresponds to series resonance of the motional equivalent capacity with the coupling coil inductance. The latter is typically one or two percent higher in frequency than the primary response and is sufficiently low in amplitude compared with the main response that oscillation will still occur at the primary response[27].

**4-4 ATOMIC FREQUENCY STANDARDS**

**4-4.1 PASSIVE RESONANT DEVICES**

**4-4.1.1 System Considerations**

The atomic resonances of cesium and

rubidium are used passively to control the frequency of quartz crystal oscillators through the action of electronic feedback circuits. The basic system arrangement is shown in Fig. 4-2. In this system, the frequency of the controlled crystal oscillator is an integral submultiple of the atomic resonance frequency. To provide standard frequency signals on the basis of the commonly defined time intervals, a frequency synthesizer is included as a part of the resonance controlled oscillator. The synthesizer may either be (1) included within the frequency control loop to relate the rational frequency of the crystal oscillator to a submultiple of the atomic resonance, or (2) part of a secondary system phase locked to the primary loop as shown in Fig. 4-2. Note that the feedback control loop attempts to match the multiplied frequency of the controlled oscillator to the frequency of the atomic resonance. The response time of the control loop determines the length of time that errors in the crystal oscillator continue before correction is made in terms of atomic resonance. In time intervals short compared with loop response period, the oscillator stability is that of the crystal oscillator, while in the case of time intervals that are long compared with the loop response period, the stability is that of atomic resonance.

An important consideration in the design of the control system is the noise associated with the atomic resonance signal. This noise will modulate the crystal oscillator frequency and degrade the short-term stability performance. To combat this, the control loop response speed may be limited by the designer and thus effectively filter the noise before it reaches the oscillator. However, disturbances caused by the environment can introduce errors that are not corrected as quickly and effectively as may be required. Compromises are, therefore, necessary to best match the application. Another frequency modulation noise may arise from the frequency multiplication process but in most practical devices, the resonance signal noise dominates[2].

The larger signal-to-noise ratio of the rubidium resonance controlled oscillators makes them more effective for short-term control of frequency errors induced by shock and vibration than those controlled by cesium resonance. Other design factors include phase control of the modulation signals, harmonic distortion of these signals, and the spectral purity of the microwave signal applied to the atomic resonator[28].

### 4-4.1.2 Cesium Beam Standards

Cesium beam standards are maintained at the National Bureau of Standards (NBS) and at the U S Naval Observatory (USNO), the two organizations chiefly involved in distributing accurate and precise time and frequency information within the U.S. NBS is responsible for the custody, maintenance, and development of the national standards of frequency and time (interval) as well as their dissemination to the general public. This mission of USNO includes the provision of accurate time as an integral part of its work associated with the publication of ephemerides in support of navigation and in the establishment of a fundamental reference system in space[29].

The National Bureau of Standards has developed several cesium standards and one thallium standard. NBS-I, the oldest machine, was originally operated using cesium but was converted to thallium in 1962. NBS-II provides a line width of 110 Hz, while NBS-III, the longest machine, provides a spectral line width of 48 Hz (see Fig. 4-15[30]). NBS-5 has an accuracy of better than 1 part in $10^{12}$.

All of the NBS devices use the Ramsey-type excitation instead of the Rabi-type. The original Rabi method of exciting the atomic transitions in an atomic beam resonance experiment uses a single oscillating field[31]. Ramsey introduced a method by which the transitions are excited by two separated oscillating fields. Advantages of this method

Figure 4-15. Cesium Atomic Beam Frequency Standard, NBS-III

buffer gas that interacts with the rubidium gas to cause a pressure shift. Commercial development of rubidium gas cell controlled oscillators has led to refined designs entirely based upon solid-state electronic circuits. The rubidium instrument is compact and operates reliably. A militarized version designed for U S Army tactical applications is illustrated in Fig. 4-16[2]. Capable of surviving rough handling and the problems involved in a wet environment, these units have been designed for operation in guided missiles and aircraft as well as in general applications for which they can be installed in a relay rack[2].

## 4-4.2 MASERS

### 4-4.2.1 System Considerations

Atomic hydrogen masers and rubidium gas cell masers are active oscillators that provide output signals directly from the quantum transitions within the atomic particles. For practical applications, it is necessary to make maser signals available at millivolt power

are[32]: (1) improved resolution of the spectrometer, (2) a high degree of uniformity of the magnetic fields not required, and (3) a higher resolution than a single field when observing very high frequency transitions. This latter advantage is gained at the expense of a reduction in the signal-to-noise ratio.

Examples of commercial cesium atomic beam frequency standards are Hewlett-Packard Models 5061A and 5062A. The former is available with two different beam tubes, standard and high performance.

### 4-4.1.3 Rubidium Gas Cell Standards

Rubidium gas cell standards require, as a component of the gas cell mixture, an inert

Figure 4-16. Tactical Rubidium Gas Cell Frequency Standard[2]

levels and synthesized to convenient frequencies. For the hydrogen maser, accessories are available to phase lock high quality crystal oscillators to the maser signal by using commercial synthesizers. The basic block arrangement is shown in Fig. 4-3.

Noise is a fundamental factor that limits stability and the spectral purity of the signal in maser frequency standards. The internal noise of the oscillator includes thermal noise and noise characteristics of the maser type and design. External circuits also contribute additional noise. The mixers are the sources of most of the noise and where very short term stability or high spectral purity are important, low noise preamplifiers are used ahead of the mixers[2].

### 4-4.2.2 Maser Devices

Techniques and design principles relevant to the construction and operation of hydrogen masers are well documented[33]. An atomic hydrogen maser clock system for installation in a satellite to be used for measuring the gravitational red shift has been developed, and performance data presented for the operation of a laboratory model, see Fig. 4-17[34]. Prototype atomic hydrogen standards for field use at tracking stations have been developed[35]. Designed for minimal operator attention, the unit contains a hydrogen maser, receiver-synthesizer, automatic cavity tuner, clock, standard output frequencies and time signals, and other system electronics.

Another type of maser that has been developed is the optically pumped rubidium

Figure 4-17. Satellite Maser With Bell Jar Removed

maser oscillator[36]. It is considerably smaller than hydrogen maser, but affected by the same factors that limit the long term stability of the rubidium gas cell standard. Short-term stability is expected to be one part in $10^{12}$ for observation times of 1 sec and long-term stability to be about one part in $10^{11}$ per month. Table 4-4[2] compares physical data of three types of atomic frequency standards.

### TABLE 4-4

### PHYSICAL DATA RELATING TO AVAILABLE ATOMIC FREQUENCY STANDARDS[2]

| Characteristic | Atomic Hydrogen Maser | Rubidium Gas Cell Controlled Oscillator | Cesium Atomic Beam (24-in.) Controlled Oscillator |
|---|---|---|---|
| Nominal Resonance Frequency | 1420.405751 MHz | 6834.682608 MHz | 9192.631770 MHz |
| Resonance Width | 1 Hz | 200 Hz (typical) | 250 Hz |
| Atomic Interaction Time $\tau_A$ | 0.5 sec | $2 \times 10^{-3}$ sec (typical) | $2.5 \times 10^{-3}$ sec Interaction length, $L = 25$ cm (typical) |
| Atomic Resonance Events per Second | $10^{12}$ | $10^{12}$ | $10^6$ |
| Principal Frequency Offsets — Magnetic | $f - f_0 = 2750\ B^2$ (gauss) $5 \times 10^{-13}$ (typical | $f - f_0 = 574\ B^2$ (gauss) $1 \times 10^{-9}$ (typical) | $f - f_0 = 427\ B^2$ (gauss) $1 \times 10^{-10}$ (typical) |
| Principal Frequency Offsets — 2nd Order Doppler | $4 \times 10^{-11}$ ($\partial f/\partial T = 1.4 \times 10^{-13}/°K$) | $8 \times 10^{-13}$ | $3 \times 10^{-13}$ |
| Principal Frequency Offsets — Collisions | $2 \times 10^{-11}$ | $3 \times 10^{-7}$ (typical) | none |
| State Selection Method | Atomic Beam Deflection in Hexapole Magnets | Optical Pumping | Atomic Beam Deflection in Dipole or Multipole Magnets |
| Resonance Detection Method | Atomic Microwave Radiation (active maser oscillation) | Optical Absorption | Surface Ionization of Deflected Atoms |
| Temperature of Resonating Atoms | 300°K | 330°K | 360°K |

### REFERENCES

1. F. D. Lewis, "Frequency and Time Standards", Proceedings of the IRE 43, 1046-68 (Sept. 1955).

2. A. O. McCoubrey, "A Survey of Atomic Frequency Standards", Proceedings of the IEEE, 54, 117-35 (February 1966).

3. Frequency and Time Standards, Hewlett-Packard Co., Palo Alto, Calif., Application Note 52, 1965.

4. O. P. Layden et al., "Crystal-Controlled Oscillators", IEEE Transactions on Inst. and Meas., IM-21, 277-86 (August 1972).

5. E. A. Gerber and R. A. Sykes, "State-of-the-Art — Quartz Crystal Units and Oscil-

lators", Proceedings of the IEEE, **54**, 103-16 (Feb. 1966).

6. Richard B. Belser and Walter H. Hicklin, "AT-Cut Resonators With Annular Electrodes", *Proceedings of the 21st Annual Symposium on Frequency Control*, U S Army Electronics Command, Fort Monmouth, N. J., April 1967, pp. 211-23.

7. D. M. Eisen, "New Developments in Glass Enclosed Crystal Units", *Proceedings of the 18th Annual Symposium on Frequency Control*, U S Army Electronics Command, Fort Monmouth, N. J., May 1964, pp. 204-216.

8. R. A. Sykes, W. L. Smith, and W. J. Spencer, "Studies on High Precision Resonators", *Proceedings of the 17th Annual Symposium on Frequency Control*, U S Army Research and Development Laboratory, Fort Monmouth, N. J., May 1963, pp. 4-27.

9. Guy Gibert, "Improvements in Sealing HC-26/U and HC-27/U Glass Holders", *Proceedings of the 22nd Annual Symposium on Frequency Control*, U S Army Electronics Command, Fort Monmouth, N. J., April 1968, pp. 155-62.

10. F. R. Brandt and G. E. Ritter, "SSB Quartz Crystal Units Utilizing Coldweld Enclosures and High Temperature Bakeout Techniques", *Proceedings of the 21st Annual Symposium on Frequency Control*", U S Army Electronics Command Fort Monmouth, N. J., April 1967, pp. 224-43.

11. W. G. Stoddart, "A New Design for Microminiature Crystals", *Proceedings of the 18th Annual Symposium on Frequency Control*, U S Army Electronics Command, Fort Monmouth, N. J., May 1964, pp. 181-91.

12. E. A. Gerber and R. A. Sykes, "Quartz Frequency Standards", Proceedings of the IEEE, **55**, 783-91 (June 1967).

13. H. E. Bommel, W. P. Mason, and A. W. Warner, Jr., "Dislocations, Relaxations and Anelasticity of Crystal Quartz", Physical Review, **102**, 64-71, April 1956.

14. W. P. Mason, Developments in Ultrasonics, *Proceedings of the 18th Annual Symposium on Frequency Control*, U S Army Electronics Command, Fort Monmouth, N. J., May 1964, pp. 12-42.

15. J. C. King, A. A. Ballman, and R. A. Laudise, "Improvement of the Mechanical Q of Quartz by the Addition of Impurities to the Growth Solution", Phys. Chem. Solids, **23**, 283-314, June 1962.

16. "Development of a High Q, BT-cut Resonator", British Journal of Applied Physics, **16**, 1341-6, September 1965.

17. G. K. Guttwein, A. D. Ballato, and T. J. Lukascek "Design Considerations for Oscillator Crystals", *Proceedings of the 22nd Annual Symposium on Frequency Control*, U S Army Electronics Command, Fort Monmouth, N. J., April 1968, pp. 67-88.

18. MIL-C-3098E, *General Specification for Quartz Crystal Unit*, Dept. of Defense, 28 November 1967.

19. J. M. Wolfskill, "Advancements in Production of 5 MHz Fifth Overtone High Precision Crystal Units", *Proceedings of the 22nd Annual Symposium on Frequency Control*, U S Army Electronics Command, Fort Monmouth, N. J., April 1968, pp. 89-117.

20. M. Bloch, et al., "Kold-Seal Thermal

Compression Bonded Crystals", *Proceedings of the 22nd Annual Symposium on Frequency Control*, U S Army Electronics Command, Fort Monmouth, N. J., April 1968, pp. 118-35.

21. Erich Hafner, "Theory of Oscillator Design", *Proceedings of the 17th Annual Frequency Control Symposium*, U S Army Electronics Command, Fort Monmouth, N. J., 27-29 May 1963, pp. 508-36.

22. Erich Hafner, *Analysis and Design of Crystal Oscillator, Part I*, Tech. Report ECOM 2473, U S Army Electronics Command, Fort Monmouth, N. J., May 1964.

23. *Quartz Crystal Oscillator Circuits, Design Handbook*, Serial TP64-1072, The Magnavox Co., Fort Wayne, Ind., Contract DA-36-039-AMC-00043(E), 15 March 1965.

24. M. Bloch, et al., "Low Power Crystal Ovens", *Proceedings of the 20th Annual Symposium on Frequency Control*, U S Army Electronics Command, Fort Monmouth, N.J., April 1966, pp. 530-43.

25. F. Destal, "The Increasing Applications of Tuning Forks and Other Vibrating Metal Resonators in Frequency Control Systems", *Proceedings of the 19th Annual Symposium on Frequency Control*, U S Army Electronics Command, Fort Monmouth, N.J., April 1965, pp. 57-78.

26. H. P. Westman, Ed., *Reference Data for Radio Engineers*, Fifth Edition, Howard W. Sams and Co., Inc., New York, March 1969, p. 22-13.

27. O. Colpen and H. E. Gruen, "Tuning Precision Oscillators", *Proceedings of the 13th Annual Symposium on Frequency Control*, U S Army Research and Development Laboratory, Fort Monmouth, N.J., May 1959, pp. 165-81.

28. J. H. Holloway and R. F. Lacey, "Factors Which Limit the Accuracy of Cesium Atomic Beam Frequency Standards", *Proceedings of the 6th International Conference on Chronometry*, Lausanne, 1964, 1964, p. 317.

29. J. A. Barnes and G. U. R. Winkler, "The Standards of Time and Frequency in the USA", *Proceedings of the 26th Annual Symposium on Frequency Control*, U S Army Electronics Command, Fort Monmouth, N.J., June 1972, pp. 269-78.

30. R. E. Beehler and D. J. Glaze, "The Performance and Capability of Cesium Beam Frequency Standards at the National Bureau of Standards", IEEE Transactions, **IM-15**, 48-55 (March-June, 1966).

31. I. I. Rabi, et al., "A New Method of Measuring Nuclear Magnetic Moment", Phys. Rev., **53**, 318 (1938).

32. R. C. Mockler, et al., "Atomic Beam Frequency Standards", IRE Transactions on Instrumentation, **I-9**, 100-32 (September 1960).

33. D. Kleppner, et al., "Hydrogen Maser Principles and Techniques", Phys Rev., **138**, A972-83 (May 1965).

34. R. Vessot, et al., "The Design of an Atomic Hydrogen Maser System for Satellite Experiments", *Proceedings of the 21st Annual Symposium on Frequency Control*, U S Army Electronics Command, Fort Monmouth, N. J., April 1967, pp. 512-25.

35. H. E. Peters, et al., "Atomic Hydrogen Standards for NASA Tracking Stations",

*Proceedings of the 23rd Annual Symposium on Frequency Control*, U S Army Electronics Command, Fort Monmouth, N. J., May 1969, pp. 297-304.

36. P. Davidovits and R. Novick, "The Optically Pumped Rubidium Maser", Proceedings of IEEE, **54**, 155-70 (February 1966).

CHAPTER 5

TIMING EQUIPMENT AND MEASUREMENT

## 5-1 INTRODUCTION

Chapters 2, 3, and 4 are concerned with the basic considerations of setting up precision reference timing systems. In this chapter we deal with the equipment and techniques available to the local user who must solve problems requiring accurate knowledge of time. For accurate time keeping — regardless of whether the required time scale is atomic time, universal time, or any other (see par. 2-1) — the local system should (1) provide a consistent time interval, (2) be initially synchronized with the master time, and (3) be checked periodically against the master time to ensure that the scales remain in correspondence. In other words, accurate time keeping requires an accurate and stable measuring system and effective means of synchronizing this system with the master. The equipment, techniques, and resources available to the local user to achieve these ends are discussed in the paragraphs that follow.

## 5-2 TIME MEASUREMENT

Precision time keeping or time indication usually requires the use of a very stable oscillator such as a quartz crystal oscillator (see pars. 3-2 and 4-2) or one that depends on the frequency corresponding to a particular energy-level transition (see pars. 3-4 and 4-4). Military Specifications for the cesium beam frequency standard are listed as Refs. 1-4.

Unless very long time intervals are being measured, the full stability inherent in these oscillators cannot be obtained with ordinary techniques. By use of the most basic method of time interval measurement, the number of complete oscillator cycles which elapse between start and stop of an interval are counted electronically. If, for example, the oscillator frequency were 100 MHz and its short-term instability were one part in $10^{10}$, then oscillator stability would be the limiting factor in the measurement accuracy only for time intervals greater than a hundred seconds. To make full use of oscillator stability for shorter time intervals, measurements can be made of the fractional cycles that occur during the time interval[5]. To measure the number of oscillations in an interval, use is made of electronic counters. The basic elements of counters include (1) decade counting assemblies, (2) a main gate that controls the time interval during which the input signal is measured, (3) a time base which supplies a reference of time for the main gate, (4) decade divider assemblies, and (5) a means of shaping the input signal to provide for proper operation of the gate. Numerical readouts are used to display the count.

## 5-2.1 GATED COUNTER

Transient or repeated time intervals often are measured using a gated counter to count the cycles of an oscillator. In its simplest form, a gated counter (Fig. 5-1[5]) forms a pulse from the input sinusoid and this pulse train is "gated" into a counter. Referring to the figure, the crystal oscillator is free-running and thus can be highly stable. Before the initiation of a time measurement, the reset multivibrator sets the counter to zero. Transmission of the crystal-controlled train to the counter is prevented by the gate. When a start pulse arrives, the gate binary switches

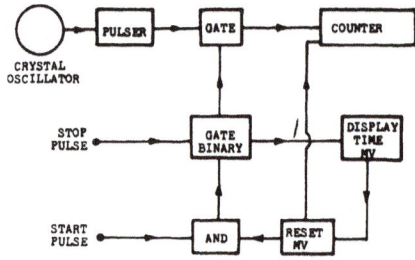

Figure 5-1. Gated Counter[5]

and opens the gate that allows subsequent time pulses to pass through to the counter. These are counted until a stop pulse arrives, and then the gate binary returns to its original state and the passage of timing pulses to the counter is stopped. The display-time multivibrator deactivates the gate binary so that the accumulated count can be viewed[5].

For transient measurements, the accuracy obviously is limited to ± 1 cycle of oscillation in a system using a simple polarized threshold detector. The accuracy of this "gated counter" system would be doubled if a bipolar threshold detection system were employed. However, any increases in order of magnitude are not assured and the counter has to operate at higher speeds. Present technology sets a counting rate limit on binary counters at approximately 100 MHz and, consequently, accuracies greater than 10 nsec cannot be obtained by the brute force application of this method[5].

A somewhat better accuracy can be obtained from the gated counter approach when doing repetitive measurements. Results of an extensive analysis indicate that 100 measurements are required for an accuracy of 1 nsec with 90% confidence with a 100-MHz counter[5]. Recycle and display typically requires several milliseconds, and thus 100 measurements might require 1 sec (Ref. 6).

## 5-2.2 VERNIER SYSTEM COUNTER

The next level of improvement in accuracy has been achieved by the use of "vernier systems"[7]. Essentially, the principle of operation is that a coarse counter counts the whole number of crystal clock oscillations while vernier counters indicate the correction due to fractional clock cycles. A block diagram of the system is shown in Fig. 5-2[6]. The correction is measured by causing a vernier oscillator to commence pulsing at the start of the time interval. The vernier oscillator frequency is slightly higher than that of the clock oscillator, and vernier oscillations are counted until the vernier oscillator comes into phase with the clock generator. A measure of this phase of the clock generator then can be obtained. This phase determination at the end of the time interval is made by another vernier actuated by the stop signal[5]. Two restrictions on this method are (1) there are stringent stability requirements for the vernier generators, and (2) there is a "dead time" because the vernier measurement is not available until the vernier and the clock coincide.

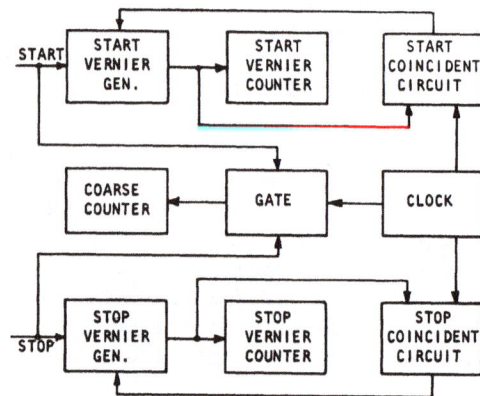

Figure 5-2. Vernier Counting System[6]

## 5-2.3 DIGITAL INTERPOLATION COUNTER

Another technique that does not have the restrictions of the vernier is the digital interpolation technique[5]. In this method, a pulse train derived from and synchronous with the clock oscillator is fed into a delay line. At the start of a time interval to be measured, a pulse is developed that is applied to the opposite end of the line. If the line has the proper length, there will be a superposition of pulses at only one point on the line. By tapping the line at regular intervals and connecting threshold detectors to each tap, the detector at or near a point of superposition will be triggered. Since the location of the detector is a measure of the time by which the start of a given time interval and a particular clock pulse differ, the detector at that point can be used to drive an indicator as in a time interval measurement. It has been shown that this method makes possible the improvement in accuracy of a conventional gated-counter time measuring system by a factor of 5 (exclusive of crystal accuracy)[5].

## 5-2.4 COMPUTING COUNTER

Computing counters combine the most precise electronic counter with a built-in arithmetic capability. With this combined system, the raw data from the measurement section can be treated mathematically to produce the desired form. For example, the computer section can be programmed to correct for fractional frequency deviation of the source. At present these counter systems can measure fractional frequency divisions as small as $5 \times 10^{-10}$ for 1 sec averaging time.

## 5-3 SYNCHRONIZATION SYSTEMS

### 5-3.1 RADIO TRANSMISSION

Time synchronization accuracy to ± 1 msec can be achieved with presently available standard time signals such as those transmitted by station WWV[7]. The method used here requires the determination of the propagation delay between the transmitter and the clock station. This delay then is applied as a correction to the clock reading[8]. In addition, HF services broadcast time of day information. Time ticks of these broadcasts may be used in much the same way. With present LF/VLF time services and the receiving equipment now available, time synchronization accuracy is a few milliseconds[9].

### 5-3.1.1 Factors Affecting Radio Transmission

The principal factors that affect the propagation delay for HF signals are (1) the great circle distance between transmitter and receiver, (2) the transmission mode, and (3) the virtual height of the ionosphere reflection layers. How these parameters are determined and how the propagation delay is calculated are discussed in the paragraphs that follow.

### 5-3.1.1.1 Great Circle Distance

A straightforward way to compute the great circle distance between two points makes use of haversines. Refer to Fig. 5-3[9],

*Figure 5-3. Great Circle Distance Calculation[6]*

suppose A and B are two points on the earth for which the latitude and longitude are known. The following relationships must then be found:

(1)  For longitude

$$Lo_{AB} = Lo_A - Lo_B \text{ , deg} \qquad (5\text{-}1)$$

where

$Lo_A$  = longitude of Point A, deg

$Lo_B$  = longitude of Point B, deg

(2)  For latitude, if the two points are on the same side of the equator

$$L_{AB} = L_A - L_B \text{ , deg} \qquad (5\text{-}2)$$

where

$L_A$  = latitude of Point  A, deg

$L_B$  = latitude of Point  B, deg

If the two points are on opposite sides of the equator,

$$L_{AB} = L_A + L_B \qquad (5\text{-}3)$$

The haversine of the great circle distance, in degrees of arc, is given by[9]

$$\text{hav } D = (\cos L_A)(\cos L_B)(\text{hav } Lo_{AB})$$
$$+ \text{ hav } L_{AB} \qquad (5\text{-}4)$$

Note that the haversine of an angle $\theta = (1 - \cos \theta)/2 = \sin^2 (\theta/2)$. Great circle distance can be determined by reference to a table of haversines versus angles[10].

## 5-3.1.1.2 Transmission Modes

The propagation path is a groundwave path for most LF/VLF transmissions. Short distance HF transmissions closely follow the great circle route from transmitter to receiver. For HF transmission over a distance of more than about 160 km, the propagation path is a sky-wave.

The maximum distance that can be spanned by a single hop (i.e., one reflection from the ionosphere) via the F2 layer is about 4000 km (see Fig. 5-4[9]). It follows then that the fewest number of hops between transmitter and receiver is the next integer greater than the great circle distance divided by 4000. Transmission modes with one or more hops greater than the minimum number of hops occur frequently as shown in Fig. 5-5[9].

## 5-3.1.1.3 Height of Ionosphere

Long distance HF transmissions usually are reflected from the F2 layer that varies in height from about 250 to 450 km. It has been determined from experience that the virtual height of the F2 layer averages about 350 km (see Fig. 5-4). This height can be used for delay estimation.

Once the transmission modes and layer heights have been determined, the transmission delay can be determined graphically by reference to Fig. 5-6[9]. This graph is plotted for a single hop, but it can be used to determine the distance for multiple hops. The

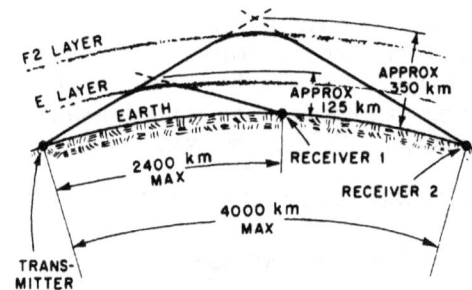

Figure 5-4. Single-hop Sky-wave Paths

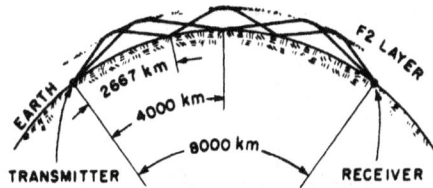

*Figure 5-5. Multiple-hop Transmission Path*

distance covered by a single hop and the delay for a single hop is determined, and this delay is then multiplied by the total number of hops to obtain the total delay.

This propagation delay is applied as a correction to the clock reading, and the clock ticks are then produced in synchronism with the transmitted master timing signal (i.e., WWV). WWV is a high-frequency station operated by the U S National Bureau of Standards at Ft. Collins, Colorado. Its time code transmission is made for one minute out of five, ten times per hour, and is produced at a 100 parts per second rate carried on a 1000-Hz modulation. Fig. 5-7[8] shows time code transmission from station WWV.

### 5-3.1.2 Normal Radio Transmission

### 5-3.1.2.1 LF/VLF Radio Transmission

Low frequency and very low frequency signals follow the earth's curvature, and are in effect guided by the ionosphere acting as a boundary. The high phase stability and long range coverage of the lower frequencies make them valuable for standard frequency transmissions. Time synchronization with the present LF/VLF services, as with HF, must include the factors of propagation and receiving system delay except for the simplification of all ground wave transmission. VLF waves propagate at about 292,000 km/sec as compared with 278,000 km/sec for HF waves[9]. Since LF and VLF transmissions are propagated for relatively great distances

by ground wave, propagation delay for these frequencies usually can be found directly, after the great circle distance is calculated.

Whereas time synchronization with LF/VLF could be accomplished to no better than a few milliseconds before, there is a new method, using dual VLF transmissions, in which the synchronization can be accomplished to an accuracy of 2 or 3 $\mu$sec or better[10]. In this method, the Dual Frequency Transmissions of Radio Station WWVL at Fort Collins, Colorado, were used as a means of time synchronization of remote clocks. A special receiving system was designed and fabricated for the simultaneous reception of either 19.9 and 20.0 kHz or 20.0 and 20.5 kHz. The system automatically combines the received signals to produce a timing pulse train with a repetition rate equal to the difference frequency and with a phase determined by the instantaneous phase coincidence of the received signals. Two receiving systems were operated simultaneously, one at NASA, Greenbelt, Maryland, and one at RMS Engineering, Inc., Atlanta, Georgia. Adequate data were obtained with the 100-Hz separation to confirm a received time accuracy of better than a few microseconds during certain hours of uninterrupted data[10].

Another system similar in concept is the NBS-NASA system[11]. The frequencies used here are the 20.0-kHz standard frequency carrier, with either 20.5 or 19.9 kHz as the single auxiliary frequency for a given test period. These frequencies were all synthesized and phase locked to the USFS. Carrier period ambiguities were 50 $\mu$sec; difference frequency signal ambiguities of the period were either 2 or 10 msec. The single transmitter and antenna at Fort Collins, Colorado was used because frequency shift keying was used. This procedure allowed 10-sec transmission time alternately for each frequency. To relate these transmissions to UTC, the phases of the two frequencies being broadcast were adjust-

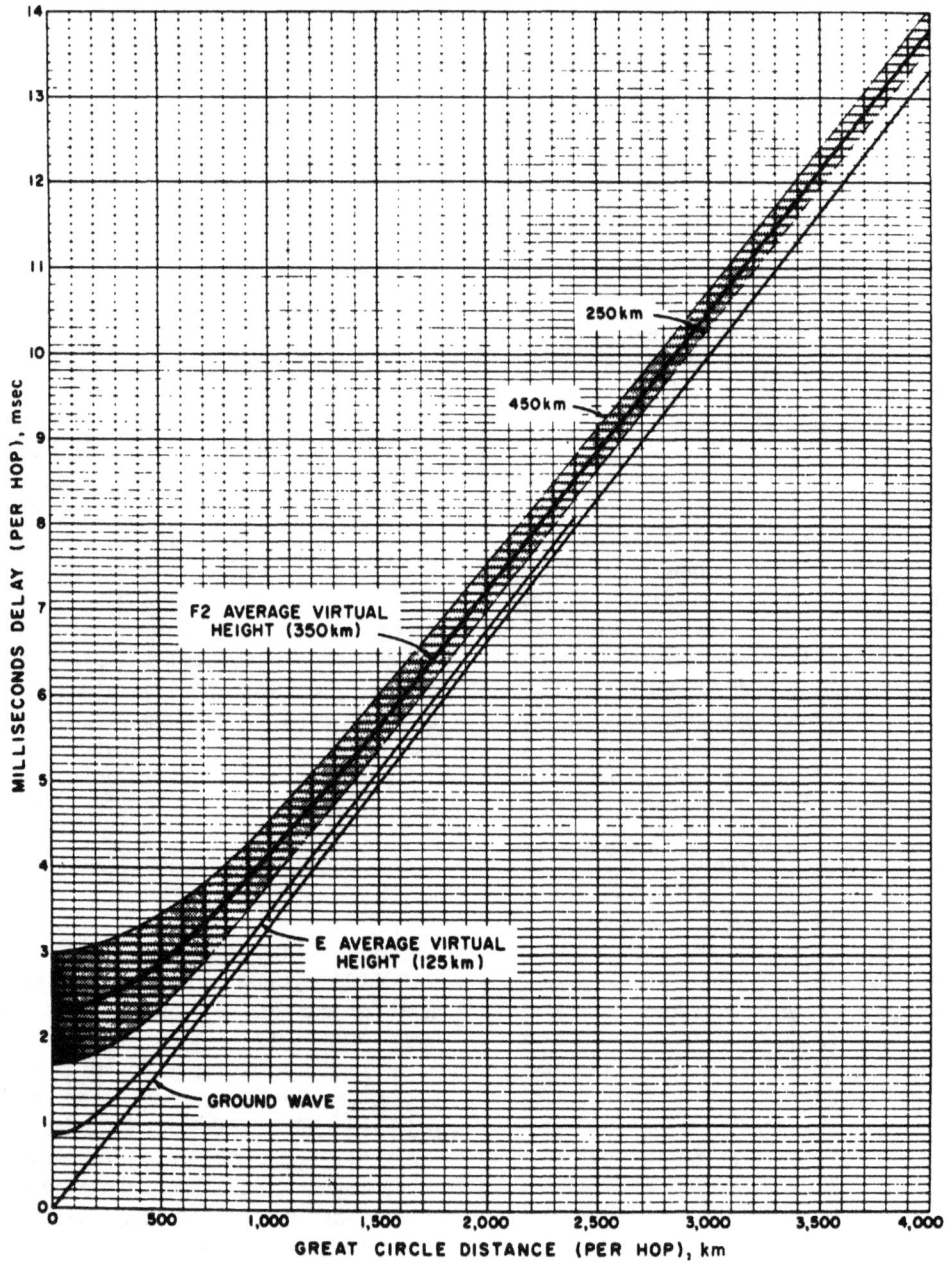

Figure 5-6. Transmission Delay Graph

FORMAT H, SIGNAL HOO1, IS COMPOSED OF THE FOLLOWING:

1) 1 ppm FRAME REFERENCE MARKERS R = (P₀ AND 1.03 SECOND "HOLE")
2) BINARY CODED DECIMAL TIME-OF-YEAR CODE WORD (23 DIGITS)
3) CONTROL FUNCTIONS (9 DIGITS) USED FOR UT₁ CORRECTIONS
4) 6 ppm POSITION IDENTIFIERS (P₁ THROUGH P₅)
5) 1 pps INDEX MARKERS

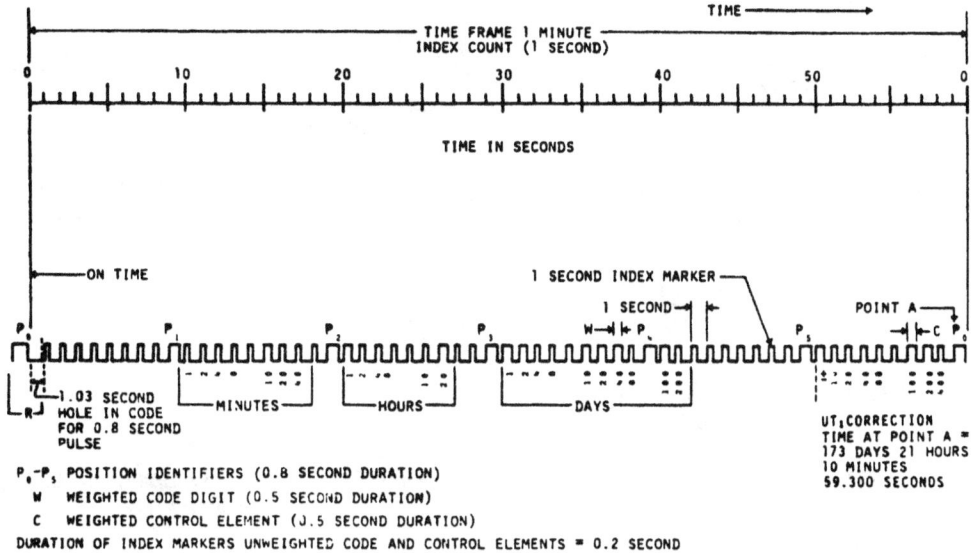

Figure 5-7. Chart of Time Code Transmissions from NBS Station WWV[9]

ed at the transmitter to go through zero simultaneously in a positive direction coincident with a second's tick of UTC. Phase relations are maintained, upon reception, with the time of simultaneous phase agreement with a delay the amount of which depends upon the phase velocity of the signals and the distance between receiver and transmitter.

To extract timing information from the signals, one or two conventional phase tracking VLF receivers can be used. To relate the phases of the received signals to the time scale operating at the receiving site for time synchronization, a calibrator was used to generate the same two frequencies as transmitted from the local time scale, with the same simultaneous phase relationship conditions. The calibrator consists of a sawtooth waveform generator that is synchro-

nized by a 100 pulse per second rate taken from the local time scale divider chain. An interchange of signals from the receiving antenna and the calibrator output permits a measurement to be made of the pair of phase differences. From the two measured quantities, the relation between transmitter time scale and receiver time scale may be calculated, provided that the carrier propagation delays are known. Propagation fluctuations on the path from Fort Collins, Colorado, to the Goddard Space Flight Center at Greenbelt, Maryland were studied by keeping the clocks at the two locations synchronized using portable clocks. Results are shown in Figs. 5-8[11] and 5-9[11] that indicate the propagation delay as derived from the difference frequency measurements. To evaluate the maximum stability of synchronization of the system, the day-to-day

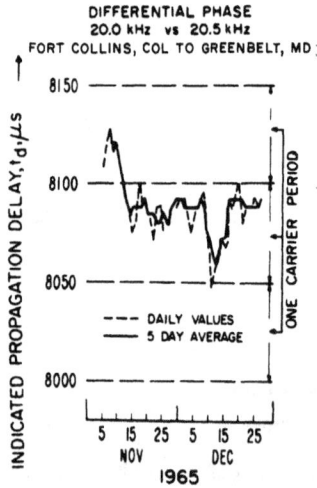

Figure 5-8. Indicated Propagation Delay,
20.0 vs 20.5 kHz[9]

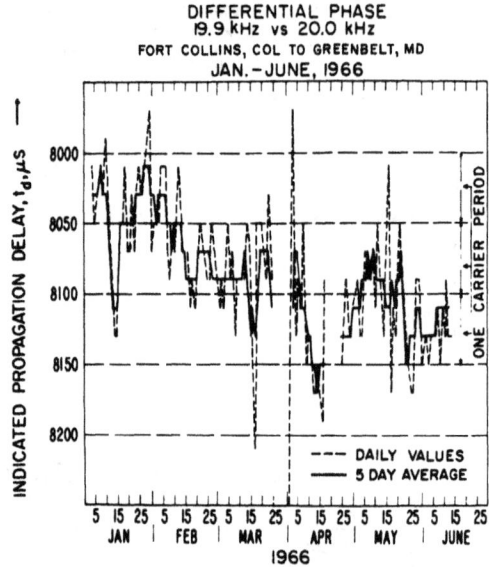

Figure 5-9. Indicated Propagation Delay,
19.9 vs 20.0 kHz[9]

time variation of the received carrier was plotted as shown in Fig. 5-10[11]. If the same cycle of the carrier had been identified each day, this figure would indicate the system capability of synchronization.

### 5-3.1.2.2 Loran-C Synchronization

A pulsed radio navigation system, operated by the U S Coast Guard, called the Loran-C system, uses a 100-kHz carrier and offers a means for precision transfers of time. Clock synchronization to ± 1 μsec is possible within the range of any station of the East Coast Loran-C chain. The master Station of the East Coast Chain, at Cape Fear, N.C., is controlled with respect to the U S Naval Observatory master clock[9].

The mechanics of station clock synchronization to Loran-C transmissions depend on the type of receiver being used and the required timing precision. Fig. 5-11[12] shows a simple visual timing system. This system can be used within a few hundred kilometers of a master chain transmitting 1 part per sec

identifiers or within that range of any station on the East Coast chain. Useful range may be expanded by photographic integration, and accuracies up to ± 20 μsec for ground waves can be achieved. Fig. 5-12[12] shows a visual timing system for use with any skywave situation and for ground waves not requiring automatic receiver accuracies. An epoch monitor system (Fig. 5-13[12]) produces a pulse rate that is reset to the delayed 1 part per sec. The phase decodes the signal. Accuracies are limited by ability to cycle-select and may be improved by photographic integration.

A phase tracking system is shown in Fig. 5-14[12]. This system provides a record of the local oscillator phase drift with respect to Loran-C. Used within the ground range of a Loran-C station, the resulting record, which is free of diurnal shift, can be used to compare the station oscillator with the cesium reference at the Loran-C station to within one or two parts in $10^{12}$ per 24-hr period.

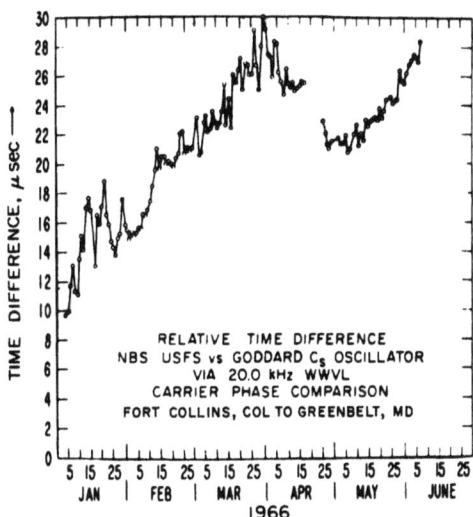

Figure 5-10. Residual Daily VLF Time Fluctuations[9]

Accuracies are generally in the range of ± 20 μsec. An automatic receiver system, Fig. 5-15[12], is used depending on the particular chain to be monitored. An automatic receiver has the capability of cycle-selection, and accuracies up to 0.1 μsec have been attained. The system can be calibrated by a Loran-C simulator.

A recent comparison among three precise timing centers in the United States was conducted for over one year[13]. These data involved a 3500-km ground wave path, and the long term precision achieved between the three remote stations was 2 μsec.

### 5-3.1.2.3 Omega Synchronization

Omega is a very low frequency navigation system. The basic navigational frequency of Omega is 10.2 kHz. In addition, navigational

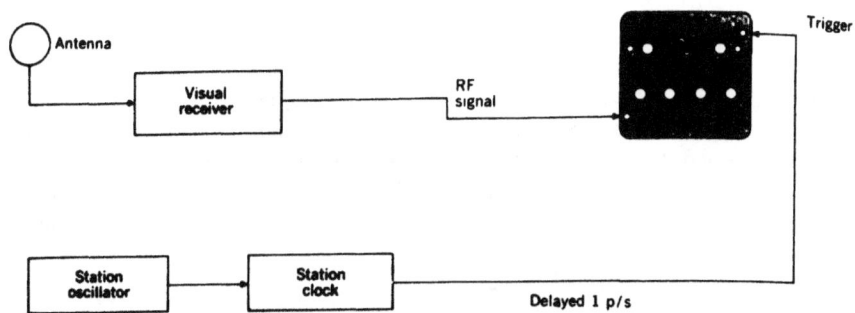

Figure 5-11. Simple Visual Timing System[9]

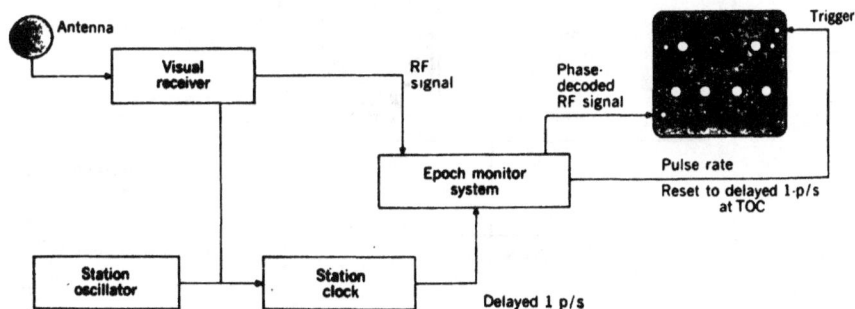

Figure 5-12. Visual Timing System[11]

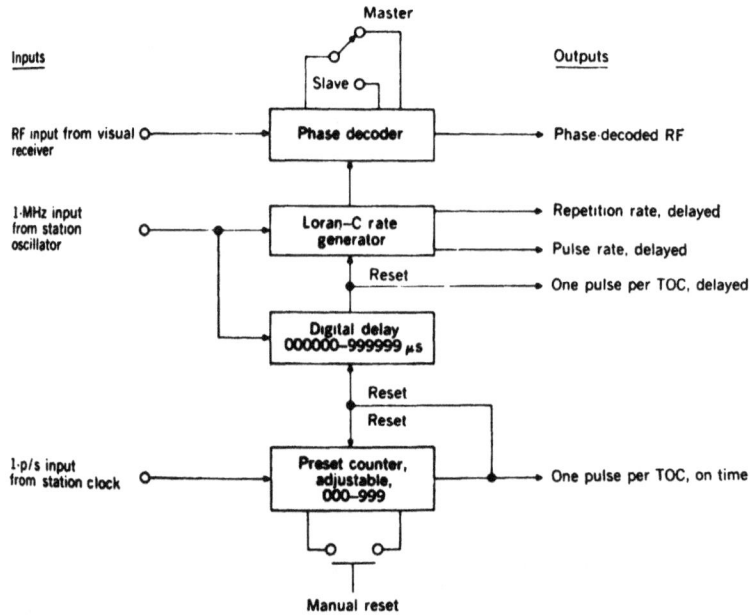

*Figure 5-13. Simplified Diagram of an Epoch Monitor System*

frequencies of 13.6 kHz and 11-1/3 kHz also are broadcast[14]. Epoch at timing sites may be deduced and maintained to better than 3 $\mu$sec, and frequency can be maintained to one part in $10^{12}$. Long bursts of continuous wave are transmitted (on the order of 1 sec or more) and repeated every 10 sec. All stations go off and on together, with 0.2 sec off between transmissions. There is no master-slave relationship, all stations being equal.

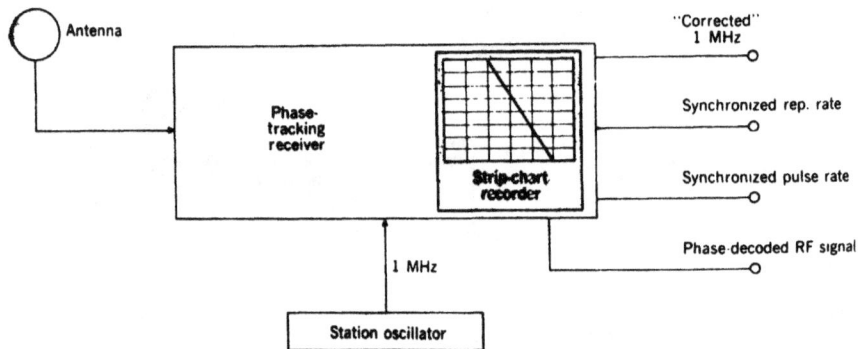

*Figure 5-14. Phase-tracking System[11]*

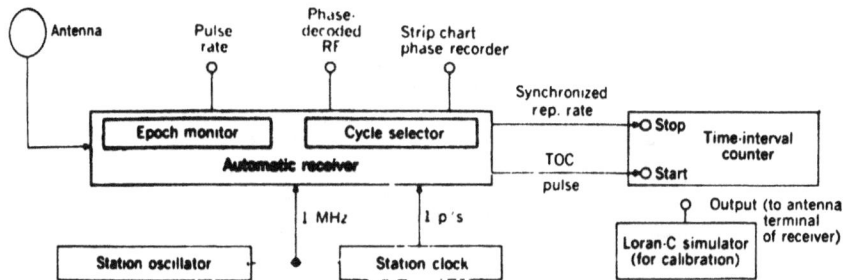

Figure 5-15. Automatic Receiver System[1][2]

### 5-3.1.2.4 TV Network Broadcasts

This method is readily available, inexpensive, and a common source for many users within the continental United States. In essence the time comparison is between the "read" from counters that are located near different clocks[13]. The counters are started by the clocks with the same tick from the local reference clocks. The counters are stopped by a "sync" pulse broadcast from a TV transmitter. The difference between any pairs of counter readings remains constant.

A system developed by NBS for use in the continental United States uses line-10 of the odd field in the 525-line system M as a passive transfer pulse. The broadcasts originate from the New York City studios of any or all of the three commercial TV networks. Carried through the TD-2 relay system, the received microwave signal is converted at a terminating station to a video signal and retransmitted by VHF or UHF to local service areas.

NBS distributes the line-10 daily measurements in terms of UTC (NBS) in the Monthly NBS Time Service Bulletin. The USNO distributes line-10 data in terms of UTC (USNO-MC) in the weekly Series 4 Time Services Bulletin. These publications also cover future changes, modifications to the TV system, or other factors affecting users in the field.

Advantages of line-10 timing include (1) simplicity, (2) low cost of comparison equipment, (3) low cost of maintaining synchronization with long range precision of 10 $\mu$sec, and (4) a method for simultaneous maintenance of microsecond synchronization of several clocks within the service area of one transmitter.

### 5-3.1.3 Satellite

Experiments in time transfers by use of satellites have shown that clocks can be synchronized to high precision at facilities equipped with satellite tracking antennas. Pulsed signals were transmitted simultaneously from ground stations over the satellite circuit with correlations to time kept by a master clock at the U S Naval Observatory[9]. Accuracy was within $\pm$ 1 $\mu$sec.

### 5-3.1.3.1 Telstar

By using narrow band signals between ground stations in the U S and the United Kingdom, a one-way transmission was achieved by means of Echo I (a passive reflector) in August 1960. The successful operation of Telstar 1 (an active relay satellite) provided an essentially symmetrical circuit for simultaneous, two-way transmission and reception having a high signal-to-noise ratio of 40 dB. The purpose of this experiment was to determine the time

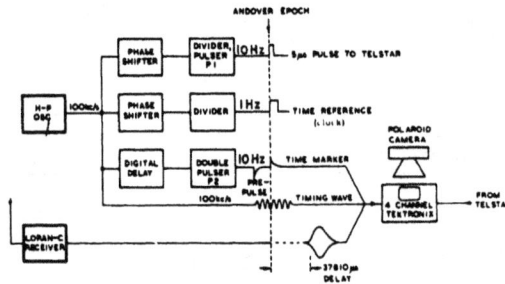

Figure 5-16. Pulse Generating and Measuring Arrangements at Andover, Me.[12]

difference between the master clocks of the U S Naval Observatory and the Royal Greenwich Observatory at Herstmonceux, U.K. Signals were transmitted simultaneously over the satellite circuit from the ground stations at Andover, Maine, and Goonhilly Downs, Cornwall, U.K. These signals were pulsed and were of a 5-μsec duration. The difference in time between the transmitted and received pulses was measured at each station, and the relative setting of the station clocks was obtained directly. The equipment used at Andover, Maine, (see Fig. 5-16[15]) included the following[15]:

(1) Hewlett-Packard (HP) quartz crystal oscillator and clock

(2) Pulser (P1) that initiated pulses to Telstar

(3) Pulser (P2) to provide two sets of pulses, one to trigger a 4-channel oscilloscope and one to provide time markers

(4) An oscilloscope camera

(5) Receivers for time signals transmitted on HF, VLF, and Loran-C.

Pulses were emitted using the Loran-C station on Nantucket Island, in groups of 8, spaced 1 msec apart, at a repetition rate of 20 groups/sec. The cycles within a pulse had a period of 10 μsec. The equipment at Goonhilly is indicated in Fig. 5-17[15]. The Post Office oscillator was a 100-kHz Essenring oscillator. Along with a chain of dividers, it gave an output signal of 1 Hz which provided the time reference. The 60-kHz transmission at MSF, Rugby, was used to relate Goonhilly to Herstmonceux. It was demonstrated by the use of Telstar that clocks at satellite ground stations on either side of the Atlantic could be synchronized with an accuracy of ± 1 μsec.

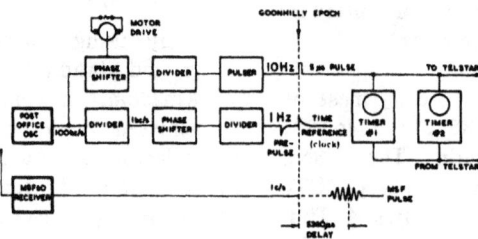

Figure 5-17. Interconnections of Generating and Measuring Equipment at Goonhilly Downs, Cornwall[12]

RELAY II SYNCHRONIZATION EXPERIMENT

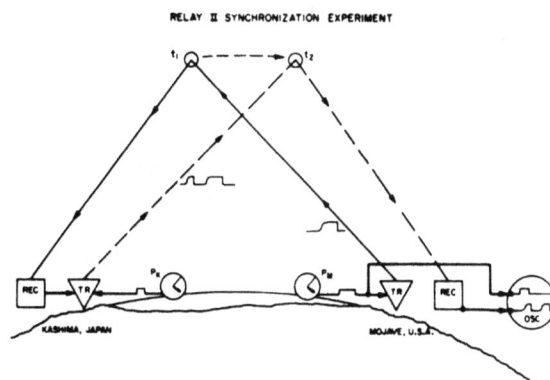

Figure 5-18. Retransmission of Pulses from Mojave by Kashima and Injection of Pulses at Kashima[12]

### 5-3.1.3.2 Synchronization Via Relay II Satellite

Experiments were carried out to synchronize clocks via the NASA Communications satellite Relay II between Mojave, U S., and Kashima, Japan in 1965. Pulses 11 μsec long were emitted from a pulse generator (Pm) at Mojave and pulses 5 μsec long were emitted from a pulse generator (Pk) at Kashima, at rates of 100, 1000, and 10,000 pulses per sec. Fig. 5-18[16] illustrates the determination of the time difference in clocks between Mojave and Kashima when the Pm and Pk pulsers were operating at 10,000 pulses per sec rate. Fig. 5-19[16] shows the pulses photographed on February 20, 1965 at Mojave. The first and fourth traces are 1-μsec time markers. The Mojave-transmitted pulse is shown in the second trace. The third trace shows the Kashima pulse transmitted on the left and the Mojave pulse retransmitted at the right. Figs. 5-20[16] and 5-21[16] indicate the oscilloscopes, pulsers, and oscillators used for generating and recording the pulses at Mojave and Kashima, respectively. By using this system, it was estimated that the clocks were related to ± 0.1 μsec on each part with a probable error, obtained independently at the two stations for each part, of about ± 0.01 μsec.

### 5-3.1.3.3 Meteor Burst Clock Synchronization

An experimental system has been developed in which the frequency offset between two remote frequency standards was determined through an exchange of timing information over a meteor burst radio propagation channel[17]. In this experiment, forward scatter propagation at VHF frequencies from a meteor burst provides a phase-stable, broadband propagation made for the transmission of timing information over

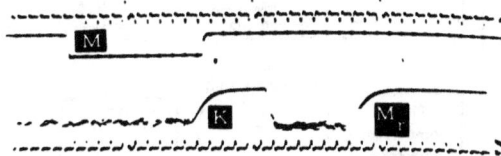

Figure 5-19. Pulses Photographed on February 20, 1965 at Mojave[12]

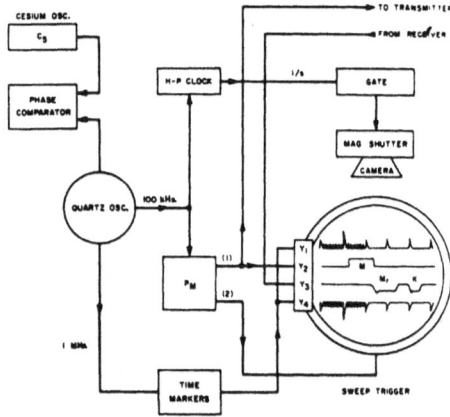

*Figure 5-20. Equipment Used at Mojave To Generate and Record Pulses*[5]

ranges up to 2000 km. A trail of electrons is used as a reflector to transmit information. The electron trail is created from the collision of the meteor with air particles in the lower E-region of the ionosphere. Techniques employed to measure instantaneous time offset were similar to those used in the Telstar experiment. Fig. 5-22[17] is a block diagram of the system used to synchronize clocks at Seattle, Washington, and Boseman, Montana. Values of $\Delta f/f$ measured by this meteor burst system, obtained by computer results, compare to within 4 parts in $10^{10}$ to the values measured by standard phase comparison methods.

## 5-3.2 TRANSFER STANDARD

To compare frequency and time between places far removed, a very satisfactory

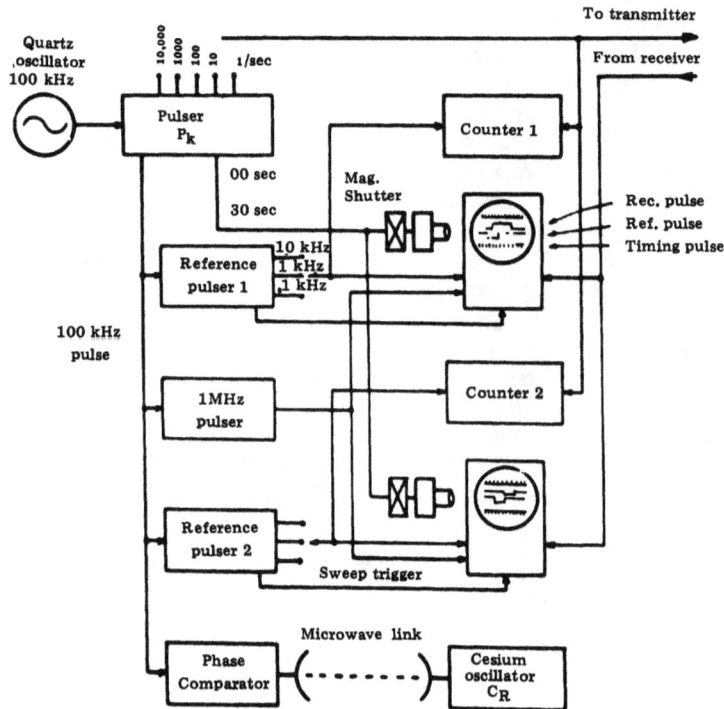

*Figure 5-21. Equipment Used at Kashima To Generate and Record Pulses*[5]

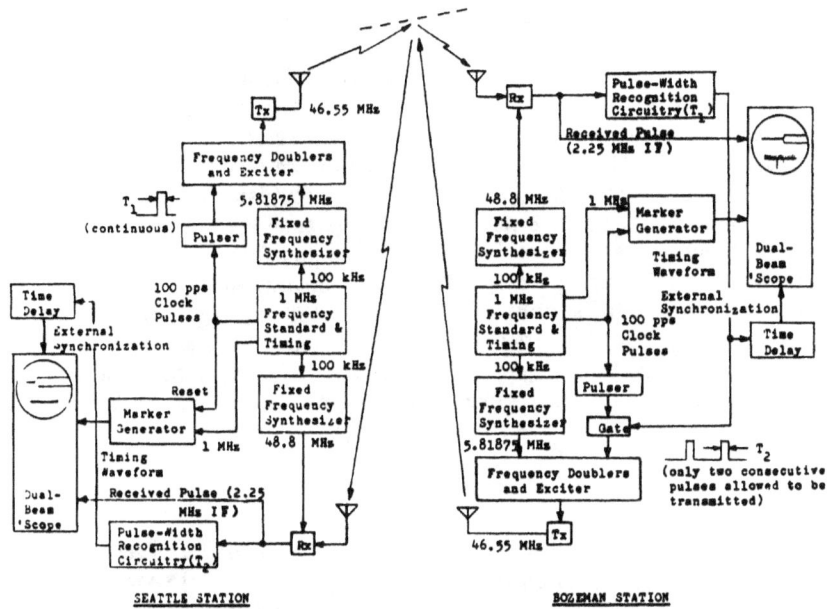

Figure 5-22. Diagram of Meteor Burst Clock Synchronization System[17]

## TABLE 5-1

### SUMMARY OF HP FLYING CLOCK EXPERIMENTS

| Date | Description |
|---|---|
| April, 1964 | Time correlated between U S and Switzerland to about 1 microsec. RF propagation time established within about 200 microsec. Two clocks (#1 and 2) operated within a few parts in $10^{12}$ of one another and within a few parts in $10^{12}$ of "long beam" cesium standards at Neuchatel, Switzerland, and NBS at Boulder, Colorado. |
| Feb./Mar., 1965 | Time (or frequency) correlated between 21 places in 11 countries to within 1 microsec. One clock (#3) accumulated less than 6 microsec time difference and the other (#4) less than 1 microsec in 23 days (compared against NBS UA). |
| May/June, 1966 | Time (or frequency) correlated within about 0.1 microsec between 25 places in 12 countries. Two clocks (#8 and 9) agreed with each other within 1 microsec after 31 days with an average frequency difference of less than 3.6 parts in $10^{-13}$. |
| Sept/Oct., 1967 | Time (or frequency) correlated between 53 places in 18 countries to about 0.1 microsec. Two clocks (#51 and 52) exhibited time differentials of 1.7 and 3.5 microsec over 41 days (compared to HP house standard), corresponding to average frequency differences of $5 \times 10^{-13}$ and $10 \times 10^{-13}$, respectively. |

method is to carry an accurate clock, usually by airplane, between them. Such a "flying clock" experiment has been conducted several times by Hewlett-Packard, the fourth of which was conducted in 1967[18]. The time span for the experiment was 41 days and the clocks traveled a total of 100,000 km. Two HP clocks were used. By comparing these clocks to the HP house standard at the beginning and end of the 41-day period, the time differences were found to be only 1.7 and 3.5 $\mu$sec. This corresponds to offsets of 5 and 10 parts in $10^{13}$ in the frequency standards that actuate the clocks. Time correlations were made to 0.1 $\mu$sec. Table 5-1[18] is a summary of the four HP flying clock experiments to date. The clock used was an atomic-controlled cesium beam standard and clock assembly that had its own standby battery power supply. The frequency in the standard of the portable clock as well as the time kept initially was compared with the national standards at the National Bureau of Standards and the U S Naval Observatory. After the trip, the portable clock was again compared with these same standards so that any slight changes could be referred back on previous measurements as a tolerance.

## REFERENCES

1. MIL-F-28734A, *Frequency Standards, Cesium Beam, General Specification For*, Dept. of Defense, April 1972.

2. MIL-F-28734/1, *Frequency Standard Cesium Beam, Type I, Detail Specification For*, Dept. of Defense, April 1972.

3. MIL-F-28734/2, *Frequency Standard, Cesium Beam, Type II, Detail Specification For*, Dept. of Defense, May 1972.

4. MIL-F-28734/3, *Frequency Standard, Cesium Beam, Type III, Detail Specification For*, Dept. of Defense, May 1972.

5. Robert Gregory, "A Digital Interpretation Technique in Time Measurement", IEEE Trans. on Instrumentation and Measurement, IM-13, 159-63 (December 1964).

6. V. R. Latorre and H. J. Jenson, "Subnanosecond Time Interval Measurements", Proc. National Electronics Conf., 22, 36-40 (October 1966).

7. R. G. Baron, "The Vernier Time-Measuring Technique", Proc. of IRE, 45, 21-30 (January 1957).

8. *NBS Standard Frequency and Time Services*, U S Department of Commerce, National Bureau of Standards, Boulder, Colorado, Misc. Publication 236, 1972.

9. *Frequency and Time Standards*, Hewlett-Packard Co., Palo Alto, Calif., Application Note 52, 1965.

10. A. R. Chi and S. N. Witt, "Time Synchronization of Remote Clocks Using Dual VLF Transmissions", *Proc. 20th Annual Symposium on Frequency Control*, U S Army Electronics Laboratories, Fort Monmouth, N. J., April 1966, pp. 588-611.

11. L. Fey and C. H. Looney, Jr., "A Dual Frequency Timing System", IEEE Trans. on Instrumentation and Measurement, IM-15, 190-5 (December 1966).

12. L. D. Shapiro, "Time Synchronization from Loran-C", IEEE Spectrum, 46, 46-55 (August 1968).

13. D. W. Allan, B. E. Blair, D. O. Davis, and H. E. Macklan, "Precision and Accuracy of Remote Synchronization Via Portable Clocks, Loran C, and Network Television Broadcasts", *Proc. 25th Annual Symposium on Frequency Control*, U S Army

Electronics Laboratory, Fort Monmouth, N. J., April 1971, pp. 195-208.

14. E. R. Swanson and C. P. Kugel, "Omega VLF Timing", *Proc. 25th Annual Symposium on Frequency Control*, U S Army Electronics Laboratories, Fort Monmouth, N. J., April 1971, pp. 159-66.

15. J. McA. Steele, et al., "Telstar Time Synchronization", IEEE Trans. on Instrumentation and Measurement, **IM-13**, 164-70 (December, 1964).

16. W. Markowitz, et al., "Clock Synchroni-zation via Relay II Satellite", IEEE Trans. on Instrumentation and Measurement, **IM-15**, 177-84, (December 1966).

17. W. R. Sanders, et al., "A Meteor Burst Clock Synchronization Experiment", IEEE Trans. on Instrumentation and Measurement, **IM-15**, 184-9 (December 1966).

18. L. N. Bodily and R. C. Hyatt, "Flying Clock Comparisons Extended to East Europe, Africa and Australia", Hewlett-Packard Journal, **19**, 12-20 (December 1967).

PART TWO — ELECTRONIC TIMERS

CHAPTER 6

INTRODUCTION

The last decade saw a gradual evolution in the field of electronics. First came the development of the semiconductor devices (transistors and diodes) and second, a technology known as *integrated circuits*. The combination of these two had a great impact on the design of electronic circuits, especially in regards to miniaturization. Moreover, this advance has changed the design of military timers.

Research and development in all phases of the technology is continuing at an ever increasing rate; consequently, new developments are arriving rapidly. This influx of new information presents a problem in writing a handbook that is supposed to contain the latest circuits and techniques because the procedures will be superseded by newer ones by the time the handbook is printed. The best that can be done is to give the designer background information and impress on him the need for reviewing the current literature before selecting a given circuit.

## 6-1 SEMICONDUCTOR DEVICES

### 6-1.1 TRANSISTORS

Like the triode vacuum tube, its semiconductor counterpart, the transistor is a three-terminal electronic device that performs many of the same functions as the vacuum tube. The transistor has replaced the vacuum tube in most areas of electronics. There are several reasons for this wide acceptance:

(1) Transistors, being small and light, are the logical choice when size and weight are important factors.

(2) They require no filament supply, which means that their power requirement is much less than that for a vacuum tube.

(3) They have a longer useful life than vacuum tubes.

(4) As long as transistors are operated in a nondestructive environment, their operating characteristics do not deteriorate with time. A nondestructive environment is one wherein temperature, humidity, and radiation levels are such that the electrical properties of the semiconductor materials are not destroyed. Vacuum tubes have a limited operating life, because of cathode deterioration with use.

(5) Finally, transistors are physically more rugged than tubes because they are solid state.

During the early development of transistors, vacuum tubes were superior to transistors but with the advance of semiconductor technology their advantages soon disappeared. Some of the advantages of vacuum devices were that they could be designed with a higher input impedance and that they could operate at higher voltage levels than semiconductor devices. For the unijunction transistor that is used as a level detector, the input impedances are now comparable. Silicon transistors have made the use of higher voltages possible, although still not as large as those for vacuum tubes.

Fig. 6-1 compares three basic transistor circuits with the equivalent vacuum tube circuits. Most of the electronic timing circuits discussed in this handbook use semiconductors rather than vacuum tubes. However, the designer should keep in mind that circuits can be constructed using either component.

## 6-1.2 INTEGRATED CIRCUITS

An ideally integrated circuit completely eliminates individual electronic parts — such as resistors, capacitors, and transistors — as the building blocks of the electronic circuit. In the place of these parts are tiny chips of semiconductor material which serve the

functions of several transistors, resistors, capacitors, and other electronic elements. The elements are interconnected to perform the task of a complex circuit, often comprising a number of complete conventional circuit stages. Within or on top of these tiny silicon chips are microscopically small depositions or growths of material layers which serve the functions of the individual discrete parts. Thus, multistage amplifiers, complex flip-flops, and dozens of other functional circuits are becoming the basic components of electronic circuits.

Basically, there are two general classifications of integrated circuits — the semiconductor monolithic integrated circuit and the thin-film integrated circuit. Each of the basic circuits can have a variety of structures with specific advantages or disadvantages for a particular application. Thin-film component arrays also have been added to semiconductor circuits in networks that promise to combine various features of the individual devices in a single configuration of maximum utility.

### 6-1.2.1 Monolithic Circuits

The basic semiconductor integrated circuit is a monolithic device; i.e., the circuit elements are fabricated on a silicon substrate. In today's circuits, this material is silicon, which serves the monolithic device as the chassis, called a substrate.

### 6-1.2.2 Thin-film Circuits

The basic thin-film circuit begins with a substrate, made of an insulating material such as glass or ceramic. Upon this substrate is deposited a pattern of the passive elements of the circuit (i.e., resistors, capacitors) as well as an interconnecting pattern of metal. Deposition is by various techniques, such as vacuum evaporation, sputtering, and silk screening. The active elements (i.e., transistors, diodes) are added separately to the thin-film pattern.

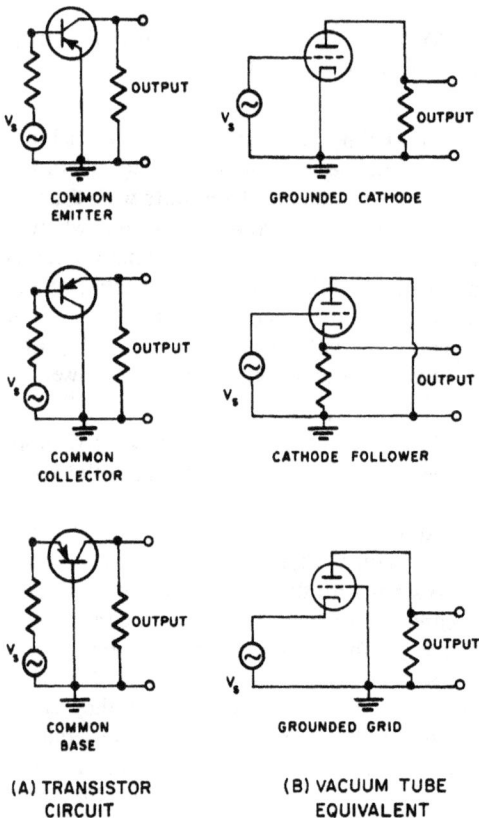

COMMON EMITTER     GROUNDED CATHODE

COMMON COLLECTOR     CATHODE FOLLOWER

COMMON BASE     GROUNDED GRID

(A) TRANSISTOR CIRCUIT     (B) VACUUM TUBE EQUIVALENT

*Figure 6-1. Basic Small-signal Circuits*

### 6-1.2.3 Multichip (Hybrid) Integrated Circuits

The multichip circuit is not a monolithic device. Rather, it is composed of individual parts, made either by film or diffusion processes. These parts are attached to a ceramic substrate and interconnected by a combination of metalization processes and wire bonding techniques. The completed circuit may be housed in packages identical to those used for monolithic circuits.

## 6-2 GENERAL TIMING SYSTEMS

Electronic timers can be classified into two general types — analog and digital — distinguished in Fig. 6-2. The circuits are discussed in detail in pars. 10-2 and 11-2 to 11-4, respectively; however, a brief description is given here to illustrate the basic difference.

### 6-2.1 THE ANALOG SYSTEM

In engineering language, the term analog usually connotes a continuous stimulus as opposed to discrete pulses that typify a digital system. The factor that determines the classification is how the time interval is generated. In an analog system, the time interval is generated by a continuous circuit action as typified by Fig. 6-3(A). In this circuit, when switch S is closed, capacitor C

charges through resistor R and the voltage across C increases with time as shown in Fig. 6-3(B).

The voltage developed across the capacitor $V_c$ at any given time is a function of the series resistor and capacitor, the voltage of the power supply, and time

$$V_c = V_b \left[1 - e^{-t/(RC)}\right], \text{V} \qquad (6\text{-}1)$$

Eq. 6-1 can be arranged to solve for time

$$t = RC \ln \left(\frac{V_b}{V_b - V_c}\right), \text{sec} \qquad (6\text{-}2)$$

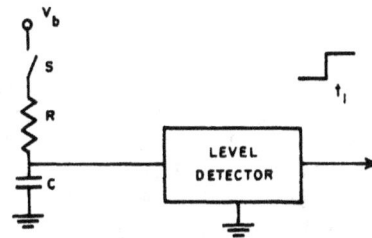

(A) ANALOG CHARGING CIRCUIT

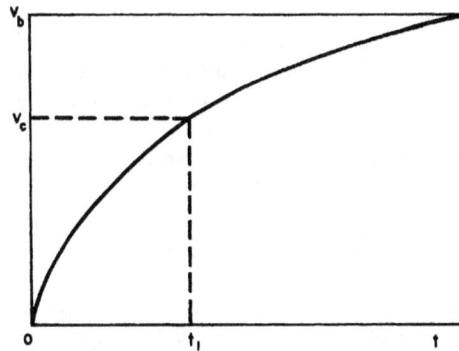

(B) ANALOG CIRCUIT CHARACTERISTIC

Figure 6-3. Simple Analog System

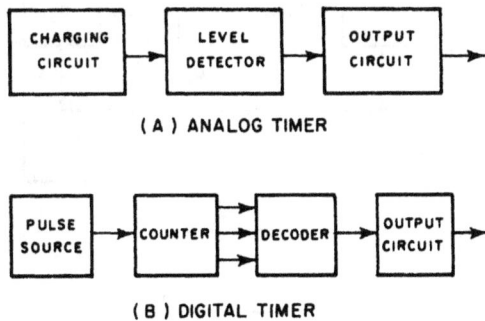

(A) ANALOG TIMER

(B) DIGITAL TIMER

Figure 6-2. Block Diagrams of Analog and Digital Timers

where

$V_c$ = voltage across capacitor, V

$V_b$ = power supply voltage, V

$C$ = capacitance, F

$R$ = resistance, ohm

$t$ = time, sec

The level detector is quiescent until $V_c$ reaches a value that causes the detector to trigger. The time delay is determined by $V_b$, $R$, $C$, and the value of $V_c$ at which triggering occurs. The simplified analog circuit illustrated in Fig. 6-3(A) is an example of a one-shot device. Modifications to this circuit which permit it to recycle and generate a succession of timing pulses are presented in par. 10-2.

The main advantage of this type of system is simplicity. Only a capacitor, a resistor, a voltage source, and a level detector are required. On the detriment side is its inaccuracy and instability. From Eq. 6-2 it can be seen that the error is a direct function of $R$ and $C$. For example, if $R$ and $C$ drift by 10%, the time will change by 10%. The error due to small changes in source and bias voltages is a function of the change in the ratio of bias to source voltage and the value of that ratio. The smaller the ratio, the smaller the error for equal percentage change in the ratio. For applications where the ratio is approximately 0.5, a 10% change in the ratio can give a time error of up to 16%.

## 6-2.2 THE DIGITAL SYSTEM

Even very simple digital timing circuits require many more individual components than analog timers. Thus, digital timing systems have benefited greatly from integrated circuit technology.

Fig. 6-4 illustrates a simple digital timing circuit. The pulse source is typically an oscillator and pulse shaper that supplies uniform time-spaced pulses to the counter when $S_1$ is closed  The counter is composed of the discrete state elements $D_1$, $D_2$, and $D_3$. The discrete state of these elements ($L_1$, $L_2$, and $L_3$) is monitored by the decoder that has a pulsed output F when some predetermined combination of the outputs occurs. The sequence of these outputs is a function of the number of pulses supplied to the counter. The decoder pulse output is converted to a continuous output signal by the output circuit.

## 6-3 MILITARY APPLICATION

This handbook emphasizes the military application of electronic timers. The main categories for these timers are

(1) High-acceleration systems (artillery)

(2) Low-acceleration systems (rockets, bombs, guided missiles)

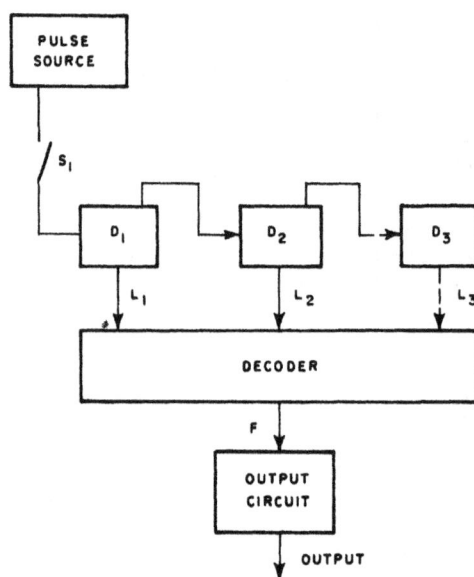

Figure 6-4. Simple Digital System

(3) Stationary timers (demolition, controls).

Each of these uses represents a peculiar environment that the timer must endure. None of these applications, however, would use what is considered to be a high-precision timer.

Electronic timers can be used to provide delay for (1) arming, (2) functioning, or (3) self-destruction or sterilization.

## 6-4 REQUIREMENTS

Timers used for military application must meet electrical, mechanical, and environmental requirements. Electronic timers are similar to other pieces of military equipment in that they must operate in all types of environment.

### 6-4.1 ELECTRICAL REQUIREMENTS

Since most of the electronic timers for military purposes are portable, the power sources are restricted to batteries. Much work has been done on trying to develop batteries that have a long shelf life, high regulation, and small size. The types of power sources available are discussed in pars. 10-5 and 11-5. Like other military timers, it is desirable that electronic timers be simple, compact, rugged, accurate, reliable and have low power consumption.

### 6-4.2 MECHANICAL AND ENVIRONMENTAL REQUIREMENTS

While these timers have an electronic time base, they are nonetheless components of mechanical systems that must withstand storage, shock, and vibration over the military temperature range. Ammunition timers may be subjected to high acceleration and spin forces. See par. 16-1.1 for a listing of typical military conditions that may be required.

Since the components of electronic timers have no moving parts, they can be encapsulated or potted in a plastic material; many ammunition timers are so packaged. Potting provides many desirable attributes: (1) increased shock and vibration resistance, (2) protection against moisture and other contaminants, (3) increased thermal resistance, and (4) handling protection. Potting, then, makes it easier to achieve the required ruggedness.

Electronic timers are more susceptible than other timers to electromagnetic radiation in the environment. See par. 7-3 for methods of radiation hardening.

## 6-5 AUXILIARY EQUIPMENT

Auxiliary equipment for military timers ranges from the calibration and maintenance equipment usually associated with any military electronic equipment to the field operated time-to-action setting devices (fuze setters) used in conjunction with ammunition timers. Auxiliary equipment usually is subject to less stringent mechanical and environmental requirements than the timers themselves; however, time-to-action setting devices for ammunition timers must be rugged and capable of working in the field. Space and power requirements for auxiliary timer equipment are also less stringent.

Ammunition timing systems usually require an integral timer plus a time-to-action setting device that introduces the desired time setting into the ammunition timer. Since only the ammunition timer is expended, considerable effort is made to incorporate all self-checking features and components necessary for setting in the non-expended and reusable setting device. Thus timers themselves in an overall ammunition timing system are likely to be of less complexity than the timer auxiliary equipment. Time-to-action setting devices for analog and digital timers are covered in pars. 10-4 and 11-5, respectively. Timer testing and calibration are treated in par. 7-5.

# CHAPTER 7

## SYSTEM DESIGN CONSIDERATIONS

### 7-1 STATIONARY SYSTEMS

### 7-1.1 GENERAL

Design goals are determined at the outset when any new military system is being developed. These goals, which are defined for each major component of the system, provide that the total system will result in a product that is safe, reliable, combines maximum accuracy with design simplicity, is constructed ruggedly for successful operation in a military environment, and has a reasonable life cost cycle.

### 7-1.2 ANALYSIS OF REQUIREMENTS

In the development of a military item, an analysis of requirements is mandatory when planning either a new concept or a modification of an existing unit. These requirements, after defining the limits within which a designer may operate, usually contain: (1) performance factors that must be met, (2) a description of associated equipment with which the item must interface, (3) approximate cost to manufacture, (4) operating parameters, and (5) probable environments.

In particular, the requirements for an electronic timer might include the following:

(1) Delay time range with ability to set in predetermined increments

(2) Accuracy given as plus or minus an allowed percentage of timing error

(3) Operating temperature range

(4) Military transportation, vibration, environmental, and storage requirements

(5) Maximum space allowable for the unit including the electrical power supply

(6) Cost goals

(7) Setting considerations, including method and resolution.

The designer must be thoroughly familiar with the stated requirements so that he is in a position to evaluate them. Should any of the requirements be too difficult, time consuming, or costly to achieve, a judgment can then be made whether or not to request relaxation of one or more requirements to reduce development time and cost risk. This judgment process is called trade-off. It is quite possible that the most economical approach would be to use an existing timer, perhaps with modifications. However, if a new design is called for, then the timer development should follow a series of logical steps.

### 7-1.3 STEPS IN DEVELOPING A SYSTEM

Stationary electronic timers are essentially of two types:

(1) Precision reference timers, typified by a precise crystal controlled oscillator followed by a binary divider network. This timer can be extended to any desired time period at the expense of providing more divider stages and the additional power to drive them.

(2) Electronic timers having an *RC* time

7

base used in relaxation oscillator circuits. These devices can be relatively simple.

Details of precision reference timers are covered in Chapters 4 and 5, while those of electronic timers are located in Chapters 10 and 11. Stationary timer systems must fulfill all of the requirements listed in par. 7-1.2, except perhaps for the following:

(1) Size and weight restrictions are not as stringent.

(2) Shock requirements, such as operability under setback and acceleration forces, are not specified.

The development of a timer used in a stationary system or one used in a mobile system (see par. 7-2) is considered successful and complete only when pilot lots have passed all tests and the timer has been accepted as a standard item ready for mass production. There are many steps between the first preliminary schematic and the final production of the standard timer. Throughout development, the designer must consider a great many details at each of the four basic stages: (1) preliminary schematics, (2) bread boarding, (3) testing and revision of model, and (4) final acceptance.

### 7-1.3.1 Generalized Process

The generalized process of time delay is illustrated in Fig. 7-1[1]. The output is triggered when the physical variable coincides with the preset reference value. Two important design factors are the repeatability of the characteristic curve, including hysteresis and independence of changing operating conditions, and the accuracy of the detector. Linearity is desired at times when a convenient readout dial is needed. In many applications, the ability to reset to zero is also of major importance. Another important physical variable in many time delay devices is temperature because of the large variations in

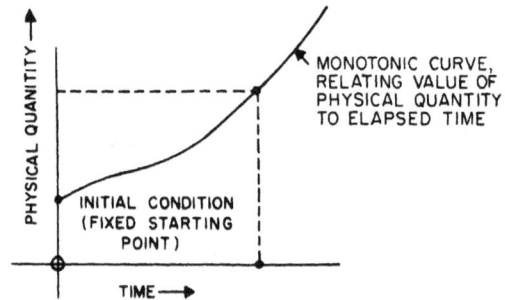

Figure 7-1. Basic Timer Diagram[1]

ambient conditions from place to place[1].

Figs. 7-2[1] and 7-3[1] show generalized behavior of a physical variable in a timing mechanism. The value of the physical variable is changed by a source of driving energy. When the timing cycle is initiated by an external signal, the change begins. Change then proceeds at a controlled rate that depends on the design and adjustment of the mechanism. The detector delivers an output signal when the value of the physical variable matches the built-in reference signal. Electrical power is used in electronic timers as the driving power that is modulated to change the physical variable[1].

The elements to be considered for use in an electronic timer are[2]:

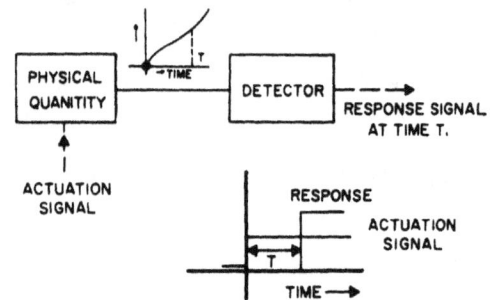

Figure 7-2. Block Diagram of an Elementary Timer[1]

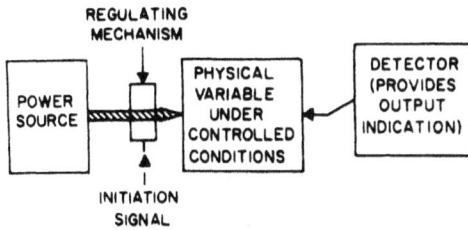

*Figure 7-3. Control of Physical Variable*[1]

(1) Power Supplies. Batteries using alkaline zinc and silver oxide, a solid-state battery, existing ground power, etc.

(2) Time base. *LC* oscillator, *RC* oscillator with gas-diode trigger, magnetic core oscillator, nuclear time base, etc.

(3) Frequency Dividers and Counters. Transistor flip-flop binary divider, transistor flip-flop ring counter, magnetic core flip-flop, magnetic core ring counter, etc.

(4) Output Circuits. To produce a signal that initiates the first element in an explosive train, for instance, or to delay the application of high voltage to a radar set in order to allow the acquisition magnetron to reach operating temperature.

### 7-1.3.2 Trade-off Possibilities

An excellent way to evaluate the various alternatives is to tabulate each class of component and then estimate the relative suitability of each factor to meet the design requirements. Such a tabulation is shown in Table 7-1[2] in which the oscillator type and accuracy, voltage and current requirements, power available, frequency, and approximate size are given for various oscillators.

Better than marginal improvements over old systems are required to justify a new design. These improvements should include such features as increased reliability, reduced cost, or greater accuracy. Production or shelf-item components that are introduced into a system must still be checked out by electrically testing the assembled item. This must be done because (1) component shelf life usually must be known or estimated, (2) mating components may introduce incompatibility, and (3) the performance of the new design may be degraded by coupling effects between old and new items.

After a review of the timer requirements has been made and trade-offs decided upon, the electronic timer designer can begin preliminary designs. At this phase of development it is quite possible that the choices of the various components of the system have been narrowed so that either available data or prototype hardware development can be used to eliminate all but one of the approaches. The prototypes can then be fabricated and prepared for testing.

### 7-1.3.3 Testing

The prototypes are subjected to tests that simulate or duplicate the forces and environments which the final version will have to survive. Most of the parameters can be measured with standard measuring equipment. The tests include[3]:

(1) Electrical Tests. Conformance of the timer with circuit requirements, dielectric strength, insulation resistance, and continuity.

(2) Environmental Tests. Vibration, operating temperatures, storage temperatures, and drop tests.

(3) Timer Performance and Accuracy. Checking after exposure to each of the preceding test groups.

### 7-1.4 COMPLETION CRITERIA

After the prototype is proven to be capable of satisfying all requirements, a testing

**TABLE 7-1**

**COMPARISON OF VARIOUS OSCILLATORS**

| Oscillator type and accuracy | Voltage, V | Current, A | Power Level, mW | Frequency, Hz | Size, in.$^3$ |
|---|---|---|---|---|---|
| LC (Including amplifier) 0.1 to 1.0% | 4 to 6 | $4 \times 10^{-4}$ | 1.6 to 2.4 | 1000 | 1 |
| RC Feedback oscillator 0.01 to 0.1% | 120 to 280 | $10^{-3}$ | 12 to 28 | 1k to 10k | < 0.5 |
| Magnetic core 0.1 to 1.0% | 15 | $1 \times 10^{-2}$ to $2 \times 10^{-2}$ | 150 to 300 | 10 to 100 | 1 |
| RC unijunction transistor 1 to 3% | 10 to 20 | $2 \times 10^{-3}$ | 20 to 40 | 1 to 100 | 1 to 2 |
| RC (SCR as trigger device; initial negative voltage on capacitor) 3% | 50 | $2 \times 10^{-4}$ | 10 | 1 | 2.5 |
| Multivibrator 2 to 4% | 10 | $2 \times 10^{-5}$ | 0.2 | 1 to 10 | 2 |
| RC Four-layer diode 5% (questionable) | 60 | $15 \times 10^{-5}$ | 9 | 1 to 100 | 2 |
| RC Gas diode 5% (questionable) | 300 | $4 \times 10^{-5}$ | 12 | 1 | 5 |

program is initiated which is based on statistical sampling. See Table 7-2[3] for a sampling plan for a typical electronic timer. The proven timer then can be produced in quantity and integrated into the weapon system. Perhaps the most important criterion is reliability.

Reliability is the probability that material will perform its intended function for a specified period under stated conditions[4]. With respect to the design of electrical timers, several important considerations have a direct bearing on device reliability:

TABLE 7-2

SAMPLING PLAN

| Lot size inclusive | Sample size | Accept lot | Reject lot |
|---|---|---|---|
| 0−25 | Use 100% Inspection | | |
| 26−35 | 20 | 0 | 1 |
| 36−50 | 25 | 0 | 1 |
| 51−150 | 30 | 0 | 1 |
| 151−300 | 30 | 1 | 2 |
| 301−over | 60 | 2 | 3 |

(1) The number of component parts should be a minimum.

(2) If discrete components are used, the reliability of each should be as high as is commensurate with total circuit reliability and cost.

(3) Parts must withstand extreme environmental conditions and still operate successfully.

(4) Components should be used whose important properties are known and reproducible.

(5) Timers should be designed so that factors that affect reliability can be detected by means of nondestructive tests or inspection.

From the reliability measurements point of view, two kinds of systems can be distinguished[5]: (1) The repairable or recoverable system designed to perform for a long time in a large number of operations, and (2) the nonrepairable or nonrecoverable system designed to perform only once. Stationary systems can be put in the former category and mobile systems in the latter. See par. 7-2 for reliability consideration for mobile systems.

The reliability of stationary systems depends on maintenance policy, i.e., on schedules for overhaul, parts replacement, and inspection. Repair maintenance is not used until parts fail during system operation. In this case, component wearout becomes the dominant factor and the system failure rate becomes much higher than it originally was. A newly built system that has been debugged is essentially free from wearout failures and its reliability is governed by the frequency of chance failures. During this period, components are replaced as they fail. After this period, wearout failures begin to show up and if only repair maintenance is used, the system failure rate would rise. However, when high reliability is required, preventive maintenance is used to prevent wearout failures from appreciably affecting system operation[5]. In other words, long-life systems must be maintained preventively if long failure-free operating periods are required, and the system chance failure rate must be very low. In regard to the probability of the first failure of the system, whether it be chance or wearout, the answer can be given only by means of a system probability analysis when the component parameters are known and not by means of system tests. For more details about system reliability measurements and testing, see Ref. 5.

## 7-2 ACCELERATED SYSTEMS

### 7-2.1 GENERAL

Timers and programmers are used by the military in virtually every type of munition from low g missiles to high g projectiles. The use of electronic timers, used in fuzes, began during and after World War II when system requirements called for more versatile and sophisticated equipment. In particular, electronic digital timers have been used extensively in more expensive large-scale weapons after many problems of size, power supply, and shock which are inherent in vacuum tubes were eliminated by the introduction of the transistor.

The use of electronic timers in accelerated ammunition requires a low-cost rugged time base. An error of less than 0.1% under all conditions of use and operating temperatures range from −50° to 125°F are typical. The use of temperature control is not feasible in applications that require an oscillator output to be available a few milliseconds after initiation. Timers must operate under high spins, after high setback accelerations during firing, and sometimes under severe deceleration forces imposed by target impact. Other requirements for the time fuze are small size (e.g., about 1 in.[3]) and a low power drain[6]. The paragraphs that follow discuss various design considerations for this type of electronic timer.

## 7-2.2 COMPONENT SELECTION AND QUALIFICATION

### 7-2.2.1 Considerations

The electric components used in the circuits of electronic timers consist of capacitors, resistors, magnetic cores, tubes and transistors, and power supplies. The accelerated environment puts stringent requirements on their ruggedness, aging, and temperature characteristics. Factors working against fuze component reliability vary with the type of fuze in which the components are used. Requirements for long inactive shelf life, extreme environmental conditions while in operation, and the limited ability to pretest for functioning just before use add to the difficulties in selection of components. For these reasons, the designer should use standard components wherever possible.

When selecting timer components, the designer may take advantage of the fact that components which are inadequate for long time applications may be entirely suitable for use in fuzing systems. Possibilities include components with a relatively short operating life or with failure rates that rise sharply with cycling. Even though some fuzes undergo many tests prior to actual use, their total operating life expectancy is normally much less than that of stationary system components. Tolerances of some components may prohibit their use in certain types of electronic equipment, but they may be used in an electronic time fuze.

Mounting of components in a preferred orientation along with potting to strengthen the entire configuration will help the component survive the severe environment. To eliminate the effects of aging and thermal changes, three solutions can be used[7]:

(1) Use components whose original properties are adequate.

(2) Use hermetic sealing to protect components from damage due to environmental conditions.

(3) Choose components so that the variation of one component property is compensated by another component.

### 7-2.2.2 Resistors

The types and characteristics of resistors that are available for military use in an electronic timer are shown in Table 7-3[8]. Available resistance varies widely. The units are capable of withstanding the required environment and have a minimum change of values under operating conditions. For additional information about resistors, see par. 10-4.2.

### 7-2.2.3 Capacitors

Capacitors currently produced that will meet the temperature and stability requirements for electronic timers are porcelain capacitors (a ceramic capacitor with a temperature coefficient of 15 ± 5 parts per million/deg C is presently in production). The porcelain capacitor is stable during storage life and is readily available with a 0 ± 25 parts per

TABLE 7-3

RESISTOR CHARACTERISTICS

| | Wire Wound | Carbon Film | Metal Film |
|---|---|---|---|
| Nominal Value, ohm | $0.1\ \Omega - 1\ \text{Meg}\ \Omega$ | $1\ \Omega - 1000\ \text{Meg}\ \Omega$ ($10^{14}$ ohm avail.) | $30\ \Omega - 15\ \text{Meg}\ \Omega$ |
| Initial Tolerance, % | 0.002% | 0.1 | 0.1 (0.05 avail.) |
| Wattage Rating, W | 0.05 – 250 | 0.1–5 | 0.1–4 |
| Temp. at which derating starts, °C | 125 Typ. | 70–120 | 70–125 |
| Zero Wattage Temp, °C | 250 Typ. | 150 Typ. | 175 Typ. |
| Temp. Coefficient of Resistance, ppm/deg C | ± 10 stand. ± 2 avail. −80 to +6000 avail. | −250 to ± 500 | 10–250 |
| Range over which TCR holds, °C | various −65 to +150 for stand. ± 10 ppm | −65 to +150 | −65 to +150 Typ. |
| Maximum Working Voltage, V | 100–1200 | 200–15,000 | 200–750 |
| Dielectric Withstanding Voltage, V | 100–2500 | 200–1000 | 200–1000 |
| Current Noise, $\mu$V/V | − − − − − − | 0.1–1 | 0.05–0.2 |
| Distributed Inductance, $\mu$H | ⩾ 1 | ⩽ 10 GHz | ⩽ 10 GHz |
| Distributed Capacitance, pF | ⩾ 0.8 | ⩽ 10 GHz | ⩽ 10 GHz |
| Voltage Coefficient of Resistance, ppm/V | | 2–20 Typ. | 1–5 |
| Load Life, Max. change, % | 0.05–1.0 | 0.1–1.0 | 0.1–1.0 |
| Short-time Overload, Max. change, % | 0.01–0.1 | 0.01–0.5 | 0.1–0.5 |
| Moisture Resistance, Max. change, % | 0.005–5 | 0.03–1.5 | 0.1–1.5 |
| Temp. Cycling, Max. change, % | 0.07–0.2 | 0.02–0.5 | 0.05–0.5 |
| Terminal Strength, Max. change, % | 0.00–0.05 | −0.2 | −0.2 |
| Low Temp. Operation, Max. change, % | 0.02–0.25 | 0.05–0.5 | 0.005–0.5 |
| Soldering, Max. change, % | | 0.01–0.5 | 0.05–0.5 |
| Shelf-life, Max. change, % | 0.002–0.1 | 0.05–0.2 | 0.05 |
| Shock, Max. change, % | 0.00–0.2 | 0.1–0.5 | 0.01–0.5 |
| Vibration, Max. change, % | 0.01–0.2 | 0.1–0.5 | 0.01–0.5 |

million/deg C temperature characteristic. The capacity shifts a maximum of 0.1% over a ten-year period with most of the drift occurring during the first few months. Temperature characteristics within ± 5 parts per million/deg C are available. To achieve 0.1% accuracy in the design of oscillators, a calibration capacitor is used. For a basic oscillator frequency of 1000 Hz, a calibration capacitor must have a tolerance of ± 2% up to 300 pF and ± 1% on values above this. High spin and shock should have no adverse effects on the capacitors[9].

### 7-2.2.4 Magnetic Cores

Magnetic cores are used in electroni counters because of their ability to maintai memory with no power. The core is made o ferromagnetic material that exhibits a squar hysteresis loop. Fig. 7-4[10] shows an idealize version of this material. The fluxes shown i the figure are

$\phi_r$ = remanent flux

$\phi_s$ = saturation flux

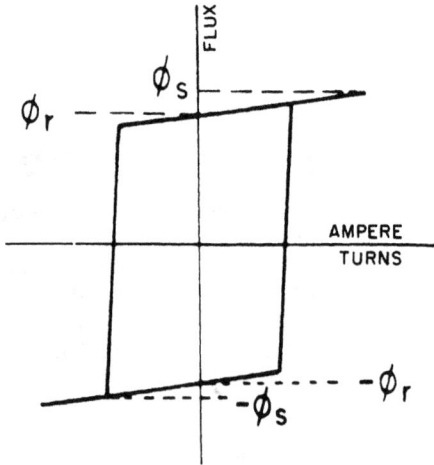

Figure 7-4. Square Hysteresis Loop Material

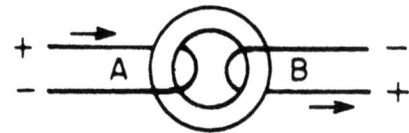

(A) Core driven to "1" state

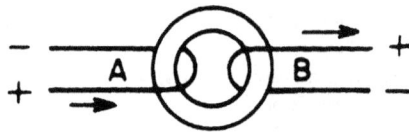

(B) Core driven to "0" state

Figure 7-5. Methods of Driving the Core

The squareness ratio is equal to $\phi_r/\phi_s$, which, for the cores used in military timers, is a minimum of 0.9. With no drive applied, the core is normally in either the $\phi_r$ or the $-\phi_r$ state. When sufficient drive is applied in the proper direction, the core switches to the opposite $\phi_s$ state; when the drive is removed, the flux changes to $\phi_r$. Fig. 7-5[10] illustrates the method of driving the core to the "0" and "1" states. The $-\phi_r$ state is designated "0" and the $\phi_r$ state is designated "1". Further information on magnetic cores and core circuits is given in Ref. 11. Fig. 7-6[2] shows the diagram of a typical magnetic core driver.

### 7-2.2.5 Tubes and Transistors

Rugged tubes are commercially available for accelerated military items. Cold cathode diodes and triodes depend on light to provide initial ionization. This problem has been solved by placing a band of radioactive material around the tube which helps to provide a consistent breakdown voltage. The choice between a diode and a triode often is made on the basis of available energy because a triode, while slightly more complicated, has more efficient energy transfer character-istics[7].

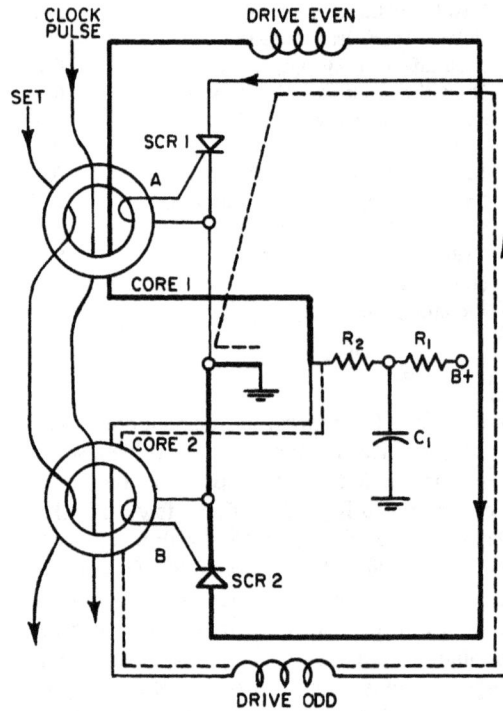

Figure 7-6. Magnetic Core Driver

Transistors at first were too expensive to use in one-shot devices such as electronic fuzes. However, with improvements in the state of the art, cost has been reduced so that more and more transistors now are being used in oscillators for electronic timers. A unijunction transistor is used as the trigger device in an *RC*-oscillator[2]. Transistors with high-gain bandwidth products, high voltage, and a guaranteed beta over three current decades are available commercially for use in amplifiers, oscillators, mixers, and other signal processing applications. They provide maximum reliability and heat transfer, and are designed to meet rugged military specifications. They are encapsulated by a welded hermetic seal and can be mounted in any position[2]. Storage and operating range is $-65°$ to $+200°F$.

### 7-2.2.6 Power Supplies

Several types of batteries are used in electronic timers. One type is the alkaline zinc and silver oxide battery. Voltage rating is 1.5 V per cell. It can be fabricated with the cells free of the electrolyte for reserve action. The electrolyte is stored in a container adjacent to cells and forced out into the cells, for example, by firing an electric squib[2]. See par. 11-6 for more information on power supplies.

### 7-2.3 CIRCUIT DESIGN AND BREAD-BOARDING

Circuit design begins with the synthesis of a circuit that meets the performance objectives. In the interest of cost and time savings, such a circuit is developed initially within the component constraints imposed by the system requirements indicated in par. 7-1.2. Since integrated circuits are being used more and more because of their packaging and reproducibility attributes, the designer also should keep in mind several present monolithic component limitations. Generally, large values of resistors and capacitors, as well as close absolute component tolerances, are to be avoided. The number of capacitors should be held to a minimum because they occupy valuable die space. Inductors should be eliminated entirely in a monolithic design because at present they are not fabricated adequately in integrated form. If a hybrid construction is planned, then discrete components are used along with the integrated circuit elements. See par. 8-2 for more information about integrated circuit techniques.

To synthesize the circuit design, a decision must be made as to the power supply, the frequency divider and counter type, and the output circuit that will attain the desired accuracy.

Compilations such as shown in Tables 7-1 and 7-4[12] show the various oscillators and their relative performance capability. Accuracy obtainable, voltage and current requirements, frequency stability, and approximate size are listed. Table 7-5[2] gives the same data for frequency dividers and counters.

Once the circuit is synthesized, a breadboard model can be built. Its performance can be evaluated with well known test procedures. Reasons for discrepancies in the anticipated performance usually can be pin-pointed by taking stage-by-stage measurements. This emphasizes the value of breadboarding because theoretical analysis cannot localize problem areas. Initial problems sometimes can be eliminated by part changes. At other times, a new circuit must be synthesized to remove the limitations of the first attempt.

Once a satisfactory breadboard model has been developed and the interconnection requirements of the system have been satisfied, the designer's initial task is ended[13]. The next step is the development of a prototype.

TABLE 7-4

RELATIVE PERFORMANCE OF VARIOUS OSCILLATOR TYPES

| Oscillator type | Frequency stability* | Resistance to shock and spin | Power drain | Complexity |
|---|---|---|---|---|
| LC parallel tuned | good | fair | low | low |
| LC series tuned | very good | fair | low | med |
| RC wien bridge | very good | good | low | med |
| RC parallel-T | very good | good | low | med to high |
| RC bridged-T | very good | good | low | med |
| RC phase shift | fair to good | good | low | med |
| multivibrator | fair to good | good | low to med | med to high |
| magnetic square loop | good | fair to poor | med to high | high |
| pulse RC discharge | fair | good | low | med |
| pulse curpistor-C | fair to good | good | very low | med |
| mechanical resonator | excellent | poor | low | med to high |

*Frequency stability, variation: excellent $< 1 \times 10^{-4}$; very good $10^{-4}$ to $10^{-3}$; good $10^{-3}$ to $10^{-2}$; fair $> 10^{-2}$.

## 7-2.4 LAYOUT AND ASSEMBLY TECHNIQUES

With the circuit breadboard performing as required, a suitable layout can be considered. The first basic rule that insures reliable circuit operation involves packaging. The circuit should be compatible with the thermal limitations of a standard package whenever possible. Since the package provides a conductive path that dissipates the heat generated internally, circuit power dissipation must be limited so that the internal operating temperature does not exceed allowable limits.

## TABLE 7-5

### COMPARISON OF VARIOUS FREQUENCY DIVIDERS AND COUNTERS

| Type of counter or frequency divider | Voltage, V | Current, A | Power Level, mW | Size, in.$^3$ |
|---|---|---|---|---|
| Magnetic core decade Input frequency– 1 Hz max. | 10 | $5 \times 10^{-5}$ | 0.5 | 0.3 |
| Magnetic core decade Input frequency– 10 Hz | 10 | $7 \times 10^{-5}$ | 0.7 | 0.3 |
| Magnetic core decade Input frequency– 10 Hz | 10 | $2.5 \times 10^{-4}$ | 2.5 | 0.3 |
| Magnetic core decade Input frequency– 1000 Hz | 10 | $2.1 \times 10^{-3}$ | 2.1 | 0.3 |
| Magnetic core ring counters or feed-back shift register– Arbitrary number of stages | 10 | Depends on input frequency as given above | Depends on input frequency as given above | Maximum size: 1/3 in.$^3$ per decade of frequency reduction |
| Transistor decade (flip–flops) Input frequency– 1000 Hz max. | 6 | $5 \times 10^{-5}$ | 0.3 | 0.3 0.01 |
| Transistor decade (flip–flop ring counter) 1000 Hz max. | 6 | $3 \times 10^{-5}$ | 0.2 | 0.7 0.03 |
| Complementary metal oxide semiconductor | 10 to 15 | $< 10^{-3}$ | $< 15$ | 0.02 |

The maximum number of permissible leads depends on package limitations. By use of integrated circuits, standard transistor can-packages are limited to 12 leads while flat and dual in-line packages will accommodate 14 leads.

A second fundamental layout objective is the size of the ultimate monolithic die. With properly designed circuits produced with high packaging yields, die size is often the controlling factor in the manufacturing cost of integrated circuits.

The circuit designer has no control over the size and geometry of active components. These are fixed by the types and specifications used in the final breadboard. In the light of the variations in manufactured components, it is advisable that the circuit designer work in close cooperation with the part supplier.

A modern aid for integrated circuit design is the computer. If the circuit design is known well enough to provide complete characterization electrically and topographically, it can be optimized by computer methods. A manufacturer's off-the-shelf circuits can be specified completely and stored in a computer library[13].

All printed circuit assemblies, even for small production quantities, are dip soldered. This method saves labor, and results in greater reliability. Generally, all conventional components are mounted on the side of the circuit board opposite the pattern, so that they are not immersed in the solder during the dipping operation. With two-sided circuits, the components are mounted on the side opposite the one to be dipped. Components mounted on the dip side are assembled after dipping and soldered in place by hand.

Preforming and precutting of axial lead components greatly increase the speed and efficiency of assembly of printed circuit boards. Other efficient methods of assembly are flow soldering, wave soldering, and cascade soldering. The simplest way to assemble components to a printed circuit board is to bend the leads of the components and insert them through holes in the printed wiring board, hand solder the connections between the component pigtail tips and the printed conductor, and then cut off the excess lead lengths. Care should be exercised in the hand-soldering operation to avoid cold-solder joints and resin joints[14].

## 7-2.5 PERFORMANCE STANDARDS

Performance standards that an electronic timer is expected to meet for accelerated systems usually are determined by the end use of the timer. Timers can be used in missile, rocket, bomb, or projectile fuzes. Existing off-the-shelf power supplies, for instance, usually will not meet the requirements for projectile fuze timers.

Setback and spin requirements vary widely. Table 7-6[15] is a general summary of these operating requirements. Electrical requirements are comparatively uniform, typical requirements being listed in Table 7-7[13]. In electronic time fuzes, accuracy is dependent upon the time error in starting as well as the running error.

### TABLE 7-6

### OPERATING REQUIREMENTS

| Fuze | Setback, g | Spin, rps |
|---|---|---|
| Missile | 25 | 5 |
| Rocket | | |
| "Nonspin" | 100 | 5 |
| Spin | 100 | 100 |
| Projectile | 27,000 | 500 |

TABLE 7-7

ELECTRICAL REQUIREMENTS

| Projectile Fuze | Voltage, V | Current, mA |
|---|---|---|
| Missile | 12–15 | 30–50 |
| Rocket | 6.7 ± 0.7 | 100 |
| Projectile* | 6–15 | 30–100 |

*Summary of three developments.

Typical performance standards for a long-delay bomb fuze include[2]:

(1) Delay time from 15 min to 100 hr settable in 15-min steps

(2) Accuracy within ± 5%

(3) Operating temperature range −65° to +160°F

(4) Operation after an impact deceleration of 20,000 g for 5 msec in any direction

(5) Usual military, vibration, and environmental requirements

(6) Maximum size: 3 in. in diameter by 2-5/8 in. long including power supply.

Oscillator frequencies of interest fall in the range of 10 to 10,000 Hz. Frequency choice is determined mainly by the incremental setting resolution desired and the technical limitations of the particular type of oscillator chosen. Setback at launch or firing will occur momentarily and will require continuity of signal, with recovery to the original signal. Spin may persist throughout the useful life of the timer. Since some rounds do not spin, it may be desirable to develop separate oscillators for spin and nonspin applications. The oscillator frequencies must remain within

specification tolerances for (1) variations of supply voltage of ± 20% from the nominal design value and (2) variations of other parameters to their extreme values[12].

In most applications, protection will be required against catastrophic failures that would increase the frequency and shorten arming or firing time, thus creating a safety hazard. To minimize timing error, most uses require a rapid buildup and stabilization of the energizing power.

A final requirement for electronic timers is accuracy after prolonged storage at temperature extremes. Crystallization, cold flow, and other deteriorating effects in the frequency determining elements must be minimized. Effects of irradiation from nuclear detonations may be of importance. To be confident that the timer will perform as required, it is necessary to subject it to extensive tests (see par. 7-5).

## 7-2.6 EXTERNAL INTERFACES

The primary external interface for electronic timers used in accelerated systems is remote selection of the time delay Electronic setting in its simplest form eliminates the need for circuit power while the selection is being done. For information on time setting methods see par. 10-4.

Provisions for external interfaces should insure electrical, mechanical, and environmental compatibility between the timer and any external cabling and selecting system. Mating connectors should pass MIL-STD specifications[16]. The connector in the timer should be capped when the cabling from the selection system is removed. A cap should act as an environmental shield to protect the timer[17].

## 7-3 RADIATION HARDENING

It is necessary for some types of timers to survive the radiation environment of a nuclear burst. The radiated energy will destroy the average electronic circuit within several miles of the burst unless radiation hardening is employed. Before discussing the techniques used to harden electronic circuits, it would be well to examine the environment that is generated by a nuclear explosion.

## 7-3.1 RADIATION FROM A NUCLEAR EXPLOSION

A nuclear explosion resembles a blackbody at a temperature of several tens of millions degrees and, in accordance with Planck's radiation formula, radiates most of its energy as soft X rays. Almost all of the energy, approximately one billion kilowatts per megaton, is released in the first microsecond in the form of X rays, gamma rays, alpha rays, beta rays, and neutrons. Note that the first two are electromagnetic while the three remaining are corpuscular. Table 7-8[18] compares four of these components.

### 7-3.1.1 X Rays

#### 7-3.1.1.1 Nature of X Rays

X rays are produced in a nuclear burst by high-velocity electrons colliding with electrons attached to an atom. When a charged body—such as an electron—is decelerated, a pulse of electromagnetic radiation is produced. For convenience, the acceleration or deceleration of an electron is measured in electron volts (eV). A 500-eV electron, for example, refers to one that has been accelerated by a potential difference of 500 V. The energy required to ionize an atom by removing an outer electron is much less than 100 eV while that required for an inner electron may be as high as 70,000 eV. X rays produced by collision with outer electrons are called soft X rays while those produced by displacement of inner electrons are called hard X rays. The main difference between these emanations is the wavelength: soft X rays range from ten to one angstrom, hard X rays cover one to 0.1 angstrom.

#### 7-3.1.1.2 Interaction With Materials

Ionization, the production of hole-electron pairs in a material, results from the absorption of photons (soft and hard X rays). Photons, which comprise up to 75% of the energy in a nuclear explosion, interact with matter in any of three ways: (1) Compton scattering, (2) photoelectric effect, and (3) pair production (see Fig. 7-7[19]).

The results of these processes are the production of hole-electron pairs that migrate to oppositely charged junctions under the influence of the applied electric field. This photocurrent in proximity to the junctions causes minority carriers to be swept across, creating an amplified secondary photocurrent due to the current gain of the device. The momentary effect is saturation of the device with a high current surge and a resultant voltage transient. Among the subsequent effects in this chain of events are transistor latchup, circuit ringing, junction breakdown and, in integrated circuits, even lead melting because current surges can range from 1 mA to 1 A. Spurious triggering is common with even small photocurrents and can disrupt the most carefully designed circuit. Photocurrents short out capacitors, turn nominal insulators into conductors, and are especially harmful in integrated circuits because they appear in the base material as well as at the site of active devices and thus form coupling and shorting paths throughout the chip.

### 7-3.1.2 Gamma Rays

Approximately 3.5% of the energy radiated from a nuclear burst occurs in the first 50 $\mu$sec as gamma rays. These rays, with a wavelength of 0.1 to 0.01 angstrom, are similar to X rays generated by several million volts.

Gamma rays are similar to X rays in that they ionize the material into which they penetrate; i.e., they remove electrons from neutral atoms. In most situations this effect is

## TABLE 7-8

## COMPARISON OF RADIATION FROM NATURAL SUBSTANCES[18]

| | X rays | Alpha rays | Beta rays | Gamma rays |
|---|---|---|---|---|
| Character | Quanta $h\nu$ (governed by waves of length $c/\nu$) | $He^{++}$ ions = He nuclei | Electrons | Same as X rays, same or shorter wavelength |
| Velocity | Velocity of light $c$ | Up to $1/15\,c$ | Up to $99.8/100\,c$ | Velocity of light $c$ |
| Energy of particle or quantum | Visible light: several eV. Technical X rays: $10^4 - 10^6$ eV | Up to $8.8 \times 10^6$ eV | Up to $2 \times 10^6$ eV | Up to several $10^6$ eV |
| Penetrating power | | 1/100 mm aluminum | Several 1/10 mm aluminum | Several cm lead |
| Absorption, observed | $I = I_o e^{-\mu x}$ | Limited range; dense ionization over the full length of range; straight paths except very rare sharp deflections | Limited, not sharply defined range; less dense ionization than by $\alpha$ rays; crooked paths | $I = I_o e^{-\mu x}$ good approximation |
| Absorption, theory | Each quantum when absorbed gives away total energy in one process; fraction of energy lost per cm path is constant | Each particle spends its energy on many ionization processes; no deflection in collisions with electrons because of large mass of $\alpha$ particle; very rare sharp deflection in collision with nuclei | Each particle spends its energy on many ionization processes; deflection occurs since $\beta$ particle is light | Each quantum when absorbed gives away its energy in one or a few processes: ionization (photo-electric effect), Compton effect or pair production |
| Relation to emitting atom | Wavelength sharply defined characteristic of electronic structure of emitting atom | Energy of $\alpha$ particle sharply defined, characteristic of emitting nucleus | *Continuous* $\beta$-ray spectra with sharp high-energy limit are characteristic of emitting nucleus; *sharp-line* $\beta$-ray spectra represent emission of K or L . . . electrons | Wavelength of $\gamma$ ray sharply defined characteristic of emitting nucleus |

*Reprinted with permission of McGraw-Hill Book Company.*

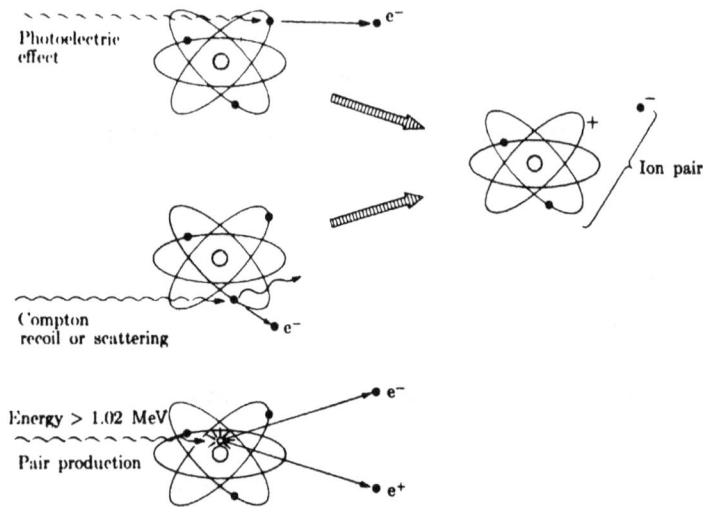

Figure 7-7. Summary of X-ray and Gamma Ray Interactions With Matter[19]

temporary so that the device returns to its original state when the radiation is discontinued. The time lapse is referred to as recovery time.

### 7-3.1.3 Neutrons

The effects of neutrons—although they are particles of appreciable mass—are similar in character to gamma rays. Neutrons can penetrate a considerable distance through the air and constitute a great hazard.

### 7-3.2 ELECTROMAGNETIC PULSES (EMP)

Under the proper circumstances a significant portion of the energy released during a nuclear detonation can be made to appear as an ElectroMagnetic Pulse (hence, EMP) having the same frequencies or wavelengths as those employed by most commercial radios and military systems[20]. Two unique properties of EMP are of crucial significance—(1) its extremely great range, EMP being capable of disabling electrical and electronic systems as

far as 3000 miles from the site of the detonation, and (2) the fact that EMP can cause severe disruption and sometimes damage when other weapon effects such as nuclear radiation blast, thermal effects, dust, debris, and biological effects are all absent. This means that a high-yield nuclear weapon, burst above the atmosphere, could be used to knock out improperly designed electrical and electronic systems over a large area of the earth's surface without doing any other significant damage. The range of EMP is diminished greatly if the weapon is detonated within the atmosphere.

An idea of the amplitude of EMP can be gained when we compare it with fields from man-made sources. A typical high-level EMP pulse could have an intensity of 100,000 V/m. This is a thousand times more intense than a radar beam of sufficient power to cause biological damage such as blindness or sterilization. It is ten million times as intense as fields created by sources in a typical metropolitan area. Recently two factors have greatly increased the significance of the EMP threat: (1) increased sophistication in nuclear strategy and weapons, and (2) increased susceptibility of electronic systems due to the broad introduction of semiconductors and the ever-greater dependence on complex operational hardware.

EMP differs from the electromagnetic environment usually encountered. The very intense EMP fields are distributed widely, whereas intense natural (lightning) or man-made fields are localized. Further, EMP occupies the spectrum through the lower microwave band. Most man-made fields are confined to only a narrow part of the spectrum. It follows, therefore, that protection practices and components for non-EMP environments—radio-frequency interference, lightning, radar, etc.—are not directly applicable for EMP problems.

The asymmetrical flow of charges caused by a nuclear explosion causes electromagnetic fields to be radiated away from the burst point. These fields cause a corresponding flow of charges, or electrical currents, in distant metallic conductors. This is comparable to the way the electromagnetic fields from a TV transmitter set up currents in a rooftop TV antenna. Any metallic object exposed to electromagnetic fields can be a collector of electromagnetic energy; i.e., act like an antenna, even though it was never intended to be that. Generally, the larger the metallic structure, the greater is the amount of intercepted EMP energy.

The level of the EMP environment can be reduced by shielding, geometric arrangement, geographic relocation, or grounding. Good shielding design is the best practice. It usually centers on pragmatic compromises related to the realities of construction and fabrication, economics, and applicational requirements. The dominant principle in circuit geometry is the avoidance of coupling configurations— most notably inductive loops. Good circuit layout practices apply equally to large cable systems or to printed circuit packages. It is generally independent of circuit dimensions. Grounding is not a panacea. If it is not applied realistically, it may make matters worse. In some systems controlled resistive grounds are used purposely to promote energy dissipation and to suppress ground loop currents.

The fraction of the collected energy which is applied to a sensitive component may be reduced by introducing an amplitude limiting device, such as a spark-gap, a filter, or a disconnect mechanism between the energy collector and the component. Protection against EMP also may be achieved by choosing less sensitive components or subsystems. For instance, vacuum tubes are more damage-resistant than transistors. Their incorporation, however, might create other problems. Increasing or selecting digital logic circuits having digital voltages and switching thresholds is preferable. Coding signals sent via long interconnecting wiring will minimize EMP "fooling" the system.

Testing is as important as planning and implementation of hardware protection. The balances between laboratory tests and full-scale simulations, and between component and full-system tests, depend on many factors, such as system size, probable threat situations, and unavoidable susceptibilities. Both analytical and empirical approaches are required to realize a hardened system. Neither can be relied upon exclusively. The number of uncertainties and unknowns is so large that one approach often is used to confirm the results of the other approach.

Today, it is considered to be sufficient to permit the design and implementation of most military hardware with high assurance of acceptable EMP hardness. But it does cost some money and effort, and the indications are that this hardness will never come "free". Typical EMP hardening costs (if incorporated early) for strategic systems can be on the order of a few percent of the system cost[20].

### 7-3.3 COMPONENTS

#### 7-3.3 1 Design Considerations

The hardening of a timing system that must operate during irradiation has two major problem areas[21]. The system must not lose its capability to perform due to permanent damage to its components, primarily from neutron irradiation. In addition, timing accuracy will be affected directly by any temporary disabling due to transient response of system circuitry to ionizing radiation, primarily photons.

Hardening the time system against permanent damage by neutron irradiation can be accomplished mainly by choosing the "hardest" components and allowing for their degradation to a level to which the weakest components can be raised by proper shielding. To minimize the transient response of the system would require extensive circuit redesign.

The degree of overdesign depends on specification level and system complexity. For example, at $10^{12}$ n/cm$^2$ with energies greater than 10 keV, an overdesign factor of 20 is reasonable; at $10^{13}$ n/cm$^2$ a reasonable overdesign factor could be 5. At $5 \times 10^{14}$ n/cm$^2$, however, an overdesign factor for safety margin is hard to obtain[22]. Some of the most sensitive semiconductor components begin to show radiation effects at about $10^{11}$ n/cm$^2$. The upper end of the range is about $10^{15}$ n/cm$^2$. See Refs. 23-26 for detailed information on hardening.

#### 7-3.3.2 Hardened Components

Hardened components are the foundation for a survivable system. Advances in semiconductor device technology have produced dielectrically isolated integrated circuits of high-gain bandwidth product with thin-film passive components for use in radiation environments. Computer aided analysis and design have been made possible by the development of codes like SCEPTRE and TRAC that use large-signal radiation equivalent circuits for semiconductor components. Reliability engineering and quality control have evolved new techniques for minimizing structural defects that might limit semiconductor performance.

The most important failure mechanism due to neutrons is the degradation of semiconductor device operating characteristics. The main aspect of this problem is $\beta$ degradation, where $\beta$ is the current gain. Various studies have shown that control of base width is needed in order to control $\beta$ degradation. This can be done more directly by controlling gain bandwidth product which is inversely proportional to the square of the base width. Selection of a gain bandwidth product that ensures the largest overdesign factor consistent with the present state of art and market availability is now possible for small signal devices. For power transistors, overdesign must be supplemented by circuit design techniques that minimize the dependence of circuit functional parameters on the gain of the power transistor. Devices such as SCR's and unijunction transistors should not be used unless a verified source of hardened devices has been established. Microcircuits, including p-n-p transistors, should be avoided if possible due to the difficulty of controlling base width in such structures[22]. Several methods have been developed to produce films that are insensitive to a radiation-induced space charge buildup. These are the silicon oxynitride films[27] and the silicon-on-sapphire isolation techniques[28]. For additional details on component hardening techniques, see Ref. 29.

While surface properties of semiconductor devices also are changed by the ionizing dose, the parameters of most transistors and modern diodes are relatively independent of changes in surface properties. The net result is that these components can be used up to total doses of $10^6$ rad. Surface field-effect transistors suffer large changes in turn-on voltage due to total dose effects and have thresholds as low as 5000 rad. These transistors should be avoided in radiation hardened systems[22]. Surface dose effects can be controlled to a degree by imposing stringent specification limits on junction leakage currents.

### 7-3.3.3 Analysis

In the design and analysis of electronics in the radiation environment, it is necessary to describe the electrical effects of radiation on semiconductor devices quantitatively. The Ebers and Moll Model[30], the simplest large signal model, can be readily modified to include radiation effects. The devices must be characterized and predictions—nominal and worst case—of device response to the nuclear environment must be made. The device must then be measured in radiation facilities. Predicted and measured responses are then compared, and any significant differences must be resolved. Analysis must cover all of the current and voltage regions that the device encounters in the system. In no case should circuit analysis be performed until satisfactory device models are defined.

### 7-3.4 CIRCUIT TECHNIQUES

Current transients from photons create noise pulses, a change of state in digital circuits, and other transient phenomena. The two main elements of a circuit hardening approach are[22]:

(1) Shielding to limit radiation levels in the electronic circuits

(2) Special circuit design techniques to minimize circuit functional response characteristics to radiation effects.

Shielding can effectively reduce X ray flux but it usually is ruled out for use with neutron and gamma flux because of severe weight penalties. When shielding is not feasible, the circuit must be redesigned.

The most cost effective way to analyze a circuit is to use a computer to predict system operating parameters as functions of radiation exposure. Proof tests can then be performed at radiation facilities and compared with the computer predictions. Agreement between predicted and measured results provides confidence in the hardened design.

A typical comparison of measured and predicted response for the microcircuit of a general purpose amplifier is shown in Fig. 7-8[22]. For further information on circuit hardening techniques, see Refs. 29 and 31.

### 7-3.5 MEASUREMENT AND SIMULATION

In order to simulate exposures and to measure their effects, a radiation facility is required. Facilities such as a TRIGA Mark F reactor at the radiation facility of the US Army Harry Diamond Laboratories can be used to irradiate components. Neutron dosimetry is used for energies greater than 10 KeV with dosimetry accuracy being 20-25%[21].

Among the results found at this facility was that reset circuitry need not be hardened against permanent neutron damage but that it

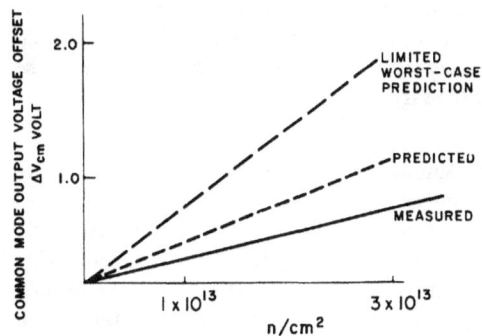

*Figure 7-8. Congruence Between Predicted and Measured Common-mode Output Voltage Offset*[22]

must remain inactive after the timing sequence has started. A burst of ionizing radiation could produce photocurrents in transistors which would act as a reset signal, thus resetting all flip-flops and magnetic circuitry. Photocurrents due to ionizing radiation in microcircuits fabricated with p-n junction isolation were almost an order of magnitude greater than those fabricated with dielectric isolation. Performance degradation of the latter type of microcircuit can be expected to follow closely the degradation of its discrete components. Thermistors irradiated at the facility were found to indicate lower temperatures after irradiation (i.e., their resistance increased); each thermistor had its own characteristic pattern of change[21].

Magnetic core dividers contained three types of component: diodes, SCR's, and magnetic cores. The SCR's failed to operate after relatively low dosages. Their principle of operation makes them extremely vulnerable to ionizing radiation even if they are built to withstand permanent damage. However, a two-transistor equivalent of the SCR is radiation resistant. Damage to the diodes of the magnetic divider showed some decrease in core switching time and loop current because of increased forward voltage drop across the diodes. Magnetic cores were not damaged. Output circuits were also vulnerable because of the presence of SCR's in the system and indicated that substantial redesign would be necessary.

Simulation of timer circuits can be done by computer. The basic information needed to model a timer circuit is the circuit schematic and its electrical performance characteristics. Gamma dose rate modeling is accomplished by the addition of a current generator and, in neutron effects modeling, the gain of the transistor is altered to simulate progressive degradation[32].

## 7-4 SPECIAL APPLICATIONS

### 7-4.1 LONG DELAY STATIONARY SYSTEMS

Electronic timers are used in mine fuzes and demolition devices. Mine fuze timers are now used for:

(1) Opening of a missile or artillery round containing a set of scatterable mines

(2) Arming of mines after ejection from the round

(3) Self-destruction of the mine after a time ranging from hours to weeks

(4) The signature analysis circuitry to help determine the functioning time against the target and to assist in resisting counter measures. These timers are made from COSMOS chips.

Examples of long-range timers are described in Refs. 33 and 34.

### 7-4.2 SATELLITE AND REMOTELY SETTABLE TIMERS

In the design of a timer for a satellite separation system, provision usually is made to provide outputs anywhere within long and short intervals by adjusting a single resistor per interval. Each interval timer is independent and may be connected in series or parallel with any number of other interval timers in any order. The timer should be able to operate reliably under normal and space temperature and pressure conditions, and also under vibration of 10 to 5000 Hz (at maximum acceleration of 54 g)[35]. Off-the-shelf components with no compensation can yield accuracies of ± 10% or better over the temperature range of −4° to 158°F. Selection of timing capacitors having a low temperature coefficient and low-leakage can yield ± 5% accuracies in this temperature range. With combinations of bias and timing current compensation, accuracies of ± 2% can be obtained. Redundancy can be accomplished easily because of the independence of the circuits that produce the timing interval and fire the explosive charge, and the fact that the oscillator for each interval is normally off. Figs. 7-9[35] and 7-10[35] show diagrams of long-interval and short-interval timers.

A remotely settable timer is an electronic system consisting of a 2-channel timer and a setting unit connected together either by a 2-wire or 3-wire link or by a 2-way radio link[10]. A typical timer system is settable from 1 to 20,000 sec in 1-msec increments. The frequency source is a crystal controlled oscillator, accurate to ± 2.5 parts in $10^7$ over a range of 21 to 31 V DC, a temperature range of 32° to 140°F, and shock and vibration environment of a missile launch[10].

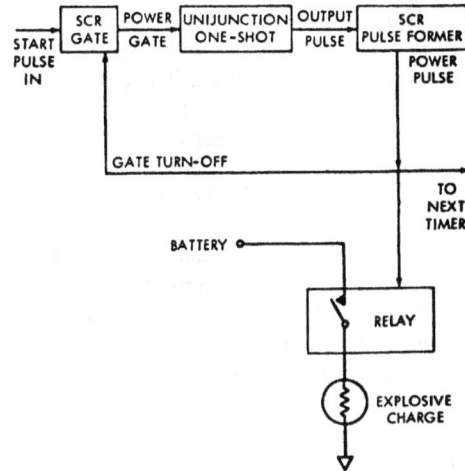

Figure 7-10. Short-interval Timer Block Diagram

Another type of remotely settable magnetic core electronic timer is shown in Fig. 7-6. This timer consists of a temperature-stabilized 1000-cycle LC oscillator accurate to 0.1%, a magnetic core counter in double line register mode, and a remote selection unit. Time delays from 1 to 200 sec can be obtained with a setting resolution to 0.1 sec. The remote selection system eliminates the need for circuit power while the time delay is being selected. The time delay selected can be stored indefinitely in the magnetic cores of the timer.

## 7-5 TIMER TESTING

### 7-5.1 GENERAL

During the design and development of an electronic timer, as with any other timer system, each component is subjected to a number of tests to determine if it performs in the manner expected. When the prototype is

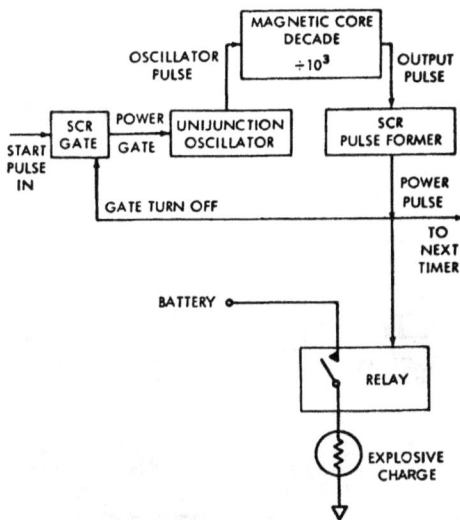

Figure 7-9. Long-interval Timer Block Diagram

built, it is subjected to performance and proof tests to see if the prototype satisfies the requirements. Since some of these tests damage the item and since the available number of items is limited, it is desirable to use special tests and it is necessary to apply special methods of data analysis.

## 7-5.2 PERFORMANCE TESTS

The components of a timer as well as the complete timer assembly must undergo tests to establish that the system functions properly. For a discussion of development and acceptance tests that are used to check out new timer designs, see par. 13-6.2.

## 7-5.3 SAFETY TESTS

Safety tests are used to determine whether performance meets safe handling requirements. There are two types: (1) destructive tests where operability after the test is not required and (2) nondestructive tests where operability is required. For a discussion of these tests, see pars. 13-6.4.1 and 13-6.4.2.

## 7-5.4 TELEMETRY

Telemetering in the broad sense is the transmission of data by any means from a remote and usually inaccessible point to an accessible location. Usually telemetering refers to electrical methods of acquiring and transmitting data, especially for transmission by an RF link from the munition to a ground station[36]. The requirement for telemetering data from fuzes using electronic timers may be quite severe, e.g., in an artillery fuze where survival of the telemetering transmitter, power source, and antenna may be necessary under accelerations in excess of 50,000 g during setback. A typical system meeting these requirements is shown in Fig. 7-11[7]. Transducers in the device being tested convert the variable being measured into an electrical signal that subsequently is used to modulate the carrier of an RF transmitter. The signal is received, amplified, and demodulated on the ground and recorded on magnetic tape or on an oscillograph for subsequent analysis.

Fig. 7-12[37] shows the sequence of events

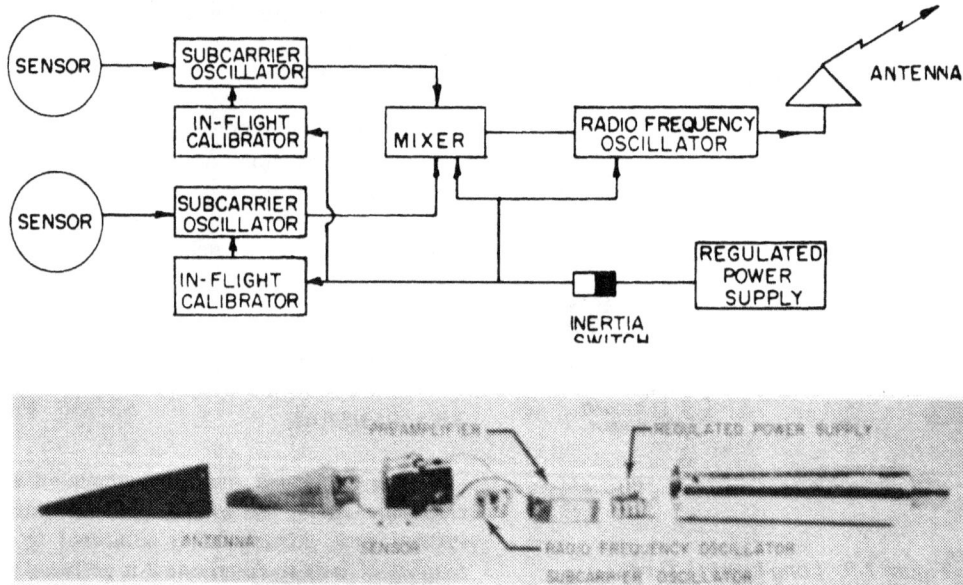

Figure 7-11. Typical VHF High-g Telemetry System

of an electronic programmer system for a projectile. The projectile was equipped with a telemetry unit that transmitted the frequency of both oscillators and the outputs from both channels. The outputs occurred at T-7, T-6, T-4, and T-0 sec. The purpose of the test was primarily to determine if the decade dividers, output logic, and firing circuit would operate properly after being subjected to gun-fired environments[37].

## 7-5.5 ANALYSIS OF DATA

It is important that the timer system designer use statistical procedures to re-enforce his conclusions. Timers may be manufactured in large lots and only a few items will be tested. These are selected randomly so that the sample will represent the lot. For more information about the analysis of data, see par. 13-6.5.

## 7-6 ADDITIONAL FACTORS IN DESIGN

In order to design a timer for use in severe environments so that it will function within the limits set forth in the specifications, a number of general design factors must be considered. These include reliability, human factors engineering, and cost.

## 7-6.1 RELIABILITY AND MAINTAINABILITY

In designing for reliability, the designer has two powerful tools at his disposal at all stages of the design. They are: (1) derating of components and (2) reliability analysis. Derating of components means operating them at substantially less than their rated values of voltage, wattage, temperature, etc. Reliability analysis is the calculation of system or equipment reliability from the failure rates of the components used. If these techniques do not reduce the component failure rate sufficiently, then redundancy or standby systems can be used. In the case of electronic timers—where high reliability, small space, and low weight are specified—redundancy and standby systems cannot always be used and therefore the first two methods commonly are employed. Techniques for reliable design are[5]:

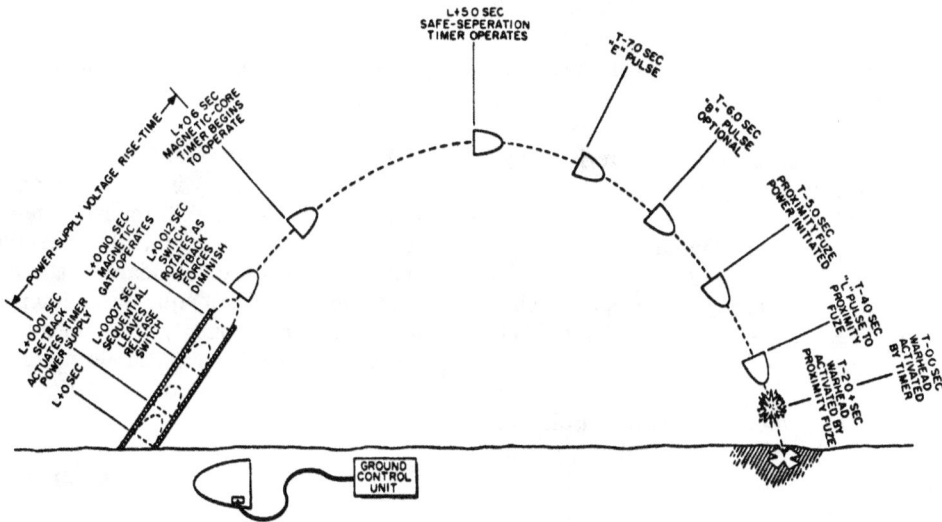

Figure 7-12. Sequence of Events

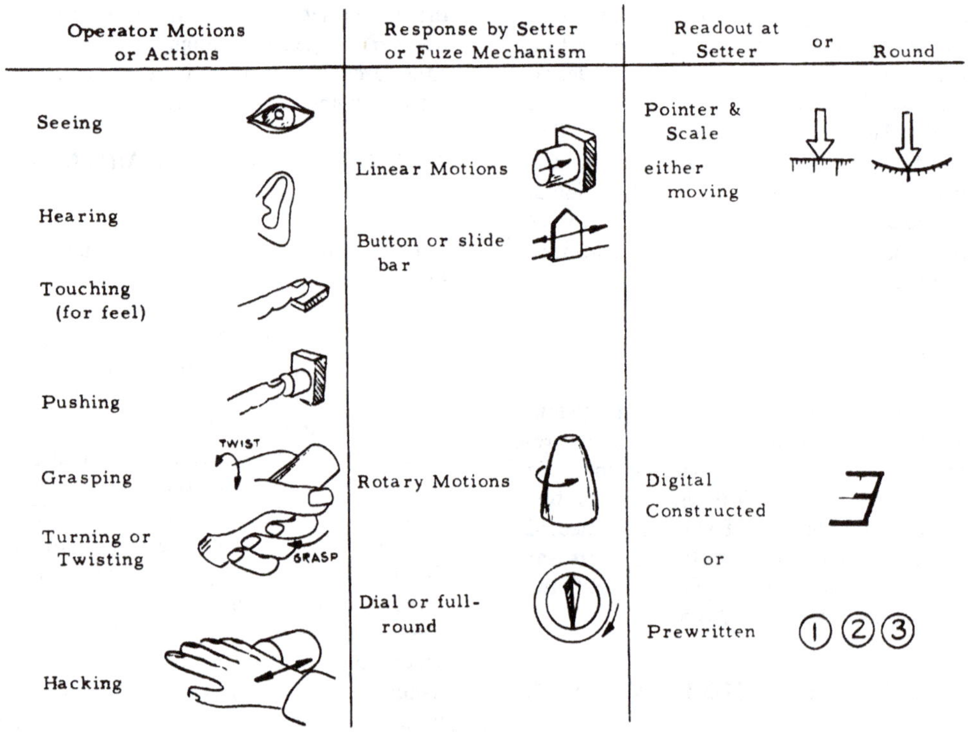

| Operator Motions or Actions | Response by Setter or Fuze Mechanism | Readout at Setter or Round |
|---|---|---|
| Seeing | Linear Motions | Pointer & Scale either moving |
| Hearing | Button or slide bar | |
| Touching (for feel) | | |
| Pushing | | |
| Grasping | Rotary Motions | Digital Constructed |
| Turning or Twisting | | or |
| | Dial or full-round | |
| Hacking | | Prewritten |

Figure 7-13. External System Building Blocks

(1) Design to a minimum of parts without degrading performance.

(2) Perform design reliability reviews by means of reliability analyses.

(3) Apply derating techniques.

(4) Reduce operating temperatures by providing heat sinks, good packaging, and good cooling.

(5) Eliminate vibrations by good isolation and protect against shock, humidity, corrosion, etc.

(6) Specify component reliability and burn-in requirements.

(7) Specify prototype tests and production debugging procedures.

In regard to maintainability, the design engineer should keep in mind the principles discussed in par. 13-2.2. Improved reliability will reduce the need for maintenance. Full use should be made of standard parts, components, tools, and test equipment. Interchangeable parts, components, and assemblies are essential to good maintainability and should be an important part of the design philosophy.

## 7-6.2 HUMAN FACTORS ENGINEERING

Human factors engineering is the scientific union of men with machines so as to enhance the man-machine system. A knowledge of the performance of both man and timer parts is required for an efficient union. An example of the use of human factors engineering is the evaluation of a setting mechanism for analog electronic timers[8]. In this study, a handbook of information and data was developed that would be useful to a designer of resistor setting switches for an electronic timer used in artillery fuzes. It contains a detailed treatment of the operator and his relations to the function of setting internal mechanisms.

A designer uses his ability to synthesize a concept from a series of known elements. An illustration of the general nature of various input mechanisms is shown in Fig. 7-13[8]. In a man-machine situation, man is considered the variable with the machine being constant. The

human factors engineer is concerned primarily with machine or product design to improve the overall efficiency. See par. 13-2.3 for further details.

## 7-6.3 COST REDUCTION

The production of electronic timers at reduced cost is subject to the same considerations as those of any other timers. The factors that the designer must keep in mind are listed in par. 13-2.4. Statements about tolerancing, as applied to electronic timers, refer to tolerances of voltage, current, or power ratings which are required to keep the final design working according to specifications and to permit the most economical manufacturing procedure.

## REFERENCES

1. R. I. Widder and C. Kaizer, Eds., "Timers", System Designers Handbook, Electromechanical Design, 9, 161-77 (January 1965).

2. R. H. Comyn, et al., *Long-Delay Timer For Air Force Bomb Fuze*, Report TR 1126, U S Army Harry Diamond Laboratories, Washington, D.C., 1963.

3. MIL-T-45139B, *Timer, Interval TD-224/M*, Dept. of Defense, 25 October 1972.

4. MIL-STD-721A, *Definition of Terms for Reliability Engineering*, Dept. of Defense, 2 August 1962.

5. I. Bazovsky, *Reliability Theory and Practice*, Prentice-Hall, Inc., Englewood Cliffs, N.J., 1961.

6. M. J. Katz, et al., *A Precision 10 Hz Clock for Military Electronic Timers*, Report TM-66-17, U S Army Harry Diamond Laboratories, Washington, D.C., 1966 (AD-645 878).

7. AMCP 706-210, Engineering Design Handbook, *Fuzes*.

8. D. Goldstein, *Engineering Study and Evaluation of Setting Mechanism for Analog Electronic Timers*, Report SME-AP-5, U S Army Harry Diamond Laboratories, Washington, D.C., 1967.

9. D. M. Anderson, "Producibility of a Twin-T Oscillator", *Proc. Timers for Ordnance Symposium*, Vol. II, U S Army Harry Diamond Laboratories, Washington D.C., November 1966, pp. 303-25.

10. *An Accurate Two-Channel Remote Settable Timer*, Report AFWL-TR-66-77, Air Force Weapons Laboratory, Kirtland, N.M., 1967.

11. A. J. Meyerhoff, *Digital Applications of Magnetic Devices*, John Wiley and Sons, N.Y., 1960.

12. J. M. Schaull, "Precision Oscillators for

Military Electronic Timers", *Proc. Electronic Timers for Ordnance Symposium*, Vol. 1, U S Army Harry Diamond Laboratories, Washington, D.C., June 1965, pp. 141-70.

13. L. Stern, *Fundamentals of Integrated Circuits*, Hayden Book Co., Inc. N.Y., 1968.

14. A. E. Linden, *Printed Circuits in Space Technology*, Prentice-Hall, Inc., Englewood Cliffs, N.J., 1962.

15. B. L. Davis, "Power Supplies for Electronic Timers", *Proc. Electronic Timers for Ordnance Symposium*, Vol. 1, U S Army Harry Diamond Laboratories, Washington, D.C., June 1965, pp. 101-14.

16. MIL-STD-454C, *Standard General Requirements for Electronic Equipment*, Dept. of Defense, 15 October 1970, Standard 10.

17. D. L. Dickerson, *A 200-Second Magnetic Core Timer*, Report TR 3225, Picatinny Arsenal, Dover, N.J., Feb. 1965 (AD-614 564).

18. O. Oldenberg, *Introduction to Atomic and Nuclear Physics*, McGraw-Hill Book Co., Inc., N.Y., 1961, p. 256.

19. J. Richards, et al., *Modern University Physics*, Addison-Wesley Pub. Co., Inc., Reading, Mass., 1960, p. 821.

20. *DNA EMP Awareness Course Notes*, Report 2772T, IIT Research Institute, Chicago, Ill., August 1971 (AD-741 706).

21. *Preliminary Nuclear Radiation Study on a High Precision, Remotely Settable, Electronic Timer System, Phase II*, Report AFWL-TR-67-34, Air Force Weapons Laboratory, Kirtland, N.M., 1967.

22. G. C. Messenger, "Radiation Hardening of Electronic Systems", IEEE Transactions on Nuclear Science, NS-16, 160-8 (1969).

23. AMCP 706-335 (SRD), Engineering Design Handbook, *Design Engineers' Nuclear Effects Manual, Volume I, Munitions and Weapon Systems (U)*.

24. AMCP 706-336 (SRD), Engineering Design Handbook, *Design Engineers' Nuclear Effects Manual, Volume II, Electronic Systems and Logistical Systems (U)*.

25. AMCP 706-337 (SRD), Engineering Design Handbook, *Design Engineers' Nuclear Effects Manual, Volume III, Nuclear Environment (U)*.

26. AMCP 706-338 (SRD), Engineering Design Handbook, *Design Engineers' Nuclear Effects Manual, Volume IV, Nuclear Effects (U)*.

27. P. F. Schmidt, et al., "Radiation-Insensitive Silicon Oxynitride Films for Use in Silicon Devices", IEEE Transactions on Nuclear Science, NS-16, 211-9 (1969).

28. R. A. Kjar and J. E. Bell, "Characteristics of MOS Circuits for Radiation-Hardened Aerospace Systems", IEEE Transactions on Nuclear Science, NS-18, 258-62 (1971).

29. *TREE Handbook*, Battelle Memorial Institute, Columbus, Ohio, 1968.

30. J. J. Ebers and J. L. Moll, "Large Signal Behavior of Junction Transistors", Proceedings of IRE, 42, 1773-84 (Dec. 1954).

31. J. G. Aiken, et al., "Design and Performance of an Integrated Circuit Flip-Flop with Photocurrent Compensation", IEEE Transactions on Nuclear Science NS-16, 177-80 (1969).

32. R. H. Dickhaut, "Simplified Microcircuit Modeling", IEEE Transactions on Nuclear Science, NS-18, 227-31 (1971).

33. M. Lazarus and D. L. Smith, *Development of a Nuclear Timer for Satellite Application*, Report TR 2564, Picatinny Arsenal, Dover, N.J., September 1958.

34. J. O. Thayer, *Long Period Curpistor-Diode Timers*, Report TR-895, U S Army Harry Diamond Laboratories, Washington, D.C., November 1960.

35. J. C. Schaffert and T. D. Glen, "A Solid State Satellite Separation Sequence Timer", *Proc. of Electronic Timers for Ordnance Symposium*, Vol. 1, U S Army Harry Diamond Laboratories, Washington, D.C., June 1965, pp. 171-96.

36. P. A. Borden and W. J. Mayo-Wells, *Telemetering Systems*, Reinhold Publishing Corp., New York, 1959.

37. O. T. Dellasanta and F. W. Kinzelman, "Electronic Programmer System for XM471 Projectile", *Proc. of Electronic Timers for Ordnance Symposium*, Vol. 1, U S Army Harry Diamond Laboratories, Washington, D.C., June 1965, pp. 21-58.

# CHAPTER 8

## PRODUCTION TECHNIQUES

### 8-1 PRODUCTION OF MICROELECTRON-IC MODULES

#### 8-1.1 GENERAL

Microelectronic digital modules are being fabricated today by a wide variety of techniques. Since each technique has strengths and weaknesses, the circuit designer should be aware of the many ways to solve a specific design problem. As in all military applications, reliability is of key importance. After reliability, the next consideration is the timer type. It can be either a control device or an ammunition component. Control devices, such as precision timers, are used repeatedly, and normally kept on the bench where they are carefully handled and can be made accessible for repair. On the other hand, ammunition components, such as fuze timers, are one-shot, low cost devices, are often subject to large forces, and must have a long storage life.

Based on the concept of packaging, the field of microelectronics may be divided into three areas[1]:

(1) Packaging of discrete, miniature component parts

(2) Printing of component parts and their interconnection pathways, *in situ*, as a series of thin films on flat wafer substrates

(3) Modification of a block of semiconductor material to obtain the varied resistive, dielectric, conductive, and junction characteristics of interconnected components.

Hybrid approaches can be employed in which combinations of two or three of the preceding techniques are used in the fabrication of a single circuit. Miniature parts are available of more or less conventional construction that can be packaged into tiny modules and subsequently into larger assemblies. A different approach, still using individual parts, involves fabrication of a component of nearly flat configuration, such as a ceramic wafer. Required wafers of a prescribed area then are connected together by means of riser wires, soldered to metallized notches at the edges of the wafers to yield a package of prescribed dimensions. This technique and another in which component parts are manufactured in the shape of tiny cylinders of predetermined dimensions and inserted into holes in a printed circuit board are designed to expedite assembly. However, they offer little size reduction over that obtained by small conventional parts. An example of this pellet or "Swiss-cheese" package is shown in Fig. 8-1[1].

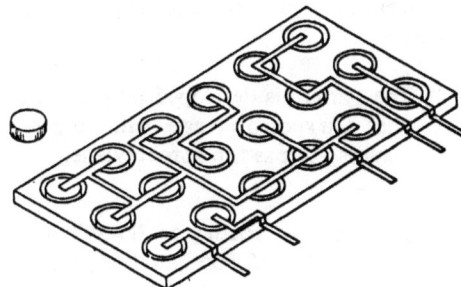

*Figure 8-1. Pellet or "Swiss Cheese" Package*

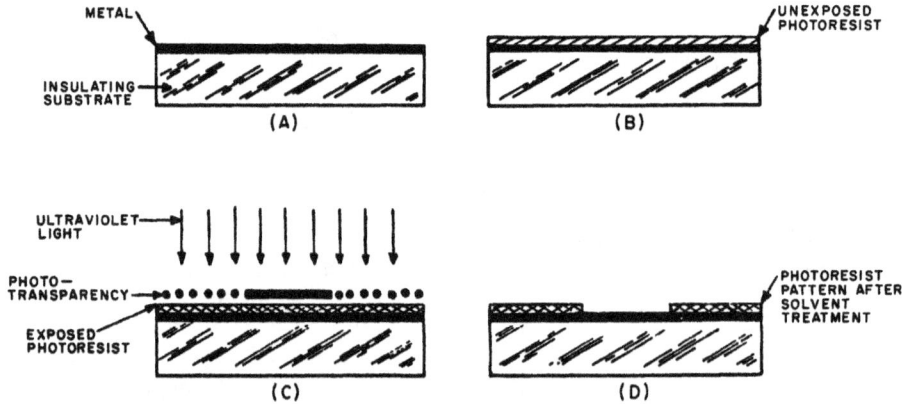

*Figure 8-2. Preparation of Pattern Using Photoresist*

## 8-1.2 PHOTORESIST MASKING

The photoresist masking technique is used extensively in thin-film and solid circuit packaging approaches. The required steps in the preparation of a pattern on a substrate are illustrated in Fig. 8-2[1]:

(1) The metal clad substrate is coated by spin-coating, dip-coating, and air brush spraying or painting.

(2) The substrate is coated with photoresist and allowed to dry.

(3) The photoresist is exposed to ultraviolet light through a negative or positive of the pattern.

(4) The pattern is developed by dissolving away the unexposed areas with organic solvent vapors, such as trichloroethylene.

The result is a resist coating on the substrate surface of the desired configuration. Equipment necessary includes a light source, an exposure frame, and fixtures to align the pattern on the substrate.

Microcircuits can be made using photoresist in a number of different ways. The photoresist can be formed into an accurate detailed mask on the surface of the metal-clad substrate. Thin layers then either can be added or subtracted from the exposed surface of the base metal. To add metallic materials, methods such as sputtering, vacuum evaporation, screen printing, electro-deposition, or chemical deposition can be used, while electroetching or chemical etching can be used to remove metal. Following the use of any of these processes, the protective photoresist can be stripped off the surface of the base material. Excellent stripping solvents include methylene chloride or hot chloroform. Fig. 8-3(A)[1] shows how photoresist protects the underlying material during an etching process. In Fig. 8-3(B) photoresist prevents the underlying metal on another plate from being electroplated. In Fig. 8-3(C), photoresist is used as a disposable mask during chemical deposition; when the photoresist is removed by solvent, the overlying deposit is also removed.

## 8-1.3 THIN-FILM MICROELECTRONICS

The more important techniques for placing conductive, resistive, and insulating thin films upon substrates include (1) screen-printing, (2) vacuum deposition, (3) sputtering, (4) vapor-phase deposition, and (5) chemical plating[1].

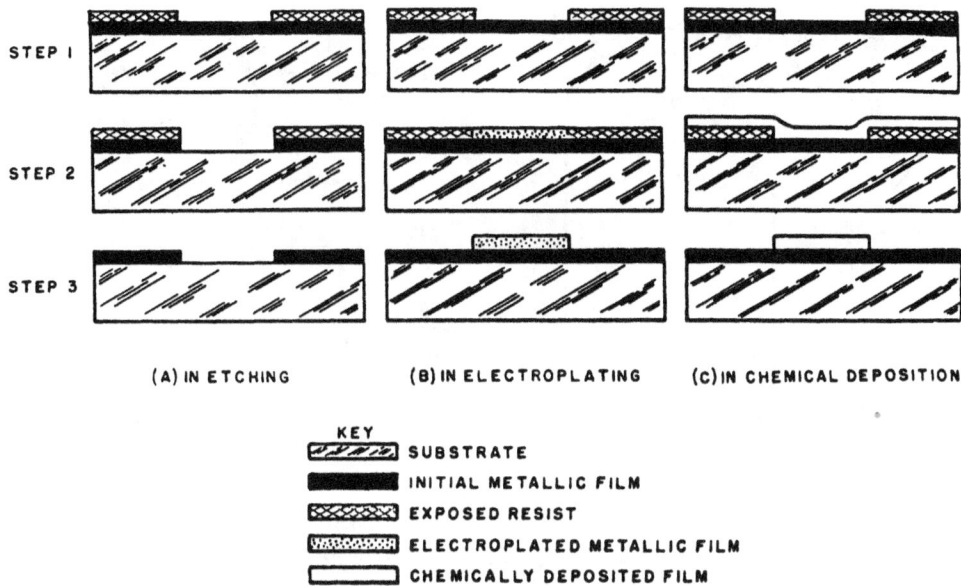

Figure 8-3. Uses of Photoresist in Various Processes

### 8-1.3.1 Screen-printing

Screen-printing commonly is used to apply ink in order to form geometric patterns on substrates. The process consists of three steps[2]:

(1) Prepare a patterned screen so that some areas of mesh are open for ink to pass while others are closed or masked.

(2) Accurately position the pattern on the substrate.

(3) Push ink down into and across the mesh of the screen with a squeegee.

While the stencil-like patterns may be prepared by both cutting and photographic methods, photography is required in microminiature circuit work because of the fine lines and intricate patterns.

### 8-1.3.2 Vacuum Deposition

In vacuum deposition, a solid material (the charge) is converted into a gas, either by evaporation or sublimation, and condensed onto a cool substrate elsewhere in the vacuum system[3]. Fig. 8-4[1] is a diagram of a typical apparatus for vacuum evaporation and deposition. Electric-resistance heating provides the energy for vaporization; other methods are induction-heating, electron bombardment, and concentrated solar radiation. A vacuum of $10^{-4}$ mm of Hg pressure or less is required to insure that the distance an atom can travel before colliding with another atom is large compared with the distance between charge and substrate.

### 8-1.3.3 Sputtering

The process in which a metal is bombarded by an ionized gas and atoms or in which

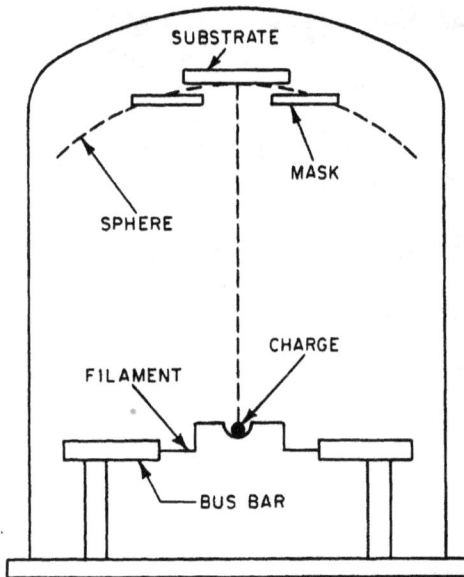

Figure 8-4. Vacuum Evaporation and
Deposition Apparatus

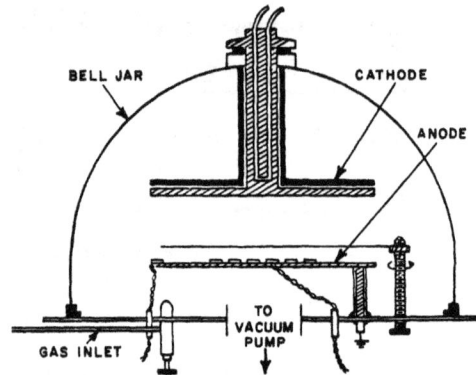

Figure 8-5. Sputtering Apparatus

groups of atoms are driven from the metal surface because of the impact is known as sputtering. Metal deposition by sputtering requires pressures as low as $10^{-2}$ to $10^{-6}$ mm of Hg and an electric field of several hundred to several thousand volts. Fig. 8-5[1] shows a typical arrangement. The metal to be sputtered is made the cathode, and the substrate is placed on a tray or rack within 1-2 in. of the cathode. The gas used (usually argon) is ionized by the electrons from the cathode. These ions are accelerated toward the cathode, which they strike, and to which they transfer kinetic energy so that surface atoms are vaporized. Some of this metal condenses on any object in its path and coats it, while some returns to the cathode after collision with gas molecules. Sputtering is slow and difficult to control but is especially useful in the deposition of refractory metals such as tungsten, tantalum, and molybdenum. Arc-melted tantalum has been sputtered in argon at a pressure of 10-20 microns and a potential of 3500 V between electrodes with

2.5-in. spacing. The sputtering rate, under these conditions, was 100 Å per min[4].

### 8-1.3.4 Vapor-phase Deposition

The process in which a substrate is coated by heating it to a high temperature in the presence of a gaseous metallic compound is called vapor-phase deposition or pyrolysis. The temperature of the substrate is high enough to decompose the metallic compound but not high enough to vaporize the deposited metal. By this means, the heated substrate decomposes the compound and the metal is deposited on the substrate. Vapor phase deposition has been used to make thin-film resistors by coating glass with tin-oxide film[5].

### 8-1.3.5 Chemical Plating

In chemical plating, metallic ions in solution are reduced by an agent absorbed on the surface of a substrate and are deposited on the substrate. The "electroless nickel" process is the most common method and is used to deposit a very adherent $Ni-Ni_3P$ alloy for electrical resistance elements[6]. The deposited nickel alloy is electroplated with copper and solder, masked with a photoresist, and thus etched in unmasked areas to yield conductor and resistor patterns for thin-film microcircuits. Fig. 8-6[1] shows a conductor and resistor pattern made in this manner.

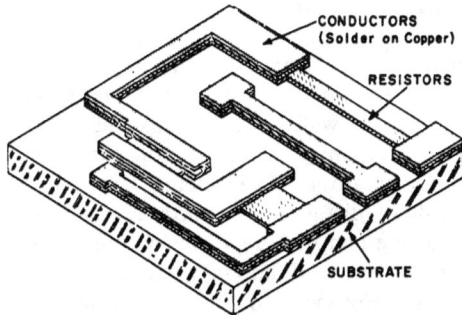

*Figure 8-6. Chemically Plated Conductor and Resistor Pattern*

## 8-1.4 TOTAL SEMICONDUCTOR CIRCUIT

The third major approach to microelectronics is that of fabricating all component parts, both passive and active, from a single block of semiconductor material. Various areas of a slice of semiconductor material are selectively exposed for diffusion by a process known as oxide-masking. An oxide is grown on the silicon slice and selectively etched away using photoresist masking. Subsequent diffusion processes then allow selected impurity atoms to enter the material at specific exposed surfaces and to continue to penetrate for a distance determined by time and temperature. This is the so-called "planar process", and is used by most of the transistor manufacturers[1]. This approach generally is classed among integrated circuits and is discussed in par. 8-2.

## 8-2 INTEGRATED CIRCUITS AND OTHER BATCH FABRICATION TECHNIQUES

The term *integrated circuits* refers to a family of devices related by advantages they have over conventionally wired circuits rather than by common design or manufacturing methods. These advantages include: (1) drastic reduction in size and weight, (2) a substantial increase in reliability, (3) promise of substantial cost reductions, and (4) improved circuit performance. Further, the integrated circuit normally is composed of

closely associated parts so that repair becomes difficult, if not impossible. In the event of trouble, the entire circuit is replaced as a single component. There are essentially two general classifications of integrated circuits: (1) semiconductor monolithic, and (2) thin-film. There are also combinations of these that could combine all of the various features of the individual devices in a complex configuration affording maximum utility. Table 8-1[7] lists the present most commonly used structures as well as those that show promise of implementation in the future.

## 8-2.1 SEMICONDUCTOR MONOLITHIC CIRCUITS

The basic structure of a monolithic circuit consists of three distinct layers of different materials, see Fig. 8-7[7]. The thick bottom layer is p-type silicon, and a thin epitaxial layer of n-type silicon is grown upon it. This p-n structure then is topped with a film of silicon dioxide. All of the component parts are built within the n-type semiconductor region. The primary function of the p-region is to act as a substrate for circuit elements formed within the n-region. The silicon dioxide film protects the semiconductor surface against contamination by external impurities and provides the means for selectively depositing various components into the n-region beneath.

Components that normally are associated with a monolithic circuit are resistors, diodes,

TABLE 8-1

TYPES OF INTEGRATED CIRCUIT

| Basic Group | Variations |
|---|---|
| Semiconductor monolithic circuits | Multichip (hybrid) circuits |
| Thin-film circuits | Monobrid circuits<br>Compatible monolithic circuits<br>Insulated-substrate monolithic circuits |

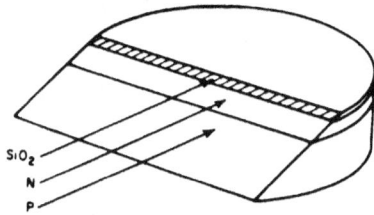

Figure 8-7. Basic Structure of Integrated Circuit Wafer[7]

transistors, and capacitors. Fig. 8-8[7] illustrates typical cross section of these components. If a number of components are to be fabricated in the n-region, means must be provided to insulate components from each other. This is done as shown in Fig. 8-9[7] where each component is bounded by a region of n-type material in contact with the p-type substrate. When the substrate is at a negative potential with respect to the n region, this p-n junction is reverse-biased and little or no current will flow through the substrate between one component and another. This form of insulation is known as diode isolation. Diffusion is used to form the isolation regions, the transistor base region, the diode anodes, and the resistor. The diffusion processes are described:

(1) Diffusion. Impurity atoms are introduced into a semiconductor wafer by diffusion to change the electrical characteristics of the wafer material. Certain types of impurity atoms will create p-type silicon while others will create n-type silicon. This

substitution process is called doping. To diffuse impurity atoms, the silicon wafer is placed in a furnace and its temperature is raised to a point between $1000°$ and $1200°C$. The heated wafer is then subjected to a gas flow containing a heavy concentration of impurity atoms. The doping level and the depth of diffusion are controlled closely by adjusting temperature, diffusion time, and the concentration of impurity atoms in the gas stream. However, to fabricate an integrated circuit, selective diffusion must be used.

(2) Selective Diffusion. Selective diffusion is used to distribute impurities only into precisely defined regions of a wafer. Selectivity is accomplished by using a series of masks that protect portions of the wafer against penetration of impurities. The masking material, silicon dioxide, easily is formed on the silicon wafer by maintaining it in an oxygen atmosphere. To diffuse the various components of an integrated circuit into the wafer of Si-$O_2$-coated silicon, areas where diffusion is to take place must be etched away. However, in order to accomplish selective etching, the $SiO_2$ layer must be masked by photoresist (see par. 8-1.2), which is used to coat one surface of the three layer wafer. A mask, which is usually glass that has an opaque printed pattern of the areas into which impurities are to be diffused, is held firmly over the wafer and the assembly is exposed to ultraviolet light. This fixes the photoresist in those areas not covered by the pattern. The wafer then is subjected to an etching solution that removes the $SiO_2$ film.

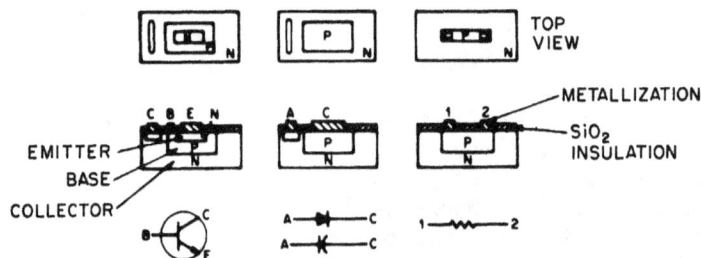

Figure 8-8. Typical Integrated Circuit Components[7,13]

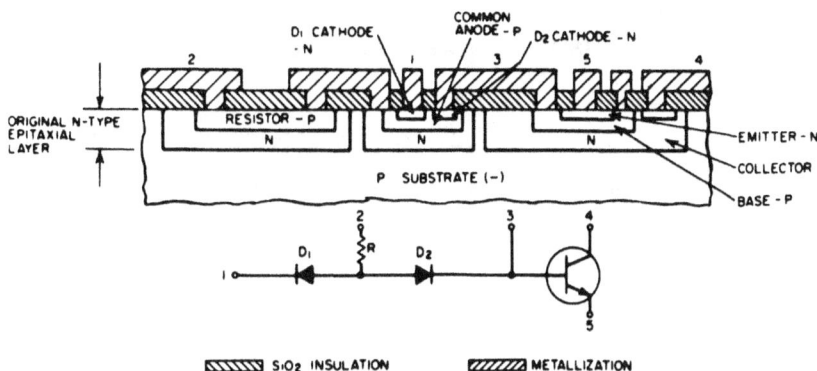

Figure 8-9. Typical Diffused Integrated Circuit Components[7,13]

After etching, the photoresist is removed. The wafer is then ready for the diffusion cycle. For more details on this process, see Refs. 7 and 8.

## 8-2.2 THIN-FILM CIRCUITS

The thin-film circuit substrate is made of a passive, insulating material such as glass or ceramic. The passive components of the eventual circuit are deposited and interconnected on this insulating surface. A large number of metallic and metal-dielectric materials (cermets), having a wide variety of electrical characteristics, are employed in these depositions. A serious drawback of present thin-film circuits is the fact that technology is unable to fabricate active components using processes that are compatible with thin-film depositions. See Table 8-2[7] for passive processes that are compatible. As a result, the cost of thin-film circuits which require a large number of active components is relatively high compared with monolithic semiconductor circuit cost.

### 8-2.2.1 Components

Properties of resistors made by thin-film methods depend on the resistivity of the material, the thickness of the deposit, and the length-to-width ratio. Some of the more

TABLE 8-2

COMPATIBLE THIN-FILM PROCESSES[7]

| Deposition Process | Material |
|---|---|
| *Metals* | |
| Vacuum Evaporation | Nichrome, aluminum, chromium, gold, nickel |
| Cathode Sputtering | Tantalum |
| Vapor Plating | Copper, gold, nickel, tin oxide |
| *Dielectrics* | |
| Vacuum Evaporation | Silicon monoxide, silicon dioxide |
| Anodization | Tantalum oxide |
| Vapor Plating | Silica, alumina, glass |

Reprinted with the permission of Hayden Book Company, Inc., Rochelle Park, N.J.

important thin-film materials for resistors are shown in Table 8-3[7]. The principal electrical characteristics of capacitors are determined by the type and thickness of the dielectric and the area of the plates. For a listing of typical dielectric materials and characteristics that can be achieved, see Table 8-4[7].

## TABLE 8-3

### THIN-FILM RESISTOR MATERIALS[7]

| Material | Suitable Process | Ohms/Sq. Range | Temperature Coefficient | Characteristics |
|---|---|---|---|---|
| Nickel-Chromium (nichrome) | Vacuum evaporation | 50–400 | ± 50–100 | Good temperature coefficient, good adhesion |
| Tin-Oxide | Vapor plating | 100–5000 | ± 100–300 | High sheet resistance, good adhesion |
| Tantalum | Sputtering | 50–500 | ± 100–200 | Good process control and temperature coefficient, good adhesion, high stability |
| Cermets | Silk screen | 10–100,000 | ± 100–300 | Wide range of resistivity, process not suitable for close tolerances |

Reprinted with the permission of Hayden Book Company, Inc., Rochelle Park, N.J.

## TABLE 8-4

### TYPICAL DIELECTRIC MATERIALS[7]

| Dielectric | Deposition Process | Capacitance at 6 Volts | Q at 10 MHz | Temperature Coefficient* | Accuracy (before trimming) |
|---|---|---|---|---|---|
| Silicon Monoxide | Vacuum evaporation | 0.013 pF/mil$^2$ at 30V | 200 | 200–250 | ± 15% |
| Alumina | Vapor plating | 0.3 pF/mil$^2$ at 12V | 10–100 | 150–400 | ± 10% |
| Tantalum | Anodic oxidation | 1.55 pF/mil$^2$ at 12V | good | 150–350 | ± 20% |
| Silicon Dioxide | Vapor oxidation | 0.25–0.4 pF/mil$^2$ | 10–100 | 100 | |

*Parts per million/deg C

Reprinted with the permission of Hayden Book Company, Inc., Rochelle Park, N.J.

### 8-2.2.2 Processing

An example of a typical thin-film circuit is shown in Fig. 8-10[7]. A typical sequence in the fabrication of a thin-film circuit includes the following steps (for more detailed information about this process, see Refs. 7 and 8):

(1) Clean substrate thoroughly to remove all contaminants.

(2) Store in a dust-free atmosphere until ready to use.

(3) Deposit the resistive pattern, usually using tin oxide.

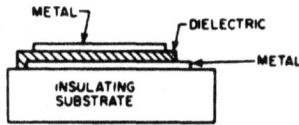

Figure 8-10. Typical Thin Film
Circuit[7,13]

(4) Deposit the bottom plates of the capacitors.

(5) Deposit the interconnecting strips.

(6) Deposit the capacitor dielectric.

(7) Deposit a conductive film over the structure to form the top capacitor plates and the final interconnecting pattern.

### 8-2.2.3 Advantages and Limitations of Thin-film Circuits

The advantages of thin-film circuits over monolithic semiconductor circuits are[7]:

(1) Thin-film passive components can be made with a wider range of values, to closer tolerances, and have better electrical characteristics.

(2) Practically any active component, made in a discrete form, can be used in a thin-film circuit.

(3) Design of circuits for conversion to thin-film assemblies is greatly simplified.

The disadvantages are[7]:

(1) Circuits are larger.

(2) Circuits are more expensive (even in large quantities).

(3) Circuits are less reliable.

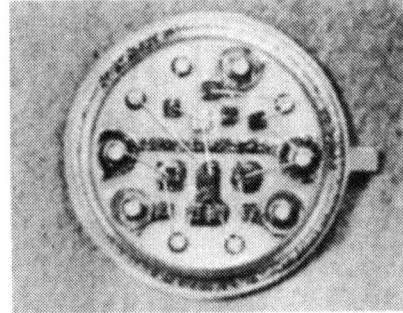

Figure 8-11. Typical Interconnected Multichip Circuit[7]

### 8-2.3 HYBRID AND OTHER INTEGRATED CIRCUIT STRUCTURES

Structures that may use both semiconductor and thin-film technologies are called hybrid and may include multichip circuits, monobrid circuits, compatible circuits, and insulated-substrate circuits.

#### 8-2.3.1 Multichip Circuits

Multichip circuits consist of components that are interconnected on a common substrate. Basically, they are more like discrete component circuits than integrated circuits; but since they possess some of the advantages of the latter, they are put in the integrated circuit family. They are practical to use in applications requiring small quantities of identical microminiature devices, or where monolithic devices cannot meet necessary performance requirements. Batch processing can be employed in multichip circuits for the various components which make up the circuit. Fig. 8-11[7] shows a typical interconnected multichip circuit.

#### 8-2.3.2 Monobrid Circuits

A monobrid circuit combines two or more

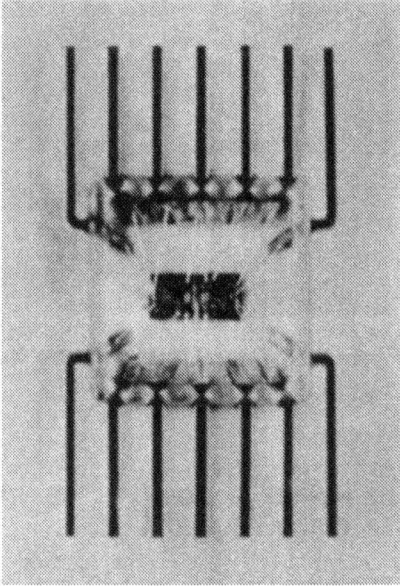

*Figure 8-12. Monobrid Circuit[7]*

monolithic circuits in a single package (see Fig. 8-12[7]). It is used when a monolithic structure cannot be employed because of complexity and large chip size. For example, transistors, which require different electrical characteristics, make processing in monolithic form difficult. Monobrid circuits are therefore used to[7]:

(1) Economically package complex circuits in a single package

(2) Optimize performance of complex circuits

(3) Use a wider range of components.

### 8-2.3.3 Compatible Integrated Circuit

The compatible integrated circuit has a basic semiconductor substrate in which diffused semiconductor components provide the basic circuit function. A pattern of thin-film passive components is placed over the monolithic substrate and insulated from it by a thin layer of silicon dioxide. Interconnections between the thin-film components and the semiconductor circuits are made through etched windows in the silicon dioxide layer. A compatible circuit requires the process steps for a monolithic circuit plus additional steps to form the thin-film components. It is, therefore, more expensive to manufacture but is a true monolithic circuit combining advantages of both monolithic and thin-film circuits[7].

### 8-2.3.4 Insulated-substrate Circuits

In order to eliminate the parasitics associated with various semiconductor junctions, a method has been developed for the construction of a parasitic-free structure that retains all the advantages of a basic monolithic structure[7]. The parasitics give rise to capacitances, diodes, and transistors that are not part of the original circuit design but are associated with the fabrication processes. Fig. 8-13[7] illustrates the following steps in the fabrication of a parasitic-free circuit:

(1) Basic substrate is prepared for insulated substrate circuit.

(2) $SiO_2$ layer is etched selectively, leaving raised $SiO_2$-covered islands.

(3) A new layer of $SiO_2$ is grown over top surface.

(4) A thick layer of polycrystal silicon is grown on top of the wafer and portions of top and bottom surfaces are removed along the dashed lines.

(5) Wafer is turned upside down, and $SiO_2$ film is grown over the surface leaving single-crystal islands completely isolated from polycrystal substrate.

### 8-2.4 DESIGN CONSIDERATIONS

Beginning either with the schematic dia-

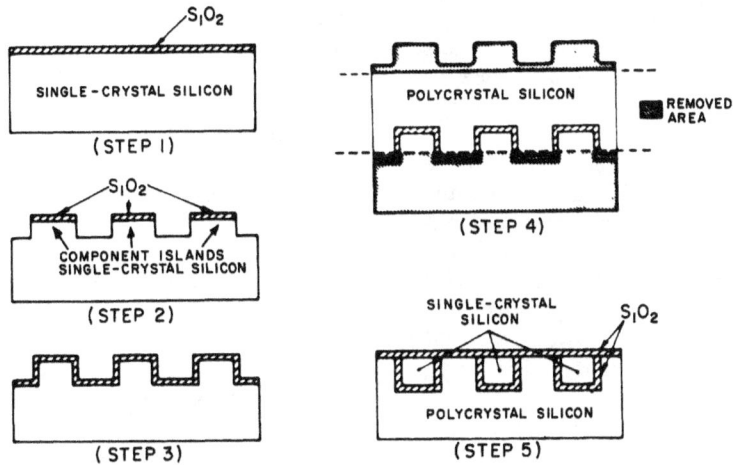

Figure 8-13. Procedure for Parasitic-free Circuit Fabrication[7]

grams of an existing system, or a block diagram of a proposed system, the following questions should be considered, each being discussed in the paragraphs that follow:

(1) What is the most economical integrated circuit that is suitable for a particular condition?

(2) What portion of a system could use the advantages of an integrated circuit?

(3) What type of packaging is best suited for a system, and how many packages are needed to house the integrated portion of a system?

(4) What are the circuit design latitudes and limitations?

### 8-2.4.1 Type of Integrated Circuit

Most off-the-shelf integrated circuits are limited to medium or low power devices. This is mainly because of the large present demand in the logic area where power requirements are minimal. Amplifiers with power dissipation of 1.8 W have been built by using integrated circuit packages[7]. If the devices

were to be mounted on a heat-dissipating surface, the power could be increased but the power limitations of the integrated elements tend to restrict the maximum limits. Integrated circuits can be made using power transistors and larger packages but the possible benefits of circuit integration must be weighed against the high development costs. Of course, in the field of low power circuits, there are few technical limitations that cannot be overcome.

### 8-2.4.2 Packaging

The type of package and how much circuit can be placed in a single package are circuit design decisions when considering layout and interconnection. Two standard types of hermetically sealed packages have been developed by the industry, a transistor-type package and a flat package. The transistor type is more compatible with component assemblies of existing equipment, while the flat pack permits a greater packing density in subsystem assemblies. Another type is the dual in-line plastic package that is less expensive but has a narrower temperature range. Special considerations include die size because manufacturing yield of integrated circuits is very dependent on it. Smaller die sizes result in more circuits per wafer, less waste material, and smaller loss because of wafer imperfections[7].

The number of components that can be placed on a given size of die depends upon the type of components needed, their composition (diffused or film), the resistivity of the material, and the value of the parts. Actual space requirements must be augmented by a 0.5-mil minimum spacing between adjacent elements within a single isolation island, and a minimum space of 2 mils between each isolation island[7]. All available die space cannot be used because of circuit layout but, with careful design, this wasted space can be kept to a minimum. Component density of a silicon chip is always a function of process refinement and is subject to improvement. Hence it behooves the designer to stay abreast of the latest techniques. See Refs. 9 and 10 for details on packaging.

### 8-2.4.3 Circuit Design

A reorientation of circuit design philosophy must be made when a change to an integrated circuit is contemplated. While it is true that many circuits can be converted part-for-part from discrete form to integrated form, it is more economical and better performance results when the circuit is designed by using integrated circuit technology. For instance, it is less expensive to fabricate integrated transistors than passive elements such as resistors and capacitors. As a result, transistors can be used freely but since passive elements are relatively large and difficult to control to close tolerances, the latter are likely to determine ultimate cost. Large values of resistors and capacitors should be avoided because the smaller values are more easily produced and more economical. Resistive and capacitive tolerances should be as open as possible to produce higher yields and lower costs.

A severe limitation in integrated circuit technology is the non-availability of inductors as integrated circuit elements. Some components have been fabricated but resulting inductance and $Q$ values have been too low. Therefore circuits requiring inductors must be externally added to an integrated circuit. A partial solution is to use extremely small toroids[7].

### 8-2.4.4 Available Component Values

There is more restriction on the range of component values for integrated circuits than for discrete components. These restrictions often prevent direct conversion of discrete circuits to integrated circuits. Tables 8-5[7] and 8-6[7] summarize integrated component values.

### 8-2.5 SEALING

All integrated circuit packages require permanent sealing. For reasons of reliability, hermetic sealing generally is required to prevent the intrusion of moisture and corrosive atmospheres that could cause an operational failure. Even when the package is encapsulated, a good hermetic seal is advisable to prevent contamination by cleaning solvents, solder flux, out-gassing of packaging, and packing material that may be forced into the package. Package sealing methods include (1) welding, (2) solder sealing, (3) glass-to-glass sealing, and (4) plastic sealing[8]. The sealing method should be established before package selection, which means that the thermal limitations of the device being packaged must be considered in light of the package sealing requirements. A discussion of the sealing methods follows:

(1) Welding. Automatic or semiautomatic production equipment is used in weld sealing. It is fast, efficient, and the internal device is subjected to the least thermal punishment. Both round can and flat packages are welded by spot stitching or continuous seam welds.

(2) Solder Sealing. Soldering is the most popular method of sealing flat packages.

**TABLE 8-5**

**VALUES FOR INTEGRATED RESISTORS[7]**

| | Diffused | | Thin Film |
|---|---|---|---|
| | **Mono** | **Multichip** | **Thin Film** |
| Usable Resistivity (ohms/sq.) | 2.5 (emitter) 100–300 (base) | 2.5–300 | 100–400 |
| Temperature Coefficient | 500–2000 ppm/deg C | 500–2000 ppm/deg C | 30–60 ppm/deg C |
| Maximum Power | 0.1 W | 0.25 W | 0.1 W |
| Tolerance | ± 10% | ±5% | ± 5% |
| Range of Values | 15 Ω 30 kΩ | 15 Ω 30 kΩ | 40 Ω 100 kΩ |

Reprinted with the permission of Hayden Book Company, Inc., Rochelle Park, N.J.

**TABLE 8-6**

**VALUES FOR INTEGRATED CAPACITORS**

| | Diffused | | Thin Film (Oxide) |
|---|---|---|---|
| | **Mono** | **Multichip** | |
| Maximum Capacitance | 0.2 pF/mil$^2$ for collector-base junction | 1 pF/mil$^2$ | 0.25–0.40 pF/mil$^2$ |
| Breakdown Voltage | 5 or 20 VDC | 5–50 VDC | 50 VDC |
| Q (at 10 MHz) | 1–10 | 10–50 | 10–1000 |
| Tolerance | ± 20% | ± 10% | ± 10% |

Reprinted with the permission of Hayden Book Company, Inc., Rochelle Park, N.J.

Packages, lids, and solder preforms are available in various configurations. Separate solder preforms should match the dimensions of the particular packages used. They are available in thicknesses ranging from 0.0005 to 0.005 in. and in several material combinations, the most typical of which is 80% gold/20% tin.

(3) Glass-to-glass Sealing. Glass-to-glass sealing has several advantages. It eliminates several glass-to-metal interfaces, and also eliminates the separate solder preform and associated gold-plating requirements. The disadvantages are limited availability of various package sizes and the need for specialized equipment to make the seal and to protect the internal circuitry from damage at high sealing temperatures (approx. 600°C)[8].

(4) Plastic Sealing. Plastics are poor sealants because they are porous and a good hermetic seal is difficult to attain with them. However, plastics are ideal for shock and vibration resistance packaging (see par. 8-2.6).

## 8-2.6 ENCAPSULATION

Potting compounds are used to encapsulate electronic parts for protection against shock, pressure, temperature, moisture, dirt, corrosion, fungus, vibration, and arcing between components. Electronic components, such as those used in fuzes, are more reliable and have longer life when properly encapsulated.

Circuits to be plastically encapsulated usually are made with chips eutectically mounted and bonded on a skeleton lead frame. These frames are gold-plated Kovar. They are made in strips or matrices so that they can be adapted to automatic, multiple unit molding. A protective resin is applied to the lead frames to protect the chip surface and to hold the gold wire leads in place. The final package is formed by the transfer molding process, and the resulting matrix of completed packages is sheared to produce the individual packaged components[8].

Potting of electronic components, however, has these disadvantages[11]:

(1) Replacement of wires of components is almost impossible.

(2) Potting compounds generally do not withstand very high or very low temperatures.

(3) Potting adds weight.

(4) Circuit design must be compatible with potting.

(5) Components must be cleaned and protected before encapsulation.

(6) Electrical circuit characteristics may be affected adversely.

The most common types of packing compounds are epoxies, polyurethanes, polyesters, and silicones. Table 8-7[11] shows characteristics typical of these materials. One major consideration is the compatibility of any packing formulation with explosives[12].

## 8-2.7 PRINTED CIRCUITS

One of the greatest benefits from the proper application of printed circuitry is product uniformity. Once procedures and processes are set up, a printed circuit from the assembly line will be the same as the previous one. This result establishes an ease of testing and inspection which is not possible with hand wiring. Design decisions are a result of trade-off studies of complexity of design versus manufacturing costs, simplicity of design versus weight, complexity of design versus skill required, and necessary tolerances versus facilities available. Factors affecting printed circuits are covered in the paragraphs that follow.

### 8-2.7.1 Design Considerations

To permit maximum efficiency of part or module placement and to facilitate the production of printed circuit boards, standard grids have been devised for the location of mounting holes. The modular units of location are based on 0.100, 0.050, or 0.025 in. and are applied to both axes of rectangular coordinates[13].

The thickness of copper foil usually is given in terms of its weight per square foot. The equivalent thickness in inches equals 0.0014 in. per $oz/ft^2$. Tables 8-8[13] and 8-9[13] give thicknesses and tolerances of copper foil and clad copper. Fig. 8-4[13] can be used to compute minimum conductor width and thickness. The estimated temperature rise above ambient temperature versus current for various cross-sectional areas of etched copper conductors is illustrated in this figure. The amount of possible undercut should be included when designing for minimum con-

**TABLE 8-7**

**COMPARISON OF PROPERTIES OF TYPICAL POTTING MATERIALS**

| Material | Linear Shrinkage | Thermal Expansion | Thermal Conductivity | Volume Resistivity | Dielectric Strength |
|---|---|---|---|---|---|
| **Epoxy** | | | | | |
| Unfilled | very low-med. | low-high | low-medium | good-excel. | very good |
| Filled (rigid) | very low-low | low | high | very good-excel. | very good-excel. |
| Filled (flexible) | low-high | low-high | medium | good-very good | very good |
| Syntatic | very low-low | very low | very low-low | very good | good |
| **Polyurethane** | | | | | |
| Foam | very low | low-high | very low | very good | (not avail.) |
| Cast | very low-high | high | very low | good-very good | good-very good |
| **Polyester** | | | | | |
| Filled (rigid) | med.-very high | low-high | medium | good-very good | very good |
| Filled (flexible) | med.-very high | high | medium | good | good-very good |
| **Silicone** | | | | | |
| Cast (filled) | low | high | very high | excellent | good |
| RTV rubber | high | very high | medium | very good | very good |
| Gel | very low | very high | medium | excellent | excellent |

**Key to Ranges**

LINEAR SHRINKAGE, in./in.: very low 0.002; low 0.0021–0.004; medium 0.0041–0.010; high 0.0101–0.010; very high 0.0201.

THERMAL EXPANSION, (in./in. deg C) $\times 10^{-5}$: very low 2.0; low 2.1–5.0; high 5.1–10; very high 10.1 (figures referenced against aluminum).

THERMAL CONDUCTIVITY, (cal/sec/cm$^2$/deg C per cm) $\times 10^{-4}$: very low 1.5; low 1.6–4.0; medium 4.1–9.0; high 9.1–20; very high 20.1.

VOLUME RESISTIVITY, ohm-cm: good $10^{11}$–$10^{12}$; very good $10^{13}$–$10^{14}$; excellent $10^{15}$–$10^{17}$.

DIELECTRIC STRENGTH, V/mil: good 225–399; very good 400–500; excellent 500.

ductor width. This normally is computed as 0.001-in. undercut per ounce of copper per side. In computing conductor spacings of either coated or uncoated boards for altitudes below 10,000 ft, use Table 8-10[13].

To increase the rigidity of the board and forestall any possible plate resonances, several practices are available. Boards between 0.031 and 0.062 in. thick should be supported at intervals of not more than 4 in. Boards greater than 0.093 in. thick should be supported at not more than 5-in. intervals. Finally, all boards should be supported within 1 in. of the board edge on at least three sides.

Moisture that enters insulating board material can greatly affect dielectric strength, dielectric constant, and the dissipation factor of the material. Conductors should not be allowed to approach the board edge within 1/8 in. If conductors are used as a board edge plug-in, the edge should be chamfered. This avoids delamination or lifting of conductors by giving the board the proper lead-in.

Terminal area annular rings should have a width of 0.020 in. for unsupported holes (Fig.

**TABLE 8-8**

**COPPER-FOIL THICKNESS AND TOLERANCE**

| Thickness | | Tolerance, |
|---|---|---|
| oz/ft² * | in. | in. |
| 1 | 0.0014 | +0.0004<br>−0.0002 |
| 2 | 0.0028 | +0.0007<br>−0.0003 |
| 3 | 0.0042 | ±0.0006 |
| 4 | 0.0056 | ±0.0006 |
| 5 | 0.0070 | ±0.0007 |

*In industry, thickness of copper foil is given in terms of its weight per square foot. Thus, 1-oz copper weighs 1 oz/ft²

8-15[13]) and 0.010 in. for plated-through holes (Fig. 8-16[13]). Fig. 8-16 shows that the ring for swaged holes should extend 0.010 in. beyond the swaged flange of a terminal or stand-off. To provide a satisfactory drill lead-in, the inside diameter of the terminal area should not exceed 0.020 in. To eliminate many soldering defects, the drill diameter should exceed the minimum inside diameter of the terminal area by at least 0.010 in.

### 8-2.7.2 Taping the Artwork

The conversion of any circuit design from schematic to a completed, printed-wiring board is a critical process for several reasons. First, no amount of touch-up can correct an inferior job of taping. Extreme accuracy and cleanliness are essential during this operation. All conductor paths and interconnections should be made with free flowing lines and rounded corners. Fig. 8-17[13] illustrates some correct and incorrect conductor patterns. In the application of the tape on the artwork, great care is necessary when forming a free-flowing bend. Tape breakage or an uneven width of a taped conductor can result if undue pressure or pull is used. No more than two layers of tape in any one area should be used because inaccuracies can result in reproduction during the photographic process. Since a static charge that attracts foreign particles is likely to build up on the polyester-film master, cleanliness should be rigorously maintained.

**TABLE 8-9**

**CLAD-COPPER THICKNESS AND TOLERANCE**

| Nominal Thickness, in. | Tolerance, ±in. | | Class II | Class III |
|---|---|---|---|---|
| | Class I | | | |
| | Paper | Glass | | |
| 1/32 (0.031) | 0.0045* | 0.0065 | 0.004 | 0.003 |
| 1/16 (0.062) | 0.0060* | 0.0075 | 0.005 | 0.003 |
| 3/32 (0.093) | 0.0075* | 0.0090 | 0.007 | 0.004 |
| 1/8  (0.125) | 0.0090* | 0.0120 | 0.009 | 0.005 |
| 1/4  (0.250) | 0.0120 | 0.0220 | 0.012 | 0.006 |

*For 1-oz copper on one side, tolerance is 0.0005 in. less than value shown.
For nominal thicknesses not shown in this table, the tolerance for the next greater thickness shown applies.

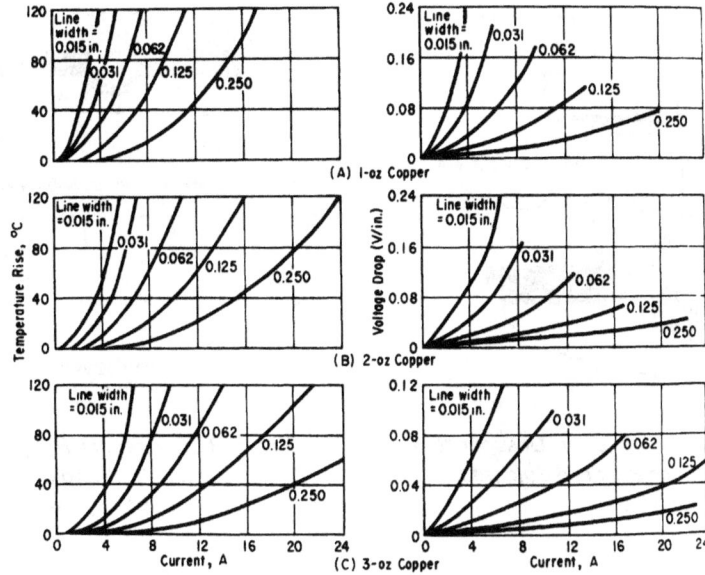

Figure 8-14. Electrical Characteristics of Etched Copper Wiring[13]

### 8-2.7.3 Laminate Materials

Printed-wiring board laminates fall into two broad categories: (1) paper-based phenolics, and (2) glass based epoxies. The phenolics are numbered x, xx, and xxx and these numbers signify that the laminate contains phenolic resin by weight of at least 35, 45, or 55%, respectively. A phenolic designated xxxp indicates that the laminate has been plasticized for punching. Phenolics have good mechanical and electrical properties, and are used except where good dimensional stability, high impact strength, arc resistance, or high insulation at high humidity are required. The most commonly used glass laminates, G-10 and G-11, exhibit good mechanical and dimensional stability at high temperature, good insulation and heat resistance, and low moisture absorption. Two disadvantages of these types are: (1) they cause excessive tool wear, and (2) they have poor punchability. A paper based epoxy (FR-3) is a compromise material. It is flame resistant, can be punched cold, and its properties are similar to the better phenolics[13].

### 8-2.7.4 Soldering Techniques

Packaging densities often make it necessary to use a double-sided board with conductors and electrical connections on both sides.

TABLE 8-10

RECOMMENDED CONDUCTOR SPACING

| Voltage Between Conductors, VDC or AC peak | Minimum Spacing | |
|---|---|---|
| | Coated | Uncoated |
| 0 to 30 | 0.010 in. | 0.025 in. |
| 31 to 50 | 0.015 in. | 0.025 in. |
| 51 to 150 | 0.020 in. | 0.025 in. |
| 151 to 300 | 0.030 in. | 0.050 in. |
| 301 to 500 | 0.060 in. | 0.100 in. |
| Over 500 | 0.00012 in./V | 0.00012 in./V |

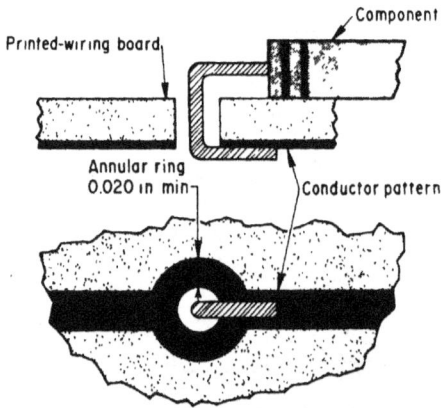

Figure 8-15. Annular Ring for an Unsupported
Hole With Component Lead Soldered
on One Side[13]

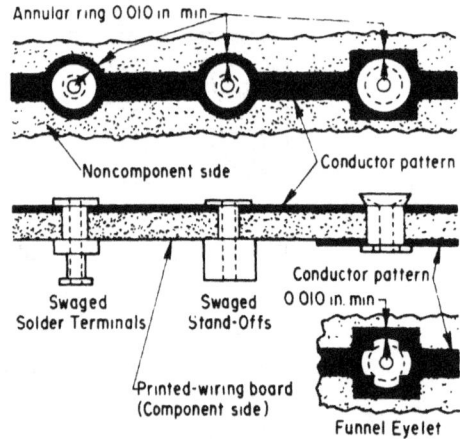

Figure 8-16. Annular Ring for Swaged
Terminals, Stand-offs, and
Funnel Eyelets[13]

Electrical, soldered connections between both sides of the board are made in one of four fundamental ways[13]:

(1) Plated-through holes

(2) Eyelets as an interface connection only

(3) Eyelets in component holes

(4) Bare holes with a terminal area on a noncomponent side, or if an electrical connection is needed on the component side, terminal areas on both sides.

For the first three methods, the board is subjected to an automatically controlled wave solder operation. Capillary action is used for the plated-through holes or eyelets when the board comes in contact with the solder wave, and this action solders the connection on the component side. The correct temperature of the solder bath and having a clean solderable surface at the right temperature are very important. For more detailed information about soldering techniques, see Refs. 15 and 16.

## 8-3 REQUIREMENTS FOR TECHNICAL DOCUMENTATION

Production requires a complete technical data package. This documentation is to ensure that military items are capable of reliable performance. See par. 14-4 for detailed information on the contents of a technical data package.

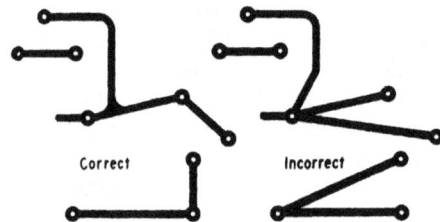

Figure 8-17. Correct and Incorrect
Conductor Patterns[13]

# REFERENCES

1. N. J. Doctor, "Status of Microelectronics", *Proceedings of the Electronic Timers for Ordnance Symposium*, Volume 1, U S Harry Diamond Laboratories, Washington, D.C., June 1965, pp. 197-226.

2. T. A. Prugh, et al., "The DOFL Microelectronics Program", Proceedings IRE, **47**, No. 5, 882-94 (May 1959).

3. *Encyclopedia of Science and Technology*, McGraw-Hill Book Co., Inc., N.Y., 1960.

4. R. W. Berry and D. Sloan, "Printed Tantalum Capacitors", Proceedings IRE, **47**, 1070-5 (June 1959).

5. R. Gomer, "Preparation and Properties of Conducting and Transparent Glass", Rev. Scientific Instruments, **24**, 993 (1953).

6. E. L. Hebb, "Microcircuitry by Chemical Deposition", Electromechanical Technology, **1**, 217-23 (July-August 1963).

7. L. Stern, *Fundamentals of Integrated Circuits*, Hayden Book Co., Inc., N.Y., 1968.

8. N. Holonyak, Jr., Ed., *Integrated Electronic Systems*, Prentice-Hall, Inc., Englewood Cliffs, N.J., 1970.

9. J. J. Staller, "The Packaging Revolution, Part I", Electronics, **38**, 72-87 (October 1965).

10. J. J. Staller, "The Packaging Revolution, Part II", Electronics, **38**, 75-96 (November 1965).

11. AMCP 706-210, Engineering Design Handbook, *Fuzes*.

12. AMCP 706-177, Engineering Design Handbook, *Properties of Explosives of Military Interest*.

13. J. Cavasin, Jr., "Printed-Wiring Design", Machine Design, **39**, 213-7 (April 27, 1967).

14. MIL-STD-275B, *Printed Wiring for Electronic Equipment*, Dept. of Defense, 24 June 1966.

15. A. E. Linden, *Printed Circuits in Space Technology*, Prentice-Hall, Inc., Englewood Cliffs, N.J., 1962, pp. 141-58.

16. MIL-P-46843, *Design and Production of Printed Circuit Assemblies*, Dept. of Defense, 22 March 1967.

# CHAPTER 9

# PACKING, STORING, AND SHIPPING PROCEDURES

## 9-1 GENERAL

Military materiel is packaged to protect it from harmful environments, including transportation. Packaging will protect the item after production, during storage, during transport, and until delivery to its ultimate user[1]. During the transportation phase, which includes both handling and carriage, the Department of Transportation (DoT) regulations must be strictly observed for movement within the U S. Since items must be stored for long periods, they must be protected against physical damage and deterioration; it also may be necessary to conduct periodic inspections of stored materiel.

From the military standpoint, good packaging methods protect the item through all phases and environments, with minimum cost. Items must be protected for one of three required levels: overseas shipment (Level A), long-term storage (Level B), or interplant shipment (Level C).

The most damaging environments during transportation by truck, rail, ship, or aircraft are usually shock and vibration. However, temperature extremes and other potentially harmful environmental factors should be considered as relevant to damage assessment or prevention.

These subjects are treated in more detail in Chapter 15; see par. 15-1 for packing, par. 15-2 for storing, and par. 15-3 for shipping.

## 9-2 DEVICES CONTAINING EXPLOSIVES

When electronic timers contain explosive charges, they must be given the special handling required to assure safety during packing, storing, and shipping. The basic reference for safety is the *Safety Manual*[2]. The applicable hazard class must be established, the device must be packed and labeled correctly, stored in magazines in accordance with quantity-distance requirements, and shipped according to transportation regulations. These topics are discussed in detail in Chapter 20; see par. 20-4 for packing, par. 20-5 for storing, and par. 20-6 for shipping.

## 9-3 SPECIAL ENVIRONMENTAL EFFECTS

In the packaging, storing, and shipping of electronic timers, particularly those containing electroexplosive devices (EED's), several environmental effects must be considered. These include stray RF energy, lightning discharge, and static electricity. It should be pointed out that packaging to reduce or eliminate these effects is possible but costly. Further, since the hazard cannot be defined precisely during all storing and shipping phases, no standard military timer makes provision for these effects in its packaging assembly. The paragraphs that follow discuss methods by which these effects can be minimized.

### 9-3.1 RF HARDENING

RF energy is particularly hazardous to EED's and to solid-state devices that are found in most modern weapon systems. This energy is limited effectively by the use of a complete Faraday shield. In general, there is a relatively large loss associated with copper and iron containers. Reflection losses account for most of the elimination of energy at the lower frequencies. Copper appears to be superior to iron for high reflection losses but the reverse is true for absorption losses[3]. Hence, a copper-flashed iron would provide an ideal material for a shielded container. Containers for sensitive devices need good electromagnetic seals at the closure junction. Gasket materials are currently manufactured that provide good contact under pressure.

Current safety practices for EED's require (1) input leads be shortcircuited, and (2) twisting together of long lead wires. For a more detailed discussion of combatting stray RF energy see Ref. 3.

## 9-3.2 LIGHTNING DISCHARGE

With respect to packaging, the safest way to take care of lightning hazard is to place the device in a steel container. However, it is much less expensive not to load and unload during storms. Storage areas are protected by lightning rods and arresters, and vehicles used to ship the devices are protected by grounding straps[2].

## 9-3.3 STATIC ELECTRICITY

Packages that tend to generate static electricity are to be avoided. Generally of the plastic film type, they are sometimes difficult to recognize because the film is deposited on the inside of a metal foil bag. The safe approach is to wrap the devices in metal foil and enclose them in a moisture resistant container if desired or necessary.

## REFERENCES

1. AMCP 706-121, Engineering Design Handbook, *Packaging and Pack Engineering.*

2. AMCR 385-100, *Safety Manual,* Army Materiel Command, April 1970.

3. AMCP 706-235, Engineering Design Handbook, *Hardening Weapon Systems Against RF Energy.*

# CHAPTER 10

## ELECTRONIC ANALOG TIMERS

### 10-1 INTRODUCTION

An analog timer derives its time base from the time interval generated by the motion of a measurable parameter between two predetermined values. An hourglass is a classical example. Electronic analog timers characteristically depend on changes in voltage or current. Their major virtues are simplicity of design and fabrication coupled with moderate accuracy and stability in military environments.

Fundamentally, all types of electronic timers are analog timers. Digital electronic timers, which are covered in Chapter 11, count the number of cycles through which an analog timer progresses. The distinction that is made in this handbook is that we will treat as analog timers those timers which are concerned primarily with a single cycle or a small number of cycles of the overall analog timing process. The distinction does not completely resolve the classification problem because some of the analog timers can and have been used as time bases for digital timers. However, most digital timing systems currently use linear oscillators as time bases. These are covered in Chapter 11.

Military electronic analog timers are designed to meet a variety of accuracy specifications, environments, and power supply conditions. Severe environmental conditions arise in projectile fuzes where analog timers may be exposed to a wide temperature range, spins of 500 revolutions per second, and setback forces of 35,000 g for 10 msec. When these environmental conditions are coupled with design requirements of $\pm 20\%$ supply voltage variability, 5 to 10 in.³ total

volume available for the timer plus power supply, and a 0.1% error limit; the design problem is difficult. A large portion of the material that follows is directed toward timer applications in projectile fuzing.

### 10-2 TIMING CIRCUITS

Many military applications require a small electronic timer or time base in the 0.5-10% stability region. The paragraphs that follow discuss networks and circuits that fulfill these requirements.

In most military high-volume timer applications, the primary engineering problem is production of a timer that meets the environmental and space requirements at minimum cost. Since high-volume applications may have widely differing environmental and space requirements, no single timer design can be a minimum cost solution for all applications. Thus, engineering design time may be considered as a good investment for high-volume applications. This contrasts with low-volume or single-usage applications where direct procurement of an overdesigned and high-cost timer may be the most economical solution of a timing problem.

#### 10-2.1 RC NETWORKS

*RC* networks are probably the most commonly used delay networks. The *RC* timer typically uses a transistor detecting element and a silicon controlled rectifier as an output driver. A basic *RC* network is shown in Fig. 10-1. When switch S is closed, an electrical charge builds up exponentially on the capacitor, thereby increasing the voltage across the capacitor as shown in Fig. 10-2.

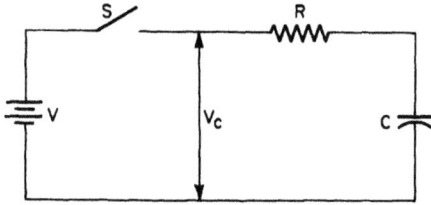

Figure 10-1. Basic RC Network

Specifically,

$$V_c = V\left[1 - e^{-t/(RC)}\right] + V_{co}e^{-t/(RC)} \quad (10\text{-}1)$$

where

$V_c$ = voltage across the capacitor, V

$V$ = applied voltage, volt

$V_{co}$ = voltage on the capacitor at $t = 0$, V

$t$ = time, sec

$R$ = resistance, ohm

$C$ = capacitance, F

The basic timing interval, assuming the switch closes at $t_o$ and the capacitor is charged to $V_{co}$ at this time, is

$$t_1 - t_o = RC \ln\left(\frac{V - V_{co}}{V - V_1}\right) \text{, sec} \quad (10\text{-}2)$$

where $V_1$ = value of $V_c$ at time $t_1$

The accuracy of the timing interval is related directly to the values of $R$ and $C$ but the dependence of the timing interval on changes in $V$, $V_{co}$, and $V_1$ is more complicated. Table 10-1 gives the percentage change in the interval for a ± 10% change in the ratio of $V_1$ to $V$ for various values of the ratio. Voltage $V_{co}$ is assumed to be zero. More detailed tables can be computed from

$$\Delta t = \left\{ \frac{\ln\left[\dfrac{1}{1 - \dfrac{V_1}{V}\left(1 + \dfrac{\delta}{100}\right)}\right]}{\ln\left[\dfrac{1}{1 - \dfrac{V_1}{V}}\right]} - 1 \right\} 100, \% \quad (10\text{-}3)$$

where

$\Delta t$ = change in the timing interval, %

$\delta$ = change in $V_1/V$, %

For example, $\delta = +10$, a 10% increase in the voltage ratio, corresponds to a $\Delta t$ value of +15.2% for a ratio of 1/2 (Table 10-1). Eq. 10-3 assumes that $V_{co}$ is zero. An extensive analysis of the accuracy of $RC$ timers is given in Ref. 1 where capacitor leakage currents; detector circuitry leakage currents; and other secondary, accuracy determining parameters are considered.

Adjustment of the timing interval is, within limits, normally provided by varying $V$, $R$, or both. A linear relationship between $V_1$ and time may be approximated by using a large $R$,

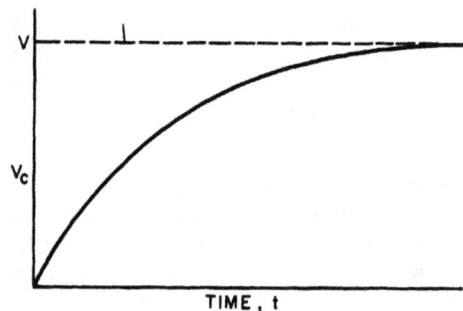

Figure 10-2. Voltage-time Curve for Basic RC Network

**TABLE 10-1**

**CHANGES IN TIMING INTERVAL FOR CHANGES IN VOLTAGE**

| Voltage Ratio $V_1/V$ | Change in Timing Interval, % + 10% change in the ratio | −10% change in the ratio |
|---|---|---|
| 0.1 | + 10.6 | −10.5 |
| 0.2 | + 11.3 | −11.1 |
| 0.25 | + 11.8 | −11.4 |
| 0.3 | + 12.2 | −11.8 |
| 0.333 | + 12.7 | −12.0 |
| 0.4 | + 13.5 | −12.6 |
| 0.5 | + 15.2 | −13.8 |
| 0.6 | + 17.7 | −15.3 |
| 0.666 | + 20.3 | −16.6 |
| 0.7 | + 22.1 | −17.4 |
| 0.75 | + 25.7 | −18.9 |
| 0.8 | + 31.7 | −20.9 |
| 0.9 | +100.0 | −27.9 |

i.e., a very long time constant circuit compared with the required range of delays. The limit to the length of time delay is established by leakage of the capacitor, which in most cases makes the *RC* arrangement inadequate for delays above several minutes. Delays of several milliseconds are attained at the lower extreme. When resetting, S must be opened and the capacitor completely discharged.

Fig. 10-3[2] shows an *RC* network incorporated in a primitive timer. The basic timing interval is the same as that given in Eq. 10-3. The voltage $V$ is applied to the emitter of a unijunction transistor (UJT). The UJT does not turn on until the voltage at the emitter exceeds the emitter peak point voltage $V_1$. When the UJT does turn on, the timing capacitor discharges rapidly through resistor $R_1$ and the emitter-to-base-one junction of the UJT. The silicon-controlled rectifier (SCR) is turned on by the conducting UJT. As the discharge current of capacitor C flowing through the UJT decreases, the UJT stops conducting. Capacitor C starts to charge again and the entire cycle is repeated, except that the silicon-controlled rectifier stays on unless deliberately unlatched, e.g., by removing the anode voltage. The load is not energized as long as the SCR is off.

A practical timer of the type described is shown in Fig. 10-4[3]. Time delays from 0.4 msec to about 4 min can be obtained with the component values shown in the figure. The timing interval is determined before a timing run by setting resistor $R_t$ and capacitor $C_t$. The timing interval is initiated by the application of the 28-V DC power supply.

A simpler example of a practical timer is shown in Fig. 10-5[3]. This single, active element circuit does not use a SCR as a load driver; instead, the UJT drives a microminiature relay directly. Time delays from 0.5 sec to approximately 3 min can be obtained with the values of $R_t$ and $C_t$ shown. The timing interval starts with the application of the DC power supply.

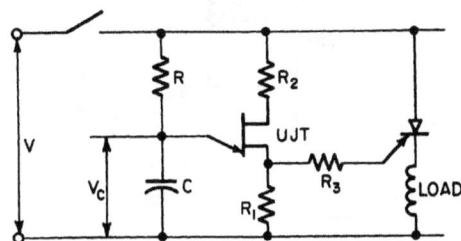

*Figure 10-3. Primitive RC Timer*

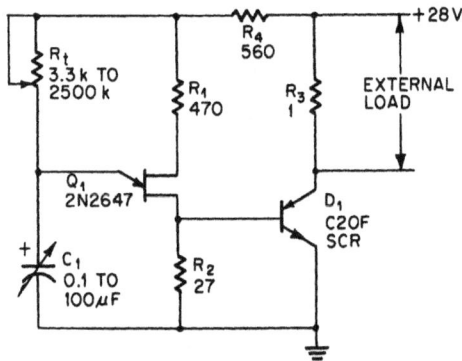

Figure 10-4. A Practical Low Accuracy Timer

The $RC$ timers discussed have used the basic $RC$ relaxation oscillator scheme of Fig. 10-6[4]. When switch S is closed, the voltage across capacitor $C_t$ increases exponentially until the trigger point of the voltage detector is reached. At this point, the input impedance of the voltage detector suddenly decreases from megohms to several ohms. The capacitor discharges through this low impedance. At the end of the discharge, the detector again assumes a high input impedance and the cycle repeats. This simple scheme is adequate for time delays up to hundreds of seconds but the values of the components that determine the primary timing interval become quite large. For example, periods of about 100 sec require capacitors of approximately 1 to 10 $\mu$F and

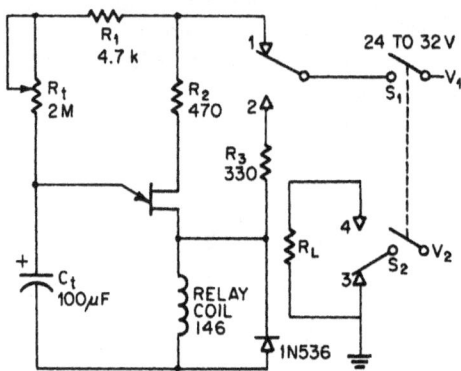

Figure 10-5. A Practical Timer Using a Single Semiconductor

resistors of approximately 10 to 100 megohms. For good accuracy these values must be quite stable.

The requirements for highly stable, high value resistors and capacitors normally limit the choice of components to those of large physical dimensions which are often prohibited by other design requirements, particularly in military fuzing applications. One way to solve this problem is to have a relaxation oscillator of fairly small period continue cycling, and then count the number of cycles with a digital counter. When the number of cycles equals the number desired, the counter "times out" and provides an output pulse. Practical digital systems of this type using $RC$ relaxation oscillators have been designed, although linear oscillators now are preferred as the time base of such systems. (Digital timers are described in Chapter 11.) Another way to avoid using large components for long period $RC$ relaxation oscillators is to open and close switch S in Fig. 10-6 at a fixed rate. The voltage seen by the detector $V_{ct}$ is shown in Fig. 10-7[4]. S is now an electronic switch that is assumed to change voltage levels by short duration ramps. Then

$$\frac{a_1}{b_1} = \frac{a_2}{b_2} = \frac{a_n}{b_n} = M, \text{ dimensionless} \quad (10\text{-}4)$$

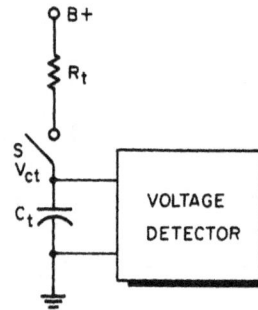

Figure 10-6. Relaxation Oscillator

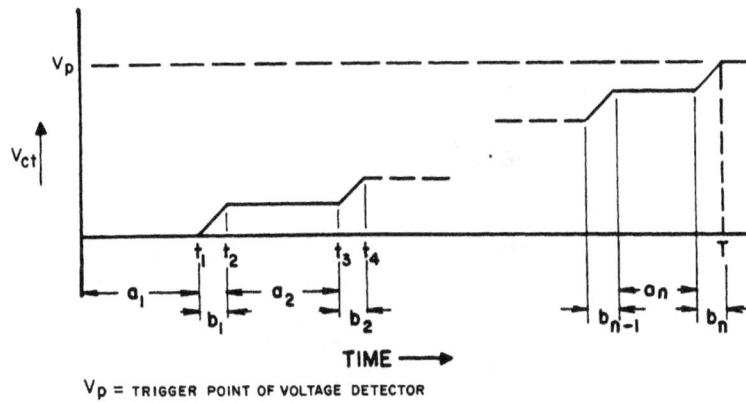

$V_p$ = TRIGGER POINT OF VOLTAGE DETECTOR

Figure 10-7. Voltage at the Dectector

where

$a$ = time between closures of S as defined in Fig. 10-7, sec

$b$ = time S is closed as defined in Fig. 10-7, sec

$n$ = number of "on-off" cycles during time $T$

$M$ = constant

$t$ = time of $C_t R_t$ to change to $V_p$, sec

then

$$T = (M + 1)t = \text{total timing interval, sec}$$
$$(10-5)$$

Thus the $RC$ time constant of the original $RC$ relaxation oscillator has been multiplied by $(M + 1)$. Fig. 10-8[4] is a block diagram of one system that has been used to achieve this multiplying effect. A complete description of this system is given in Ref. 4. Note that this system uses another $RC$ relaxation timer in the triggering circuit to control the period of the flip flops that switch the electronic switch formed by diodes $D_1$ and $D_2$; therefore, the overall circuit can be recognized as one that allows one $RC$ timer to control the total time interval in which the charging path of another $RC$ timer is complete.

Long period $RC$ timers that use unijunction transistors as detection elements, either with large value resistors and capacitors or using a time constant multiplying scheme, have a characteristic problem. The UJT usually requires that a current of about 1 $\mu$A flow into the unijunction emitter before the trigger voltage is reached. This characteristic becomes a serious problem when the charging current for the capacitor that determines the timing interval—at the end of the time interval—is of the same order of magnitude as the emitter current required before triggering. Under these conditions, the timer will function erratically, if at all. The common solution is to provide a sampling pulse that lowers the trigger voltage of the UJT. The sync circuit of Fig. 10-8 performs this function. Fig. 10-9[3] shows a general usage, long timing interval $RC$ timer. The basic timing interval is determined by $R_1$ and $C_1$. If high quality, low leakage capacitors and a low leakage diode are used, intervals up to 2 hr can be obtained from this timer. Transistor $Q_2$ and $R_2$, $C_2$, and $R_5$ form a free running $RC$ oscillator whose period is less than 2% of the basic timing interval of the timer. This free running oscillator supplies negative pulses to the upper base of voltage detector $Q_1$ to lower the triggering voltage and bypass the triggering current/UJT problem. The timing interval is started by application of the 28-V DC power supply.

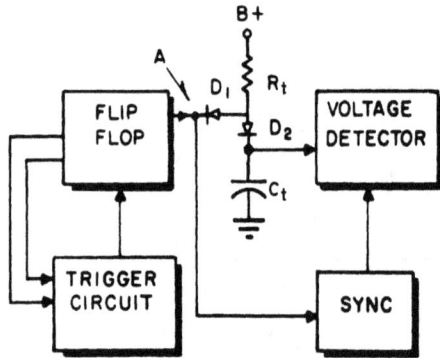

Figure 10-8. Block Diagram of Multiplier
System

## 10-2.2 RL NETWORKS

Fig. 10-10 shows the basic $RL$ network. When the switch is closed (assumed at $t = 0$), the current increases as shown in Fig. 10-11. Specifically

$$I = \frac{V}{R} \left( 1 - e^{-tr/L} \right), A \qquad (10\text{-}6)$$

where

$I$ = current, A

$L$ = inductance, H

The detector, which is frequently the coil of an ordinary relay inserted in series with the $RL$ components, responds to current. There-fore, when the current builds up to a predetermined value, the relay actuates. By varying $R$ or $V$ the delay time can be adjusted. However, $R$ directly affects the steady-state dissipation in the circuit. Several variables make this delay mechanism (which can be low-cost) incapable of very high accuracy. The copper coils of the inductor are sensitive to ambient temperature changes as well as the resistance $R$. Friction in relay pivots depends on wear as well as ambient temperature; differential expansion can change the friction by changing the fit on the pivots. Lastly, spring-supported relay elements suffer changes in spring constant over a temperature range.

Time delays may be increased by intentionally introducing eddy currents. A conducting slug or shading coil will link a significant portion of the magnetic flux. As the flux builds up, a current is generated in the shading coil that generates a bucking flux. The bucking flux slows down the buildup of the magnetic field and the coil current. Shading coils are temperature sensitive, which is a disadvantage. The build-up in inductance, coinciding with the closing of a relay, causes a momentary drop in current, thus producing arcing due to inductive energy buildup when switch S is opened. Time delays vary from milliseconds to several seconds. In suitable applications, this economical circuitry operates output contacts in a simple and direct

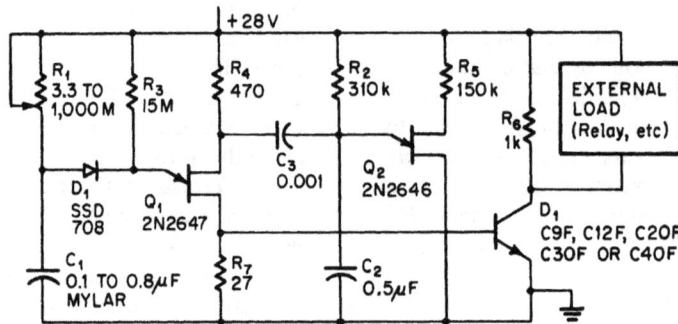

Figure 10-9. General Usage for Long Period RC Timer

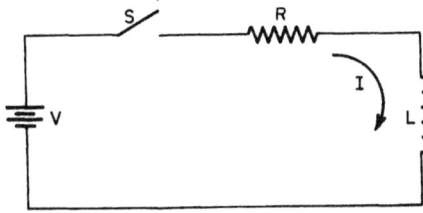

Figure 10-10. Basic RL Network

manner. Instantaneous resetting is possible by bringing the current to zero.

## 10-2.3 MILLER INTEGRATORS

The Miller integrator circuit, shown in Fig. 10-12[5], can attain longer time delays than a single $RC$ circuit because it provides a timing voltage that varies essentially linearly with time. The standard timing formula for the circuit of Fig. 10-12 is

$$t_1 - t_o = CR_2 (1 + \beta) \ln$$

$$\left\{ 1 + \left[ \frac{V_o}{V_1} \left( \frac{R_1}{R_2} \right) \left( \frac{1}{\beta} \right) \right] \right\}, \text{sec} \qquad (10\text{-}7)$$

where

$C$      = timing capacitor value, F

$R_1, R_2$    = resistor values, ohm

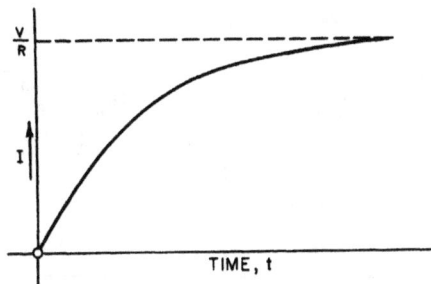

Figure 10-11. Current-time Curve for Basic
RL Network

$V_s$      = DC supply voltage, V

$V_1$      = applied input voltage, V

$V_o$      = resulting final amplitude of the output ramp,

$\beta$      = large signal current gain of transistor Q, dimensionless

The equation simplifies to

$$t_1 - t_o \approx CR_1 \left( \frac{V_o}{V_1} \right), \text{sec} \qquad (10\text{-}8)$$

if $\beta \gg 1$ so that $1 + \beta \approx \beta$ and if the natural logarithm argument is much smaller than one so that

$$\ln (1 + a) \approx a \qquad (10\text{-}9)$$

The standard timing equations, Eqs. 10-7 and 10-8, ignore transistor leakage current, transistor input impedance, and capacitor leakage current. Detailed analysis of these variables place limitations on the usefulness of the timing equations. Stable timing to within a few percent error as predicted by the standard timing equations is possible under the following conditions:

(1) Silicon transistors are used exclusively.

(2) Circuit resistances are kept approximately equal, and large (in excess of 50 k ohms).

(3) Transistor current gain is as high as possible, and in any case not less than 20.

(4) Time magnification $K$ (achieved by making $V_o = KV_i$) is no more than 3.

(5) Timing capacitor leakage resistance (in ohm) is greater than 10 times the quantity given by $\dfrac{R_1 R_2 (1 + \beta) V_o}{R_1 V_o + \beta R_2 V_i}$ (refer to Fig. 10-12 for identification of voltages and resistances).

10-7

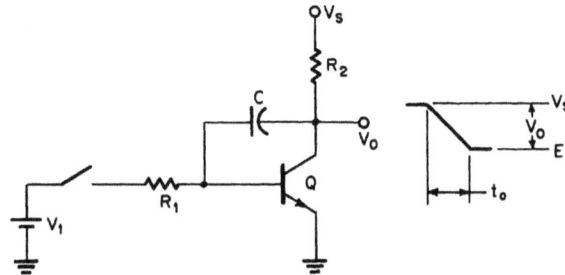

Figure 10-12. Miller Integrator

Temperature compensation for timing stability is difficult. The most important parameters are:

(1) Transistor leakage current is, within limits, associated with a negative temperature coefficient, i.e., time delay decreases with increasing temperature.

(2) Capacitor leakage current is associated with a positive temperature coefficient.

(3) Capacitor variation can produce either a positive or negative temperature coefficient depending on capacitor type. These parameters can be used together to obtain some measure of compensation. A ± 2% stability has been obtained in this manner for a temperature range of 0° to 40°C.

### 10-2.4 CURPISTOR TIMERS

Long term single cycle $RC$ analog timers can be constructed from constant current sources that have small output currents. The curpistor is a constant-current ion chamber that will pass constant currents from $1 \times 10^{-12}$ to $5 \times 10^{-9}$ A, depending upon the degree of ionization in the chamber. The constant current is obtained by collection of all the ions produced by the radium foil cathode when a potential from 50 to 300 V is applied. This current is determined mainly by the activity of the cylindrical radium foil (cathode), its distance from the center rod (anode), and the nature of the gas inclosed in the tube.

Fig. 10-13[6] shows an experimental configuration using a curpistor and XD-1C cold cathode diode. The capacitor C is charged by the curpistor current at a rate $dq/dt$ and the resulting "perfect" time integral of the charging current, assuming ideal components, is defined by

$$E = \frac{1}{C} \int I \, dt \qquad (10\text{-}10)$$

in differential form $I = C\left(\dfrac{dE}{dt}\right)$

for a linear voltage increase $I = C\left(\dfrac{\Delta E}{\Delta t}\right)$

where

$E$ = voltage, V

$I$ = constant curpistor current, A

$C$ = capacitance, F

$t$ = pulse time, sec

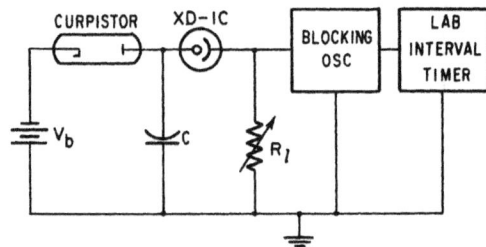

Figure 10-13. Curpistor-diode Oscillator

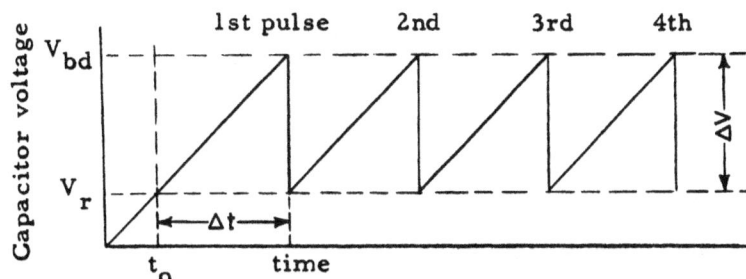

Figure 10-14. Sawtooth Charge-discharge Curve

The capacitor voltage $V_C$ rises linearly until it reaches diode breakdown voltage $V_{bd}$ at which time there is an arc discharge and about 25 percent of the energy $CV^2/2$ stored in C is transferred to the load resistance $R_L$. If C initially has zero charge, then $\Delta E = V_{bd}$. However, after the initial pulse, there is a residual voltage $V_r$ on the capacitor C which is used as a reference in all of the succeeding pulses. In this case, $\Delta E = V_{bd} - V_r$, as is illustrated in Fig. 10-14[6]. Hence $V_r$ is dependent largely upon the impedance of the diode during the time of discharge plus the load resistance. It is not significant for single cycle operation.

The primary advantage of the curpistor timer over conventional $RC$ timers is the length of delay that can be achieved easily. A serious disadvantage is the relatively high voltage necessary for curpistor operation. For example, the experimental configuration of Fig. 10-13 requires a source of at least 165 V. This is balanced partially by the fact that the regulation of the source is not a problem. Any source of 165 to 300 V that can supply $10^{-8}$ A is adequate. The use of nuclear batteries for curpistor voltage sources is possible, and this in itself is an advantage.

Other disadvantages of the curpistor timer are those associated with any long period $RC$ timer. The leakage current through the capacitor, capacitor mounting, and detector must be very small in relation to the constant charging current. Ref. 6 describes the

evaluation of experimental curpistor timers.

## 10-3 LEVEL DETECTORS

Electronic analog timers require a device to detect the arrival of the analog parameter (voltage or current) at the predetermined level that indicates the end of the timing interval of the timer. In the examples of $RC$ timers shown in Figs. 10-3, 10-4 and 10-5, the level detecting function is performed by a unijunction transistor. In the $RL$ timer described in par. 10.2-2, the detection function is accomplished by means of the magnetic flux sensitivity of the relay. The Miller integrator of Fig. 10-12 uses an N-P-N transistor as a voltage detector.

The ideal characteristics of a level detector are that: (1) it impose no load on the circuit it monitors, at least until the analog parameter being monitored has reached the predetermined level, and (2) it provide a high level output instantaneously at that time. In addition, it also should be repeatable in a given circuit, inexpensive, identical from item to item, and unaffected by environmental changes.

Various voltage level detectors for $RC$ timing circuits have been used including cold gas diodes, conventional heated cathode diodes and triode vacuum tubes, conventional silicon and germanium transistors, silicon-controlled rectifiers, UJT's, and field effect transistors.

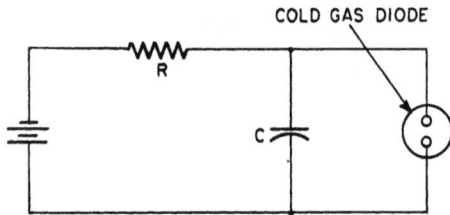

Figure 10-15. Relaxation Oscillator Using a
Cold Gas Diode

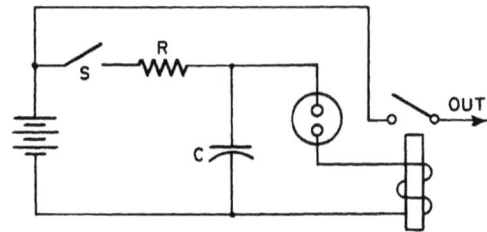

Figure 10-16. Cold Gas Diode Detection
Timer

## 10-3.1 COLD GAS DIODES

Perhaps the earliest and most used voltage level detector for electronic analog timers is the cold gas diode. The diode itself is usually a hermetically sealed envelope containing two electrodes and an inert gas such as neon or argon. Glass is the usual envelope material. Fig. 10-15 shows a primitive relaxation oscillator using a cold gas diode. Capacitor C charges through the resistance R until the diode breaks down and discharges the capacitor to some low voltage at which the diode turns off and the cycle repeats. This is essentially the circuit of many "winking" light strings for Christmas trees. Fig. 10-16 shows the same circuit adapted as a timer. The capacitor charges, after switch S is closed, until the diode breaks down and activates the relay. The relay, of course, can be replaced with a low-value resistor, and the voltage drop across the resistor can perform other output functions. Ref. 7 describes the overall behavior of cold gas diode relaxation oscillators. Cold gas diode voltage detectors have very low loading characteristics for $RC$ timer applications and, since breakdown time is very short, the output of the detector can be considered for most applications to be instantaneous. The disadvantages of cold gas diode detectors are:

(1) Variation from item to item in breakdown voltage is high.

(2) Glass envelopes usually used in cold gas diode construction allow ambient light to impinge on the gas. The varying intensity of ambient light thus causes changes in the breakdown voltage.

(3) The breakdown voltages usually obtainable are too high to be compatible with solid-state circuitry.

## 10-3.2 HEATED CATHODE DIODES AND TRIODES

Standard vacuum tube detectors normally are not used. Their primary disadvantage is the need of a filament power supply and the relatively high voltages necessary for vacuum tube operation. Loading of the $RC$ timing circuit is usually very small. If large voltages were used in the $RC$ charging circuit, relatively quick repeatable output response could be obtained.

## 10-3.3 BIPOLAR TRANSISTORS

Bipolar transistors were used widely as voltage detectors for $RC$ circuits in spite of the considerable loading effect on the $RC$ changing circuit. Variation from transistor to transistor, however, sharply influenced timer period. The appearance of unijunction transistors (UJT) and field effect transistors (FET), with their very high input impedance characteristics, makes the use of bipolar transistors in high accuracy voltage detectors a rarity except in special purpose designs.

## 10-3.4 UNIJUNCTION TRANSISTOR DE-TECTORS

The UJT is a three-terminal, solid-state device having a single P-N junction. This fact leads to characteristics that are quite different from those of conventional transistors. The primary characteristic of the UJT for application as a voltage detector for *RC* timer circuits is its relatively high input impedance. Ref. 8 summarizes the theory and characteristics of the UJT while Ref. 9 describes in detail its use in typical relaxation oscillators and timing circuits.

The primary disadvantage of the UJT as a detector occurs with long period timers. The UJT requires a small current before trigger voltage can be reached. This current, for long period *RC* networks, can be a significant part of the capacitor charging current. Hence, the UJT loads the *RC* circuit just before time out of the timer. A sampling pulse applied to the UJT to momentarily lower the triggering voltage is the common solution to the loading problem. See par. 10-2.1 for additional detail.

UJT's are available from many manufacturers at reasonable prices. It is recommended that a designer specifically consult the manufacturers' literature on any device to be used because specifications vary widely.

## 10-3.5 FIELD EFFECT TRANSISTORS

Field effect transistors are available in two types: (1) the junction field effect transistor (JFET), and (2) the metal-oxide semiconductor field effect transistor (MOSFET). For long period *RC* timer voltage-level detector functions, the difference of concern is that the leakage current of the MOSFET is considerably lower than that of the JFET so that the MOSFET can be used as a normally "off" device.

Since the leakage current of the MOSFET can be typically as low as 0.5 pA, the device is suitable for long period *RC* timer application.

Further, the MOSFET does not have the UJT characteristic current requirement previous to triggering. The primary disadvantage for voltage level detection of the MOSFET is temperature compensation and price (compared with UJT's).

## 10-4 SETTING METHODS

For analog timers, the basic timing interval of the analog process is the parameter that is to be "set". This contrasts with digital timer setting in that digital counters are usually "set" by changing the number of basic analog intervals which must occur before output.

Electronic analog timers usually initiate the timing process upon application of voltage to the timer or on the closing of a contact pair. The contact pair closing itself often applies a specific voltage to a portion of the timer which influences the basic timing interval. Thus timers often can be considered to be voltage controlled, and the basic timing interval also can be considered to be voltage controlled. However, setting of the timing interval by controlling a voltage applied to the timer from a device external to the timer is difficult, inconvenient, and perhaps, in some applications, hazardous. In practical applications analog timers usually are set by varying a resistive element. The voltage applied to the timer to initiate the timing sequence is obtained ideally from a well regulated supply.

## 10-4.1 SETTING MECHANISM

The internal setting mechanism of a time setting system is that portion which responds to the operator's input motion and translates it into a single, discrete, electrical configuration of the internal resistor package. The resistor package design, however, has a marked influence upon the internal setting mechanism. Analog resistors (potentiometers) may, for certain reasons, be limited to subranges below the total required resistance range. Here, a switching arrangement would be required in order to tie subrange resistance

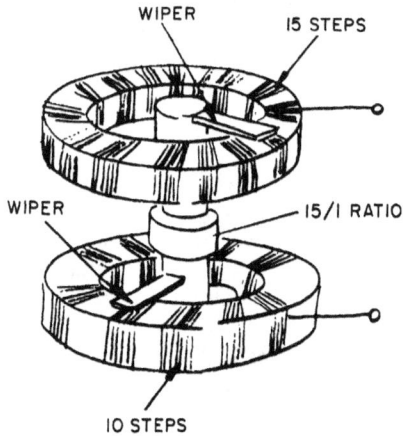

Figure 10-17. Switching for Two-section Resistor Package

together as required. Fig. 10-17[10] depicts a resistor package for use over a 150-bit range. It has two analog resistor subranges, one of 1—9 ohms in 1-ohm steps and a second resistor range of 10—150 ohms in 10-ohm steps.

If a binary count form of analog resistor package is desired (i.e., 1-, 2-, 4-, 8-, 16-, 32-, 64- and 128-ohm step values), the logic switching pattern would have to be provided either at the wiper or at the contact point resistor interface. A schematic representation

of such a system is presented in Fig. 10-18[10]. Other setting schemes are described in Ref. 10.

## 10-4.2 RESISTOR PACKAGE

The selection of a resistor package is, like the external and internal mechanism, dependent upon a series of input and output requirements or characteristics. The inputs are comprised of environmental loads or conditions coupled with the form of the internal switching and the space available. The output requirements dictate resistor accuracy, linearity, base and range of values, capacity (wattage), voltage rating, operational drift, and inductive/capacitance impedance.

The primary division of resistors into types is based on the resistance element. If the element is a metallic wire, wound on a form, the resistor is a "wire-wound" type; if the resistive element is a conducting film deposited on a nonconducting base, the resistor is a "film" type. High cost and a low upper limit of resistance have restricted wire wound resistor usage. The film resistor is a newcomer, and uses less expensive materials and processes. It is capable of extreme upper values of resistance. A reference chart that shows the general ranges of parameters for each major type of conventional resistor is shown in Table 7-3.

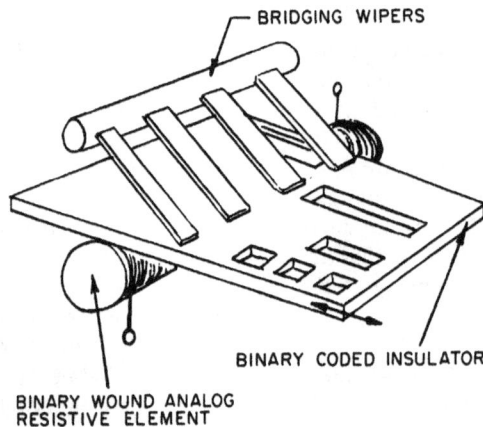

Figure 10-18. Binary Switching System for an Analog Resistor

There are distinct advantages to the selection of thin film networks as the resistor package for analog electronic timers. With respect to separate components, these pre-packaged networks are superior for the following reasons:

(1) Size. The package density is greater than available with individual components.

(2) Performance. The capacitive coupling is constant from network to network.

(3) Reliability. The number of external connections is reduced, and internal connections are stronger. The single package concept reduces the probability of mislocated components. The rugged construction is well suited to gun launch environment.

(4) Cost. Assembly and inspection time are reduced, and inventory control is simpler.

With respect to potentiometer type resistive elements, the prepackaged network offers the advantages of lower cost and smaller size. Wire wound potentiometers have an upper resistance limit below the capability of the network approach. Further, discrete valued resistances allow additional leeway on posi-tion control of the input while maintaining any specific set resistance value.

## 10-5 POWER SUPPLIES

Unless internal voltage regulation is incor-porated in an analog timer, timing error usually is affected by power supply variation. Since digital timers rely on a form of analog timer for their counting time interval, they too usually are subject to error as a result of poor voltage regulation. For general usage of most timers, a regulated supply is recom-mended. Power supply design problems for analog timers under general usage conditions differ little from those of other electronic equipment. Off-the-shelf power supplies usu-ally can be incorporated to solve regulation, environmental, space, and weight constraints. The special problems of projectile fuzing design, however, present difficulties of power supply design which are seemingly unique to this application. Since these problems are common to both analog and digital projectile fuzing timers and, in addition, are usually more complex for digital timing systems, power supply selection for these applications is covered in par. 11-6.

## REFERENCES

1. W. E. Ryan and I. R. Marcus, *Error Analysis of RC Circuits Commonly Used for Timing*, Report TR-1323, U S Army Harry Diamond Laboratories, Washing-ton, DC, July 1966 (AD-640 022).

2. Ante Lujic, "Time Delay Relays", Instru-ments & Control Systems, **40**, 86-90 (1967).

3. D. V. Jones, "Quick-on-the-trigger De-sign", Electronics, **38**, 105-10 (1965).

4. R. J. Reyzer, *An RC Time Constant Multiplying Circuit*, Report TR-1234, U S Army Harry Diamond Laboratories, Washington, DC, 10 June 1964.

5. G. W. R. Hole, "Stable Miller Oscillators With Very Long Periods", A.W.A. Tech-nical Review (Amalgamated Wireless Australasia, Inc.), **13**, 1-14 (1964).

6. J. O. Thayer, *Long Period Curpistor-Diode Timers*, Report TR 895, U S Army Harry Diamond Laboratories, Washington, DC, 29 November 1960.

7. A. L. Ward and L. G. Schneekloth, *Calculations of Relaxation Oscillations in Gas Tube Circuits,* Report TR-1166, U S Army Harry Diamond Laboratories, Washington, DC, 28 August 1963.

8. T. B. Bergersen, *Theory and Characteristics of the Unijunction Transistor,* Note AN-293, Motorola Semiconductor Products, Inc., Phoenix, AZ (no date).

9. T. B. Bergersen, *Unijunction Transistor Timers and Oscillators,* Note AN-294, Motorola Semiconductor Products, Inc., Phoenix, AZ (no date).

10. D. Goldstein, *Engineering Study and Evaluation of Setting Mechanism for Analog Electronic Timers,* Study Report SME-AP-5, U S Army Harry Diamond Laboratories, Washington, DC, March 1967.

# CHAPTER 11

# ELECTRONIC DIGITAL TIMERS

## 11-1 INTRODUCTION

Electronic digital timers establish time delays by counting how often an analog timer has cycled. Their chief advantage for general application is that their accuracy for extremely long time delays can approach that of the analog timer that drives them. Their general application to military systems has paralleled their commercial use, and virtually every large-scale weapon system uses them in some form of programming application. The application of electronic digital timers to projectile fuzing presents a special set of stringent environmental, weight, and volume restrictions on such timers. The latter applications are the principal concern in this chapter.

Although any analog timer can be used as a time base for digital timers, harmonic oscillators are particularly well suited for this use. Hence, they are thus covered in this chapter rather than in Chapter 10.

Electronic digital timer systems for projectile fuzing applications can be divided into several subsystems:

(1) Power supply

(2) Time base

(3) Counter

(4) Output circuit

(5) Internal setting and checkout equipment.

In operation, a particular timing interval is selected by the user and set into the output circuit by means of the interval setting and checkout equipment. The power supply usually is activated also at this time. At the first motion of the projectile, the time base is activated or enabled and begins stepping the counter. As the counter steps, it advances the count previously set into the counter toward the counter's maximum possible count. When this count is reached, the output circuit produces the proper output for fuze activation.

The overall accuracy of any digital timing system is limited by the accuracy of the time base, and system accuracy is close to that of the base. Hence the selection of the time base is of considerable importance. Therefore, the subject of time bases for projectile fuzing is covered in detail in par. 11-2. This subject has received considerable study at the U S Army Harry Diamond Laboratories from which the information is excerpted at length. Each of the oscillator types used has certain advantages and capabilities and may be used over a wide range of operating parameters. Highly stable *LC, RC,* and saturable core oscillators can be fabricated to perform satisfactorily over the military temperature range with variations of ± 20% in supply voltage.

In general, the setting and checkout equipment is reusable while the power supply, time base, counter, and output circuit are expended. Therefore, efforts are made to include as much as possible of the necessary sophistication in the setting and checkout equipment.

The particular advantages of electronic digital timers for projectile fuzing applications, in contrast with other types of timers, lie in the possibility of mass producing a very small number of projectile digital electronic timer types that will cover practically all projectiles and applications at costs far below those attainable with other timers. Integrated circuits with their high quality control, physical ruggedness, and small number of required connections make the electronic digital timer the best candidate for such a major cost reduction approach. Another, and from past experience more likely, consequence of the availability of electronic digital timers is the possibility of increased sophistication of fuze missions and functions. This is due to the relative ease with which more logic circuitry could be incorporated into the integrated circuits of the timer.

## 11-2 TIME BASES[1]

Oscillators may be grouped under two general types: (1) linear, or quasilinear type, having essentially a sine-wave output; and (2) relaxation, or pulse type, having a discontinuous recurring waveform. Most precision oscillators are of the first type, although some types of pulse oscillator have achieved comparable accuracy. For additional information about oscillators, see also Ref. 2.

To obtain high accuracy and to permit precise calibration, it is necessary to isolate the oscillating circuit from changes in power-supply voltage and impedance and also from impedance changes in the load or output circuit. Dependent on the type of circuit and environment, it may also be necessary to shield against electromagnetic and electrostatic interference or proximity effects. Where the highest accuracy is required, thermal control or isolation and calibration at the time of use may be necessary.

## 11-2.1 HARMONIC OSCILLATORS

### 11-2.1.1 Oscillator Characteristics

Harmonic, or sine-wave oscillators, are composed of an amplifier to provide power gain and a resonator, or filter network, to determine the frequency of oscillation. The amplifier must be provided with a positive feedback path so that enough of the output power can be returned to the input in proper phase to sustain oscillation. The resonator must be connected at some point in the feedback loop, or in the amplifier, to control the frequency of oscillation by its phase and attenuation properties. For a nearly linear system, the gain around the loop will be unity and the total phase shift will be zero, or a multiple of 360 deg. If gain of greater than unity exists, the output will increase until the loop gain is reduced to unity because of limiting in the amplifier or other controlling element which is often provided. It can be seen that the phase shift around the loop must add up to zero if a single coherent frequency is being generated. If a phase change takes place in any part of the amplifier or feedback loop, the frequency of the oscillator will shift so that the resonator will generate an equal but opposite phase shift.

Hence, the phase properties of the resonator with respect to frequency determine the frequency of oscillation. The amplitude response characteristics are also important in suppressing spurious modes of oscillation and in reducing the amount of noise amplification at other frequencies. It is desirable to have a maximum $d\theta/df$ for the resonator which maintains the frequency stability, but a minimum $d\theta/df$ for the amplifier and coupling networks which are inherently unstable.

Transistor amplifiers are characterized by rather large changes in input and output resistance and capacitance with changes in temperature and operating power levels. To minimize the effects of these impedance changes with respect to amplifier phase shift, the input and output coupling impedances should be kept as low as possible consistent with required gain. Changes in transit time within the transistors will cause a variation in phase shift. This effect can be reduced by using high-frequency transistors, which will also have less input and output capacitance and, therefore, less change in capacitance.

The reactive elements of the resonator should be loaded by the amplifier as lightly as possible to achieve a maximum phase change versus frequency. The optimum coupling impedances should be achieved by tapped reactive elements or by an impedance transformer to avoid unnecessary dissipative loading of the resonator. The oscillator should be isolated by a buffer or output stage and not designed to supply appreciable load power, as this will result in additional resonator loading and degradation of frequency stability.

The amplitude of oscillation in simple transistor oscillators is limited by hard clipping or rectification at the collector as its voltage swings beyond zero and forward biases the collector junction, or by a shift in the conduction angle toward a condition of increased cutoff as the result of base-emitter rectification. The latter condition generally is avoided in linear amplifiers by eliminating the base circuit capacitor. A resistor of several thousand ohms in series with the base usually improves the wave form and, in some cases, the frequency stability of low-frequency oscillators, where more precise amplitude limiting is not used.

An excessive amount of positive feedback usually produces considerable distortion, while only slightly more than unity feedback, required for oscillation, may result in sudden stoppage under adverse temperature, voltage, or loading conditions.

The distortion resulting from even an adequate drive signal can cause a tendency toward less stable frequency output. This results from a change in phase shift at the fundamental frequency as more or less of the fundamental component is made available from the intermodulation with the harmonic frequencies present. It is thus desirable, when highest stability is required, to provide an external limiting circuit to regulate the loop gain to unity automatically and maintain the amplifier in the linear operative region. In laboratory equipment, this usually is accomplished by a thermistor or lamp, which is heated by the output power, and is coupled into a bridge or attenuator circuit to control the loop gain.

To avoid distortion, these limiting devices must have a time constant long in comparison with the period $(1/f)$ of the oscillator. Lamps and thermistors are not suitable for use as amplitude-sensing regulators where high shock, wide temperature range, and rapid stabilization must be considered; the filaments or thermistor beads must be isolated thermally to function properly, which makes them vulnerable to shock. Also, these devices stabilize too slowly for short-interval timers requiring rapid start-up and functioning. A thermistor, however, is useful to help stabilize gain as the temperature varies, thus reducing the dynamic range required in the automatic level control.

Transistors and diodes, having nonlinear properties when operated with large control voltages, but performing as variable linear resistors with small signal voltages, may be used as amplitude limiters with a small increase in distortion. In such use, the oscillator amplitude is sampled by a rectifier-filter arrangement having a rapid charging time constant to achieve rapid stabilization and a much longer discharging time constant to achieve maximum signal filtering. The output of the filter is applied to the nonlinear control device, which adjusts to keep the amplifier operating at a constant level in the linear region.

Field-effect transistors also may be used as low-distortion voltage-controlled resistors by applying the signal between source and drain, and the control voltage between source and gate.

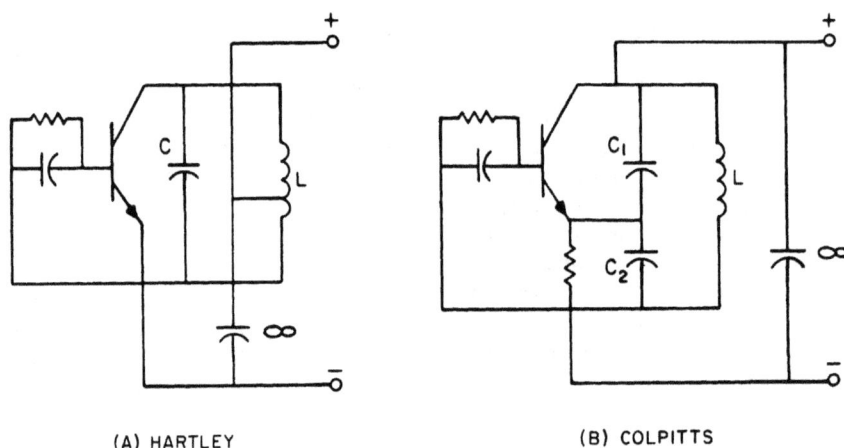

(A) HARTLEY                    (B) COLPITTS

*Figure 11-1. Parallel Resonant Oscillators*

### 11-2.1.2 Inductance-capacitance Oscillators

This type of oscillator is perhaps the oldest and most widely used. In most cases the inductor and capacitor are connected in parallel to give a maximum impedance and zero phase angle at the resonant frequency. In considering a simple transistor oscillator (Fig. 11-1[1]), the emitter usually is connected to a tap on the coil (Fig. 11-1(A)) or a tap between two capacitors used in series across the coil (Fig. 11-1(B)). The base and collector are connected at opposite ends of the *LC* parallel combination through appropriate coupling elements to permit application of DC biases to the transistor. A great variety of connection configurations are possible, which may be treated analytically as one of these two basic circuits.

The two most important considerations are preservation of as high a resonator *Q* as possible and adjustment of the amount of positive feedback for optimum performance. To minimize resonator loading by the amplifier, the collector as well as the emitter may be connected to a tap on the inductor. This permits a larger inductor and a smaller

capacitor for a given frequency and usually results in a higher operating *Q* value. An equivalent approach, known as the Clapp circuit, makes use of three capacitors in series across a two-terminal inductor with the transistor connected across the capacitive divider. At the lower frequencies, the larger of these capacitors, for a proper impedance match, becomes somewhat bulky if of high quality.

The inductor, for use in the audio-frequency range, must contain a permeable core to be of reasonable size and *Q* value. Molybdenum permalloy powdered cores are most suitable although powdered iron and ferrite cores also are used. These cores have an appreciable temperature coefficient of permeability, and must be specially designed and stabilized for use over a wide temperature range. Specially treated molyperm cores having a permeability of about 200 have been fabricated, which have a temperature coefficient about equal and of opposite sign to the temperature coefficient of polystyrene capacitors. By a matching technique, it is possible to achieve a near zero temperature coefficient for the resonator over a wide temperature range. A lower permeability core and polycarbonate

11-4

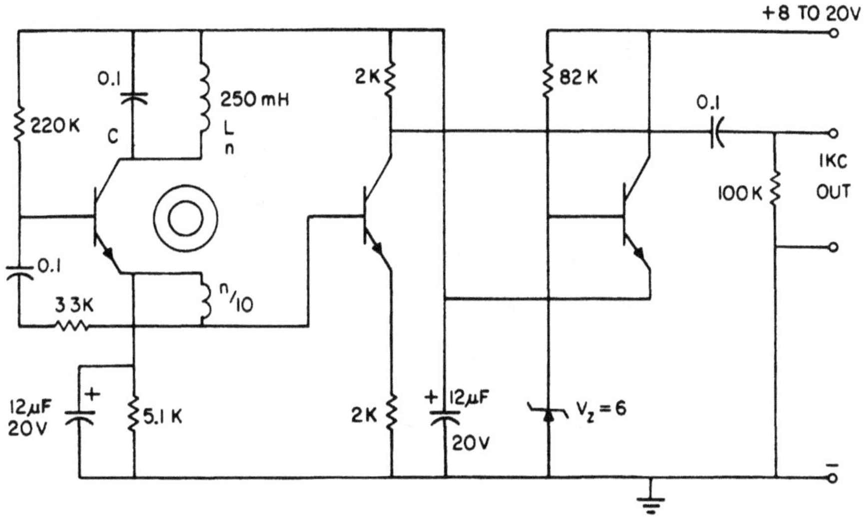

Figure 11-2. Tuned-output Oscillator

capacitor combination has also been used with good results, although generally the polycarbonate-dielectric capacitor has a more S-shaped temperature coefficient, making a wide range match more difficult.

The value of inductance is also a function of the magnitude of the direct and alternating currents present. It is, therefore, important to keep these currents very constant or at very low amplitude to maintain a stable frequency. For circuits working over an appreciable range of supply voltage, it is best to keep the direct current out of the inductor by use of a shunt-feeding resistor and capacitor. The shunt-feeding resistor, however, should not load the inductor heavily, and the coupling capacitor should be large enough to have very low phase shift if maximum frequency stability is to be preserved. Series feeding of the direct current through the inductor and operation on a low, zener-stabilized voltage seem preferable to the difficulties of maintaining a high $Q$ with shunt feed of the supply current. Such a circuit diagram is shown in Fig. 11-2[1], including voltage-regulating and output amplifiers. The regulator stage shown

is useful in that it maintains a fairly constant current drain over a large voltage range. Shunt regulators using only zener diodes draw excessively high current at the higher voltages unless the regulated voltage is low in comparison with the supply voltage. A push-pull circuit with a center-tapped primary may be used to eliminate the DC component in the inductor and reduce distortion if the stability of the resonator elements is good enough to warrant the additional complexity.

An alternate circuit arrangement makes use of a series $LC$ resonator, which has a minimum impedance and zero phase shift at the resonant frequency. When used as a series element in the positive feedback loop, it may be used to control the frequency of an oscillator. To preserve resonator $Q$, in this case, it is necessary to couple into and out of the circuit with amplifier impedances, which are low compared with the series resistance of the resonator. The series resonator capacitor conveniently isolates the DC bias voltages of the amplifier. This circuit is readily adaptable to bridge feedback arrangements permitting negative feedback for DC, which aids bias

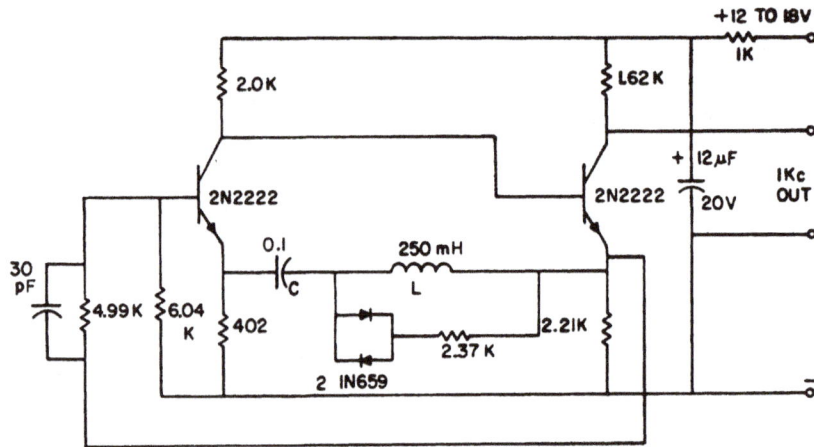

Figure 11-3. Series-resonant Oscillator With Diode Limiting

stabilization and also provides a magnification of the phase characteristic of the resonator proportional to amplifier gain. Maximum bridge phase change occurs with a unity ratio in the bridge arm pairs, although the curve is rather broad and deviations in ratio for biasing convenience up to three to one cause only a slight loss in overall stability.

Fig. 11-3[1] shows a circuit diagram of a series $LC$ bridge oscillator with limiting diodes connected across the inductor. The silicon diodes and resistor shown limit the peak-to-peak voltage across the inductor to about 1 V for supply voltages between 10 and 20 V. With some matching in the temperature coefficients of the polystyrene capacitor and the molyperm inductor, a stability of better than ± 0.1 percent has been obtained from −55° to +71°C with these voltage variations. As shown, the resonator $Q$ is degraded from about 30 to 8 by the limiting diodes, but the measured phase magnification in the bridge of 3.3 brings the equivalent $Q$ back to about 25.

A somewhat similar arrangement using an automatic level control (ALC) is shown in Fig. 11-4[1]. The series impedance of the resonant circuit will change about 40% over

the temperature range because of the temperature coefficient of the inductor wire. If the emitter resistor in the bridge is made of copper wire (noninductively wound), the bridge will stay nearly balanced over the temperature range, thus reducing the dynamic range required of the ALC transistor. A field effect transistor is used as a voltage-variable resistor to achieve the bridge balance. This transistor is controlled by a peak-detection diode connected to the amplifier output. The frequency constancy of this oscillator is determined almost entirely by the stability of the $LC$ resonator components.

### 11-2.1.3 Resistance-capacitance Oscillators

The most popular $RC$ oscillator type, which has been used in a variety of audio-frequency signal generators, is known as the Wien-bridge circuit. The two $RC$ sections forming the positive feedback arms of the bridge (Fig. 11-5(A)[1]) perform equivalent to a high-pass and a low-pass section in cascade and give maximum output with zero phase shift at a common cutoff frequency. At this frequency, the output at the center lead of the $RC$ bridge arms is about one-third of the

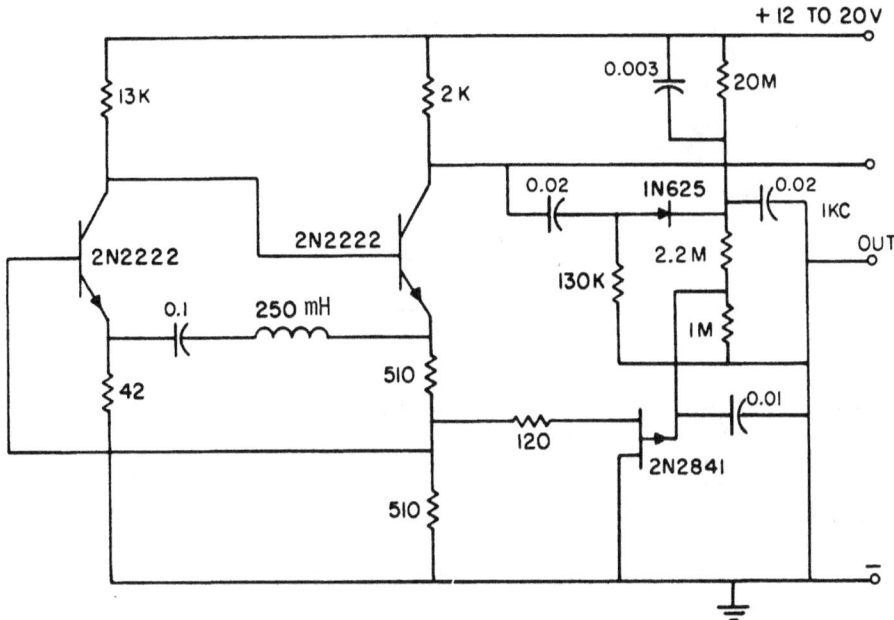

Figure 11-4. Series-resonant Oscillator With Automatic Level Control

input, and the amplitude and phase response are roughly equivalent to those of an *LC* resonator with a *Q* of one-third. When used with a high-gain amplifier coupled to the bridge to form an oscillator, the phase-versus-frequency response can be enhanced by a large amount. The Wien bridge requires, however, extremely high or extremely constant input resistance and extremely low or extremely constant output resistance for the amplifier. Both of these conditions are difficult to realize with transistor amplifiers. By optimum choice of *RC* impedance and special amplifier design, a highly stable oscillator can be achieved.

Another *RC* oscillator circuit in wide use employs a parallel-T null network as the frequency-determining element. This network, shown in Fig. 11-5(B), has a high value of $d\theta/df$ near balance and is in this manner similar to the Wien bridge. Much confusion exists in the literature because of its many possible methods of use in oscillator circuits. The parallel-T, or twin-T as it is often called,

may be symmetrical as shown, or may have the input and output sides dissimilar. The shunt arms may bear the impedance ratios of 1/2 with respect to the series arms, or may be varied by allowing the factor *n* of Fig. 11-5(B) to vary and readjusting the series arms to obtain $f_o$. Maximum phase sensitivity results with *n* equal to 1/2. In this case, as with the Wien bridge, the null frequency is given by *W* = 1/(RC) (where *R* and *C* are the values for the series elements). At null, with all elements in the proper ratio, the phase shift through the network approaches 180 deg. If the shunt resistance is now reduced by a small amount, the phase shift at minimum transmission will be 180 deg and will be of positive and high slope versus frequency in a symmetrical manner for about ± 80 deg. However, if the shunt resistance is made larger than the balance value, a somewhat similar but negative slope in the transfer characteristic will occur centered around 0 deg.

An oscillator may thus be made by using a single-stage amplifier with high gain and 180

11-7

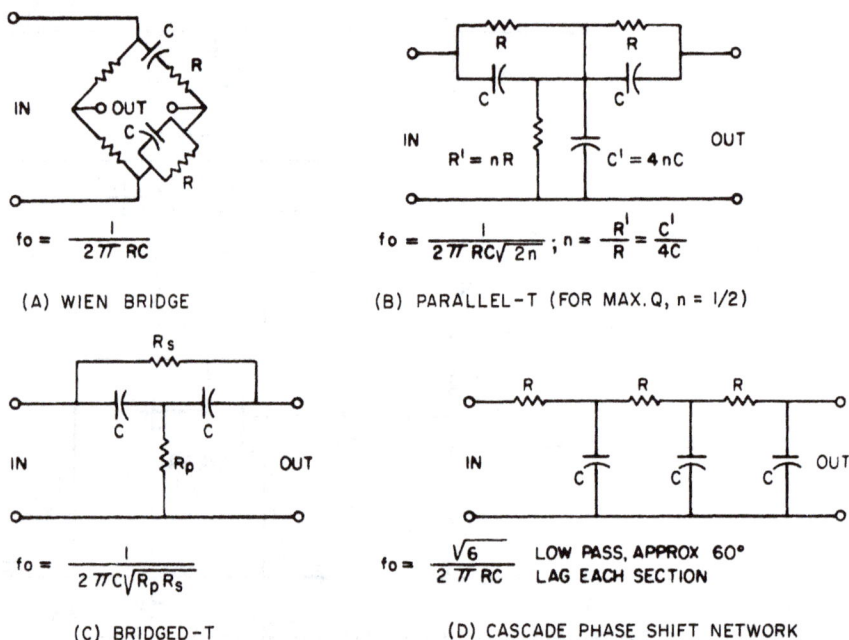

$$f_o = \frac{1}{2\pi RC}$$

(A) WIEN BRIDGE

$$f_o = \frac{1}{2\pi RC\sqrt{2n}} \; ; \; n = \frac{R^1}{R} = \frac{C^1}{4C}$$

(B) PARALLEL-T (FOR MAX. Q, n = 1/2)

$$f_o = \frac{1}{2\pi C\sqrt{R_p R_s}}$$

(C) BRIDGED-T

$$f_o = \frac{\sqrt{6}}{2\pi RC}$$ LOW PASS, APPROX 60° LAG EACH SECTION

(D) CASCADE PHASE SHIFT NETWORK

*Figure 11-5. RC Frequency-selective Networks Used in Oscillators*

deg of phase shift along with a network phase shift of 180 deg, or by using a zero phase shift amplifier along with a zero phase shift network in a single feedback loop. Two-loop feedback oscillators have been built with a precisely nulled network in the negative feedback loop and an amplitude control device in the positive feedback loop which maintains the total loop gain at unity.

Still another twin-T arrangement makes use of the network in which the common and output terminals are reversed. The new output terminal is worked into a common-base transistor stage (ideally a short circuit), and oscillation occurs by having zero phase shift in both the network and amplifier. An interesting property of this network is that it has an input-to-output voltage gain of slightly greater than unity, with a maximum occurring when $n$ (defined in Fig. 11-5(B)), is about 5. In this case any transistor having a $\beta$ greater than

11 will sustain oscillation. The network $R$ should be much larger than the emitter input impedance, and the network $C$ should be much larger than the transistor capacitance.

Other oscillator circuits using the bridged-T, or the phase shift network shown in Fig. 11-5(C), or variations of these networks, are possible. For the bridged-T, a high ratio (5 to 25) of R-series to R-parallel is desirable to give a high phase shift factor in the network, which makes the oscillator insensitive to transistor changes with temperature and voltage.

The L-section phase shift networks of Fig. 11-5(D) frequently are tapered (increased in impedance toward the output end) to reduce loading between sections. A three-section filter with $R$ and $C$ elements reversed also will oscillate, but the high-pass type is likely to give increased distortion and therefore may

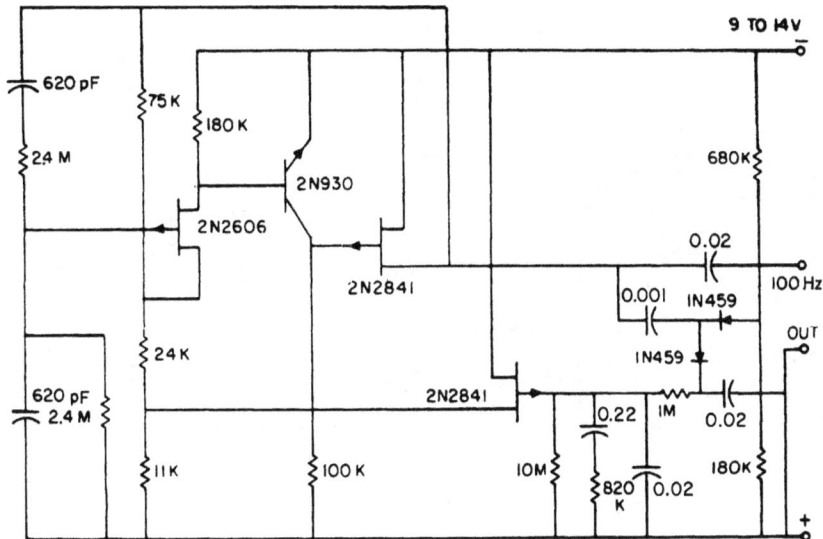

*Figure 11-6. Wien Bridge Oscillator*

not be so stable. Oscillators using these networks have been reported giving stabilities comparable to those of the other types discussed. This circuit is sensitive to changes in amplifier input and output impedances. A three-phase oscillator using three field-effect transistors (located between the L-network) has given good results.

Fig. 11-6[1] shows a 100-Hz Wien-bridge oscillator developed by Sann of U S Army Harry Diamond Laboratories. Field-effect transistors are used in the amplifier input to obtain a very high input impedance and in the ALC circuit to obtain a linear resistor that is controlled by a peak-to-peak rectifier on the output. This circuit has been tested from $-55°$ to $80°C$ with supply voltage from 9 to 14 V. Combined frequency deviations were $\pm 2$ parts in $10^4$ for the amplifier alone. Wirewound precision resistors and monolithic ceramic capacitors were used, but these were kept at room temperature to evaluate the amplifier. Compensation to 1 part in $10^3$ (0.1%) is not difficult, although the shock stability and long term drift are not evaluated so easily.

Fig. 11-7[1] shows the circuit diagram of an oscillator using a phase inverting parallel-T network in the positive feedback loop. A thermistor is used across the negative feedback resistor to stabilize the gain over the temperature range. The thermistor may be replaced with a controllable resistor as part of an ALC circuit similar to the one used in Fig. 11-4, to give better amplitude control and frequency stability. The shunt resistor in the parallel-T network is reduced by shunting with a much larger resistor, to obtain a network loss slightly less than the gain of the amplifier. Amplifier limiting (or an ALC circuit) maintains the oscillation amplitude constant. A circuit of this type has performed over the temperature range of $-55°$ to $80°C$ and $\pm 30\%$ variation in supply voltage with frequency changes of $\pm 55$ parts in $10^4$. The frequency determining network consisted of monolithic ceramic capacitors and metal film resistors with near zero temperature coefficients.

**11-2.1.4 Mechanical Resonator Oscillators**

Electromechanical resonators have been

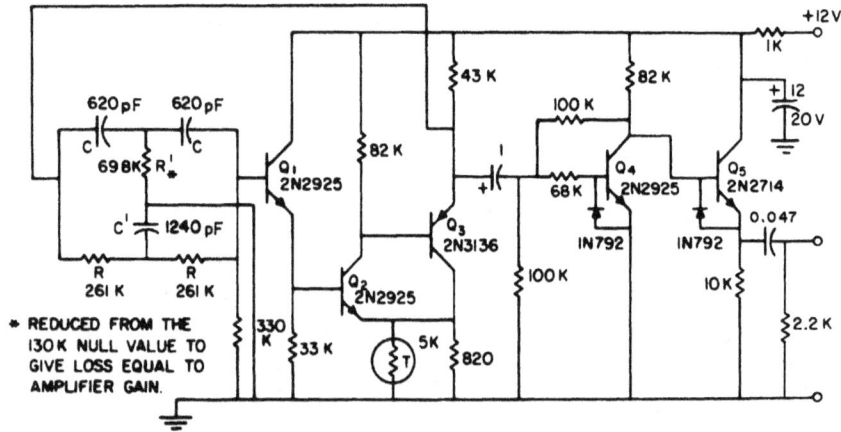

Figure 11-7. Twin-T, Unbalanced-network Oscillator

used in oscillators for a number of years, especially where highest accuracy is required. Electrically driven tuning forks have achieved accuracies of 1 part in $10^6$, and quartz crystals can be obtained with a daily stability of 1 part in $10^9$. Lower-precision resonators, operating on magnetostrictive or torsional modes, are also available. These devices are sensitive to shock and vibration and to temperature variations. High-frequency quartz crystals of the overtone types have been used under conditions of fairly heavy shock. Transistor-driven balance wheels are also operable under occasional high-shock conditions but are not considered reliable under conditions of continuous severe vibration.

## 11-2.2 RELAXATION AND PULSE-TYPE OSCILLATORS

Nonsinusoidal oscillators of a variety of types have been used in timing applications of moderate precision. The most common of these is the free-running multivibrator. A compensated unit of this type has been built that was constant to 0.1% from −55° to 75°C. Similar laboratory precision has been obtained using unijunction transistors or pnpn switches. The extremely high input imped-

ance of the insulated-gate, field-effect transistors may increase the capabilities of this type of relaxation oscillator.

Another type of relaxation oscillator makes use of a resistor, or a small ion chamber or curpistor, which contains a controlled amount of radium compound, and serves as a very low-current source for charging a capacitor. A special gas diode and sensitive counting relay are connected in series across the capacitor, which partially discharge the capacitor each time it reaches the diode breakdown voltage. The ion chamber eliminates the need for regulating the supply voltage, but requires about 75 V above diode breakdown. Principal causes of error are variations in the diode firing voltage and in the residual voltage left on the capacitor after each pulse. Accuracies of 1% have been achieved. The system is suited to very long-period timers because of the extremely small average current drawn. The radioactivity associated with the ion chamber is a slight disadvantage.

Saturable-reactor oscillators have become increasingly popular in recent years as timing devices. These oscillators make use of the sudden change in impedance or voltage of the

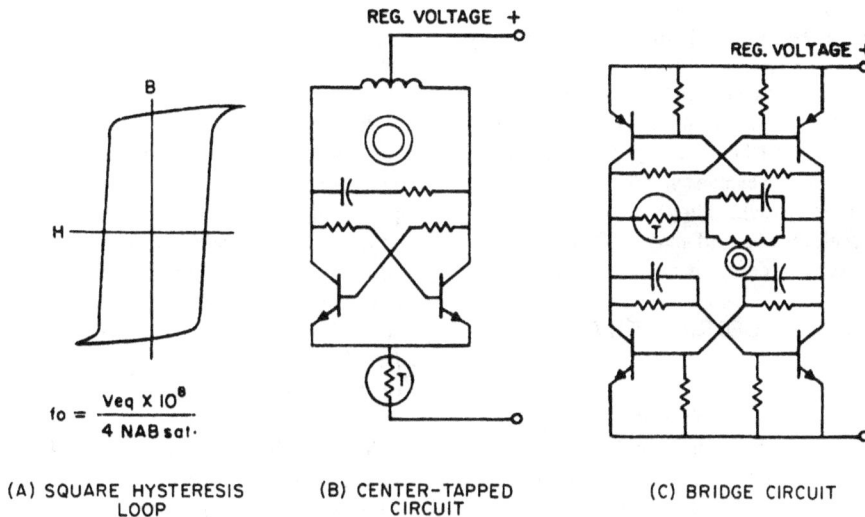

$$f_0 = \frac{V_{eq} \times 10^8}{4\, NAB\, sat.}$$

(A) SQUARE HYSTERESIS LOOP     (B) CENTER–TAPPED CIRCUIT     (C) BRIDGE CIRCUIT

*Figure 11-8. Saturable Core Oscillators*

winding on a square loop core to trigger flux reversing and readout functions as the core reaches saturation. The timing action of this device with its abrupt discontinuity is much more favorable to precise timing than the slowly changing exponential waveform of an *RC* circuit (see Fig. 11-8[1]).

Fig. 11-8(A) shows a hysteresis loop for a typical square-loop inductor. In use in an oscillator, the inductor is connected into a multi-vibrator-type circuit that switches the flux continuously from a condition of positive saturation to negative saturation and back again each time the core saturates. Fig. 11-8(B) shows a push-pull circuit for use with a center-tapped coil while Fig. 11-8(C) shows a complementary flip-flop circuit for use with a two-terminal coil. The latter has an advantage where low-frequency operation is required, since the period of oscillation is proportional to the number of turns on the coil. The center-tapped coil uses only one-half of the turns during each half cycle. Frequencies as low as a few hertz have been obtained, but the inductors become large for frequencies below 100 Hz. Other circuit

arrangements also have been employed, using multiple feedback windings to achieve oscillation.

It is necessary to regulate precisely the switching voltage applied to the core winding because frequency is directly proportional to applied voltage. Total flux change of a typical core may vary as much as 10% over a temperature range of −55° to 75°C, decreasing with increasing temperature. This variation is compensated partially by the positive temperature coefficient of the coil winding since the effective driving voltage is equal to the input voltage minus the *IR* coil drop. During the cycle a low current is drawn while the core is switching, with high current flowing as the core saturates. The residual temperature coefficient of frequency may be further reduced by use of a positive temperature coefficient thermistor in series with the inductor, or thermistor, or diode compensation of the voltage regulator. In this manner, accuracies of better than 0.1% have been achieved over the usual military temperature range.

The square-loop cores usually are made of very thin grain-oriented material, such as Orthonol or Supermendur tape wound in toroidal form, and supported by a nonmagnetic metallic spool. These cores are sensitive to shock and must be specially packaged and treated to avoid permanent frequency shifts resulting from shock. They are also sensitive to spin accelerations and may shift frequency several percent under high spin.

## 11-2.3 PRESENT OSCILLATOR CAPABILITIES

Laboratory models of all of the types of oscillators shown have been built which gave better than 0.1% accuracy over ± 20% supply voltage variation and a temperature range of −55° to 75°C.

A number of military-type, saturable-core oscillators have been produced and are commercially available from several sources. These items perform well over the temperature range and under typical missile shock environment. They are, at present, not so accurate under the high accelerations associated with gun-fired rounds. Spin accelerations of 6000 g may shift the frequency as much as 5%. Specially designed cores have reduced this variation to as low as 1%. Firing shocks of about 15,000 g have introduced permanent changes up to 1%. Operation in strong magnetic fields is to be avoided. A large bar magnet 6 in. away shifted the frequency 0.4%. When brought within 1 in., the frequency increased 14%. Residual shift in frequency was very slight when the field was removed.

A limited number of inductance-capacitance oscillators of ruggedized type have been shock and centrifuge tested. In the shock tests, permanent changes from 0.1 to 1% have been observed for accelerations of 15,000 g. The units were of a potted type but not specifically designed for high shock. Newer units presently under development are expected to show improved shock resistance. Two

units of improved type recently fired at 20,000 g showed permanent changes of 0.05 and 0.2%.

In the centrifuge tests, *LC* oscillators were spun up to 10,000 g, and showed frequency increases of 0.05 to 0.4% with the force perpendicular to the toroidal axis. Another unit with the force along the toroidal axis showed a decrease of 0.06%. If the toroidal core can be kept within less than 1 in. of the spin axis of the projectile, it is believed that frequency changes of less than 0.1% can be obtained in most uses.

The toroidal core of a 1-kHz *LC* oscillator was exposed to the same magnetic fields as were applied to the saturable-core oscillator. At 6 in., the change in frequency was about 0.05% and at 1 in., about 2%. Residual shift on removal of field was negligible.

Shock and spin tests are planned for the several types of *RC* oscillators that have been laboratory tested. It is believed that *RC* circuits are likely to prove more stable in high spin and shock environments than those circuits containing inductors because of the inherent shift in the permeability of magnetic cores when under stress. Resistors and capacitors also can be expected to change in value when subjected to force fields but it is believed that improved packaging and construction techniques can be used to reduce these changes considerably.

The problems of temperature stability and aging of the frequency-determining components must be considered. From a production standpoint it is better, insofar as possible, to place the burden of obtaining low temperature coefficient upon the component suppliers where quality control can be more closely and precisely monitored. Aging control must be placed in the area of component design. At present, wire-wound resistors with temperature coefficients as low as 10 parts per million per deg C (ppm/deg C) are readily available and 2 ppm/deg C can be obtained. They also may be obtained with specified nonzero temperature coefficients which may be used to compensate for variations in other components. Metal-film precision resistors are available from a number of sources with coefficients of ± 25 ppm/deg C.

Of the various film-dielectric capacitors, polystyrene film units offer the more nearly linear temperature coefficients, the nominal value being −120 ppm/deg C. Variations of ± 15 ppm/deg C in this value may be expected. Stability of the polystyrene capacitors under high shock at 70°C is expected to be a problem. These capacitors usually are not used above 85°C. Polycarbonate dielectric capacitors, which have a lower but less linear temperature coefficient, may be more useful at these temperatures. Monolithic glass-incased capacitors are available with a coefficient of 115 to 165 ppm/deg C. Also, monolithic ceramic capacitors are available with coefficients of ± 25 ppm/deg C. The monolithic types are said to have excellent retrace and aging properties, and are expected to be highly shock resistant. Silvered mica capacitors have excellent stability and retrace characteristics, and temperature coefficients of 40 to 70 ppm/deg C. Their shock stability depends, to a large extent, on how they are fabricated and packaged.

Permalloy powder cores are now available for which inductance will vary less than ± 0.25% from −65° to +125°C. Cores of this type are also available that have coefficients of approximately 120 ppm/deg C and give a near-zero $LC$ coefficient when used with polystyrene capacitors.

The flux storage capacity of the square-loop inductors decreases with temperature, but this is partially compensated by an increase in wire resistance with temperature when used in a saturable-core oscillator. All magnetic cores are stress sensitive and, for stability, must be packaged to keep mechanical stress to a minimum.

To obtain highest thermal frequency stability, it is likely that the resonators of any of the oscillator types will require some temperature-coefficient adjustment or parts matching on individual oscillators after performance over the temperature range has been observed. A final frequency calibration adjustment probably will be required after partial assembly and temperature cycling tests. A comparison of the relative performance capabilities of the various oscillator types is given in Table 11-1[1].

## 11-3 COUNTERS

### 11-3.1 GENERAL

Most counting circuits use an electronic or magnetic circuit having two stable states as their basic logical element. While schemes using multistable basic elements are being investigated, binary basic elements and binary logic elements are those generally available. The basic logical element function is that of a scaler or divider. The element has an output after a predetermined number of input pulses are supplied. If it has an output for every two inputs, it is a binary scaler or binary logical element and usually is termed a flip-flop. If the basic element requires three input pulses for one output, it is called a ternary scaler; one output for ten inputs is the characteristic of a denary scaler. Fig. 11-9 shows a basic binary element (Fig. 11-9(A)) and three of these elements connected to make a three-stage counter (Fig. 11-9(B)). Note that for binary counters, the scaling factor between input and output is $2^n$ where $n$ equals the number of stages.

Fig. 11-10[3] shows a typical transistor binary flip-flop constructed of discrete elements. A semiconductor integrated circuit can be used to perform the same function as the discrete elements shown.

Fig. 11-11[3] shows a block diagram of a complete electronic digital timer system and indicates how the counting stages fit into the

TABLE 11-1

RELATIVE PERFORMANCE OF VARIOUS OSCILLATOR TYPES

| Oscillator Type | Frequency Stability* | Resistance to Shock and Spin | Power Drain | Complexity |
|---|---|---|---|---|
| LC parallel tuned | good | fair | low | low |
| LC series tuned | very good | fair | low | medium |
| RC Wien bridge | very good | good | low | medium |
| RC parallel—T | very good | good | low | med. to high |
| RC bridged—T | very good | good | low | medium |
| RC· phase shift | fair to good | good | low | medium |
| multivibrator | fair to good | good | low to med. | med. to high |
| magnetic square loop | good | fair to poor | med. to high | high |
| pulse RC discharge | fair | good | low | medium |
| pulse curpistor—C | fair to good | good | very low | medium |
| mechanical resonator | excellent | poor | low | med. to high |

*Frequency stability, variation: excellent $< 1 \times 10^{-4}$; very good $10^{-4}$ to $10^{-3}$; good $10^{-3}$ to $10^{-2}$; fair $> 10^{-2}$.

overall function of the timer. A 1280-Hz *LC* oscillator is inhibited by an initiate gate. On receiving an initiation signal, a simple 7-stage ripple-through counter provides the basic time base of 10 Hz. Divide-by-ten circuits are used in the time accumulator for tenths of seconds, units of seconds, and tens of seconds. A single binary divider serves as the 100-sec accumulator. A negative-going signal from the 100-sec binary will latch a controlled rectifier in the firing circuit and energize a detonator. Varying time delays are selected by setting the counter stages to a preselected count before the initiate gate is enabled.

Electronic digital timers have been con-

structed using both transistor flip-flops and transistor driven magnetic core flip-flops. Each has its particular advantages. Examples are given in the paragraphs that follow. Other types of multistable basic elements are being investigated in relation to projectile fuzing environment, and these also are discussed.

## 11-3.2 TRANSISTOR TYPE COUNTERS

Almost all types of counter use transistor junctions or discrete transistors as the active elements to perform the electronic functions of the timer and, in that sense, all electronic digital timers for projectile fuzing are transistor counters. However, in those coun-

(A)   One Stage

(B)   Three Stage

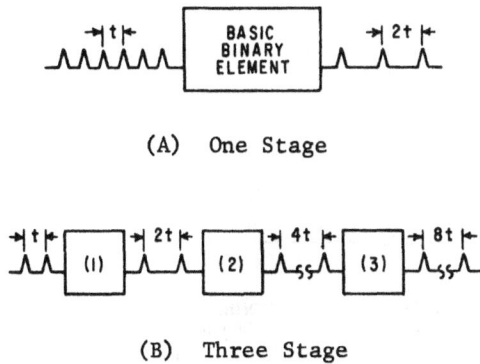

*Figure 11-9. Primitive Counters*

ters referred to as transistor type counters, only transistors are used as information storage in the basic logical element. Thus the basic logical stage of this type counter is similar in electrical configuration to that shown in Fig. 11-10.

This type of basic logical element suffers from a practical drawback because the element must be supplied with power throughout the complete time that it must preserve the count. This can be a serious

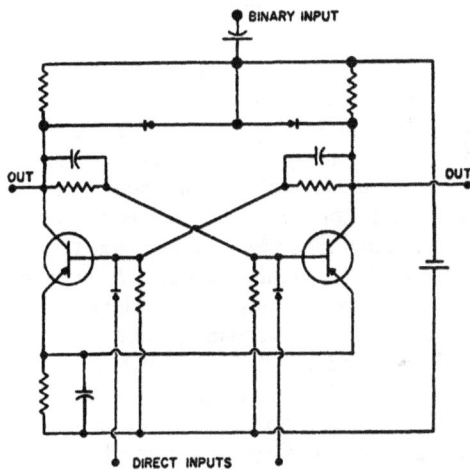

*Figure 11-10. Typical Multivibrator*

disadvantage. Hence, this type of counter seldom is used. Fig. 11-12[4] is a schematic diagram of an electronic digital timer that uses basic logic elements of the magnetic core type. Note, however, the four binary divider stages, A4 through A7, in the center of the figure. These stages are used purely as dividers, and the time interval selection information does not concern them. Thus the timer uses a transistor counter as a divider, and memory core logic to store the set time interval.

Some experimental digital timers for projectile fuzing were developed in early research using transistor counters exclusively. Fig. 11-13[5] shows the block diagram of such a system. Note that the basic logical elements are ternary dividers of all transistor construction. A schematic of one of the dividers is shown in Fig. 11-14[5].

Although pure transistor counters are not preferred at present due to their power drain during the interval from set to launch time, they do present the best prospects for miniaturization. With sufficient advance in miniaturization, power demands are sure to decrease radically so that transistor junction counters must be considered as possibly and probably desirable components for projectile fuzing applications.

## 11-3.3 MAGNETIC CORE TYPES

Cores have many advantages over strict electronic counters for military application. Aside from the ability to store information without power expenditure, they are resistant to nuclear radiation, relatively inexpensive, and have volumes comparable with discrete electronic components. Their principal disadvantage—from a long-range viewpoint—is the fact that they are relatively difficult to miniaturize and they require multiple connections. The magnetic cores used for memory operation are discussed in par. 7-2.2.4. Additional information on magnetic cores and core circuitry is contained in Ref. 6. Their important property is, of course, that

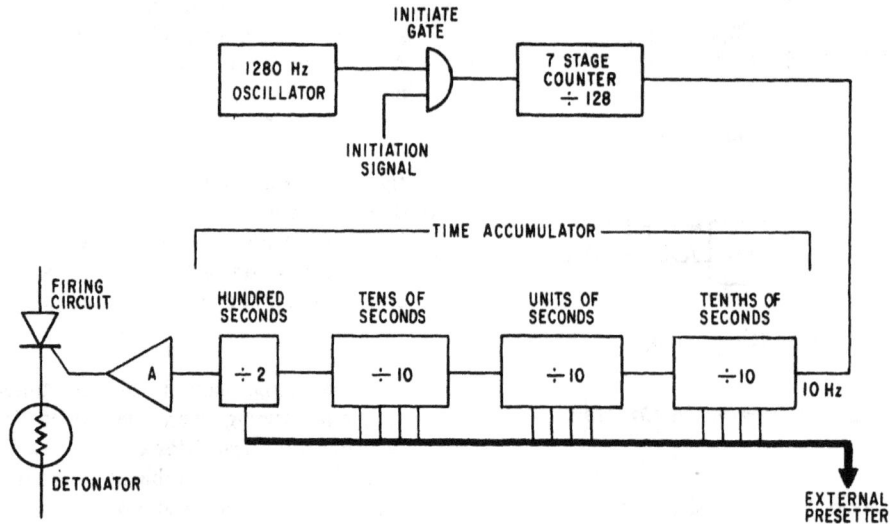

Figure 11-11. Block Diagram of Basic Timing Circuit

the core "remembers" the direction of the last significant change in the applied magnetic field.

There are many different configurations of cores and drivers that can perform counting, scaling, and memory functions. Fig. 11-15[7] shows cores connected to form a scaler that could have an output pulse for every fourth clock pulse. The odd and even drive pulses are generated from the clock pulses by the circuit of Fig. 11-16[7].

Suppose core "0" of Fig. 11-15 were put into the "1" state and cores 2, 3, and 4 were put into the "0" state. Now, a drive current of sufficient magnitude and duration is applied to the even drive winding as shown. Core 2, in the "0" state and being driven to the "0" state, does not switch (but a small voltage is generated due to the $\phi_s - \phi_r$ flux change). Core 0 in the "1" state switches to the "0" state. The output of core 0 drives core 1 to the "1" state. The diode in the 1-2 loop isolates core 2 from core 1, so that core 1 is unloaded when it switches to the "1" state and it is completely switched before

core 0 is completely switched to the "0" state. The next drive pulse appears in the odd drive line, driving cores 1 and 3 to the "0" state. This transfers the "1" from core 1 to core 2. In this manner the "1" is transferred from core to core, and since core 3 is connected to core 0, the "1" will circulate. If an output were taken from the loop joining cores 3 and 0, the output would appear once for every four clock pulses.

In Fig. 11-16, core A is initially set to the "1" state and core B is set to the "0" state. The clock pulse is derived from an oscillator or from a previous decade, and drives both cores A and B to the "0" state. Core A switches to the "0" state, turning on SCR A, which allows capacitor C to discharge through the even-drive winding and core B, putting a "1" into core B. Resistance R is large enough so that (B+)/R is less than the holding current of SCR A and SCR B. The second clock pulse from the oscillator again drives cores A and B to the "0" state; now core B switches, and C discharges through the odd-drive winding and core A, putting a "1" into core A. In this manner, the clock pulses generate alternate drive pulses to the scaler.

11-16

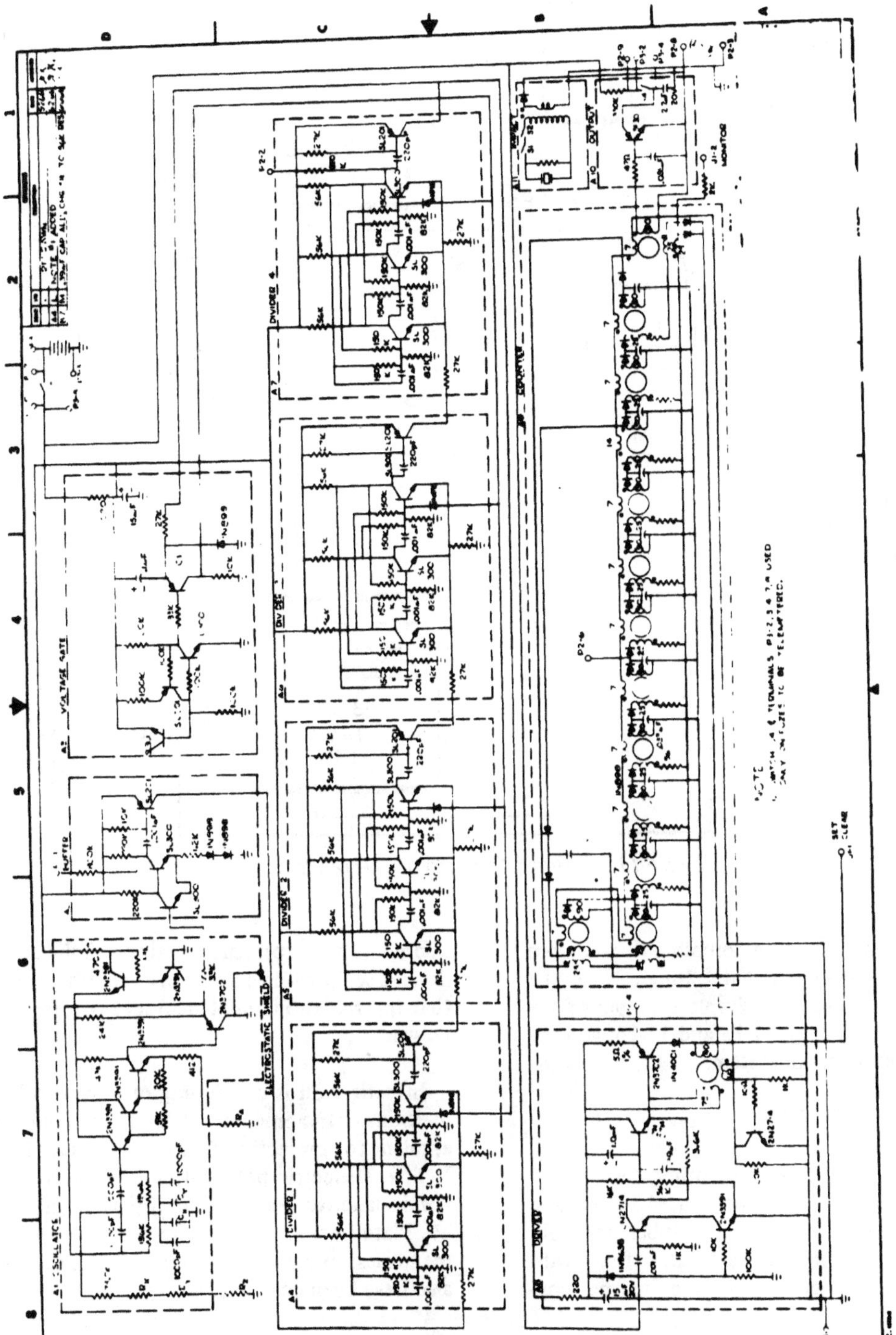

Figure 11-12. Schematic Diagram of Fuze, Electronic Time, XM587

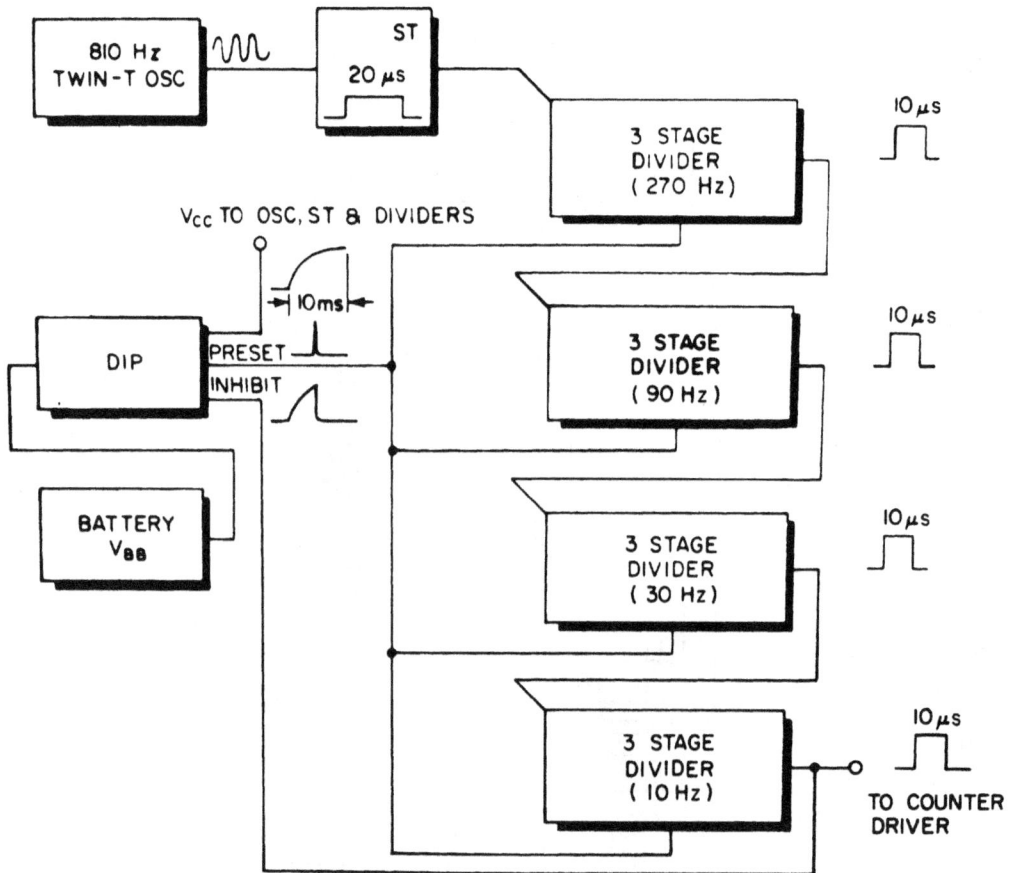

Figure 11-13. Block Diagram of 10-Hz Clock

Fig. 11-17[8] shows a complete decade core scaler for an experimental timer. Note the similarity between the simple example of the scaler given by Figs. 11-15 and 11-16, and the completed operational item.

## 11-3.4 OTHER TYPES

There are a number of other devices and phenomena that have been used to develop counters or scalers. Vacuum tubes and cold gas diodes were perhaps the first with sufficient frequency response to be con-sidered for projectile fuzing. Solid-state devices seem to have most of their advantages (barring radiation resistance) and require less power.

Investigation is continuing on several other methods. Incremental flux switching as applied to projectile fuzing is discussed in Ref. 9. Dielectric polarization memories with nondestructive readouts are discussed in Ref. 10. Plated wire memories, deposited magnetic memories, and magnetic domain propagation also have been considered.

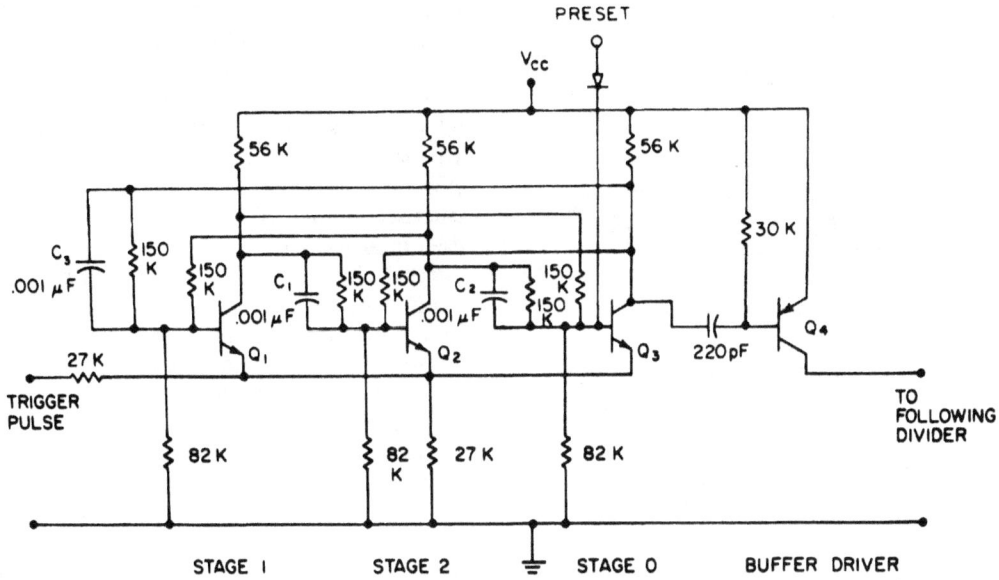

*Figure 11-14. Ternary Divider*

## 11-4 OUTPUT CIRCUITS

### 11-4.1 GENERAL

The primary function of the output circuits of digital electronic timers used for projectile fuzing applications is to produce an electrical signal on occurrence of the "timing out" of the counter circuitry which will cause detonation of the projectile warhead. The secondary function—of almost equal importance—is the positive prevention of electrical initiation of the warhead under any circumstances except when functioning is desired. In

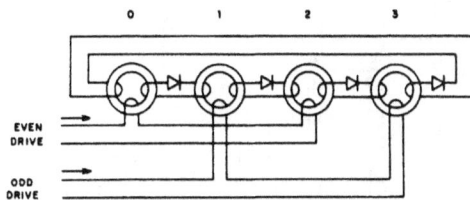

*Figure 11-15. Magnetic Core Scaler*

particular, the output circuits must not produce an electrical initiation stimulus on counter "time out" unless the projectile is armed.

In missile and rocket applications, the output circuits of the timer may produce a relatively low power signal that is transmitted to sophisticated decision making electronics in the warhead. In contrast, the output circuits for electronic timers used in artillery projectiles must provide a relatively high energy output and provide all electrical logical interlock functions. The output circuit design is heavily dependent on the system mission in the case of projectile fuzing applications. Refs. 3, 11, and 12 provide examples of output circuits for particular electronic digital timers.

### 11-4.2 OUTPUT PULSE GENERATION

Since output circuits for electrical timers used in missile and rockets need only supply an electrical signal to warhead logic circuits, they generally use a solid-state active element

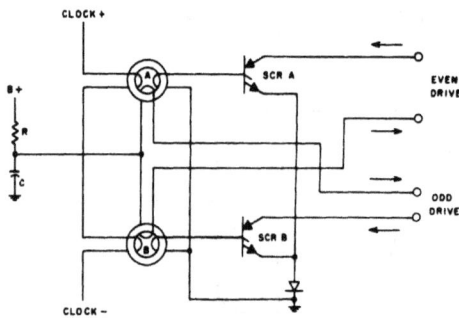

*Figure 11-16. Two Phase Driver*

for output pulse generation. The electrical requirements are seldom severe. Hence, a transistor, silicon-controlled rectifier (SCR), or unijunction transistor that is identical with those used in the counter circuits often is employed to reduce the number of different components in the overall timer and thereby reduce overall cost.

In artillery projectile fuzing and in other instances where the output circuits must directly initiate detonation, special means are employed to insure reliable functioning. Detonation initiation usually requires the output circuits to function an electroexplosive device (EED). The DC resistance of the bridgewires commonly used in hot wire EED's varies from a few tenths of an ohm to a few ohms. Currents of fractions of an ampere to several amperes are required to provide reliable initiation which usually occurs within 10 or 20 msec from application of the current. Especially sensitive EED's—requiring currents as low as 100 mA—sometimes are used in artillery projectile fuzing to minimize output circuit current requirements. Such sensitive items require special safety precautions during assembly and handling.

The basic problems for output circuit design are the relatively high power required by the EED and the impedance mismatch between the EED's resistance wire of a few ohms and the operating impedance level of solid-state circuitry which in general is much larger than 50 ohms. A typical solution to the output circuit/EED interface problem is shown in Fig. 11-18. The +12 V DC is not applied to the initiation circuit until after the rocket or missile has been launched or, in the case of artillery applications, the projectile has been fired. At this time the 150 $\mu$F capacitor charges rapidly through the transistor and the 1.5 k ohm resistor. The transistor acts as a self-biasing device to maintain the capacitor charge. If the timer "times out" and the logic decision circuits provide a positive "initiate signal" to the SCR, the 150 $\mu$F capacitor discharges through the SCR and the EED resistance wire providing initiation of

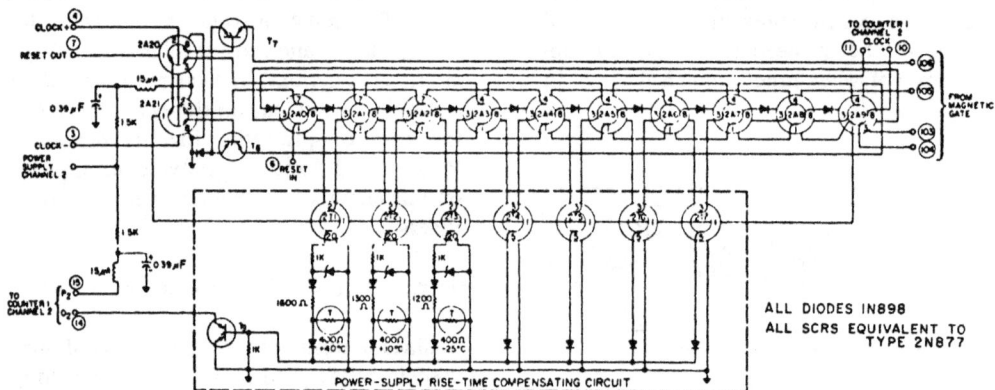

*Figure 11-17. Decade Frequency Divider*

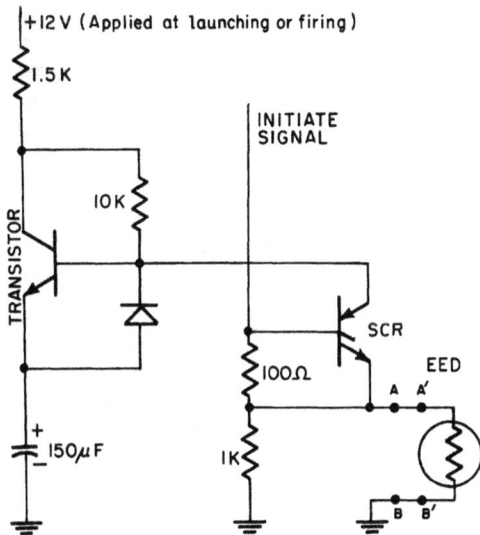

*Figure 11-18. Typical Initiation Circuit*

the explosive charge. Many output circuits are designed along these general lines. In many applications, however, a set of switch contacts are inserted between the SCR output and the EED (AA' or BB' in Fig. 11-18). The switch often is closed by a mechanical arming function in missiles and rockets, and by centrifugal force in spinning projectiles.

## 11-4.3 LOGICAL INTERLOCK FUNCTIONS

In missile and rocket application, the electronic timer output circuits may have little to do except pass on the signal from the last stage of the counter to the elaborate warhead electronic decision making circuits. These circuits, which may have inputs from elaborate sensors, make the detonation decision. The artillery projectile, in contrast, usually requires that all the electronic decision making capability be incorporated directly into the output circuitry of the electronic digital timers. The primary safe and arm function for artillery projectile fuzing applications, however, is accomplished mechanically.

## 11-5 SETTING METHODS

### 11-5.1 GENERAL

The most common method of setting a digital timer for fuzes is to preset a count into the counter such that the counter reaches its maximum count after the desired number of input pulses from the time base. When the counter has reached this maximum count, detonation occurs. Thus, if we desire detonation after 593 clock pulses have stepped in a counter with a maximum count of 1000, we would preset the counter to 407. Presetting of the counter can be accomplished by setting the counter to zero and stepping it to the correct setting by inserting single pulses or by "jamming" the stages to the desired setting. The first of these methods is called serial setting, the second parallel setting. Refs. 3, 11, and 13 provide examples of both methods.

### 11-5.2 SERIAL METHODS

Using a serial set method for counters involves:

(1) Applying power. External power is supplied if the counter can retain its reading without power. The fuze power supply must be actuated for counters that cannot do so.

(2) Clearing the counter to zero

(3) Counting the required number of pulses supplied to the counter input

(4) Checking the operation of the counter, if necessary.

Serial setting usually can be accomplished with few electrical contacts with the fuze, and this is a decided advantage. The primary disadvantage of serial setting arises when the number of setting pulses supplied must be large; for, then setting time can become intolerable. The setting pulses can be supplied at a rate much greater than those normally

11-21

supplied by the time base to overcome the disadvantage but this approach is limited by the ability of the counter to respond to the higher repetition rate pulses. In general, high response rate for the counter means a more costly counter. Hence, this approach is limited by economics.

## 11-5.3 PARALLEL METHODS

Using a parallel set method for counters involves:

(1) Applying power. External power is supplied if the counter can retain its reading without power. The fuze power supply must be actuated for counters that cannot do so.

(2) Applying the correct pulse to every stage of the counter

(3) Checking the operation of the counter, if necessary.

Parallel setting can be accomplished in very short times. The primary disadvantage is that many electrical contacts must be made with the fuze. Electrical contacts are particularly troublesome in rugged military equipment.

## 11-5.4 COMBINATION METHODS

Compromises between the serial and parallel methods involve treating the counter as if it were several smaller counters and setting each by the serial method. The disadvantages of more electrical contacts are not as serious because (1) there are less contacts than in the parallel method, and (2) the setting time will not be overly high because a much smaller number of pulses can be applied simultaneously to the smaller counters than would be necessary for the undivided counter.

## 11-6 POWER SUPPLIES

## 11-6.1 GENERAL

Digital timers rely on some form of analog

timer as a time base and hence the overall accuracy of the digital system can be no better than that of the analog timer base count. Although the electronic digital counting methods employed in digital timers are relatively insensitive to voltage fluctuations, the overall system—which includes the voltage sensitive analog timer—is thus still subject to error due to poor voltage regulation.

For general application timers, the design problems for the power supply are essentially the same as those of other electronic logic circuits. Off-the-shelf power supplies are recommended for these applications because they are available in great variety and can be obtained economically to provide well regulated voltage for the analog time base, and less well regulated outputs for the remaining, less critical portions of the system. Study of the specifications for off-the-shelf supplies from several manufacturers before selecting a final unit can be very helpful to the timing system designer if he does not have an intimate knowledge of power supply design. The specifications may point out problem areas—such as transient protection or radio frequency interference vulnerability, which the off-the-shelf power supply manufacturers have solved—which are directly applicable to the timing system problem under consideration.

Special problems arise in power supply design for projectile fuzing systems. The remainder of this discussion is aimed at these more stringent applications.

## 11-6.2 POWER SUPPLIES FOR PROJEC- TILE FUZING APPLICATIONS

In projectile fuzing design it appears that oscillators, timing circuits, safing and arming mechanisms, and arming switches sometimes are considered first and laid out. Then, almost as an afterthought, the remaining space is allocated to the power supply with a catch-all form factor. We here enter a strong plea to give the power supply more consideration during development. Designers rely on manufacturers' claims and theoretical values for

TABLE 11-2

TIMER POWER SUPPLIES (VOLUME AND SIZE)

| Fuze | Max. Volume, in.$^3$ | Typical Dimensions, in. |
|---|---|---|
| Missile | 45 | 2.5 X 3.0 X 6.0 |
| Rocket | 21 | 3 diam. X 3 long |
| Projectile | 4 | 1-5/8 diam. X 2 long |

power per unit weight or per unit size without giving adequate consideration to the overdesign needed to insure operation at temperature extremes, and ruggedization at high setback and spin forces. Most existing off-the-shelf power supplies will not meet the requirements for projectile fuze timers.

Projectile fuzing design timers can be divided into the following categories:

(1) Timers for missile fuzes (approximately 300-mile range)

(2) Timers for rocket fuzes

(3) Timers for projectile fuzes.

Maximum flight times of the fuzes are 3 min. The volume available for the power supply depends on the size of the military item. Based on past requirements for proximity fuzes, Table 11-2[14] lists these sizes and presents typical power supply dimensions. The setback and spin requirements vary widely for these vehicles, while the range of temperatures for operation and storage is the common military range. Table 11-3[14] summarizes these environmental requirements while Table 11-4[14] shows the typical electrical requirements and indicates that the power requirements are relatively uniform.

Most power supplies for fuzing applications are activated on fuze setting. It would appear

TABLE 11-3

ENVIRONMENTAL REQUIREMENTS

| Fuze | Setback, g | Spin, rps | Temperature Range, °F |
|---|---|---|---|
| Missile | 25 | 5 | −65 to +160 |
| Rocket | | | |
| "Nonspin" | 100 | 5 | −65 to +160 |
| Spin | 100 | 100 | −65 to +160 |
| Projectile | 27,000 | 500 | −65 to +160 |

sensible to lighten the load on the power supply by activating it at the time of launch or firing, thus requiring a shorter life. However, typical supplies do not have well controlled turn-on times, and they would thus influence accuracy of the overall fuze timer. For a very slow missile with a velocity of 200 fps, a 50-msec uncertainty in turn-on time would amount to a 10-ft uncertainty in position. With higher velocity projectiles the problem gets worse.

The required active life of the power supply is of major concern. Two major types of digital timer are in use or under development, each having different life requirements for power supplies. One type requires that power be supplied continuously to the timer from the time of fuze setting until projectile detonation. Integrated flip-flop circuit timers have this requirement. The other type, principally timers of the memory core type, requires power only from firing or

TABLE 11-4

ELECTRICAL REQUIREMENTS

| Fuze | Voltage, V | Current, mA |
|---|---|---|
| Missile | 12–15 | 30–50 |
| Rocket | 6.7 ± 0.7 | 100 |
| Projectile* | 6–15 | 30–100 |

*Summary of three developments.

launch time until detonation. Between setting time and launch or firing time, the only requirement on the power supply is open circuit voltage.

Thus timers using memory cores instead of flip-flop circuits are preferred because power supply requirements are less stringent. In addition, electronic timer systems that require power from time of fuze setting until projectile detonation are not compatible with several types of artillery missions in which time from fuze setting until firing is not well controlled, e.g., sustained barrage and surveillance firing at specific targets. Missions of this type often lead to the problem of determining when the power supply has adequate power to provide detonation on target. Fuzes usually are designed to permit ready replacement of the power supply thus providing a means of insuring adequate power if a low power condition is suspected or indicated.

The power supplies for projectile fuze timers are single electrochemical or stacks of such cells. In practice, the number of stacked cells required for a given electronic timer depended, in the past, on the electrochemical potential difference of the individual cell type and the voltage required to operate the electronic timer. For instance, an electronic timer requiring 5.5 V for operation would need at least four stacked cells for a 1.5-V cell and at least five cells for a 1.1-V cell. The recent development of transistorized DC to DC converters makes a second approach to the problem possible. With such a converter, a single cell can supply the total required power at the correct voltage. Although this approach is presently more expensive, it is considered superior from a reliability viewpoint because, in general, it has fewer parts and is comparatively easy to construct.

Electrochemical systems for power supplies usually are subdivided into two groups, active and reserve. The active group includes the common flashlight battery and the common automobile battery. The distinguishing characteristic is that the electrochemical system of the battery is assembled or mixed during manufacture. In the reserve systems, in contrast, the mixing of assembling of the necessary components is accomplished at the time when voltage is required. The reserve system is best represented by thermal batteries in which the heat necessary for the proper electrochemical reaction is provided by a pyrotechnic mix at battery activation time. Another good example of the reserve system is the available car battery in which the electrolyte is not added until the battery is sold and placed in the customer's car. The primary advantage of the reserve systems is long shelf life.

A composite of manufacturers' literature and laboratory data is shown in Tables 11-5[14] and 11-6[14] giving the characteristics of various electrochemical systems.

As a note of caution it should be observed that although recent development in printed circuits, transistors, solid-state junctions and components, and integrated circuitry has allowed performance of fuzing functions with less and less power per component, added sophistication of fuzes and the tasks they are called upon to perform have resulted in keeping the power requirements the same. Hence, the power supply specification problem is as important as ever and seems destined to remain so.

TABLE 11-5

PRIMARY POWER SUPPLY SINGLE CELLS

| Electrochemical System | Temperature Range, °F | Size | | Volume, in.³ | Commercial Designation | Setback, g | Spin, rps | Remarks | Shelf Life at 70°F, yr |
|---|---|---|---|---|---|---|---|---|---|
| | | L, in. | D, in. | | | | | | |
| Leclanche (carbon-zinc) | 35 to 140 | 1.83 | 1.03 | 1.56 | C | Unknown | Unknown | Marginal at high extreme | 1 |
| Alkaline-MnO₂-Zinc | 0 to 140 | 1.83 | 1.03 | 1.56 | C | Unknown | Unknown | Marginal at extremes | 1 |
| Mercury | 0 to 160 | — | — | 2 | None available | ca 20,000 | Probably 250 | Marginal at low extremes | 2 |
| Thermal (molten-electrolyte) (Reserve) | −65 to 160 | 3 | 3 | 21 | None available | 15,000 | 100 | Life marginal | 20 |
| Pb-PbO₂-Fluoboric Acid (Reserve) | −20 to 160 | — | — | 2 | None available | 20,000 | 400 | — | 10 |

**TABLE 11-6**

**PRIMARY POWER SUPPLY MULTICELLS**

| Electrochemical System | Temperature Range, °F | Size, in.³ | Commercial Designation | Setback, g | Spin, rps | Remarks | Shelf Life at 70°F, yr |
|---|---|---|---|---|---|---|---|
| Mercury | −20 to 140 | 2 | None Available | 20,000 | probably 250 | Marginal at extremes | 2 |
| Pb-PbO₂-Fluoboric Acid | −20 to 160 | 2 | None Available | 20,000 | 400 | | 10 |
| Ag₂O-Zn-KOH (commercially available) | 0 to 160 | 2 | None Available | 20,000 | 250 | Marginal at high extremes | 2 |

# REFERENCES

1. J. M. Shaull, *Precision Oscillators for Military Electronic Timers,* Report TR-1235, U S Army Harry Diamond Laboratories, Washington, D.C., 15 June 1964.

2. Report, U S Army Harry Diamond Laboratories, Washington, D.C., 1973.

3. D. N. Shaw, *A Description of the Use of Semiconductor Integrated Circuits for Artillery Timers and Other Ordnance Material,* Report TM-1355, Picatinny Arsenal, Dover, N.J., February 1964.

4. G. W. Kinzelman, "Accurate Electronic Time Fuze (XM587)", *Proceedings of the Timers for Ordnance Symposium, Volume II,* U S Army Harry Diamond Laboratories, Washington, D.C., 15-16 November 1966, pp. 231-45.

5. M. J. Katz, J. M. Shaull, and J. W. Miller, Jr., "A Precision 10-Hz Clock for Military Electronic Timers", *Proceedings of the Timers for Ordnance Symposium, Volume II,* U S Army Harry Diamond Laboratories, Washington, D.C., November 1966, pp. 209-30.

6. A. J. Meyerhoff, *Digital Applications of Magnetic Devices,* John Wiley and Sons, Inc., New York, 1960.

7. *An Accurate Two-Channel Remote Settable Timer,* Report AFWF-TR-66-77, Air Force Weapons Laboratory, Kirtland Air Force Base, N.M., May 1967 (AD-814 432).

8. O. T. Dellasanta and G. W. Kinzelman, *Electronic Programmer System for XM471 Projectile,* Report TR-1242, U S Army Harry Diamond Laboratories, Washington, D.C., 20 July 1964.

9. Edward W. DeKnight, "Modular Electronic Timer", *Proceedings of the Electronic Timers for Ordnance Symposium, Volume I,* U S Army Harry Diamond Laboratories, Washington, D.C., 30 June 1965, pp. 59-72 (AD-469 661).

10. A. Kaufman and H. Newhoff, *Ceramic Memory for Ordnance Fuzing,* Report TR-250(D)-1, U S Army Harry Diamond Laboratories, 26 February 1966 (AD-479 362).

11. M. J. Katz and G. W. Kinzelman, *A Magnetic Core Arming Programmer for the Copperhead II Fuze System,* Report TR 1215, U S Army Harry Diamond Laboratories, Washington, D.C., 20 March 1964 (AD-600 445).

12. O. T. Dellasanta and G. W. Kinzelman, *Electronic Programmer System for XM471 Projectile,* Report TR 1242, U S Army Harry Diamond Laboratories, Washington, D.C., 20 July 1964.

13. D. L. Dickerson, *A 200-Second Magnetic Coretimer,* Report PATR 3225, Picatinny Arsenal, Dover, N.J., February 1965 (AD-614 564).

14. B. L. Davis, "Power Supplies for Electronic Timers", *Proceedings of the Electronic Timers for Ordnance Symposium, Volume I,* U S Army Harry Diamond Laboratories, Washington, D.C., 30 June 1965, pp. 101-13 (AD-469 661).

# PART THREE — MECHANICAL TIMERS

## CHAPTER 12

## INTRODUCTION

### 12-1 GENERAL SYSTEMS

Mechanical timing systems are used extensively in military applications in the function of programmers, sequencers, clocks, and fuzes. These military systems use timers in two forms: (1) the purely mechanical systems in which the time base is a mechanical device such as a tuned or untuned escapement; and (2) hybrid systems that use crystals, tuning forks, or electronic circuits as the time base to control the beat rate of the escapement. It is a basic premise of this handbook that timing systems be classified according to the functional nature of their time base; therefore, the latter hybrid systems are discussed elsewhere in this handbook, and this chapter will limit its discussion to those systems having a mechanically operated time base.

### 12-1.1 BASIC MECHANICAL TIMER ELEMENTS

The basic mechanical timer systems are comprised typically of various combinations of four components:

(1) The power supply or source of energy

(2) An escapement (the time base) that periodically releases the energy, maintains oscillations, and in some systems actuates the computing elements

(3) The computer or counting element which totals the oscillations of the escapement and provides an output signal when a predetermined number of counts has been received

(4) An output such as a settable readout, a switch which closes, a cam which rotates, or a means of producing a firing signal.

Fig. 12-1 diagrams a typical example of a military timer containing these four components.

### 12-1.2 FUNCTIONS OF MECHANICAL TIMERS

Mechanical timers perform control functions in many military applications, and methods of performing these control functions are almost as varied as the requirements themselves.

#### 12-1.2.1 Interval Timers

Interval timers are the class of timers which control the time that elapses before an event takes place. An example of this type is an escapement timer as shown in Fig. 12-2[1]. The power supply is a mainspring driving the output shaft; the time base is a tuned two-center escapement; the counter is a gear train; the output is a cam-operated switch; the activator holds the escapement until removed.

A specialized group of interval timers are mechanical time fuzes. Mechanical time fuzes are placed in a separate category because they represent a major usage class, and because they are required to operate under a much more severe set of environmental conditions than are the other types of mechanical timers. Examples of these types of timer are shown in Figs. 12-3[2] and 12-4[3].

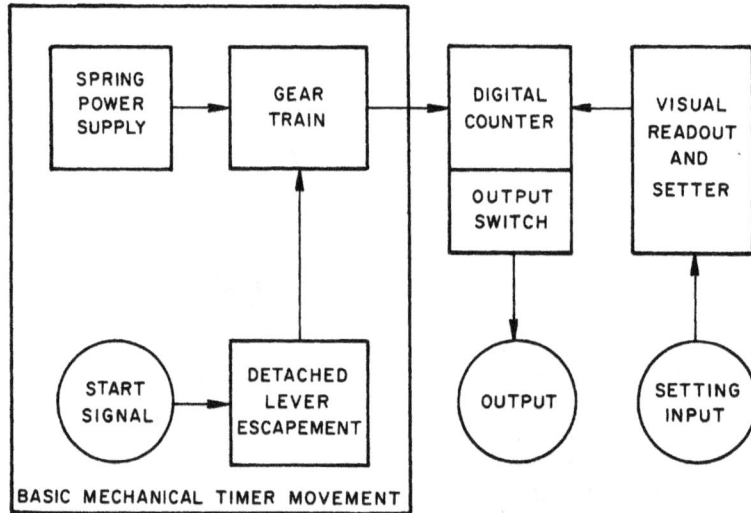

Figure 12-1. Military Timer Concept

### 12-1.2.2 Programmers and Sequencers

Programmers and sequencers are mechanical timers that control the time interval relating a series of events, control the sequence of events, or control the duration of the events. The typical function of these timers is to provide for the completion or interruption of electrical circuits at a predetermined time or for a predetermined interval.

### 12-1.2.3 Clocks

Clocks are a class of timing devices which are operated so as to have some absolute reference to time. As a general rule, the mechanical clocks that are used by the U.S. Army have as their output some type of read-out device. The clock may be used to relate to absolute time, to indicate elapsed time, or to control electrical circuits with relation to absolute time.

### 12-1.3 ACCURACY

In the use of an interval timer for a military or space application, accuracy is an important characteristic that must be given serious consideration. For field use the highest accuracy ($\pm 0.1\%$) is achieved by the crystal, the $LC$ or the tuning fork oscillators. Some mechanical timers (the spring-mass systems) provide better than $\pm 0.5\%$ accuracy. The mechanical verge and the $RC$ oscillator provide accuracies of $\pm 5\%$ in common applications. Mechanical systems under the proper conditions will supply fairly good accuracy. Electronic timers have a reputation for accuracy, usually based on the use of crystal controlled oscillators for the time base or on the $LC$ oscillator (in which higher accuracies than are possible with the common spring-balance are obtained). The spring balance, however, frequently approaches the accuracy of the $LC$ oscillator in actual use because drift in diode discharge devices degrades the $LC$ accuracy.

Increased accuracy generally costs more due to the need for precise components.

Errors in mechanical timers which contribute to lack of accuracy are those of the escapement, counter, setter, and output.

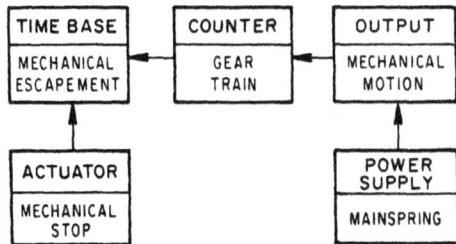

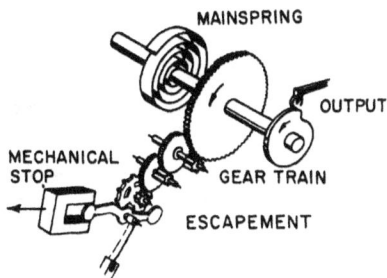

Figure 12-2. Mechanical Interval Timer[1]

Overall accuracy of the escapement will depend upon the sum of the effects that tend to destroy isochronism.

One of the principal sources of error is the interaction between the spring-mass and the ratchet during the period when energy is transferred between them to maintain oscillation. The disturbance should be as small as possible, should occur at the point of maximum kinetic energy, and should take place as quickly as possible in order not to impair accuracy.

Environmental conditions also affect timer accuracy. Special modifications must be made when the timer is to operate at extremely low or high temperatures, or under vibration or shock. Generally, operation over a temperature range of −10° to 120°F is no problem; however, a much wider operating range often is required.

Fluid timers, wherein the motion is regulated by a gas or liquid (see par. 16-7), are

not as accurate as clockworks but have found application in arming delays of fuzes.

## 12-1.4 EXAMPLES OF MECHANICAL TIMERS

### 12-1.4.1 Interval Timers

Interval timers are mechanical timers in which the escapement is used to provide for measurement or control of a single time interval. A diagram illustrating this type of timer is shown in Fig. 12-2. The mechanism may be arranged so that the interval to be measured is variable. The timer may be one-shot or resettable.

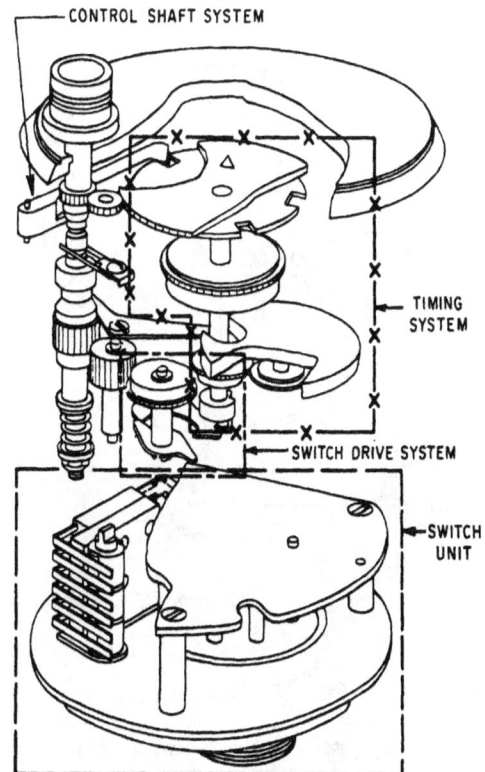

Figure 12-3. Diagrammatic View of XM4 Interval Timer

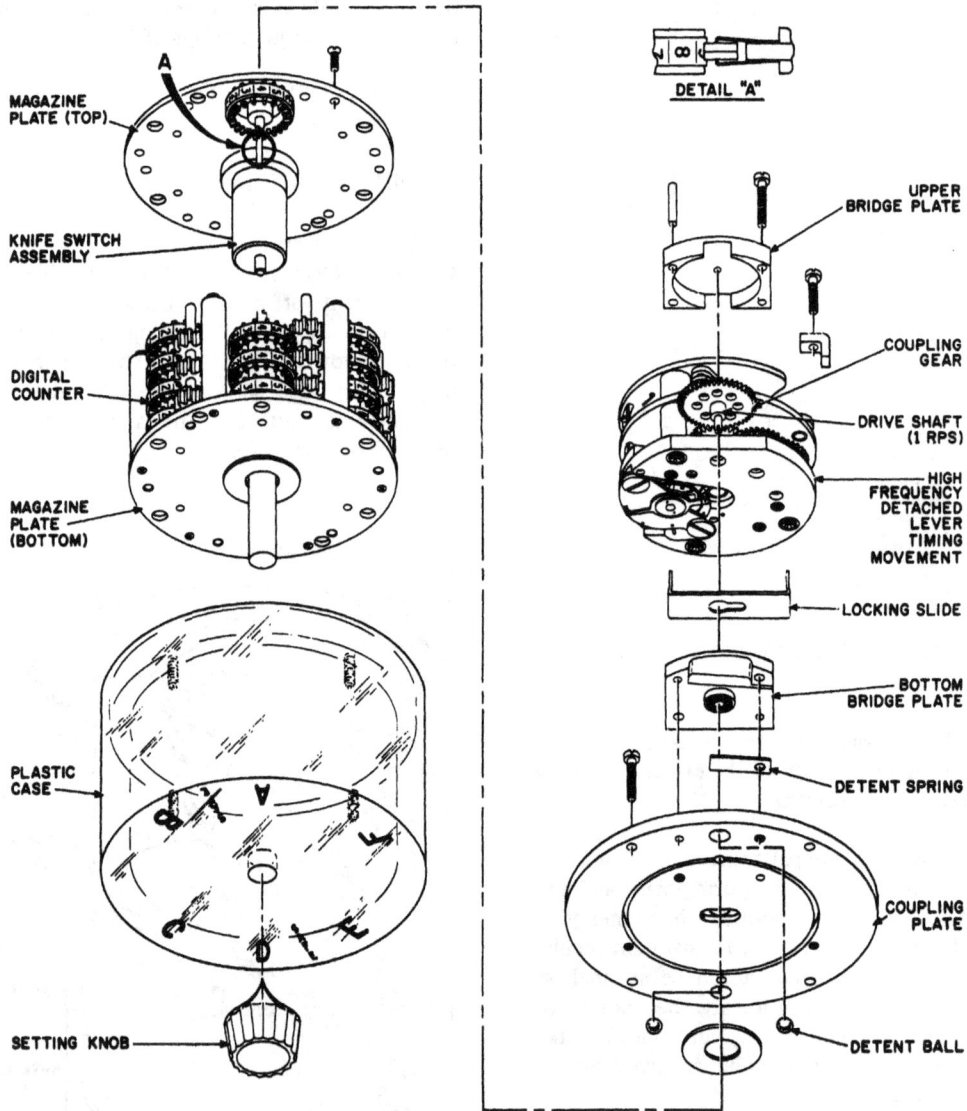

DETAIL "A"

MAGAZINE PLATE (TOP)

KNIFE SWITCH ASSEMBLY

DIGITAL COUNTER

MAGAZINE PLATE (BOTTOM)

PLASTIC CASE

SETTING KNOB

UPPER BRIDGE PLATE

COUPLING GEAR

DRIVE SHAFT (1 RPS)

HIGH FREQUENCY DETACHED LEVER TIMING MOVEMENT

LOCKING SLIDE

BOTTOM BRIDGE PLATE

DETENT SPRING

COUPLING PLATE

DETENT BALL

*Figure 12-4. Exploded View of a Mechanical Timer*

## 12-1.4.2 Fuzes

A fuze is a device with explosive components designed to initiate a train of detonation and to function an item of ammunition at the time and under the circumstances desired. A mechanical time fuze is activated by a clocklike mechanism preset to the desired time. It performs this mission when attached to ammunition that may be traveling towards a target or that may be awaiting the approach of a target. In either case, a stimulus is required to trigger the operation of the fuze.

12-4

Figure 12-5. Electromechanical Program Timer for Adaption Kit of Ballistic Missile[4]

It is obvious that the foundation of the mechanical time fuze system is the accurate measure of time. Timing mechanisms used in fuzes are predominantly tuned two-center escapement and gear-train combinations. An example of a mechanical time fuze is shown in schematic form in Fig. 12-9.

### 12-1.4.3 Programmers

An example of a mechanical programmer is shown in Fig. 12-5[4],* with a schematic shown in Fig. 12-6[4] *. It originally was developed for U S Naval Ordnance Laboratory as part of a ballistic missile adaption kit. Once triggered by a momentary electrical pulse at an appropriate point in the missile trajectory, this device runs for approximately 1 min, operating electrical switches in accordance with a predetermined program. Once initiated, the timer requires no further external power, performing its function with considerable accuracy through all anticipated flight environments.

The mechanism consists of a tuned two-center escapement and gear train which controls the rate of a spring-driven output shaft to which electrical circuit disks are secured.

---

*Reprinted by courtesy of ORDNANCE Magazine, Washington, D C.

Most mechanical timers have a major disadvantage; if they are to be tested, some sort of physical access must be provided during component and system test to allow for rewinding of the power spring. Such a requirement is not only inconvenient, but it also prevents the device from being hermetically sealed against humidity and external contamination.

To overcome this problem, a unique

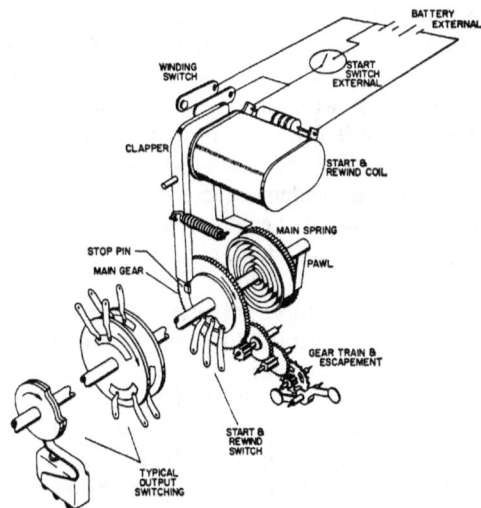

Figure 12-6. Schematic of Mechanical Timer With Electrical Rewind Circuit[4]

12-5

feature was developed. The timer is started by electrically activating an integral solenoid that releases the main shaft. Power for testing is supplied through a rewind circuit to this same coil through an internal flip-flop switch. During the timer duty cycle, one of the circuits programs this power to the solenoid for several seconds, rewinding the spring for the following cycle. The power spring has enough energy for several successive cycles without such rewinding and with no loss of accuracy. Other features of this timer are typical of most military timers:

(1) Exceptionally rugged, having a high tolerance for shock, vibration, and acceleration

(2) Almost totally unaffected by large doses of nuclear radiation

(3) Unaffected by temperature or pressure variations

(4) Small, light and inexpensive.

### 12-1.4.4 Sequential Timers

In sequential timers a mechanism is provided which will permit a time controlled series of events to occur after a specified time delay. Most sequential timers provide a series of outputs with independently adjustable time intervals from one output to the next. The initiation of each succeeding time element is dependent upon completion of the preceding output. Length of the total cycle is the sum of the individual elements.

Accuracy of short intervals following long delays can be increased by connecting timers in tandem. For example, suppose a 30 ± 3 sec interval once every 24 hr ± 15 min is required. By using a mechanical timer with 1/2% accuracy, the delay specification can be satisfied within 8 min, comfortably below the requirement. To generate with the 24 hr timer 30 ± 3 sec requires a minimum interval accuracy of 0.035%. A simpler way to achieve

this accuracy is to actuate a 30-sec interval timer of 10% accuracy. The load switch of the 24-hr delay timer is used to start the motor of the 30-sec interval timer that in turn activates the load.

### 12-2 MILITARY APPLICATIONS

Mechanical timer systems are used extensively by the US Army in fuzes for various weapons. They are used to accomplish the following functions in a fuze:

(1) Provide an arming delay; where the arming process is controlled so that the fuze cannot function until the projectile is a safe distance from the launching site.

(2) Provide an impact delay; where it is required that the fuze function within a specified time interval after striking a target.

(3) Provide a functioning delay; where it is required that the fuze function a given time interval after launching or emplacement.

(4) Provide self-destruction or sterilization; where the end of the useful life of a munition can be recognized by the time interval after launching or after some other signature.

In most fuze applications, the final action of a timing mechanism will either release another mechanism, or line up an explosive train, and set off a detonator. For details on the design of fuzes see Ref. 5 and for details on mechanical timers, Chapter 13.

### 12-2.1 GUIDED MISSILE FUZES

Another application of a mechanical timer is in guided missile application where fuzes perform an arming function and a firing function as do other fuzes. The term for the separable arming device is the Safing and Arming (S&A) mechanism.

Since a guided missile is a large, expensive item, a high probability of functioning is

FLIGHT DIRECTION

LATCH — — SLIDER

OPENING — — DETONATOR
TO BOOSTER

STOP

FROM
CLOCKWORK — RETURN SPRING

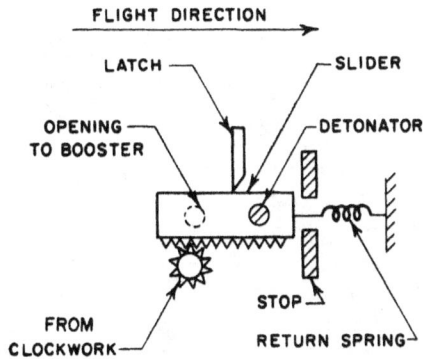

*Figure 12-7. Safing and Arming Mechanism*

required. Accordingly, multiple fuzing often is employed. The advantage of the multiple units is that the probability of functioning (success) increases exponentially.

The environmental and operational conditions found in guided missiles require the design of fuzes, S&A mechanisms, power supplies, and other components which are peculiar to missile applications. For example, the acceleration of missile upon firing is limited to a value of about 60 g; therefore, the arming mechanism must be designed to operate with this acceleration. Although a wound spring might be used as a source of power to provide for arming a fixed time after launch, as a general rule, any arming system that uses stored energy is considered undesirable from the safety standpoint. Perhaps the best type of arming mechanism for these low accelerations involves a time-acceleration integrator that senses the true launching environment.

Suppose an arming device is required for a hypothetical missile that has the following characteristics: (1) it will arm when under an acceleration of 11 g if this acceleration lasts for 5 sec, and (2) it will not arm when under an acceleration less than 7 g for a period of 1 sec. Consider the arming device shown in Fig. 12-7[6]. Setback forces encountered during acceleration of the missile apply an inertial

force to the slider. Thus after a specified time, the detonator will be aligned with the booster and the latch will drop down to lock the slider in the armed position. If at any time during this process the acceleration drops below 7 g, the slider will be returned to its initial position by a return spring. Because of its weight (if not damped), the slider would move too fast under these accelerations; hence, a restraining force is necessary. Therefore, an untuned escapement is used to regulate the motion of the slider. It turns out that this arrangement results in a double integrating accelerometer; i.e., the displacement of the slider is related to the distance the missile has travelled.

An example of this type of mechanism is the Safing and Arming Device, GM, M30A1 shown in Fig. 12-8[5]. This device is, of course, much more refined than the example cited, although the performance parameters for the example were taken from this device.

## 12-2.2 ARTILLERY FUZES

The mechanical time fuze provides a preset functioning time for an artillery projectile. Fig. 12-9[6] shows the complete timing movement for Fuze, MTSQ, M564. This fuze can be used only in spin-stabilized projectiles because centrifugal force is used to release the escapement lever and permit operation of the timing mechanism. The fuze could be used, for example, on a projectile fired from a 105 mm howitzer.

When a projectile with the M564 Fuze is fired, setback force resulting from the acceleration of the projectile drives the hammer weights of the hammer spring assembly back against the setting lug on the timing disk of the movement. This frees the timing mechanism from the setting pin. Centrifugal force, caused by rotation of the projectile in the weapon, moves the arbor-stop lever outward and unlocks the arbor of the timing mechanism. Simultaneously, the detents move to free the escapement lever.

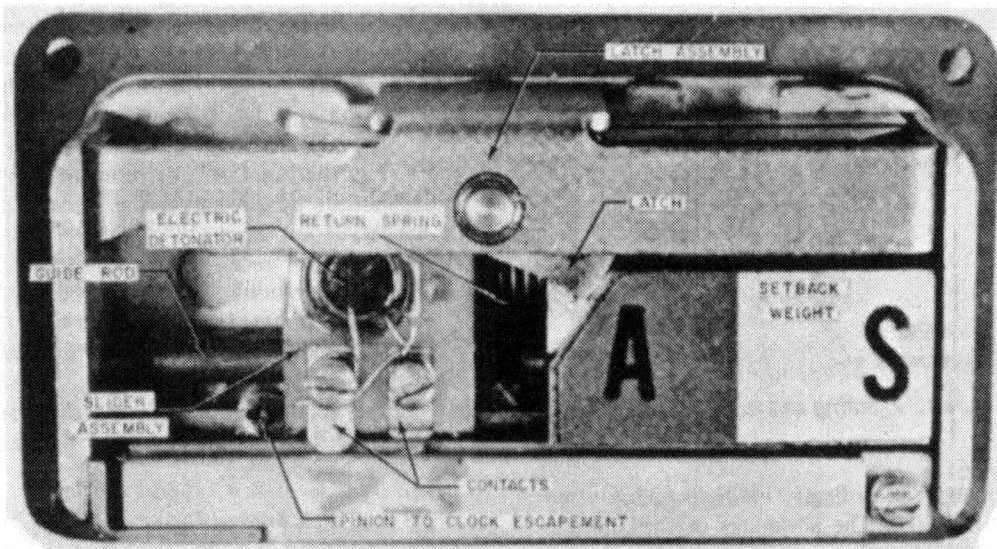

*Figure 12-8. Safing and Arming Device, GM, M30A1*

The timing movement begins to operate.

The movement is a spring-driven clockwork, which can be set to control a mechanism for releasing the firing pin. When the clockwork is assembled and stored, the driving spring is locked in the wound position by a spring-loaded arbor-stop lever that swings outward against the resistance of its spring when acted upon by sufficient centrifugal force. With the stop lever in the outward position, the arbor is free to turn, and the spring can unwind to drive the gear train. At the set time, the firing arm drops into the firing notch on the timing disk, releasing the firing pin that strikes the detonator. The detonator fires, igniting the explosive train.

### 12-2.3 ARMING DELAYS

A typical fuze for 81 mm mortar ammunition is Fuze, PD, M525, as shown in Fig. 12-10[5]. The M525 is a superquick, point-detonating fuze that has been quite successful because of its relative simplicity. It consists of two major parts:

1. A head assembly that contains a striker, firing pin, and a clockwork for delayed arming. The striker with tapered striker spring is designed especially to permit the fuze to be fully effective when impact is at low angles.

2. A body that contains the arming mechanism (a slider), detonator, lead, and booster pellet.

The fuze has two pull wires, connected by a cord for easy withdrawal, that remove two setback pins that lock the fuze in the unarmed position to insure safety during storage, transportation, and handling. The wire is pulled with the round immediately over the mortar tube and just ready for release.

Operation is as follows: upon firing, acceleration of the projectile produces setback forces that cause the setback pin to move to the rear. The safety pin is released as

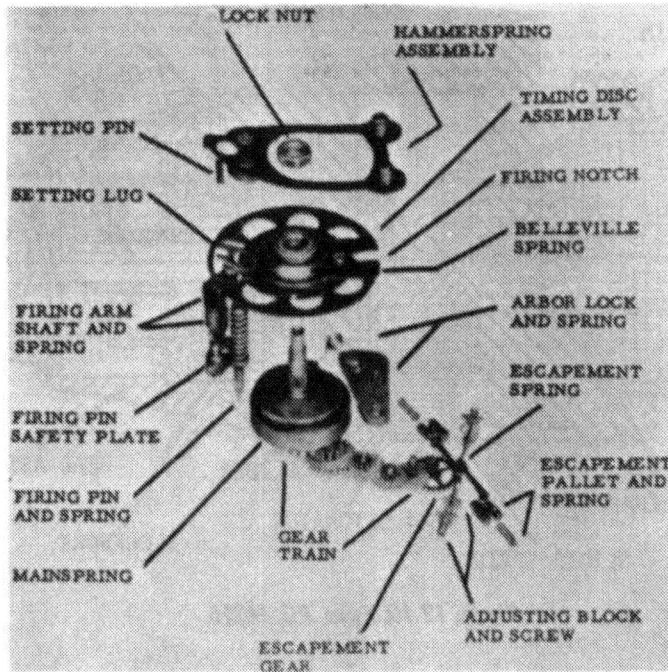

*Figure 12-9. Timing Mechanism of Fuze, MTSQ, M564*

a result of this motion so that the spring on the safety pin pushes it outward. Provided the projectile is within the mortar tube, the pin rides on the bore. Since the slider therefore still is retained from moving, the fuze is bore safe. The pin is thrown clear of the fuze when the projectile emerges from the muzzle. The firing pin in its rearward position is in the blank hole of the slider so as to act as a second detent on the slider.

Setback also frees the escapement to start the clockwork in the head assembly. At the end of the 3-sec arming delay, a spring causes forward motion of the firing pin, causing it to withdraw from the slider. The slider, then, is prevented from moving until both (1) the projectile clears the tube, and (2) the clockwork runs down.

When the slider is free to move, the detonator in the slide is aligned with firing pin and lead. Upon target impact, the stricker pushes the firing pin into the detonator. The detonation sets off the lead and booster.

## 12-3 REQUIREMENTS

Some of the basic requirements for a military timer are compactness, low power consumption, low cost, high accuracy, and resistance to shock. Other related factors include simplicity, ruggedness, reliability, environmental compatibility, and resistance to vibration.

Simplicity and ruggedness are really sub-headings under the factor of reliability. Simplicity means nonintricate and few components parts, and a minimum need for precision assembly and adjustment. Ruggedness usually follows from simplicity. It means

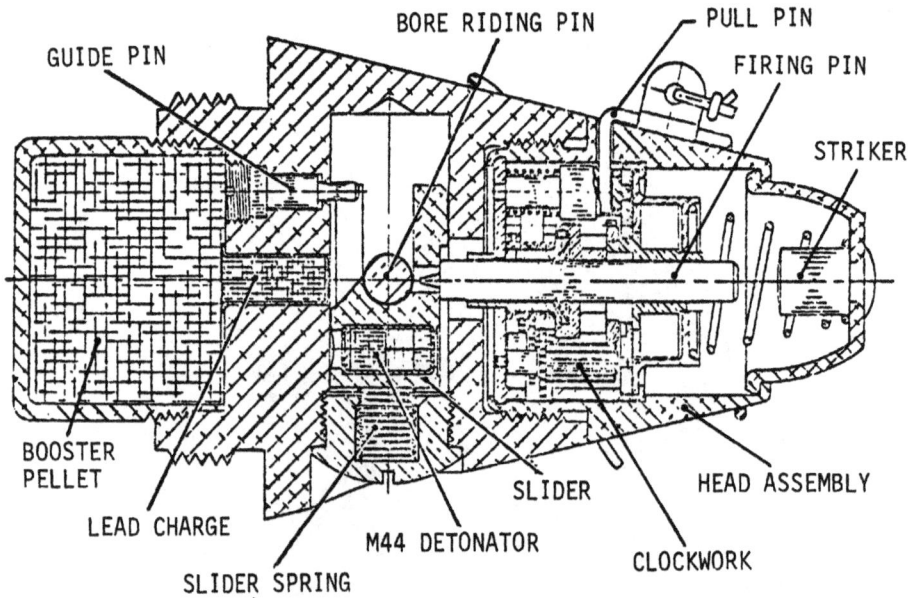

*Figure 12-10. Fuze, PD, M525*

the ability to withstand and still operate successfully after prolonged storage and exposure to extremes of temperature, humidity, shock, vibration, etc. without the need for periodic maintenance. In this respect, a rugged timer would also be a reliable timer. In regard to ruggedness, the detached lever (tuned three-center) escapement—with its delicate pivots and precision jeweled bearings (if used) that require periodic lubrication, and dirt-free environments and the fine hairspring—does not tend to be very rugged. Shock-mounted bearings and dry lubricants can make this type of escapement more adaptable to military requirements and it has been used with moderate success in bomb fuze timers. In addition, in order to survive high shock loadings, components having a high strength-to-weight ratio often are used.

Actual performance requirements that the various mechanical timers must meet vary with the system in which the timer must operate. A typical set of requirements for an experimental mechanical interval timer are:

(1) Unit to be designed to provide output with an accuracy of ± 0.2 sec

(2) Output selected from six possible firing intervals between 40 and 120 sec with one-knob control

(3) All-mechanical system with volume approximately 6 in.$^3$

(4) Operate over a temperature range −40° to +160°F.

Requirements for a typical time fuze (M564) are:

(1) The unit is to be designed to provide an output between 2 and 100 sec. Timing error is not to exceed 0.18% while spinning at the specified rpm.

(2) Time Range Setting:

Method               fuze setter or wrench

| Maximum | 100 sec |
| Minimum (prescribed) | 2.0 sec |
| Divisions | 1.0 sec |
| Setting accuracy | 0.1 sec |
| Setting torque | 90-110 lb-in. |
| Setting direction | clockwise |

(3) System: all mechanical

(4) Temperature Limits: $-40°$ and $52°C$

## 12-4 AUXILIARY EQUIPMENT

### 12-4.1 TORQUE REGULATORS TO COMPENSATE FOR MAIN SPRING RUN DOWN

The beat rate of mechanical timing mechanisms varies to some degree with input torque. This effect is known as torque sensitivity[7]. Because of this torque sensitivity, inherent to some degree in all mechanical escapements, many devices for regulating the torque delivered to the escapement have been developed. Among them are the gravity escapement, the fuzee, and the remointoire or constant-force device. They are attempts to improve on accuracy by minimizing inherent torque variation in the power supply.

A gravity escapement is an accurate clock escapement in which the mainspring or driving weight lifts a small lever, which drops to give an impulse to the pendulum or balance, and is thus independent of the driving force. The fuzee is a conical, spirally-grooved pulley in a time piece from which a cord or chain unwinds onto a barrel containing the spring, and which by its increasing diameter compensates for the lessening power of the spring. The remointoire is a device by which the clock train does not drive the escape wheel directly but rather

through an auxiliary mechanism—such as a small spring, or weight, or a DC motor which drives the wheel of the ordinary escapement. Except for the remointoire, these regulators are attempts to compensate for the torque variation due to mainspring run-down.

### 12-4.2 TORQUE REGULATORS TO COMPENSATE FOR VARIATION IN OUPUT LOADING

In many military applications, instead of having a small fixed load (such as minute and second hands), the timer is asked to throw switches, to turn cams, or to bring rotors into alignment. These functions not only impose relatively large loads on the timing movement but also the loads are often variable throughout the cycle and from one timer operation to another. Therefore, the escapement sees a relatively large variation in torque resulting from the difference between the torque available and the torque applied to the load.

An example of a torque regulator that allows the power supply to deliver varying amounts of torque for external work while maintaining a constant torque to the escapement is the remointoire. The principle of operation is based on the age-old capstan (Fig. 12-11[7]). The input torque is applied to one

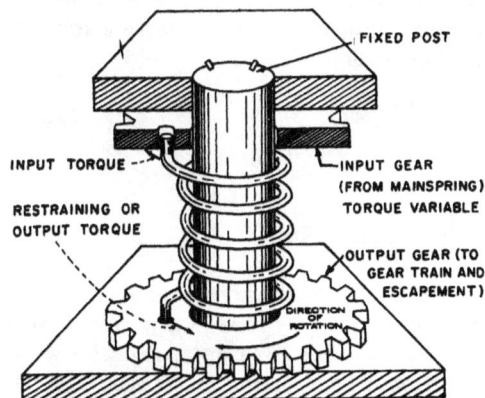

Figure 12-11. Capstan Torque Regulator

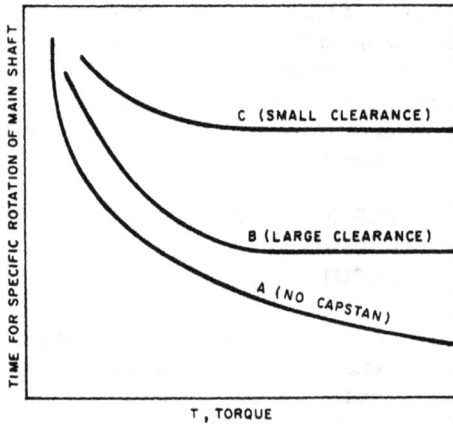

*Figure 12-12. Effect of Capstan Regulator on Torque-Time Curves for Runaway Escapements*

end of the torsion spring, and the output is obtained from the other end. The ID of the coil is only slightly larger than the OD of the fixed shaft. If the output gear is restrained by an increased load, the input torque causes the coil to deflect angularly, thereby decreasing the ID until contact is made on the fixed shaft, at which time the difference between output and input force follows the torque amplification behavior characteristic of the capstan. For example, in a runaway escapement the time $t$ required for a specific angular rotation of the main shaft is approximately inversely proportional to the square root of the torque, or

$$t \approx \sqrt{\frac{1}{G}}, \text{ sec} \qquad (12\text{-}1)$$

where

$G$ = torque, oz-in.

This is plotted as curve A of Fig. 12-12[7] and represents the case when no torque regulation is used. If a capstan regulator were added with large clearance between coil ID and shaft OD, the curve marked B would result. If a small initial clearance were used,

curve C would result. The flat ranges of both B and C indicate those portions where the capstan coil is tightly wrapped about the fixed shaft. These curves show that the running rate is constant for relatively large ranges of torque. The capstan regulator is simple, rugged, and inexpensive, and thus is suitable for military applications.

## 12-4.3 AUTOMATIC DYNAMIC REGULATION MACHINE

Some form of escapement adjustment is required to correct for escapement beat rate variations due to small construction differences. In the pallet type escapement (shown in Fig. 16-27) the stiffness of the escapement spring can be changed by movement of the adjusting nut along the spring. This is done by turning small adjusting screws. Another possible complication is that a particular beat rate is required not at zero spin, but at some rpm close to what the escapement will have when operating in a spin-stabilized projectile. To speed up this adjustment operation of the production phase, an automatic dynamic regulating machine has been designed and is presently in operation at Frankford Arsenal. The equipment works in the following manner[8]:

(1) The fuze movement is held securely in a suitable chuck and allowed to run while being rotated about its geometric center at a rotational velocity simulating free flight conditions.

(2) As the movement is being rotated at constant speed, a beam of light is intercepted by the escapement lever. This beam of light emanates from a light source located directly above the chuck and on the same center line. The light passes through an opening in the movement, is reflected from an inclined surface on one of the movement plates, and leaves in a radial direction. In passing through the movement, the beam is interrupted by one wing of the escapement lever as the lever oscillates. The light leaving the movement

becomes a blinking signal that is pulsating at exactly the same frequency as the escapement. This radial beam then strikes an annular mirror that surrounds the rotating chuck and is reflected to a photoelectric cell. The signal generated by the photocell has a frequency identical to the movement frequency.

(3) The photocell signal is amplified and sent to the printer where it is electromechanically compared with a signal received from a crystal controlled frequency standard. When the movement frequency is correct, the printer records a straight vertical line of dots on a paper tape.

The automatic dynamic regulation machine provides for adjustment of the movement while it is rotating in the chuck. In this sense the automatic machine is a true regulator because it both performs the operation and gives a visual indication of the result. The automatic dynamic regulator is basically the same as the standard machine, except for the addition of a servo control and a special chuck.

The machine operates at spin speeds up to 13,500 rpm and has been tested safely up to 26,000 rpm.

Some advantages of this regulating process are:

(1) Increased output with fewer machines

(2) Increased output with less labor

(3) Improved quality through elimination of human bias

(4) Lower level of skill required because the machine performs the adjustment

(5) Increased capability with substantially lower capital investment.

## REFERENCES

1. M. Saltsman, Jr., "Timer Types and Features", Instruments and Control Systems, 38, 83-6 (October 1965).

2. *Development of Interval Timer, XM4*, Army Materiel Command, Technical Info. Report 8-11-1B1(1), November 1962 (AD-446 929).

3. D.L. Overman and D.S. Bettwy, *Experimental Mechanical Timer With Detached Lever Escapement and Digital Readout System*, U S Army Harry Diamond Laboratories, Washington, D.C., Report TM-65-44, 31 August 1965.

4. M. Saltsman, Jr. "Improved Timing Devices", Ordnance, L, 86-8 (July-August 1965).

5. AMCP 706-210, Engineering Design Handbook, *Fuzes*.

6. *Development of Mechanical Time Superquick Fuze, M564 (T197E2)*, Army Materiel Command, Technical Info Report 8-1-1B2(3), January 1963 (AD-444 612).

7. D.S. Bettwy, *A Capstan Type Torque Regulator for Timing Movements*, U S Army Harry Diamond Laboratories, Washington, D.C., Internal Report R-450-62-16, October 1962 (Note—no external distribution).

8. F. Luke, "Development of Firing Device Demolition, XM70," *Proceedings of the Timers for Ordnance Symposium*, Vol. I, U S Army Harry Diamond Laboratories, Washington, D.C., November 1966, pp. 35-43.

# CHAPTER 13

## MECHANICAL TIMER SYSTEMS

### 13-1 SYSTEM DESIGN CONSIDERATIONS

When a new military system is being developed, various design goals are determined at the outset. These goals then are defined for each component so that the total system will result in a product that is reliable, combines maximum accuracy with design simplicity, is ruggedly constructed for successful operation in a military environment, and has a reasonable life cost cycle.

### 13-1.1 DESIGN SPECIFICATIONS

The design specifications are the basis for all development work, whether they relate to an entirely new concept or a modification of an existing unit. By defining the limits within which a product designer may operate, these specifications usually contain: (1) performance factors that must be met, (2) description of associated equipment with which the product must interface, (3) approximate cost to manufacture, (4) operating parameters, and (5) probable environments.

In particular, the requirements for a timer might include:

(1) Delay time range with ability to set in predetermined increments

(2) Accuracy within plus or minus a certain percentage

(3) Operating temperature range −50° to 125°F

(4) Ruggedness, including required operation after impact deceleration of a given number of g's for a given duration in any direction

(5) Military transportation, vibration, environmental, and storage requirements

(6) Maximum space allowable for the unit including the mechanical power supply

(7) Cost goals

(8) Setting considerations, including methods and resolutions.

### 13-1.2 PRELIMINARY CONSIDERATIONS

The designer of a timer must be thoroughly familiar with the bases for the stated requirements. He then is in a position to evaluate the requirements. Should any of the requirements be too difficult, time-consuming, or costly to achieve, he could then give an intelligent proposal to relax one or more of them in order to reduce the development time and cost risk.

Often the most economical approach to the problem would be to use an existing timer—with perhaps slight modifications—to satisfy the present requirements. Failing this, the next step would be to review various types of timer (literature search, patents, in-house designs, etc.) so that a decision could be made as to which type holds the most promise for the design goals.

Many elements can be considered for use in a mechanical timer:

(1) The time base types include balance

wheel and hair spring, pallet mass and reed spring, inertial governor, timing fork, counter oscillating rims, and vibrating beam (wire).

(2) Escapements including detached lever and escape wheel; nondetached lever and escape wheel; runaway escapement; crank and cam; cylinder type, nonlinear gear train; and, for electromechanical timers, mechanical switch and electronic armature, and electronic switch and armature.

(3) Power supplies include the mainspring (either spiral power spring or Negator B type spring motor); the remointoire (a means of rewinding the mainspring to regulate torque to the escapement); and, for electromechanical timers, the battery.

(4) Computing systems include gear train counters that can be either straight spur and pinion sets, epicyclic or hypocyclic train, and the harmonic drive and stepping wheel counter.

(5) The outputs (for display or activation of another device) can be either analog or digital.

An excellent way to evaluate the various alternatives is to tabulate each class of component and then estimate the relative suitability of each factor to meet the design requirements. Such a table for the comparison of time bases is shown in Table 13-1[1]. This table was prepared for the development of a long-delay timer for a bomb fuze, and includes electromechanical and electronic time basis. In that study, each of the timer elements was rated by a weighting factor to aid in judgment. This method will give a clear picture of which type of time base, escapement, etc. most likely will fit the design requirements.

Obviously, new designs must provide more than marginal improvements over old systems or they are economically not worth the effort. These improvements would include increased accuracy, higher reliability, greater capability, decreased cost, greater ease of handling, improved safety, etc.

Even though production or shelf-item components are phased into the design, a good designer will conduct all necessary development tests on the assembled item.

TABLE 13-1

COMPARISON OF TIME BASES

| Time base | Weighting Factor, % | Frequency range | Accuracy | Simplicity | Ruggedness | Size | Cost | State of development | Total |
|---|---|---|---|---|---|---|---|---|---|
| | | 15 | 15 | 25 | 25 | 5 | 10 | 5 | 100 |
| Balance wheel & hairspring | | 85 | 70 | 70 | 60 | 80 | 60 | 90 | 70.3 |
| Pallet mass & reed spring | | 70 | 60 | 75 | 65 | 75 | 65 | 75 | 68.5 |
| Inertial governor | | 70 | 30 | 85 | 70 | 75 | 80 | 70 | 69.0 |
| Tuning fork | | 50 | 80 | 80 | 75 | 65 | 70 | 60 | 71.5 |
| Counter-oscillating rims | | 60 | 80 | 80 | 75 | 70 | 70 | 35 | 72.0 |
| Vibrating beam (wire) | | 50 | 70 | 40 | 75 | 40 | 40 | 10 | 53.3 |
| RC circuit | | 100 | 60 | 70 | 85 | 50 | 60 | 90 | 75.8 |
| LC circuit | | 45 | 70 | 70 | 85 | 40 | 55 | 90 | 68.0 |
| Crystal oscillator | | 20 | 90 | 30 | 75 | 30 | 30 | 85 | 51.5 |

(Reliability spans Simplicity and Ruggedness columns)

This should be done for several reasons:

(1) Different components have different shelf lives.

(2) Mating components may possibly be incompatible.

(3) Coupling effects between old and new items may degrade the performance of the new assembly.

## 13-1.3 ARMING AND FUNCTIONING DE-LAYS

One of the basic decisions the designer must make is whether to use similar or different methods for arming and functioning. Arming is the changing from the safe condition (that of storage, transportation, and handling) to that of readiness for functioning. Functioning is the succession of normal actions from initiation of the first element to delivery of an impulse from the last element of the explosive train.

The selection is based on the same considerations discussed in par. 13-1.2. For some applications, a combination of mechanical arming and electrical functioning may be optimum and has in fact been used. As far as timers are concerned, the combination of an arming clock and impact functioning is common where no functioning delay is desired.

Timing accuracy is the important characteristic that distinguishes these two applications. For functioning, timing must be precise or else the terminal effects of the munition will be degraded.

The usual tolerance on functioning delay timers is usually less than ± 3%. Arming delays are far less critical, provided they insure safety before and functioning after arming. Common arming delay tolerances are usually greater than ± 10%.

## 13-1.4 PROTOTYPE DEVELOPMENT

After a review of the requirements has been conducted and decisions as to trade-offs completed, the designer is ready to begin preliminary designs. The choices might typically be narrowed down to two approaches by this time. When available data are not sufficient to eliminate one of the approaches, prototype hardware development could be tested to aid in the selection of the best approach. The prototypes then can be fabricated and made ready for testing.

## 13-1.5 TESTING

The tests that the prototypes are subjected to simulate or duplicate the forces and environments that the final version will have to survive. The test program is described in par. 13-5.

The failure of either of the two prototypes in any of the tests will entail its elimination from the program. If the performance of either type is marginal, rework or possible complete redesign may be required. The tests then are rerun and, if successful, the design then is reviewed for production methods and tooling requirements.

## 13-1.6 FINAL ACCEPTANCE

After the prototype is proven to be capable of satisfying all requirements, a testing program is initiated which is based on statistical sampling. The proven timer now will be produced in quantity and integrated into the weapon system.

## 13-2 GENERAL FACTORS IN DESIGN

The timer used in military applications is a complex device. Certainly, engineering knowledge is required to design the timer so that it will function within the limits set forth in the specifications while working under severe environmental conditions. There are a number of general design factors that must be considered including reliability, maintainability, human factors engineering, and methods of cost reduction.

## 13-2.1 RELIABILITY

Reliability is the probability that materiel will perform its intended function for a specified period under stated conditions[2]. It is generally defined as a statistical probability at some confidence level. For example, a system may be described as having a reliability of 99 percent with a confidence level of 95 percent. This means that for a specified period of time we would be 95 percent confident that there would be no more than one failure out of 100.

With respect to the design of mechanical timers, several important considerations have a direct bearing on the reliability of the device. These are:

(1) The escapement system has a self-starting capability.

(2) The number of components parts is minimum.

(3) Parts are simple in design and manufacture.

(4) The need for precision assembly and adjustment is minimum.

(5) Parts must withstand and still operate successfully under extreme environmental conditions.

(6) Materials should be used whose important properties are reproducible and well known to the designer.

(7) As far as possible, timers should be designed so that factors which affect reliability can be detected by means of nondestructive tests or inspection.

The usual practice of providing redundant components to increase overall reliability is not always practical, particularly in military applications. Hence, when the overall system reliability is specified, the components must have higher reliabilities because the overall system reliability normally is the product of the individual component reliabilities. Designers of timers should be aware of the characteristics of the system in which the timers will be used and the desired system reliability so that they can determine the limits in which the timer must perform. Tests to determine reliability sometimes are carried out by a group independent of the design organization. Difficulties in interpreting results can be more easily resolved if the designer is familiar with the techniques used by evaluators, uses similar techniques to assure himself that his designs are reliable, and designs devices in which reliability is as nearly inherent in the design as possible.

## 13-2.2 MAINTAINABILITY

In regard to maintainability the design engineer should keep the following objectives in mind[3]:

(1) Improve reliability to reduce the need for maintenance.

(2) Reduce the frequency of preventive (cyclic) maintenance.

(3) Improve maintainability to reduce downtime.

(4) Reduce the logistic burden by making full use of standard parts, components, tools, and test equipment.

(5) Make parts, components, and assemblies interchangeable.

(6) Consider modular replacements.

(7) Reduce the requirements for highly trained specialists.

Maintainability must be designed into a system; it just doesn't happen. This places the burden of solving a large part of the maintenance problem squarely on the shoulders of the design engineer. Every designer should attempt to view the maintainability requirements from the standpoint of the maintenance technician.

In selecting components, the designer first should consider the use of proven parts in order to reduce development time and cost.

Current procedure in regard to handling a defective clockwork in a timer is to remove the defective unit, and substitute a new one; thus design for ease of removal of the clockwork is of paramount importance. In many instances, it does not pay to repair a defective clockwork. Interchangeability then becomes significantly important.

## 13-2.3 HUMAN FACTORS ENGINEERING

Human factors engineering is the scientific union of men with machines so as to enhance the man-machine system. An efficient union requires a knowledge of the performance of both man and the timer parts. The first attempt at understanding a man-machine relationship was made about 1880 and became the "time and motion study" branch of industrial engineering. In a man-machine situation, man is considered the variable with the machine constant. The human factors engineer primarily is concerned with machine or product design to improve the overall efficiency.

When properly applied, human factors engineering reduces the cost of the final product, makes it easier to use in the field, and often means the difference between success and failure. Guidelines for such factors as maximum torque settings, minimum visibility requirements, and other physiological properties have been compiled in reference handbooks[4,5].

For an illustration of human factors engineering applied to time fuzes in tank-fired ammunition, see par. 16-5.2.

## 13-2.4 COST REDUCTION

There are a number of factors which must be considered in the continuing effort to reduce the cost of manufacture of mechanical timers. These include:

(1) Standard parts

(2) Manufacturing tolerances

(3) Mass production

(4) Material substitution

(5) Corrosion preventive finishes

(6) Ease of maintenance

(7) Simplicity

(8) Ruggedness vs accuracy.

The paragraphs that follow review these factors and are aimed to pinpoint the philosophies involved.

### 13-2.4.1 Standard Parts

The decision as to whether to adopt an existing timer to a system design or to design a new timer is often one of the most difficult decisions a designer has to make. A new item often has been developed because, in the layout stage, it took less effort to sketch in something that fit the dimensions than to learn what was available. On the other hand, the hard and fast resolution to use only shelf items often has resulted in systems that are appreciably inferior to the best obtainable with regard to reliability, effectiveness, or compactness, and in the perpetration of obsolete items.

As a general rule, the standard item always must be given first preference and must be considered carefully.

### 13-2.4.2 Manufacturing Tolerances

All timer parts must be properly toleranced following good design practice. Every length, diameter, angle, and location dimension must be given and defined in tolerances as broad as the performance of the part can tolerate to permit the most economical manufacturing procedure. Particularly in high-volume parts, costs rise rapidly as tolerances are made tighter. All fits must be stipulated. These fits should be chosen with primary concern for function and accuracy. Tolerance stackups indicate whether parts can be assembled properly and whether an assembly will operate as expected. Expected user environments, temperature extremes, and their effects upon critical interference and clearance fits must be considered. It is imperative in the development of mechanical timers that tolerance stackup determinations be complete before manufacture of hardware is begun.

Tolerancing affects the interchangeability of parts. Complete interchangeability is desirable whenever feasible. However, in complex mechanisms, such as timers, where components are small and tolerances are critical, complete interchangeability is often impractical. Selective assembly may be used to conform to the tolerance specifications.

### 13-2.4.3 Mass Production

Design for mass production using modern machines, tools, jigs, and fixtures has definite advantages in cost reduction. This, in combination with streamlined methods, have proven to be the greatest factor in making it possible to produce large quantities of parts at substantial savings (see par. 14-2).

### 13-2.4.4 Material Substitution

New material technology is increasing all

the time, and the possibilities of substituting materials such as plastics for metals in gear design (i.e., molded gears) warrants continuing review. It must be borne in mind, however, that requirements of ruggedness in timer parts to resist setback and acceleration forces will be difficult to meet with plastics unless through good design these forces are resisted in a manner such that the gearing is not affected adversely. Savings also can be affected by purchasing materials in large quantities.

### 13-2.4.5 Corrosion Preventive Finishes

The use of corrosion preventive finishes is very important, particularly, for timer assemblies that are stored for future use. If these assemblies are stored for any period of time, the possibility of rust or corrosion build-up can inhibit free running of the timer so that additional work is required to make them ready for use. This increased time and money is prohibitive, and can be avoided by adequate finish specifications.

Closely associated with corrosion prevention is the subject of lubrication. This is discussed in par. 14-2.5.

### 13-2.4.6 Ease of Maintenance

If the timer assemblies are so designed that they can easily be maintained, a great saving will result (see par. 13-2.2).

### 13-2.4.7 Simplicity

Simplicity means nonintricate design, few component parts, and a minimum need for precision assembly and adjustment. Overall cost is a function of simplicity and, of course, of the materials used.

### 13-2.4.8 Ruggedness vs Accuracy

There are a number of accurate timing movements that are manufactured by various clock and watch manufacturers. However, to adapt these to the stringent environmental conditions that military timers require means that various components have to be strengthened, tolerances have to be reduced for cost savings, etc. These decisions require a series of design trade-offs that must be made by the designer so that the timer will perform at the required accuracy. Only through thorough analysis and testing can these problems be resolved.

## 13-3 SYSTEMS FOR HIGH ACCELERATION

A timer used in an artillery projectile experiences high forces due to projectile acceleration. Maximum acceleration occurs while the projectile is still in the gun tube. Setback acceleration can be large, up to 50,000 g, and rotation in spin-stabilized rounds reaches 100,000 rpm. Spin introduces centrifugal, Coriolis, and precessional forces. In addition to setback, the projectile also is subjected to linear accelerations in forward and sideways directions.

A typical time fuze designed to operate under setback and spin forces is the Artillery Fuze M564 (Fig. 12-9). Mortar Fuze M525 (Fig. 12-10) is subjected primarily to setback forces.

### 13-3.1 SETBACK

The force of setback $F_s$ is expressed as[6]

$$F_s = \left(\frac{W_p}{g}\right) a, \text{ lb} \qquad (13-1)$$

where

$W_p$ = weight of the part, lb

$g$ = acceleration due to gravity, ft/sec²

$a$ = linear acceleration of the projectile, ft/sec²

The high setback accelerations (occurring in the gun tube) affect the clock mechanism by causing a "stamping" action in the region near the bearings; this may introduce added friction in the clock mechanism. To overcome this, some clockwork manufacturers subject new mechanisms to very high accelerations to "prestamp" the mechanism beforehand. Since clockworks used in fuzes are not started until the linear acceleration is reduced considerably, the effect of high g forces on the escapement and mainspring need not be considered, assuming of course that the setback did not cause any permanent deformation in either the mainspring or escapement spring.

### 13-3.2 FORWARD AND SIDEWAYS ACCELERATION

Setforward—a negative setback—is an acceleration in the direction of projectile travel[6]. It occurs when projectiles are rammed into an automatic weapon and after firing. Present time fuzes will withstand about 1000 g setforward.

A sideways acceleration occurs when, upon firing, the projectile aligns itself with the gun tube. For example, the 175 mm field gun and the 120 mm tank gun have such high lateral forces that fuze ogives have broken off[6]. These forces have not been measured; however, the damage was simulated in drop tests by accelerations of 10,000 g.

### 13-3.3 CENTRIFUGAL FORCE

The centrifugal force vector $F_c$ is directed on a radial line orthogonal to the spin axis[6]

$$F_c = \left(\frac{W_p}{g}\right) r \omega^2, \text{ lb} \qquad (13-2)$$

where

$r$ = radial distance from the spin axis to the center of gravity of the part, ft

$\omega$ = angular velocity, rad/sec

Since $F_s$ and $F_c$ are forces applied at the center of gravity of the part that experiences both types of acceleration and since they act at 90 deg to each other, the maximum force $F_{max}$ occurs when the quantity $(F_s{}^2 + F_c{}^2)^{1/2}$ is a maximum.

## 13-3.4 CORIOLIS FORCE

If a part within the timer has a radial component of velocity as the unit spins, it experiences a Coriolis force. The condition is illustrated in Fig. 13-1[6], where a part within a radial slot is allowed to move outward as the body is subjected to a constant angular velocity. The Coriolis force $F_{co}$ is determined by [6]

$$F_{co} = 2 v \left(\frac{W_p}{g}\right) \omega, \text{lb} \qquad (13-3)$$

where

  $v$ = radial component of velocity of the part, ft/sec

## 13-3.5 TORQUE

Torque, the product of a force times its lever arm, occurs within the projectile. If the projectile is increasing or decreasing its spin rate (i.e., an angular acceleration is present), a torque $G$ will be produced on every internal part such that[6]

$$G = I_p \, \alpha, \text{lb-in.} \qquad (13-4)$$

where

  $I_p$ = moment of inertia of the part, slug-in.[2]

  $\alpha$ = angular acceleration, rad/sec[2]

Torque also may be produced from centrifugal effects when the pivot point of the part does not coincide with its center of gravity. Under this condition both a force and a lever arm are present.

## 13-3.6 PRECESSIONAL FORCE

If a part experiences a torque about any axis other than its spin axis, it will precess, i.e., it will turn about still another axis. It can be shown (from the equation governing rotating bodies) that the part will turn about an axis perpendicular to both the spin axis of the projectile and the torque direction. The precessional angular velocity $\Omega$ is expressed as

$$\Omega = \frac{G}{I\omega}, \text{rad/sec} \qquad (13-5)$$

## 13-3.7 EFFECT OF SPIN ON TIMERS

In general, spin will affect a spring-mass clock mechanism so that the timer will run fast. Centrifugal forces acting on the ends of the escapement spring tend to stiffen the spring, thus resulting in a higher escapement-lever frequency as spin speed increases.

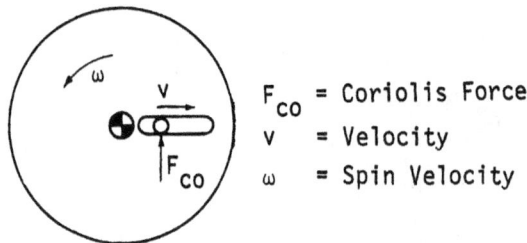

$F_{co}$ = Coriolis Force
$v$  = Velocity
$\omega$  = Spin Velocity

*Figure 13-1. Coriolis Force*

Effects of spin on two types of cylinder escapement are discussed in Ref. 7.

In a study at Frankford Arsenal, the clockwork movements of 56 mechanical time fuzes (T316) were spun at various rates on a production type regulation machine and their rate variation was noted[8]. The movement center was concentric with the center of rotation. Thirty fuzes were initially regulated dynamically to 77.52 Hz (the natural frequency of this movement) at 6000 rpm. The remaining group of 26 was regulated to 77.52 Hz at 12,000 rpm.

Operation of the movement at spin speeds other than their regulation speed caused deviations in timing accuracy. High spin speeds caused the movements to run on the fast side whereas low spin speeds resulted in slow running times. A movement of the type tested regulated at 12,000 rpm and set for 20 sec would have the following errors in running time:

(1) Fired from a 155 mm howitzer in Zone 7 (11,700 rpm), the movement would run approximately 0.025 sec slow.

(2) Fired from a 90 mm gun (17,000 rpm), the movement would have a timing error of approximately 0.500 sec fast.

(3) Fired from the 155 mm howitzer in Zone 1 (5400 rpm), regulated at 6,000 rpm and set for 20 sec, the movement would produce an error of approximately 0.0242 sec slow.

## 13-4 SYSTEMS FOR LOW ACCELERATION

### 13-4.1 MUNITION CLASSES

In this category we are dealing with such military items as rockets, guided missiles, grenades, and some mortar projectiles. The components of these munitions experience much lower forces from acceleration, being exposed to about 10,000 g as an upper limit. Setback forces may be calculated from Eq. 13-1. The magnitude of the acceleration is often relatively low, and its duration is much longer than that for high-acceleration systems. For some rockets, the duration may be as long as 2 to 4 sec. Arming and functioning of the fuze system by ballistic forces must, therefore, be dependent on this different acceleration-time profile (low acceleration-long time).

A second important characteristic of the munitions in this low-acceleration category is that the munitions usually are fin-stabilized; thus spin forces are negligible.

Bomb fuzes are a special class in this category. Being dropped from aircraft, bombs do not experience setback forces. Like the other munitions in this category, bombs usually have negligible spin.

### 13-4.2 TIMERS FOR GUIDED MISSILE FUZES

Consider the following typical requirement for a guided missile fuze:

(1) It shall arm when under an acceleration of 11 g, or higher if this acceleration continues for a minimum of 5 sec.

(2) It shall not arm when under an acceleration less than 7 g for a period of 1 sec or more.

A schematic drawing of a possible arming mechanism is shown in Fig. 12-7. The slider can move to the left after overcoming the force of the return spring. After the slider has moved a certain distance, the detonator will be aligned with the booster and the latch will drop to prevent return of the slider. The mechanism is then "armed". If the latch is not permitted to drop in back of the slider, the slider will return to its initial starting position when the acceleration is removed. This would be the case when an acceleration of 7 g or less is applied. Note the presence of the clockwork gear which restricts the slider movement. It has been found that the clockwork can provide the time-dependent restraint on the slider to satisfy the given condition (1).

Clockwork devices, in addition to being used for arming, can be used for self-destruct mechanisms. Self-destruction devices are added to rockets and guided missiles to prevent armed ammunition from causing damage if they impact on friendly territory and to prevent capture by the enemy. Mechanical time devices may be used to detonate the bursting charge if the target is missed—i.e., a firing stimulus is not received—and a certain delay time has been exceeded.

The guided missile Safing and Arming Device, GM, M30A1 is shown in Fig. 12-8.

### 13-4.3 TIMERS FOR BOMB FUZES

A bomb fuze, like other fuzes, must arm at an appropriate time after release and function at or near the target. Time fuzes are used when an air burst is desired. Timing devices for both arming and functioning are used in some of the newer bomb fuzes.

Fig. 13-2[6] illustrates Fuze, Bomb Nose, M198 that contains a timer for both arming and firing processes. See Ref. 6 for a description of its action.

### 13-4.4 TIMERS FOR GRENADE FUZES

#### 13-4.4.1 Launched Grenades

Grenade launchers, designed to propel grenades farther than they can be thrown by hand, have a range of about 400 m. Since they impart spin to the grenade, both spin and setback forces can be used for arming.

Fuze, PD, M551 (Fig. 13-3[6]) is used for grenades launched from the 40 mm XM79 Grenade Launcher. It functions by impact or graze action. See Ref. 6 for a description of its action. The timer permits arming times of 66 to 132 msec, corresponding to 60 to 120 ft in range of the temperature extremes.

#### 13-4.4.2 Rifle and Hand Grenades

The current standard service rifle is not designed to accommodate a rifle grenade. However, if a rifle grenade is desired, the required timer would be designed like any other low-acceleration system. Rifle grenades are not spun but subjected to setback accelerations of 500 to 1000 g. This setback force in combination with an escapement timer can serve for arming safety. The setback forces may be calculated from Eq. 13-1.

Hand grenades commonly use a pyrotechnic delay element, see par. 21-2.

### 13-5 STATIONARY AND LONG-DELAY TIMERS

Stationary timers perform in essentially fixed locations; hence, they need not withstand acceleration or spin forces. Very often, the delay desired of stationary timers is long (minutes to hours) compared with that of other military timing systems. Most stationary timers are used for demolition purposes. They occasionally may be used for mines although most mines are triggered by foot or vehicle impact. If the timer output is a switch closing or if the timer is located remotely from the explosive charge, the timer may be used for repeated operations.

There is an occasional need for long-delay timers that can be delivered by a projectile. Naturally, these timers must withstand the acceleration and spin forces imparted by the particular ammunition, and the setback forces upon impact with the ground.

### 13-5.1 DEMOLITION FIRING DEVICE, XM70

The XM70 (Fig. 13-4[9]) is a mechanical

REMOVE WITH COTTER PIN
AFTER ARMING WIRE HAS
BEEN INSERTED IF BOMB IS
NOT DROPPED, REPLACE
COTTER PIN AND SEALING
WIRE BEFORE REMOVING
ARMING WIRE.

SECTION AA

END VIEW

| | |
|---|---|
| 1 — ARMING PIN | 8 — ARMING STEM |
| 2 — FIRING PIN | 9 — SLIDER |
| 3 — VANE | 10 — PRIMER |
| 4 — GOVERNOR DRUM | 11 — DETENT |
| 5 — GOVERNOR SPRING | 12 — DISK ASSEMBLY T5 |
| 6 — GEAR REDUCTION | 13 — TIMING DISK LEVER |
| 7 — ARMING GEAR | 14 — FIRING PIN SPRING RETAINER |

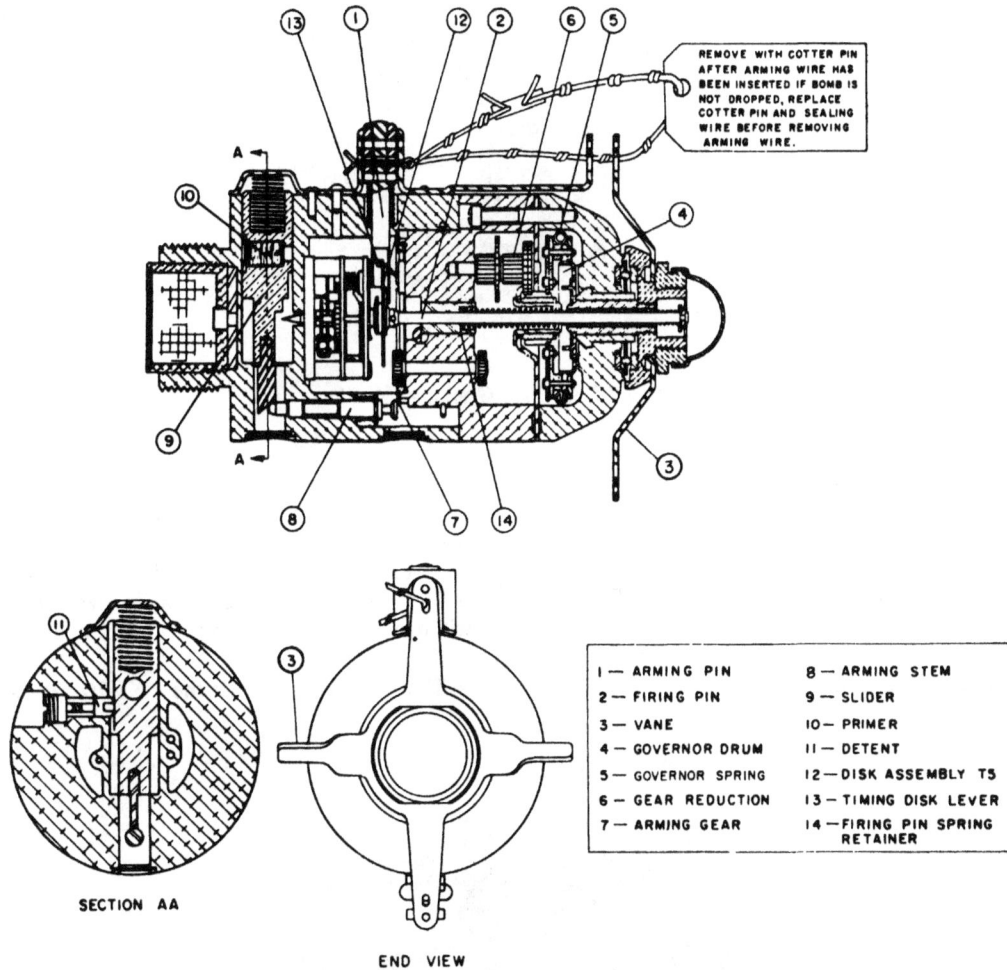

*Figure 13-2. Fuze, Bomb Nose, M198*

timer. It releases a spring-loaded firing pin after any desired time interval between 5 min and 24 hr. While the timer does not contain explosives, it permits access to its firing pin through a tapped hole in the case (not shown). This hole fits a standard coupling for attachment of primacord, RDX, or other demolition charge. If desired, the firing pin motion can be used to operate a switch in an electric circuit.

The power source for the XM70 is a constant-torque spring motor (negator). The spring also provides two other functions. To set the unit, the spring must be transferred from the storage spool to the output spool. Thus, the amount of spring transferred is an indication of the time interval desired, obviating a separate setting indicator. The constant-torque spring is printed with graduations that represent 5-min intervals when the

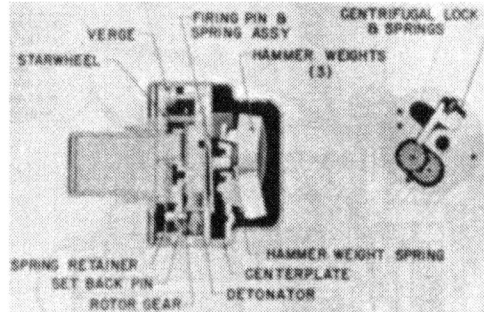

Figure 13-3. Fuze, Grenade, M551

spring is permitted to return to the storage spool at the controlled rate. These graduations are visible through a window in the case. The other function performed by the spring is to serve as a safety latch. A notch in the output spool of the spring serves as the release point for the firing pin. In the unarmed condition the release is held off the spool by the arming mechanism. In setting the XM70, the spring is wound onto the output spool covering the notch. The notch is not uncovered again until the end of the elapsed time at which time the spring-loaded release will drop and permit firing pin motion.

The clockwork of the XM70 is an

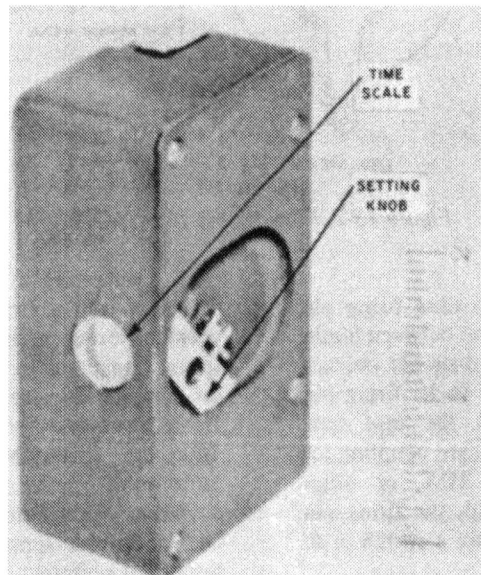

Figure 13-4. Demolition Firing Device, XM70

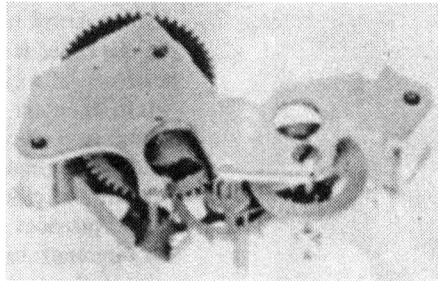

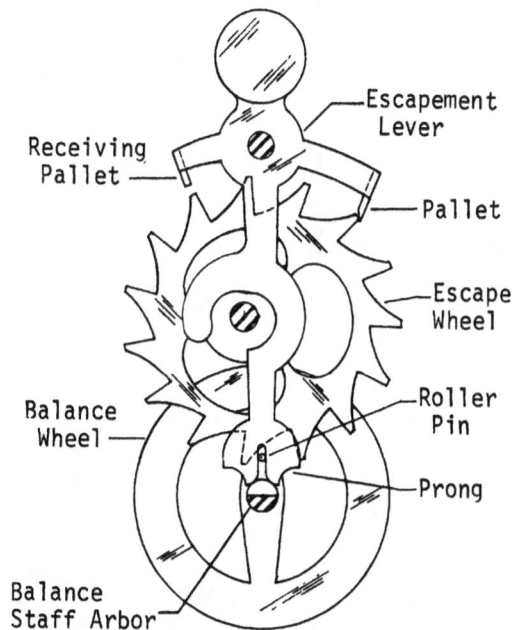

Figure 13-5. Clockwork for Demolition Firing Device, XM70

inexpensive, self-starting detached-lever escapement (Fig. 13-5[9]). The assembly includes —in addition to those linkages required for arming, disarming, and recocking—linkages that prevent the unit from being reset to a different time until it is disarmed. These interlocks insure user safety.

## 13-5.2 INTERVAL TIMER, XM4

The XM4 is a 30-min delay timer designed to switch three circuits at any time that is set from 5 to 30 min, in intervals of 5 min each. With a luminous dial, it can be set either in daylight or darkness. It is sealed in an olive-drab cylindrical case. A small threaded electric connector extends downward from the lower end of the case while the knob for winding and setting the timer extends upward from the top of the flange. After the case is sealed, the XM4 is completely waterproof. The timer is shown in Fig. 13-6[10] and the

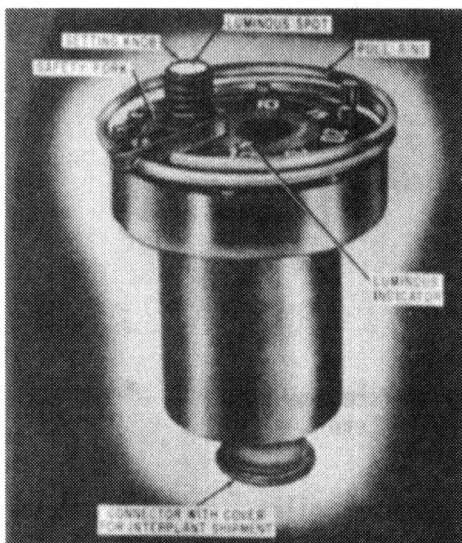

Figure 13-6. Interval Timer, XM4

clockwork in Fig. 13-7[10]. The clockwork is a modified, standard, 15-jewel watch movement. A small torque, supplied by the timing mainspring, is transmitted through the main timing shaft to a release mechanism that controls a switch drive system. The switch drive system is powered by a separate mainspring that gives the larger torque needed to turn the drum of switch. Both clockwork and switch are powered by Dynavar mainsprings that are wound when the timer is set. This makes the XM4 a self-contained unit requiring no external power source.

The XM4 passed all required tests and performs with a standard deviation in timing error of less than ± 1% at all delay settings. Its principal characteristics are shown in Table 13-2[10].

### 13-5.3 180-HR AIRBORNE TIMER

As an example of a long-delay timer consider the 180-hr airborne timer shown in Fig. 13-8[11]. This timer was developed for use in the Army lunar-probe missile. The chassis of the timer is made up of four plates, and is enclosed in a cylindrical cover measuring 2.75 in. in diameter by 2.06 in. in length. A T787 bomb-fuze clockwork is mounted on the third plate. The mainspring panel is mounted on top of the clock. Since the T787 clockwork only provides a 78-hr running time, the requirement for a 180-hr timer was met by use of the mainspring from the T908 Bomb Fuze.

Three separate electronic switches are controlled by the programmer which is made up of two concentrically mounted commutators. The programmer is suspended between the top and second plates. A dimple motor is mounted in a housing on the third plate. The motor actuates a switch that closes the first circuit and gives a starting kick to the clock mechanism.

Suspended under the third plate is a restarting clock with a free-running verge escapement. It runs during the acceleration period of the first stage of the rocket and gives the main clock a restarting kick if it has been stopped by the acceleration forces. The restarting clock is released under an acceleration of 40 g by the motion of a g-weight that travels on two shafts fixed between the second and bottom plates.

### 13-6 TIMER SYSTEM TESTING

### 13-6.1 GENERAL

During the design and development of a timer system, each component is subjected to a number of tests to determine if it performs in the manner for which it was designed. When the prototype is built, it is subjected to performance and proof tests to determine if the prototype satisfies its requirements. Since some of these tests damage the item and, since the available number of items is limited, it is desirable to use special tests and necessary to apply special methods of data

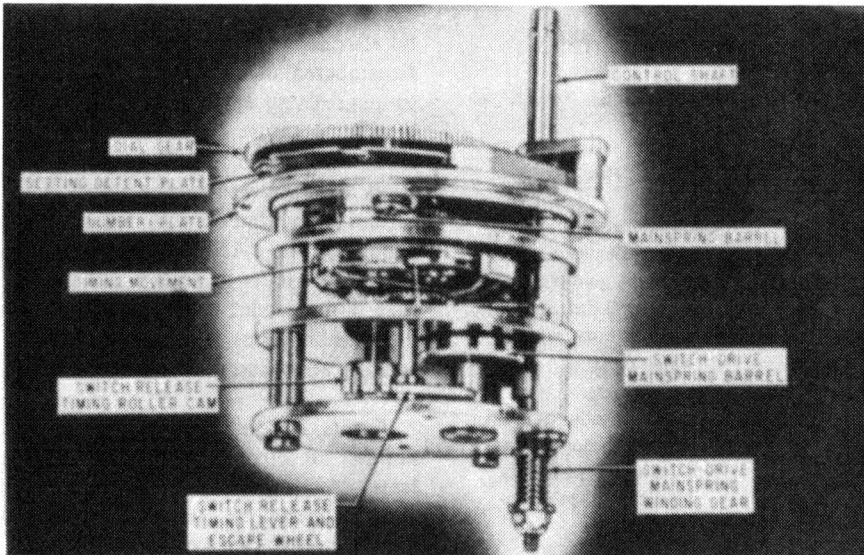

*Figure 13-7. Clockwork of XM4 Timer*

analysis. There is a definite trend toward test standardization. The pertinent standard tests are indicated in the paragraphs that follow.

## 13-6.2 PERFORMANCE TESTS

The components of a timer as well as the complete timer assembly must undergo tests to establish that the unit functions properly and does its intended job. *Development* tests are performed to evaluate the designer's latest effort; *acceptance* tests are performed to evaluate the final design and often are called approval tests or evaluation tests. Development tests seek an answer, while acceptance tests confirm it. The tests differ in three respects:

(1) Development tests are applied to individual components; acceptance tests are applied to the entire timer only.

(2) Development tests are severe and run until failure; acceptance tests stop at a predetermined acceptance level.

(3) Development tests are specified by the designer; acceptance tests are specified by a Service Board.

Timers are tested to determine timing accuracy by comparing performance with a timing instrument that is accurate enough to assure compliance with the accuracy requirements in the specification.

If the timer is to be used in a spin-stabilized projectile, the effects of spin upon the timing rate must be investigated to determine the extent of the error. In this test, timers are mounted in a production-type regulation machine that simulates the spin of the projectile. Movements first are regulated dynamically at predetermined frequencies and speeds. Then the movements are spun at various speeds, and timing deviations are recorded.

An important characteristic of many timers is the ability to start or restart by themselves. This feature is tested by subjecting the timer

13-15

**TABLE 13-2**

**CHARACTERISTICS OF XM4 TIMER**

| Type | Interval timer |
|------|----------------|
| Materials | |
|   Case | aluminum |
|   Setting system | aluminum, steel, stainless steel |
|   Clockwork | aluminum, steel, stainless steel, 12% nickel silver, Dynavar, gold-plated brass, drill rod, red ruby, commercial watch parts |
|   Switch drive | steel, Dynavar |
|   Switch | aluminum, rhodium-plated beryllium copper, glass-filled diallyl-phthalate plastic, solid-gold contacts, lead |
| Diameter of case | 1.5 in. |
| Diameter of flange | 2.2 in. |
| Length below flange | 2.1 in. |
| Weight | 161 g |
| Time range | 5-30 min |
| Time intervals | 5 min |
| Temperature limits | $-55°$ and $75°C$ |

to accelerations or other forces and noting the starting performance of the timer.

## 13-6.3 TEST PROGRAMMING

It is important to plan a test schedule ahead of time. Planning will avoid wasting units in overtesting, and will permit sequential and combined testing when desired. A typical test plan that covers safety and surveillance is shown in Table 13-3[6]. For a given timer design, some of these tests may be omitted while other more appropriate ones may be added—i.e., tests should be consistent with the intended use of the timer. It is important to use enough samples, so that valid conclusions may be drawn, i.e., establish an acceptable confidence level.

The order of testing should be considered carefully. It often is desired to perform sequential tests where the same timer is subjected first to one test, then to another. In this way, cumulative effects, but not synergistic effects, may be evaluated. It also must be remembered that a timer never encounters a simple isolated environment during its service life. Temperature, humidity, and pressure are always present in the atmosphere, and each alters and is altered by the others; it is almost certain the timers will experience rough handling. The JANAF Fuze Committee has suggested a series of tests which evaluates the fuze design under conditions closely simulating combinations of environments as they actually are encountered[1,2].

Timers for fuzes are required to meet a number of environmental and performance tests, described in MIL-STD-331[3]. Tests are either destructive (see par. 13-6.4.1) or nondestructive (see par. 13-6.4.2).

## 13-6.4 SAFETY TESTS

Safety tests are designed to investigate the requirements for safe handling. There are two types:

(1) Destructive tests where operability after the test is not required.

(2) Nondestructive tests where operability is required.

### 13-6.4.1 Destructive Tests

In destructive tests, the timer assembly usually is installed in its end item such as a fuze. Forty-foot drop, jolt, and jumble tests are designed to test the ruggedness of the fuze[3]. It is advisable also to conduct these tests at extreme temperature ($-50°$ to $125°F$) in order to find out whether the materials or components are vulnerable at these temperatures.

Safing and arming devices for projectile fuzes must pass muzzle impact and impact safe distance tests; for bomb fuzes they must withstand a number of jettison and accidental

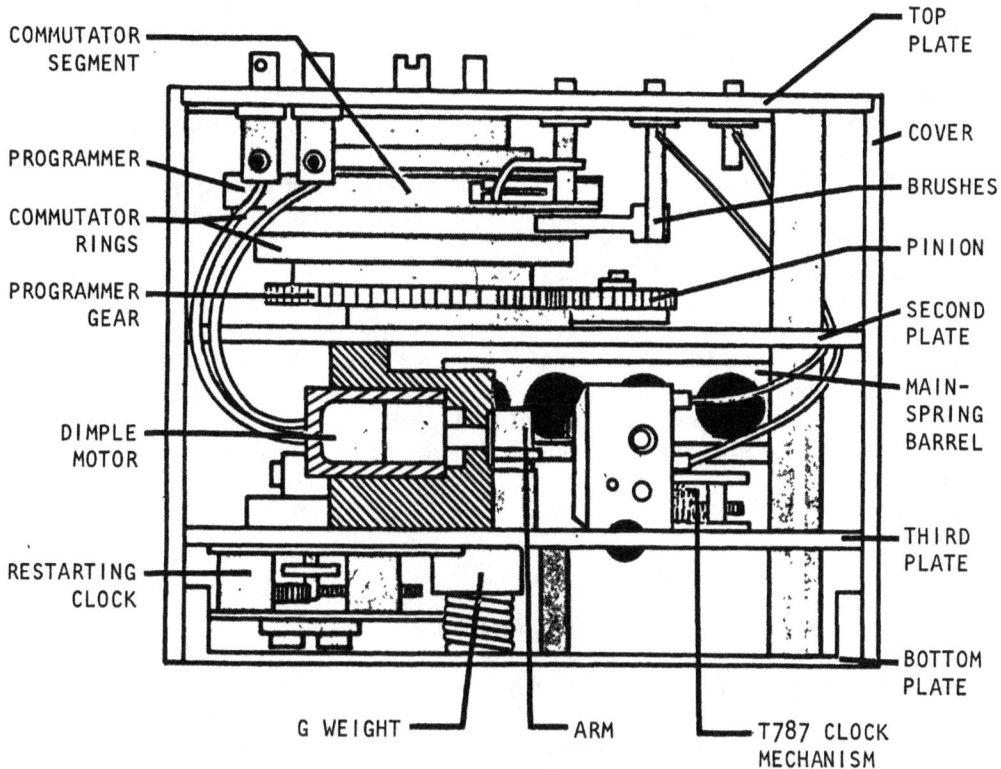

*Figure 13-8. Cutaway View of 180-hr Timer*

release tests; and for air delivery the devices must be safe in parachute delivery and arrested landing tests[3].

### 13-6.4.2 Nondestructive Tests

These tests check the performance, ruggedness, and reliability of the timer assembly by simulating actual conditions in which the timer will be expected to perform.

A large group of environmental tests has been standardized to evaluate the performance of timers under various conditions that simulate handling, transportation, and atmospheric conditions. Applicable tests for fuze timers are selected from among transportation-vibration, temperature-humidity cycling, salt spray, five-foot drop, thermal shock,

extreme temperature storage, rough handling (packaged), sand and dust, and fungus resistance[3].

**TABLE 13-3**

**SAFETY AND SURVEILLANCE TESTS**

| Test | MIL-STD-331 Test No. | Typical Quantity |
|------|------|------|
| Jolt | 101 | 6* |
| Jumble | 102 | 6 |
| Five-foot drop | 111 | 10 |
| Forty-foot drop | 103 | 10 |
| Transportation vibration | 104 | 10 |
| Temperature and humidity | 105 | 5 |
| Vacuum-steam pressure | 106 | 5 |
| Waterproofness | 108 | 5 |
| Salt spray | 107 | 2 |

*Sequential in 3 positions

13-17

To check the effect of acceleration upon the timing movement, units are placed in a centrifuge, air gun, or shock machine and subjected to simulated accelerations of the prescribed magnitude[6].

## 13-6.5 ANALYSIS OF DATA

It is important that the timer system designer use statistical procedures to make certain that his conclusions are valid.

Timers are manufactured in huge lots and only a few are chosen to be tested from the lots. These are selected carefully "at random" so that the sample will represent the lot faithfully. Costs of procuring timers and running tests will limit the sample size. Since economy requires that a small sample be tested, confidence in high reliability cannot be assured. However, statistical principles make it possible to attribute a certain degree of confidence to the results obtained with a given sample size. The designer must determine what compromise between accuracy and economy should be adapted to his particular case.

The concepts of random sampling, frequency distributions, measures of reliability, statistical significance, and practical significance should all be understood by the designer so that he can at least recognize the need for a professional statistician when it arises. The subject of military application of statistics is covered in other handbooks[14-18].

## REFERENCES

1. R. H. Comyn, et al., *Long-Delay Timer for Air Force Bomb Fuze*, U S Army Harry Diamond Laboratories, Washington, D.C., Report TR-1126, 29 April 1963.

2. MIL-STD-721A, *Definition of Terms for Reliability Engineering*, Dept. of Defense, 2 August 1962.

3. AMCP 706-134, Engineering Design Handbook, *Maintainability Guide for Design*.

4. Morgan, et al., *Human Engineering Guide for Equipment Designers*, McGraw-Hill Book Co., N.Y., 1964.

5. MIL-STD-1472, *Human Engineering Design Criteria for Military Systems, Equipment and Facilities*, Dept. of Defense, 9 Feb. 1968.

6. AMCP 706-210, Engineering Design Handbook, *Fuzes*.

7. F.K. Soechting, *Influence of Spin Rate on Dynamic Behavior of Fuze Elements*, Picatinny Arsenal, Dover, N.J., Report TM 1109, Feb. 1963.

8. E.M. Escandarian, *Spin Sensitivity of Mainspring Driven Movements Used in Mechanical Time Fuzes*, Frankford Arsenal, Phila., Pa., Report R-1509, 1959.

9. F. Luke, "Development of Firing Device, Demolition, XM 70", *Proceedings of the Timers for Ordnance Symposium*, Vol. 1, U S Army Harry Diamond Laboratories, Washington, D.C., 15-16 November 1966, pp. 35-43.

10. *Development of Interval Timer, XM4*, Army Materiel Command, TIR No. 8-11-1B1(1), November 1962 (AD-446 292).

11. D.S. Bettwy, et al., *180-Hour Airborne Timer*, U S Army Harry Diamond Laboratories, Washington, D.C., Report TR-767, 4 March 1960.

12. *Combined Environments Testing*, Journal Article No. 53 of the JANAF Fuze Committee, 12 April 1968 (AD-835 813).

13. MIL-STD-331, *Fuze and Fuze Components, Environmental and Performance Tests for*, 10 January 1966.

14. AMCP 706-110, Engineering Design Handbook, *Experimental Statistics, Section 1, Basic Concepts and Analysis of Measurement Data*.

15. AMCP 706-111, Engineering Design Handbook, *Experimental Statistics, Section 2, Analysis of Enumerative and Classificatory Data*.

16. AMCP 706-112, Engineering Design Handbook, *Experimental Statistics, Section 3, Planning and Analysis of Comparative Experiments*.

17. AMCP 706-113, Engineering Design Handbook, *Experimental Statistics, Section 4, Special Topics*.

18. AMCP 706-114, Engineering Design Handbook, *Experimental Statistics, Section 5, Tables*.

AMCP 706-205

CHAPTER 14

PRODUCTION TECHNIQUES

**14-1 DESIGNING FOR PRODUCTION**

A knowledge of production tools, machines, and processes is imperative in working toward the goals of improved production methods and lower manufacturing costs. Manufacturing processes are the primary processes used in the fabrication of engineering materials, and must be kept in mind when considering alternative production procedures and attempts to optimize the production process.

In general, the quantities involved in the manufacture of timer components for military use are large, and hence the philosophy of production is based primarily on the mass production process. In designing for mass production, the designer should be aware of how mass production techniques are applied and the manufacturing methods that are available in modern machine shops. In addition, production of these large quantities must be carefully documented and controlled to assure quality production.

**14-1.1 THE PRODUCTION PROCESS**

The totality of operations that transform raw materials and blanks that enter the plant into finished products is known as the production process.

The selection of a process for a given operation is influenced by several factors, including the product quality desired, the cost of labor needed, the number of units to be produced, and the materials of construction. When an item can be produced by one of several methods, there is often one best method for a given set of requirements[1].

There are three types of production: (1) unit production, (2) series production, and (3) mass production, namely:

(1) **Unit Production** is the production of single products or of small batches of products with large time intervals between successive runs. Since parts of various dimensions and diverse configurations may be required, universal equipment and general-purpose tools are used. Because of the diversity of the operations, personnel engaged in unit production must have high qualifications. As one might expect, costs may be high and production slow but unit production is optimum for small quantities.

(2) **Series Production** is the manufacture of several products in small batches. One or several operations are carried out at each workplace, and either the entire batch or a smaller group of parts moves together from operation to operation.

(3) **Mass Production** is the steady production of uniform products. The elements of work are of specialized character, and wide use is made of automatic and mechanized tools and devices. For details, see par. 14-2.

**14-1.2 IMPORTANT FACTORS IN PRODUCTION**

Accuracy as well as reliability are of utmost importance in the production of mechanical timers for military use. Making the parts rugged helps to meet both of these requirements. Processing accuracy is understood to refer to the degree to which the finished parts correspond to the form and dimensions specified in the drawings. Variations in the dimensions of parts produced can result from

14-1

many factors that influence the manufacturing process. A few of these are tool wear, inaccuracy in the machine or cutting tools, nonuniformity of the material being processed, and deformations in the machine tool or part during processing.

## 14-1.3 STEPS LEADING TO PRODUCTION

During the development of a timer for military use, successful testing of prototypes is necessary before production of the timer. This may require several stages of development, leading to full-scale field tests. At the other extreme, there may be a product improvement and limited testing because the item may be based on a previously successful timer design. The design engineer should collaborate with the production engineer while the timer is still in the design phase to insure that the timer can be mass produced with readily available materials.

## 14-2 DESIGNING FOR MASS PRODUCTION

### 14-2.1 THE MASS PRODUCTION PROCESS

Mass production has as its aim the steady production of uniform products. Each workplace is assigned a specific operation, and the elements of work have a narrowly specialized character. In mass production, wide use is made of specialized and automatic machinery, special cutting tools, measuring instruments, and a variety of automatic and mechanized instruments and devices. The product design, technological process study, and equipment design should be complete and thorough because of the large investments required in the preparation for mass production. Details of production are well documented in the literature[2].

The materials, the processes for forming and protecting them, and the skilled hands to use them are available but it is up to the production engineer to integrate these into the optimum combination, not only for a

reliable product but also for efficient and economic production. The problem is to establish—for each component of a mechanical timer—the optimum combination of materials and processes to meet the functional requirements of shape, strength, and resistance to chemical and physical hazards.

Other considerations are: (1) timer components must be chemically compatible over the military temperature range with other metals, plastics, lubricants, explosives, or other materials with which they may be exposed, particularly in storage[3]; and (2) the production process must not adversely affect ease of maintenance (see par. 13-2.2).

The choice of the manufacturing process is influenced by the material chosen, the quantity of parts involved, and the compatibility of the parts with the process. Each component is considered as an individual design, with alternate methods of manufacture being assessed on the basis of the processes and materials involved to obtain an effective assembly. Decisions such as the use of punch and shave techniques rather than milling is an example of this weighing of alternatives with respect to economy. Materials and processes also must be considered together because of the possible effect upon each other.

One method that can aid in the choice among different solutions is value engineering analysis. In this method, ways of achieving the objectives are defined and cost or resource consequences are estimated for each way. Basically, a choice must be made between maximizing product capability for a given cost or minimizing the cost for achieving a given objective[4].

### 14-2.2 COOPERATION WITH THE PRODUCTION ENGINEER

The designer must have a working knowledge of the aspects of production but need not be an expert in detailed production techniques. The designer should consult with the production engineer. This should be done at the stage when the mechanical principles have been established and coordinated to meet the specifications but before the design is consolidated to the point where all details are frozen. The production engineer's knowledge of the various processes for forming and fabrication is invaluable to the designer, and his suggestions often can improve the design or its cost-effectiveness.

## 14-2.3 REQUIREMENTS FOR MECHANICAL TIMERS

The materials, parts, and assemblies used in timers must comply with the requirements stated in the applicable drawings, referenced standards, and specifications. Those for Fuze, MTSQ, M564E3 (see Fig. 12-9) are typical of the requirements for mechanical time fuzes and therefore can serve as an example. (For further details on the data package, see par. 14-4.) The M564E3 is required to

(1) Be made as shown on assembly drawing F10 543 194

(2) Meet all conditions stated in its specification, MIL-F-60802 (MU)

(3) Pass the specified fuze, fuze component, environmental, and performance tests of MIL-STD-331

(4) Be manufactured within the quality assurance provisions of MIL-STD-109, with critical, major, and minor defects as defined

(5) Have surfaces with roughness values according to MIL-STD-639

(6) Be accompanied by a data card according to MIL-STD-1167

(7) Be packaged as specified

(8) Meet the following requirements for workmanship:

(a) Metal Defects. All components shall be free from cracks, splits, cold shuts, inclusions, porosity, or any similar defect.

(b) Threads. All threads shall be full and undamaged for the length required by the applicable drawing.

(c) Burrs. No parts shall have a burr that might interfere with the assembly or function of the item or that might be injurious to personnel handling the item.

(d) Foreign Matter. No part or assembly shall contain dirt, chips, grease, rust, corrosion, or other foreign matter.

When the assembly to be manufactured contains explosive components, proper safety precautions must be taken, see par. 19-2.

## 14-2.4 METALS FOR MECHANICAL TIMERS

Part uniformity, tool wear, and manufacturing economy are all affected by the material being used in mass production. It is therefore important that all materials, particularly metals, be uniform in composition. The mechanical properties of metals used in the production of timer components must satisfy the applicable specifications. It is expedient for the designer to use qualified materials when at all possible.

All metals should be either corrosion-resistant or protected against corrosion by appropriate application of coatings[5]. Corrosion due to galvanic action resulting from dissimilar metals also must be considered. General properties and characteristics of the metals used for components of mechanical timers are covered in the following Military Handbooks:

**TABLE 14-1**

**COMMON TIMER LUBRICANTS**

| Type | MIL. Spec. | Composition | Comments |
|------|-----------|-------------|----------|
| Oil | MIL-L-3918[7] | Specified synthetic ester mixture and additives | Low temperature (−40°F), non-spreading lubricating oil |
| Oil | MIL-L-11734[8] | Specified mixture of C9-C10 dibasic acid esters and additives | Standard fuze oil; used in many mechanical time fuzes over military temp. range |
| Solid film | MIL-L-8937[9] | $MoS_2$, graphite, etc. in resin binder | Bonded solid-film lubricant; resin cures at 300°F for 1 hr |
| Solid film | — | Same as above | Same as above except resin cures at 400°F |
| Solid film | MIL-M-7866[10] | Powdered $MoS_2$ min. purity 98.5% | Unbonded; applied by tumbling or burnishing |
| Solid film | — | TFE Telomer | Introduced recently |

(1) MIL-HDBK-723(MR), *Steel and Wrought Iron Products*

(2) MIL-HDBK-693(MR), *Magnesium and Magnesium Alloys*

(3) MIL-HDBK-694A(MR), *Aluminum and Aluminum Alloys*

(4) MIL-HDBK-697(MR), *Titanium and Titanium Alloys*

(5) MIL-HDBK-698(MR), *Copper and Copper Alloys.*

For example, the metals used in the timing movement of the M564E3 Fuze are aluminum, stainless steel, and brass, in addition to an alloy steel mainspring.

## 14-2.5 LUBRICATION

A lubricant is expected to perform the jobs of minimizing friction, wear, and galling between sliding or rolling parts. It must do these jobs under two types of condition: (1) those which are inherent in the component element itself — such as load, speed, geometry, and frictional heat; and (2) those which are imposed from external sources — such as temperature and composition of the surrounding atmosphere, nuclear radiation, inactive storage, vibration, and mechanical shock. The imposed conditions are usually the more restrictive ones for lubricant selection[6]; low temperature is particularly significant.

Lubrication itself as well as the selection of the proper lubricant is critical because it affects timer reliability. For example, improper lubrication will prevent parts from moving as intended. A proven lubricant should be selected and employed at the time of material and part design.

A wide variety of lubricants with proven military properties are available. The most commonly used lubricants for escapements, gears, bearings, and linkages are listed in Table 14-1[6]. Modern timer lubrication trends may be summarized as follows:

(1) Timer components have been successfully lubricated with various oils and dry solids.

(2) Of the oils, the most commonly used is the standard fuze oil (MIL-L-11734). This is as expected since the oil has been standardized for over twenty years.

(3) Solid film lubricants now predominate over the oils because they have better storage characteristics.

(4) The TFE telomer solid lubricant, introduced only a few years ago, appears to have proven itself as an effective and versatile lubricant for timers.

## 14-3 MANUFACTURING METHODS FOR TIMER COMPONENTS

### 14-3.1 ESCAPEMENT

The escapement consists of a number of small parts (see Fig. 13-5) that are produced by conventional production techniques[2,11]. In manufacturing the escape wheel the following operations may be used: blanking, tooth milling, punching and shaving, hardening and tempering, face grinding and polishing. Similar operations, used for pallets and levers, are blanking and shaving, milling, piercing, deburring, grinding, and polishing.

The escape wheel must satisfy the following requirements[12]:

(1) Surface finish of at least 63 rms on the working surfaces of the tooth

(2) Accurate geometrical tooth form

(3) Minimum run-out with respect to outside diameter.

### 14-3.2 GEARS

Production methods for the gears used in the gear train include milling, hobbing, piercing, and blanking. Both involute and epicyloid tooth shapes are used, the selection depending on the factors discussed in par. 16-3.5. The Wickenberg gear tooth design allows greater radial tolerances because of the larger root depth. A minimum of six teeth is used on small pinions in current practice[3].

### 14-3.3 HAIRSPRING

The hairspring is manufactured by repeatedly drawing wire, flattening it, cutting it to blanks of the required length, coiling the blanks in a barrel to give them the shape of an Archimedean spiral, followed by heat treatment of the coiled springs in order to stabilize their shape (form fixing), and lastly separation of the stabilized hairsprings[12]. Special machines used for these operations are a drawing machine, a rolling mill, a coiling machine, and heat treatment furnaces.

In military timers that use the tuned two-center escapement (see par. 16-4.2), the coiled hairspring is replaced by a straight metal bar that acts as a flexing beam. Manufacturing method is similar except for coiling.

### 14-3.4 OTHER PARTS

The balance wheel is made by blanking and shaving followed by stress relief using heat treatment furnaces.

Pins and shafts usually are made on automatic screw machines. The plates to complete the assembly are drilled on an automatic jig-borer or the holes can be punched, depending upon the accuracy of dimensions required between holes.

### 14-3.5 TOLERANCES

Most military components are toleranced to permit complete interchangeability of parts over the entire temperature range. However, in timers, where components are small and tolerances critical, complete interchangeability is not practical. Hence, selective assembly methods are used (see par. 13-2.4.2).

## TABLE 14-2

### SUPPORTING DATA PACKAGE

**Military Specifications**

| | |
|---|---|
| MIL-I-45208 | Inspection System Requirements including Appendix A, Supplemental General Quality Assurance Provisions |
| MIL-HDBK-204 | Inspection Equipment Design |

**Purchase Descriptions**

| | |
|---|---|
| TL-PD-18 | Purchase Description for Manufacturing Process Report |
| TL-PD-37 | Purchase Description for Drawing Package |
| TL-PD-124 | Purchase Description for Evincive Data Reports |
| TL-PD-125 | Purchase Description for Preparation of Item Purchase Description |
| TL-PD-328 | Purchase Description for Inspection Equipment Lists |

**Other Documents**

| | |
|---|---|
| AR 380-32 | Security Classification Guide for Proximity Fuzes and Components |
| AR 380-130 | Armed Forces Industrial Security Regulations |
| AR 754-10 | Conservation of Materials |
| HDLR 380-10 | Thermal Batteries, Security Classification and Controls |

## 14-4 REQUIREMENTS FOR TECHNICAL DOCUMENTATION

Production requires complete documentation. The required documents should be comprehensive and carefully written, and then assembled into a technical data package. A technical data package should comprise the documentation needed to ensure (1) that military items are capable of reliably performing the functions for which they were designed, and that (2) they can be produced promptly and economically by commercial contractors[13].

The content of the technical data package varies depending on the nature of the item being manufactured. A purchase description or specification always is included as are drawings of the item to be constructed. The only exception to the requirements for a drawing is for items that do not commonly require one, such as a can of lubricant. A manufacturing process report is included for the production of items that are unique or require special instructions beyond common manufacturing techniques. A typical data package may contain the following documents:

(1) Specification or purchase description

(2) Drawings and parts lists

(3) Manufacturing process report

(4) Test equipment manuals and drawings

(5) Gage drawings and inspection equipment list

(6) Instructions for preparing a data report.

Other documents sometimes used include security requirements check list, and depot maintenance work requirements. A typical supporting data package used by US Army Harry Diamond Laboratories is shown in Table 14-2[13].

# REFERENCES

1. H.B. Maynard, *Industrial Engineering Handbook*, McGraw-Hill Book Co., Inc., N.Y., 1963.

2. G.B. Carson, Ed., *Production Handbook*, Ronald Press Company, N.Y., 1958.

3. AMCP 706-210, Engineering Design Handbook, *Fuzes*.

4. R.N. Grosse, *An Introduction to Cost-Effectiveness Analysis*, Research Analyses Corporation, July 1965 (AD-662 112).

5. MIL-HDBK-721(MR), *Corrosion and Corrosion Protection of Metals*, Dept. of Defense.

6. *The Lubrication of Ammunition Fuzing Mechanisms*, Journal Article No. 49.0 of the JANAF Fuze Committee, 3 May 1967 (AD-829 739).

7. MIL-L-3918, *Lubricating Oil, Instrument, Jewel Bearing, Nonspreading, Low Temperature*, Dept. of Defense.

8. MIL-L-11734A, *Lubricating Oil, Synthetic (for Mechanical Time Fuzes)*, Dept. of Defense.

9. MIL-L-8937 (ASG), *Lubricant, Solid Film, Heat Cured*, Dept. of Defense.

10. MIL-M-7866A (ASG), *Molybdenum Disulfide Powder, Lubricant*, Dept. of Defense.

11. AMCP 700-1, Logistics, *Machining Data*, Army Materiel Command.

12. S.V. Tarasov, *Technology of Watch Production*, Israel Program for Scientific Translations, Ltd., 1964 (OTS 64-11110).

13. G.R. Keehn, *Purchase Description for Technical Data Package*, U S Army Harry Diamond Laboratories, Washington, D.C., Report TL-PD-68, Rev. B, 26 October 1967.

# CHAPTER 15

# PACKING, STORING, AND SHIPPING PROCEDURES

## 15-1 PACKING

### 15-1.1 PURPOSE

The basic purpose of military packaging is to insure that items are fit to perform their intended functions when the time comes for their use. Packaging will protect the item after production, during storage, during transport, and until delivery to its ultimate user. During the transportation phase, which includes both handling and carriage, the Department of Transportation (DoT) regulations must be strictly observed. Since items must be stored for long periods in protected or unprotected storage areas, they must be protected against physical damage and deterioration; means also may be required to conduct periodic inspections of the stored material.

### 15-1.2 OBJECTIVES

The objective of military packaging include: (1) uniform, efficient, and economical packaging; (2) similar items packaged and marked in a similar way; and (3) packaging requirements and materials kept to a minimum.

### 15-1.3 PACKAGING ENGINEER

The preceding objectives are the responsibility of the packaging engineer. In addition, the packaging engineer is responsible for the development and engineering execution of all aspects pertaining to the preservation, packaging, packing, processing, and marking. His functions and responsibilities are related closely to activities of supply management, field service maintenance, procurement, and

production. The role of the packaging engineer and his relation to the design engineer are shown in Fig. 15-1[1].

### 15-1.4 METHODS

From the military standpoint, good packing methods are those that protect the item through all phases and environments, with a minimum cost. Some questions to be answered as a guide to the packing problem are:

(1) Does the method afford the required protection?

(2) Does the method result in the lowest cost consistent with required protection?

(3) Does the method use noncritical materials?

(4) Are materials readily available?

(5) Does the method require special facilities and equipment?

(6) Does the method permit minimum depreservation effort at time of use?

(7) Does the method result in the least cube and weight, consistent with required protection?

(8) Can the method be easily adapted for mechanization, when conditions such as quantity to be procured warrant?

(9) Does the method provide for reuse of containers for recoverable, repairable items?

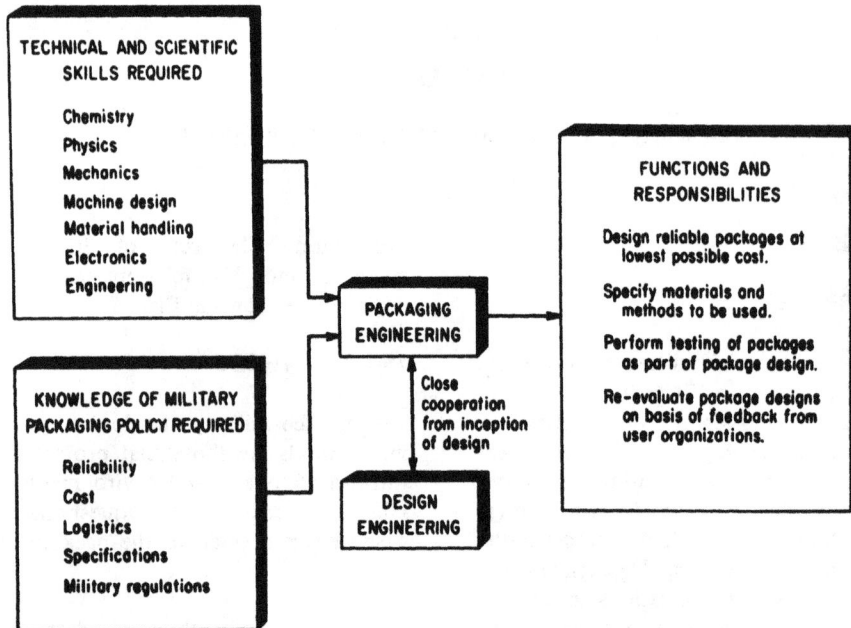

*Figure 15-1. Role of Packaging Engineering*

(10) Is the method consistent with quantity end-use?

(11) Does the method provide continued protection for more than one like item in a unit pack until the contents are depleted?

(12) Does the method promote uniformity in the application of packaging methods?

(13) Does the method meet statutory requirements?

## 15-1.5 PROTECTION LEVELS

There are three levels of military protection. The levels of protection are described in terms of the performance expected of the package and must be translated into specific technical or design requirements for individual items or categories of items. The performance criteria are:

(1) *Level A, Military Package.* Preservation and packaging which will afford adequate protection against corrosion, deterioration, and damage during world-wide shipment, handling, and open storage.

(2) *Level B, Limited Military Package.* Preservation and packaging which will afford adequate protection against known conditions that are less hazardous than Level A is designed to meet. This level provides a higher degree of protection than afforded by Level C. The design of Level B is based on firmly established knowledge of the shipment, handling, and storage conditions to be encountered and on the determination that the costs of preparation are less than Level A.

(3) *Level C, Minimum Military Package.* Preservation and packaging which will afford adequate protection against corrosion, deterioration, and damage during shipment from

SPECIFIED PROCEDURES    CLEANING    DRYING    PRESERVING

WRAPPING    CARTONIZING    MARKING    READY FOR PACKING

(A) BASIC STEPS IN MILITARY PACKAGING

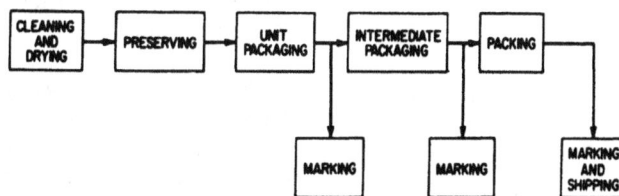

CLEANING AND DRYING → PRESERVING → UNIT PACKAGING → INTERMEDIATE PACKAGING → PACKING

MARKING    MARKING    MARKING AND SHIPPING

(B) NORMAL SEQUENCE OF PACKAGING STEPS

*Figure 15-2. Basic Steps in Military Packaging and Their Normal Sequence*

supply source to the first receiving activity for immediate use or for controlled humidity storage.

### 15-1.6 PACKAGING STEPS

The normal sequence of packaging and the steps which may be followed are shown in Fig. 15-2[1]. The materials and methods which can be used in carrying out each of the basic steps are well documented in specifications and standards.

For cleaning, drying, preserving, and unit packaging the basic governing specification is MIL-P-116.

Standard markings (an example is given in Fig. 15-3[1]) for use on military supplies and

equipment to be shipped or stored are given in MIL-STD-129C.

### 15-2 STORING

All military material must be packaged so that it is capable of withstanding the effects of extreme environmental conditions during storage. The various classes and types of storage for which the packaging engineer must provide are given in par. 15-2.1.

### 15-2.1 TYPES OF STORAGE

(1) *Class A, Dormant Storage*. The packaged item is protected against the elements by preservation, sealing, covering, or placing in shelter and buildings, either dehumidified or nondehumidified. Items in dormant storage

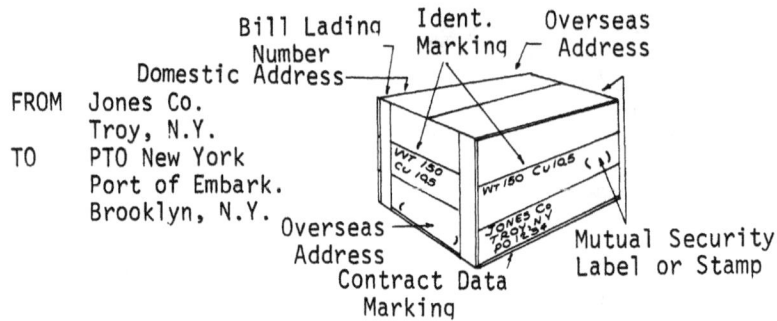

*Figure 15-3. Markings for Boxes*

are not operated between reprocessing cycles.

(2) *Class B, Active Storage.* The packaged item is protected by the same basic measures as dormant storage except that certain preservation requirements are replaced or supplemented by specific periodic exercising, either by running the equipment or by operating the equipment with an external power or driving source.

(3) *Type 1, Outside Storage.* The storage area is exposed to the extremes of local, natural environments. The stored item must be protected from all weather elements as well as fungus, pests, dust, pilferage, and the unpredictable results of idle curiosity. For open storage, maximum use of known packaging methods and materials is specified to insure the protection of the material. Surveillance and maintenance are required to minimize deterioration.

(4) *Type 2, Sheltered Storage.* The material is stored in ventilated or unventilated, heated or unheated buildings and shelters. The stored material is not protected from atmospheric changes of temperature and humidity. Surveillance and maintenance are required to keep deterioration to a minimum.

(5) *Type 3, Dehumidified Structural Storage.* The material is stored in structures in which the atmosphere is maintained at a

relative humidity of 40% or less. Controlled humidity provides the highest degree of protection, and little surveillance and maintenance are required.

(6) *Type 5, Dehumidified Nonstructural Storage.* The material is stored in completely or partially sealed packages with a mechanical or static dehumidification of each item, single or in series, with the relative humidity of the atmosphere within the interior areas not exceeding 40%. Inspection and surveillance are required to ensure constant protection.

## 15-2.2 STORAGE OF EXPLOSIVES

The storage of devices containing explosive material is controlled by the potential hazards of the explosive. Explosives must be stored in standard ammunition magazines designed for this purpose and in special areas. These areas generally are not heated nor wired for electricity. The package must be labeled to identify the contents as explosives and to indicate the explosive classification. Periodic surveillance and inspection are required. For additional details, see par. 20-5.

## 15-3 SHIPPING

### 15-3.1 TRANSPORTATION ENVIRONMENTS

During transportation the most damaging

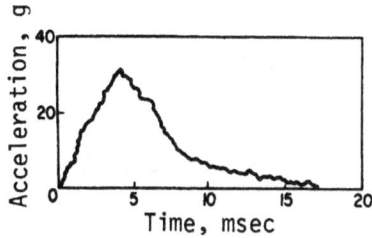

Figure 15-4. Plot of Typical Shock Motion

environments are from shock and vibration. Vibration is an oscillating motion which can be either repetitive or transient. Shock connotes impact, collision, or a blow, usually caused by physical contact. It denotes a rapid change of loading or a rapid change of acceleration with a resultant change of loading.

Although an equation may be developed to define a particular shock motion, actual shock motions are usually complex and can best be described graphically. A typical example is given in Fig. 15-4[1] where acceleration is used as the parameter. Velocity or displacement also could have been used as the parameter. There are a variety of possible shock motions because pulse shape, duration, and peak acceleration may vary[2,3].

## 15-3.2 TRUCK TRANSPORT

When considering truck transport, we are concerned mainly with the effect of vibration on the cargo. Vibration frequencies in motor trucks are determined by:

(1) The natural frequency of the unsprung mass on the tires

(2) The natural frequency of the spring system

(3) The natural frequency of the body structure.

The amplitude of vibration depends on

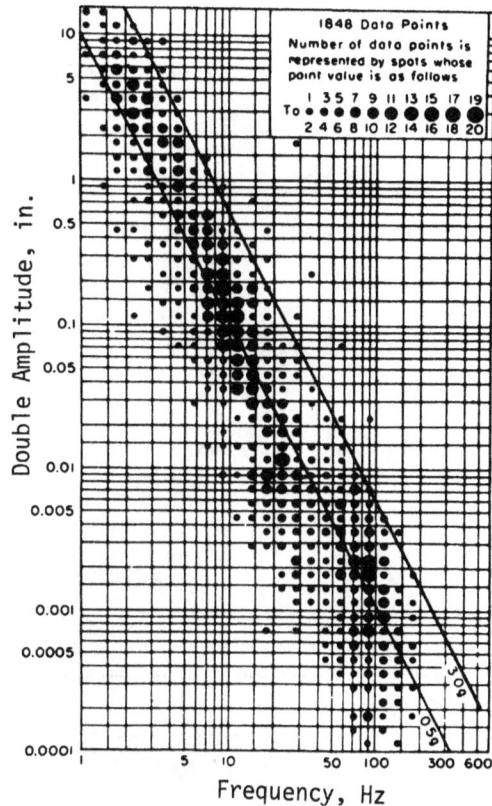

Figure 15-5. Truck Transportation Vibration Data

road conditions and vehicle speed. Intermittent road shocks of high magnitude can occur, with resultant extreme truck body displacements. If the cargo is not secured, it may be severely damaged. The predominant natural frequencies of various military transport vehicles, as measured in the cargo space, are given in Table 15-1[1]. Fig. 15-5[1] presents vibration data measured in the cargo space of trucks and trailers.

## 15-3.3 RAIL TRANSPORT

Freight car vibrations are caused by track and wheel irregularities, and occur principally in the lateral and vertical directions. The most damaging phases of rail shipment are those

TABLE 15-1

PREDOMINANT FREQUENCIES MEASURED IN CARGO SPACES OF
VARIOUS MILITARY TRANSPORT VEHICLES

| Type of Vehicle | Direction of Acceleration | Predominant Frequencies, Hz | | |
|---|---|---|---|---|
| | | Springs | Tires | Body |
| Truck (2-½ tons) | Vertical | 2 to 4 | 8 to 13 | 70 to 180 |
| | Longitudinal | – | 10 to 20 | 70 to 100 |
| | Lateral | 2 | 10 to 20 | 100 to 200 |
| Truck (3/4 ton) | Vertical | 2 to 3 | 5 to 10 | 60 to 110 |
| | Longitudinal | – | – | 70 to 100 |
| | Lateral | – | – | 60 to 70 |
| Trailer (1 ton) | Vertical | 3 to 5 | 8 to 10 | 50 to 100 |
| | Longitudinal | – | – | 50 to 100 |
| | Lateral | 2 | – | 50 to 120 |
| M-14 Trailer | Vertical | 1 to 4 | 7 to 10 | 50 to 70 |
| | Longitudinal | 3 to 4 | 8 to 10 | 200 and greater |
| | Lateral | 2 to 4 | – | – |
| M1, 2T Trailer | Vertical | 2.5 to 5 | 7.75 to 10.5 | 100 to 150 |

shock and transient vibrations that occur during coupling, starting, and stopping[4].

Fig. 15-6[1] presents data on the number of occurrences of shock of various levels recorded on the floor of a freight car during a 700-mile trip. Lateral and vertical shock occurrences are plotted against car speed. Table 15-2[1] shows the percentage of travel time in each speed range during the 700-mile trip. The duration of the shock impulses was estimated to be between 10 and 50 msec.

By observing a representative number of railroad operations, Fig. 15-7[1] was constructed. It shows the velocity of impact during switching versus the percentage of occurrence at each velocity. From this figure, 7 mph is obtained as the approximate mean speed of impact. This is above the 5-mph limit for which the switching gear provides cushioning protection. The maximum longitudinal acceleration at various impact velocities from 1 to 7 mph is plotted in Fig. 15-8[1].

## 15-3.4 AIR TRANSPORT

The loadings that are important during air transport are those that occur during flight in rough air. The in-flight shock and vibration environment is generally not too severe. The dynamic loadings are differentiated from shock loadings in that they consist of fairly high magnitude acceleration imposed for a prolonged period of time. These accelerations can be as high as 2 to 3 g during normal operation of large transport aircraft.

The most damaging conditions encountered during air shipment are the shocks resulting from handling operations. This is shown in Fig. 15-9[1] which is a plot of the maximum shocks recorded during a test shipment by a major airline. For the test, two impact recorders were placed in a wooden box whose total weight was 73 lb. Both longitudinal and vertical shocks were recorded.

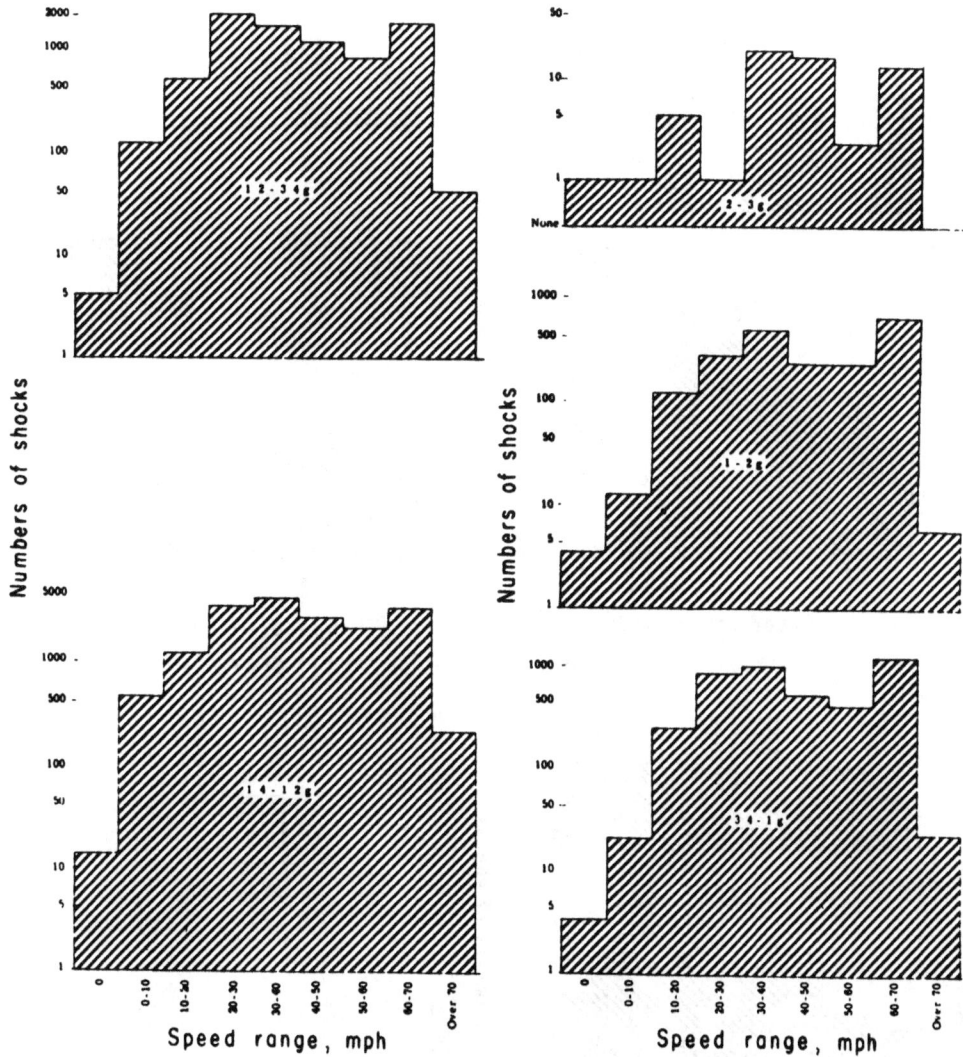

(A) VERTICAL SHOCKS MEASURED ON FREIGHT

*Figure 15-6. Freight and Freight Car Shock Measurements*

## 15-3.5 SHIP TRANSPORT

Vibrations are produced in ship structures from an imbalance or misalignment of the propeller shaft system and the flow of water from the propellers around the ships' hull. The frequency range of these vibrations is about 5 to 25 Hz, with acceleration reaching a maximum of about 1 g.

During normal voyages ship cargos do not experience damaging shocks. The highest shocks occur during loading and unloading operations.

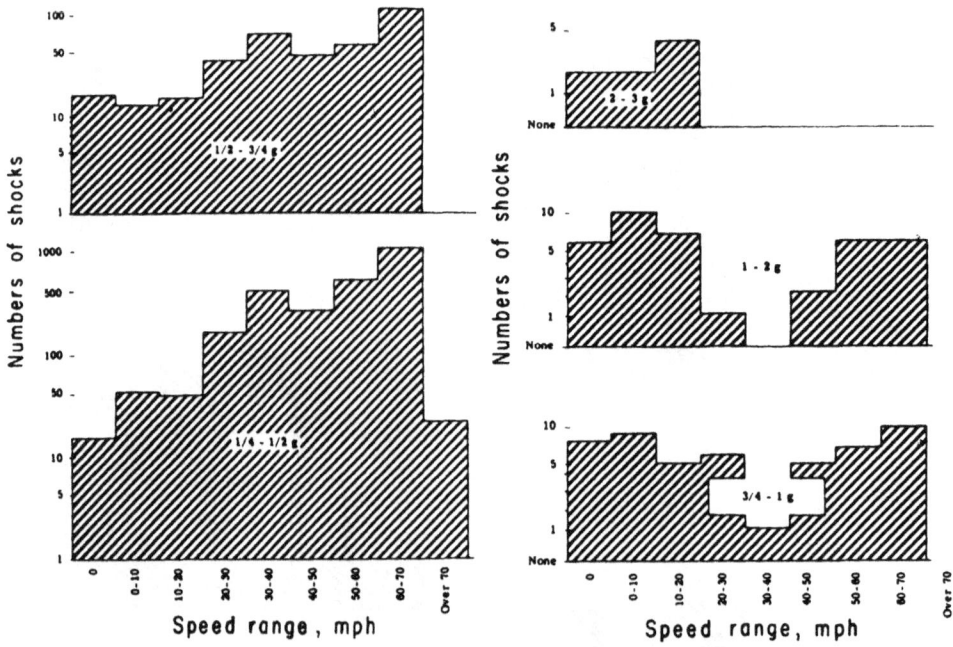

(B) LONGITUDINAL SHOCKS MEASURED ON FREIGHT CAR FLOOR

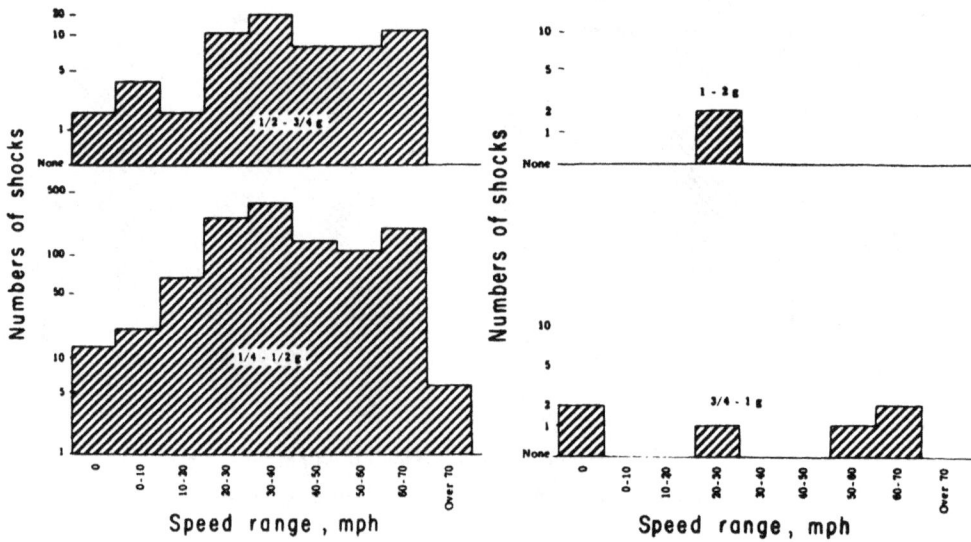

(C) LATERAL SHOCKS RECORDED ON FREIGHT CAR FLOOR

Figure 15-6. Freight and Freight Car Shock Measurements (Cont'd)

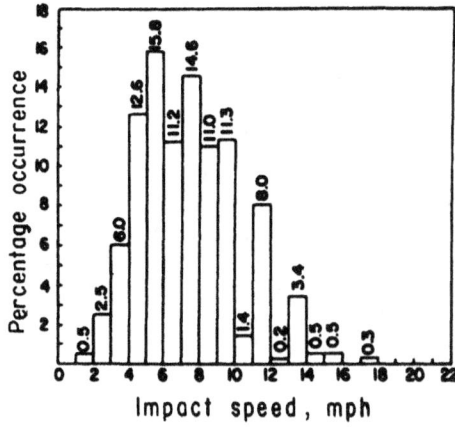

Figure 15-7. Impact Speed During Freight Car
Switching Operations

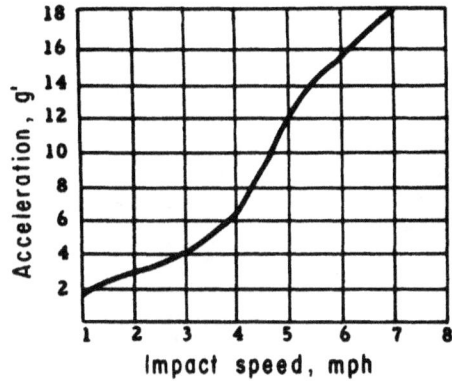

Figure 15-8. Maximum Longitudinal Acceleration of Freight Car Body vs Switching Impact
Speed

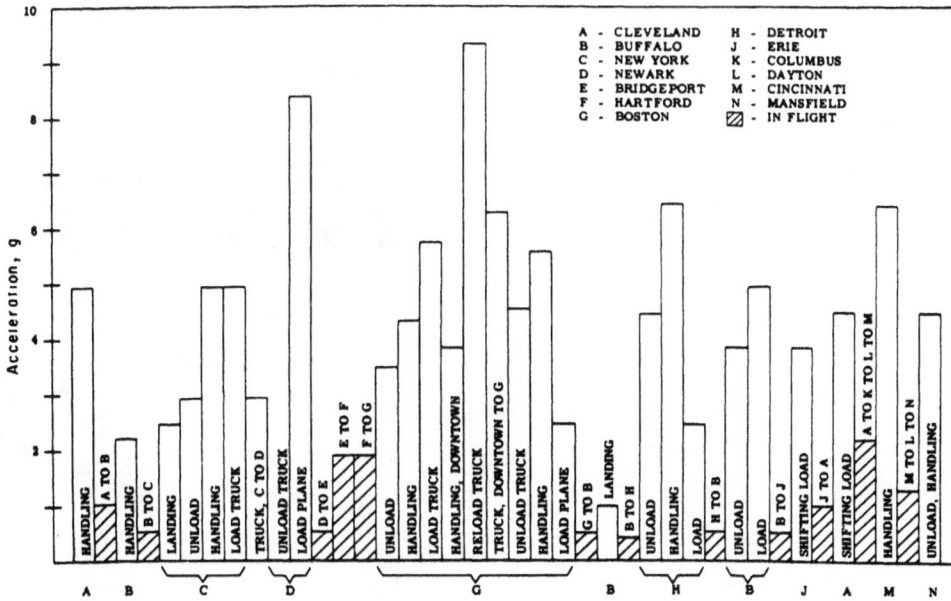

Figure 15-9. Maximum Shocks Recorded During Airline Test Shipment

15-9

## TABLE 15-2

### PERCENTAGE OF TRAVEL TIME VS
### SPEED RANGE FOR FIG. 15-6

| Speed Range, mph | Travel and Recording Time, % |
|---|---|
| 0 | 9.3 |
| 0-10 | 14.3 |
| 10-20 | 9.4 |
| 20-30 | 20.3 |
| 30-40 | 21.2 |
| 40-50 | 10.8 |
| 50-60 | 6.1 |
| 60-70 | 8.3 |
| over 70 | 0.3 |

## REFERENCES

1. AMCP 706-121, Engineering Design Handbook, *Packaging and Pack Engineering*.

2. R.E. Barbiere and Wayne Hall, *Electronic Designer's Shock and Vibration Guide for Airborne Applications*, Aeronautical Systems Division, Wright-Patterson Air Force Base, Ohio, Report WADC TR 58-363, December 1958 (AD-204 095).

3. E.C. Theiss, et al., *Handbook of Environmental Engineering*, Aeronautical Systems Division, Wright-Patterson Air Force Base, Ohio, Report ASD TR 61-363, 1961.

4. C.M. Harris and C.E. Crede, Eds., *Shock and Vibration Handbook*, Vol. 3, McGraw-Hill Book Co., Inc., N.Y., 1961.

# CHAPTER 16

## COMPONENT DESIGN

### 16-1 INTRODUCTION

All timers can be considered to be composed of 5 separate components:

(1) An *Actuator* to start the timer

(2) A *Power Supply* or source of energy

(3) An *Escapement* (Frequency Converter) that serves as the time base, periodically releases energy, maintains the oscillations, and in some movements activates the counting system

(4) A *Counting System* that counts a predetermined number of oscillations

(5) An *Output* such as a settable readout, or switch closure, or cam rotation which provides an output pulse.

Using different combinations of these components, many types of timers can be constructed. Some of the factors that must be considered in the design of a mechanical timer are:

(1) Timing accuracy required

(2) Duty cycle required

(3) Spin environment (max rpm)

(4) Acceleration environment (max g's)

(5) Allowable gross weight

(6) Allowable gross volume

(7) Temperature extremes (operational and storage)

(8) Number and type of outputs required

(9) Pressure environment

(10) Allowable impact shock (direction and amplitude)

(11) Allowable vibration.

A good timer design has the following features:

(1) Reliability of action

(2) Resistance to damage in handling and use

(3) Resistance to deterioration in storage

(4) Simplicity of construction

(5) Adequate strength in use

(6) Compactness

(7) Ease of manufacture

(8) Economy of manufacture.

The last item, economy of manufacture, deserves special note because it affects each of the preceding features. The best way to achieve economic mass production is to use as many as possible proven components from previous timer designs. The designer therefore should be acquainted with the characteristics of military timers. New designs require extensive tests before economy can be assured.

The most common military usage of mechanical timers is in the mechanical time fuze. The timer clockwork is an integral part of a mechanical time (MT) fuze, and thus is subjected to and must pass all tests that fuzes are subjected to. The development of the clockwork will occur concurrently with the development of the fuze. The clockwork for MT fuzes used in projectiles may be called upon to operate in the time interval between the instant of firing to the initiation of the primer. Thus the forces existing within the realm of interior ballistics, exterior ballistics, and terminal ballistics must be considered. Disturbing forces are created on delicate clockwork mechanisms from setback, spin, temperature variations and extremes, and impact decelerations. These forces will cause the timepiece to deviate from its at rest behavior. Therefore, materials used for the clockworks should be strong, light as possible, and properly lubricated.

Fig. 12-2 shows a mechanical timer and a block diagram of its five basic components. The *Time Base* is a mechanical escapement; the *Counter* is a gear train; the *Output* is a cam operated switch. The *Power Supply* is a mainspring driving the output shaft; the *Actuator* holds the escapement until removed.

## 16-1.1 DESIGN CONSIDERATIONS

### 16-1.1.1 Tactical (Fuze Timers)

Mechanical fuze timers are designed for different tactical situations[1]. They may be used with various ammunition items such as artillery projectiles, aircraft bombs, sea mines, small arms projectiles, rockets, and guided missiles. Each series has its own tactical requirements and will experience different environmental forces during operation and storage. A mechanical timer may be designed for one specific round or it may be functional and applicable to a given series of one type of projectile. For example, it may be used with all HE projectiles that can be fired from howitzers and guns ranging from 75 mm to 155 mm. The latter type timer will experience muzzle velocities ranging from 420 to 3000 fps. Some of the tactical situations that the timer will be subjected to are: ground detonation (surface or after penetration), or air burst.

### 16-1.1.2 Environmental

Environmental conditions will influence type of materials, method of sealing, design of component parts, and method of packaging.

(1) Operating Temperature. The timer must withstand temperatures ranging from an air temperature of 125°F (ground temperature of 145°F) in hot-dry climates to an air temperature of −50°F (ground temperature of −65°F) in cold climates. Temperatures can drop to −80°F in bomb bays of high-flying aircraft, and aerodynamic heating can raise the temperature of missiles launched from high speed planes above 145°F[2].

(2) Storage Temperature. −70° to 160°F, operable after removal from storage.

(3) Relative Humidity. 0 to 100%.

(4) Water Immersion. For certain applications the timer may be required to be waterproof. For this test, the timer must be operable after immersion in water at 70°F ± 10 deg F under a pressure of 15 ± 5 psi for one hour.

(5) Rough Treatment. The timer must withstand the rigors of transportation and rough handling.

(6) Fungus. The timer must not support fungus growth.

(7) Surveillance. Operable after 10-20 yr storage.

### 16-1.1.3 Safety

Another of the designer's important considerations is safety in manufacture, in loading, in transportation, in storage, and in assembly to the operating system. This applies to timers in fuzes which initiate an explosive train.

## 16-1.2 TIMER ACCURACY

It has been reported that accuracies under 0.1% can be achieved with mechanical timers employing tuning forks or crystals for their time base. The best spring-mass systems can approach 0.5% in timer accuracy. Increased accuracy generally costs more.

Spring-mass systems are best employed for long time durations where the frequency of the spring-mass can be 5 Hz or lower. For time intervals of fractions of seconds, the tuning fork or crystal time base operating at 1000 Hz or higher may be used. Spring-mass systems operating at high frequency (above 100 Hz) can provide for good timer accuracy for short time intervals[3].

All timers have some inherent error in their operation. Proper adjustment can reduce this error to a lower limit depending on the type of mechanism. Spring-mass timers commonly available may have errors ranging from 30 sec per day to 0.1 sec per day, with the cost of the timer increasing as the inherent error decreases[4].

Environment also will affect timer accuracy. Special modifications must be made when the timer is to operate at extremely low or high temperatures, or under vibration or shock. Operation in the military range of −50° to 125°F requires special design and assembly techniques. Table 16-1[4] shows environmental temperature effects on various types of timepieces.

## 16-2 POWER SOURCES

The power supply provides the torque required to operate a mechanical timer. The five types of power supplies which may be applied are spring motors, centrifugal drives,

AC motors, DC motors, and ram air.

## 16-2.1 SPRING MOTORS

There are basically two types of mainspring motors*, the spiral spring and the "constant-force" negator. These are shown in Figs. 16-1 and 16-2, respectively. The spiral spring can be made more compact for small outputs because only one spring panel is required, whereas the negator spring must use a storage drum and a separate output drum. The negator spring is capable of producing more usable energy from a spring of equal dimension, because it supplies a constant usable torque throughout its entire length. Another advantage of the negator spring is that no interleaf friction is present since the spring uncoils with a rolling motion. The constant torque negator is especially desirable with torque-sensitive escapements. Compared with battery supplies, mainsprings generally have the advantage of lower cost, greater ruggedness, larger storage life, and better low-temperature operating capability.

### 16-2.1.1 Torque

For the spiral spring the factors of torque, stress, width and thickness are related to each other

$$G = \frac{\sigma b h^2}{6} , \text{lb-in.} \qquad (16-1)$$

where

$G$ = torque, lb-in.

$\sigma$ = stress, psi

$b$ = width, in.

$h$ = thickness, in.

---

*When a strip of spring steel is wound around itself so that each turn touches the next, these are known as power or motor springs. Small spiral springs when wound with space between the coils are called hairsprings. Hairsprings can be analyzed theoretically more easily than power springs because the disrupting factor of surface friction between coils is not present.

**TABLE 16-1**

**APPROXIMATE ACCURACIES OF TIMEPIECE TYPES**

| Timekeeper | Temperature Range, °F | | | | | |
|---|---|---|---|---|---|---|
| | −30°/150° | 0°/120° | 30°/120° | 50°/90° | 72° | Constant Temp |
| Chronometer | D | C | B | A | A | A |
| Chronometer Watch | D | D | C | B | A | A |
| Balance-Driven Clock (adjusted)* | D | D | C | B | B | B |
| Balance-Driven Watch (adjusted) | D | D | D | C | C | C |
| Balance-Driven Clock (unadjusted) | E | E | D | D | C | C |
| Balance-Driven Watch (unadjusted) | E | E | D | D | C | C |
| Spring-Wound Watch (adjusted) | D | D | D | C | C | C |
| Spring-Wound Clock (adjusted) | D | D | D | C | C | C |
| Spring-Wound Watch (unadjusted) | F | F | E | E | D | D |
| Spring-Wound Clock (unadjusted) | G | G | F | F | F | F |
| Electric-Rewind Clock | F | F | E | D | D | D |
| Stalled-Motor Clock | G | G | F | E | E | E |

| | |
|---|---|
| A ± 1  sec/day | E ± 1 min/day |
| B ± 5  sec/day | F ± 2 min/day |
| C ± 10 sec/day | G ± 5 min/day |
| D ± 30 sec/day | |

*Adjusted means timed in more than one position for 24 hr or more. Rate of watch in all positions is equal.

In Eq. 16-1 $G$ and $\sigma$ are the maximum values attainable with any particular spring, taken with the spring let back one revolution from fully wound. Stress has a fairly definite relation to thickness (as shown in Fig. 16-3[5]). This figure is valid for springs with a length $\ell$ to thickness $h$ ratio of about 7000. With higher ratios the curve would shift downward; with lower, upward. Also as the ratio of arbor diameter $d_1$ (Fig. 16-1) to

thickness $h$ increases, the curve would shift downward.

*Example:* Find the maximum torque delivered by a spring 0.5 in. × 0.05 in. × 30 ft wound on a 1 in. arbor. From Fig. 16-3 the stress corresponding to a thickness of 0.05 in. is about 210,000 psi. Substituting in Eq. 16-1 yields

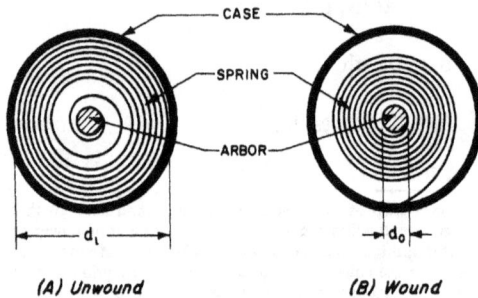

*Figure 16-1. Spiral Spring Power Motor[4]*

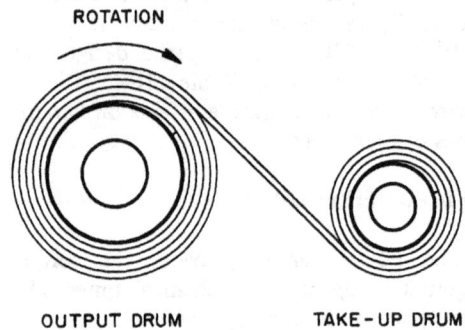

*Figure 16-2. Negator Spring Motor*

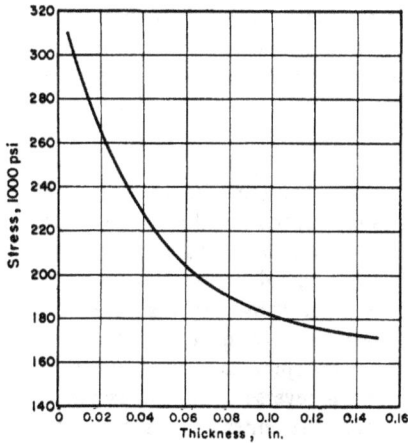

*Figure 16-3. Average Maximum Solid Stress in Power Springs (for ℓ/h approximately 7000)*[5]
*Reprinted with permission of McGraw-Hill Book Company*

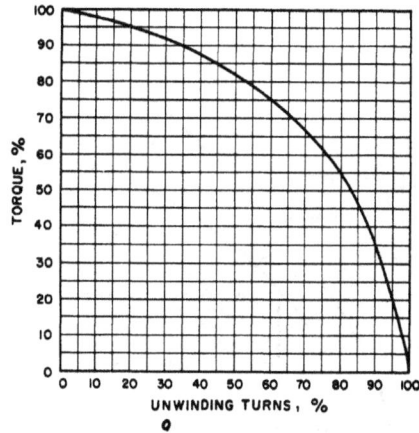

*Figure 16-4. Typical Torque Curve for Power Spring*[5]
*Reprinted with permission of McGraw-Hill Book Company*

$$G = \frac{210,000 \times 0.5 \times 0.05^2}{6} = 43.7 \text{ lb-in.}$$

In order for a spring to contain the maximum number of turns in a given volume, the value ($\ell \times h$) should equal 1/2 the available volume (drum volume minus arbor volume). Expressed mathematically in two dimensions, (the spring width $b$ being constant)

$$\ell h = \frac{1}{2}\left(\frac{\pi d_2^2}{4} - \frac{\pi d_1^2}{4}\right) \quad (16\text{-}2)$$

$$\text{or } \ell = \frac{d_2^2 - d_1^2}{2.55h} \quad (16\text{-}3)$$

$$\text{or } d_2 = (2.55\ell h + d_1^2)^{1/2} \quad (16\text{-}4)$$

where

$\ell$ = length, in.

$h$ = thickness, in.

$d_2$ = case drum diameter, in.

$d_1$ = arbor diameter, in.

With reference to the arbor size, it has been found, both experimentally and practically, that the arbor diameter should be from fifteen to twenty-five times the spring thickness $h$.

The number of turns $N$ can be expressed by the equation

$$N = \sqrt{\frac{\sqrt{2(d_1^2 + d_2^2)} - (d_1 + d_2)}{2h}} \quad (16\text{-}5)$$

In the foregoing paragraph the torque and stress are at a maximum. In order to determine the torque at any intermediate position of the spring from solid to free, the curve of Fig. 16-4[5] may be used. The shape of this curve is typical and fairly constant for springs of well proportioned design.

*Example:* A 15-turn spring has a maximum torque of 20 lb-in. Find the torque when 6 turns are let out. From Fig. 16-4 at 40%, the torque is 88%. Thus 88% of 20 is 17.6 lb-in.

16-5

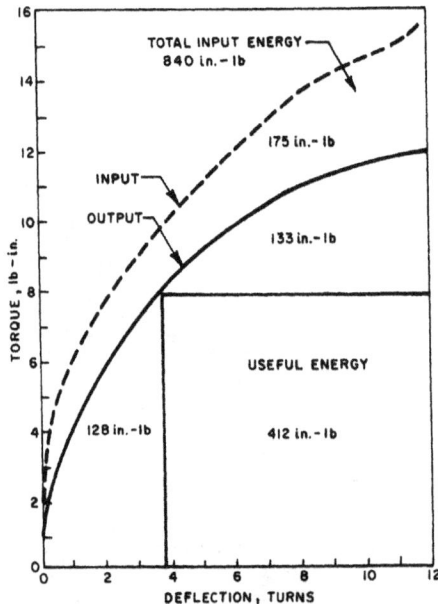

Figure 16-5. Input and Output Curves of a
Conventional Spiral Power Spring[6]

The spiral spring motor is relatively inexpensive, compact, and convenient. The moment delivered by the spring decreases as it unwinds—from maximum at release, to zero at full rundown. The spiral spring can produce a specific torque at only one point; at other times it is either too great or too small. Further, all associated components—transmission elements, escapements, governors, etc.—must be designed to take the maximum full wind torque. Most spiral springs operate through a maximum of only about 17 useful revolutions. One way to increase the number of turns is to lengthen the spring and use thinner material. The practical length-to-thickness ratio ($\ell/h$) is about 15,000:1. Additional length always increases intercoil friction. Gearing will be needed if the driven mechanism must operate through more revolutions than the spring.

## 16-2.1.2 Efficiency

The useful output of a spiral spring is about

1/2 of the input as shown in Fig. 16-5[6]. The number of turns at high torque also is limited. To increase total output, it is necessary to design for higher torque and use gearing to obtain the required length of run.

A special form of the spiral spring is the prestressed spiral spring (spirator). The spirator—a compact, single-axis mechanism—resembles a spiral spring in appearance, but contains a *prestressed* spring. The spirator is prestressed to form a tight coil. After forming, the spiral is turned inside out; the outer end is reverse bent, and fastened to the arbor; and the inner end is secured to the case. This action results in:

(1) High resisting moment at the start of winding

(2) Gradual increase in stress and torque during winding

(3) Use of thinner material and greater length of prestressed material

(4) Activity of a greater proportion of length during run-down.

The outstanding characteristic of the spirator is its ability to sustain delivery of useful torque through many turns. The input-output curves for a typical spirator is shown in Fig. 16-6[6]. The spirator is obviously superior to the conventional spiral spring. Furthermore, the spirator produces more total useful energy for the same weight of material, making it much more efficient in terms of useful energy compared with the energy potential of the material.

## 16-2.2 NEGATOR MOTORS

With the constant-torque negator motor (also known as a negator B motor) the motor is wound from drum B to A (Fig. 16-2). On unwinding as indicated by the arrow, almost constant torque is developed due to the tension produced by the spring plus the

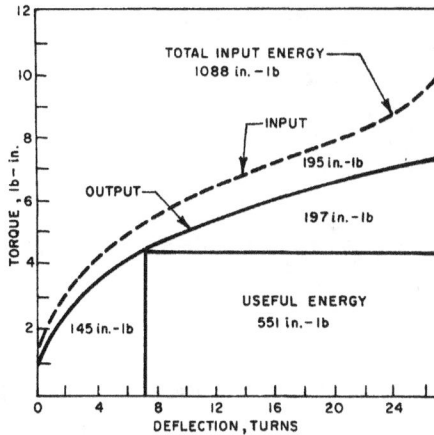

Figure 16-6. Input and Output Curves of a Prestressed (Spirator) Spring of the Same Weight as Power Spring in Fig. 16-5[6]

moment developed due to reverse bending of the strip as it is unwound from the output drum. Normally the strip is attached to the A drum, but is merely allowed to wind about the B drum. Besides constant torque, the B motor has the advantage of no intercoil friction.

About 60 revolutions is the upper practical limit for the B type motor if constant torque is required. If a slight linear gradient can be tolerated, type B motors can be made up to 100 turns. The B type motor can be designed to deliver the exact number of turns required, and thus eliminate or reduce required gearing.

The constant torque or low linear gradient of the B motor makes it easy to control the output. Constant torque often is misinterpreted as constant speed, which it is not. Constant speed requires a governor, the complexity of which can be reduced greatly by coupling to a B motor.

Since there is no intercoil friction, there is no need for lubrication. Optimum B motor design to produce maximum total energy in a given space, accepting a slight torque variation, should be considered unless there is no need for requiring absolutely constant output torque.

Fatigue should be considered in a B type motor use due to high repetitive stresses. A motor rating should consider both torque required and required life. Fig. 16-7[6] shows the input-output curve for a B-type negator.

## 16-2.3 COMPARISON OF SPRING MOTORS

An energy comparison of the three types of motor—spiral, spirator, negator—for the same material weight as shown in Fig. 16-8[6]. A tabular comparison is given in Table 16-2[6].

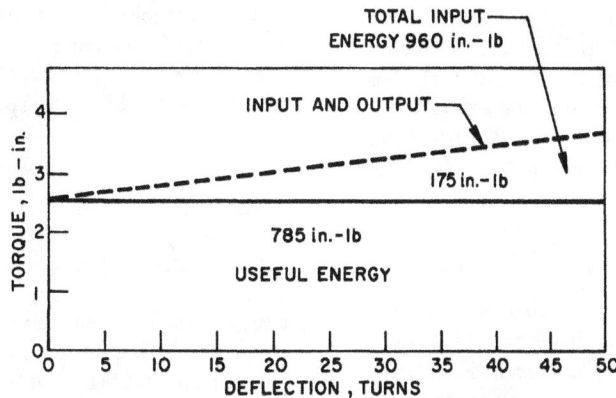

Figure 16-7. Input and Output Curves of a B Type Negator Motor of the Same Weight as the Spiral Spring Motor in Fig. 16-5[6]

16-7

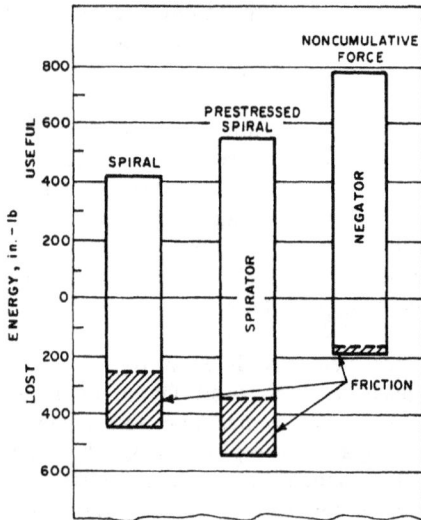

Figure 16-8. Energy Comparison of Three Types of Motor Spring Having the Same Weight of Material[6]

For a thorough analysis of spiral spring motors under spin and nonspin environments in both axisymmetric and nonsymmetric conditions consult Ref. 6. For a detailed description of spiral spring motors consult Ref. 7.

The diminishing torque characteristic of the spiral spring drive as it unwinds can be minimized to a certain extent through combination of an electric drive motor and the spiral spring drive. The timer drive mechanism is designed so that when a selected amount of rotation of the output shaft has been made under power from the spring drive, a set of contacts is closed which permits an electric motor to rewind the spring to a selected torque level[7].

## 16-2.4 CENTRIFUGAL DRIVE

Centrifugal force can be used as the power source. Although this drive has no present application it is described here to illustrate a design approach. Note, for example Fuze MTSQ, M502A1 that is fired from a 105 mm howitzer. As the projectile spins, internal weights move outward radially thereby applying torque to the main pinion that is geared to the escapement wheel and lever. Power is supplied to the escapement as long as the projectile is spinning. The torque $G$ on the centrifugal gear section shown in Fig. 16-9[1] can be expressed as

$$G = \frac{m\omega^2}{12} r_s r_p \sin \phi, \text{ lb-in.} \qquad (16\text{-}6)$$

where

$m$     = mass of the gear segment with its center of mass at A, slug

$\omega$     = angular velocity, rad/sec

$r_p$ and $r_s$ = radii shown on Fig. 16-9, in.

$\phi$     = angle through which the gear could be turned by this torque, deg

For this gear, the mass $m$ is 0.014 slug, $r_s$ and $r_p$ are 0.48 and 0.16 in., respectively and $\phi$ is 135 deg. Let us assume this projectile and fuze are fired from a 105 mm howitzer with a velocity of 2200 fps at a spin of 210 rad/sec. This produces an applied torque of 3.95 lb-in. The gear ratio is 275 so that the torque on the escapement shaft is decreased to 0.0144 lb-in. However, because there are friction and bearing losses within the gear train, assume 28% of the theoretical torque will appear at the escapement shaft or 0.0040 lb-in. Since two centrifugal gears are always used in a drive system of this type, all torque values should be doubled, or approximately 0.008 lb-in.

## 16-2.5 AC MOTORS

AC motors operate on a supply of alternating current. They are usually synchronous, with a speed exactly related to line frequency. The relation among rpm, frequency, and the number of poles, can be expressed as

$$\omega = \frac{120f}{p} \qquad (16\text{-}7)$$

## TABLE 16-2

### TABULAR COMPARISON OF MOTOR SPRINGS[6]

| | Spirator (Spiral) | Prestressed Spiral | Negator (Noncumulative Force) |
|---|---|---|---|
| **Physical Characteristics** | | | |
| Mounting | On a single arbor with restraining ring or case. | Same as spiral spring. | On two spindles: one output, one take-up. |
| Relaxed Configuration | Coils expanded against retaining case. | Same as spiral spring. | Tightly coiled spiral on take-up spool. |
| Wound Configuration | Tight coil on arbor. | Same as spiral spring. | Tightly coiled on output drum. |
| Number of Rotations | Seldom greater than 15. | Readily, 20 turns. As many as 50 turns possible at reduced energy storage efficiency. | Limited only by space available (100 turns present practical limit). |
| Forming Process | Wound directly on arbor, or some reverse winding of longer springs. | Prestressed through sequential bending and thermal processing then reverse wound and fastened to arbor. | Prestressed through sequential bending and thermal processing. |
| **Performance** | | | |
| Charging | Stressed cumulatively as a unit by winding to a tight spiral. | Same as spiral spring. | Stressed uniformly by reverse wrapping onto second drum. |
| Useful Energy Storage | Limited by number of effective revolutions, moderate nonuniform working stresses. | More useful energy because of high working stresses. | Highest energy storage because of higher working stresses, nearly uniform stress distribution throughout material. |
| Torque Output | Optimum value available only momentarily. Diminishes rapidly from maximum throughout cycle. | Diminishes on delivery stroke less rapidly than power spring from a beginning maximum. | Initial torque maintained throughout cycle.* |
| Efficiency | Energy wasted by friction and in compensating for nonuniform rate of release. | Same as spiral spring except much lower percentage of input energy need usually be wasted. | Negligible friction loss and little energy wasted because it is released at uniform rate.* |
| Input Torque | Effort increases during winding. Torque input required to fully wind reaches a high peak. | Effort also increases but less rapidly. Lower input required to fully wind. | Winding effort remains low because restraining torque does not build-up. |

*For constant-torque B motor. Low-gradient B motor produces very slight torque increase which decreases gradually at a constant rate during rundown.

16-9

**TABLE 16-2**

**TABULAR COMPARISON OF MOTOR SPRINGS[6] (Cont'd.)**

| | Spirator (Spiral) | Prestressed Spiral | Negator (Noncumulative Force) |
|---|---|---|---|
| **Application Characteristics** | | | |
| As a Drive Unit | Limited number of turns. Gearing usually required. | More turns usually possible. Simplification of associated gearing possible. | Greatest number of turns. Sometimes applied without gearing. |
| To Deliver Linear Force | Requires gearing and linkage to provide linear displacement. | Can provide longer linear displacement, with less linkage than power spring. | Large number of turns available for conversion to long-displacement linear force. |
| Adjustability | Wide output torque range available for output adjustment. | Narrow output torque range available for adjustment. | Output nonadjustable once mounted unless bands are specially formed. |
| Associated Components | Must be able to handle high maximum output torque. | Lower maximum torque allows use of lighter duty parts. | Constant output allows use of light-duty parts. |
| Output Control | Governor escapement must operate over wide torque range. | Governor operates over narrower torque range. | Governor required only for very close control. |

where

$\omega$ = motor shaft speed, rpm

$f$ = frequency of electric supply, Hz

$p$ = number of poles

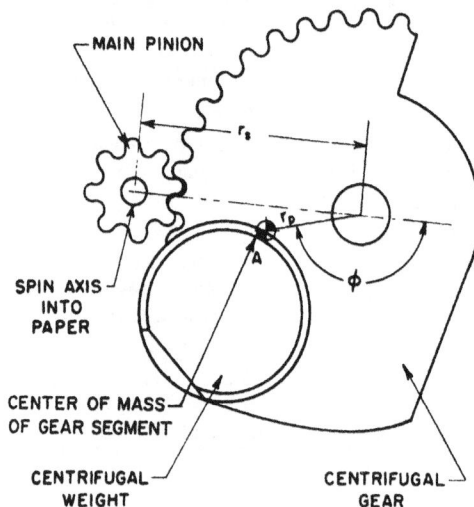

*Figure 16-9. Centrifugal Drive Employed in the MTSQ, M502A1 Mechanical Time Fuze*

A synchronous motor will, on the average, be exactly as correct as the line frequency, rotating through an angle corresponding to a pair of poles during each cycle. Instantaneous speed variation of a synchronous motor is called "hunting".

Some of the more salient features of the synchronous motor are:

(1) Exact relation exists between speed and line frequency.

(2) Speed is independent of line voltage and wave form distortion as long as synchronism is maintained.

(3) Pull-out torque and starting torque vary as the square of the applied voltage.

(4) Input power increases as square of applied voltage where shaft load is negligible.

(5) Hunting is a superimposed low frequency modulation of instantaneous speed, with average speed held constant. Hunting may be important where output is driven at high speed.

(6) Very special designs are available where it is economical to produce a unit with a large number of poles (low synchronous speed) in a small size. These units are at their best where load demand for shaft power is almost negligible. Low speed means fewer gear reduction stages and longer bearing life.

(7) Most economical starting means is generally by a shading coil, which does not have high starting torque. Thus bearing friction must be very low, even at low temperatures. Capacitor starting provides more starting torque in larger units.

(8) Hysteresis motors employing a motor of permanent magnet material commonly are employed to develop synchronous torque.

(9) Reluctance motors using a notched rotor often are used as synchronous drives. These develop relatively poor starting torque that can be boosted by providing a small amount of induction motor action on top of the synchronous behavior.

(10) With very many poles, synchronous speed is so low that the motor readily synchronizes with a minimum of starting torque.

(11) Extremely high accuracy can be obtained from synchronous motors driven from a temperature controlled crystal oscillator source. Somewhat lesser accuracy is obtainable from a tuning fork supply. However, these supplies are often cumbersome.

The electrical drive, as compared with weight and spring drives, has the great advantage of obviating the winding and running time limitation. Also the motor can be kept small since the speed of the working shaft or shafts usually is smaller than the motor speed, as a result of which the torque reduced to the motor shaft also becomes small. Although the electric motor represents the ideal form of drive, since the need for a regulator is also obviated, its applications are limited. When constructed as an asynchronous motor, it responds to voltage variations; and when designed as a synchronous motor, it responds to frequency variations (disadvantage of synchronous clocks that are connected to a nonfrequency-stabilized line). If frequency stability is assumed, the synchronous motor is an ideal driver.

## 16-2.6 DC MOTORS

Governor controlled DC motors can provide timer accuracies to 0.1%. A means for speed control, however, must be provided because the speed of DC motors is usually voltage dependent. A convenient method is to provide a set of centrifugally operated contacts on a plate fixed to the DC motor shaft, with a pair of slip ring brush contacts. The centrifugally operated contacts make and break armature current. When the required speed is reached, the contacts open, which may open the armature circuit thereby slowing down the motor or introduce a dropping resistor that also will decrease armature current. A dropping resistor is more common and can result in finer speed control.

For good accuracy the springs associated with the centrifugal governor must be free of internal strains that may gradually shift the governing point. The design must take into account such environmental factors as shock, vibration, and acoustic noise which may cause faulty operation. Contact ratings must be adequate for the life of the unit and protection must be provided against salt spray and humidity in service environments.

An example of a commercial timing motor manufactured by the A. W. Haydon Co. is the chronometric governor shown in Fig. 16-10(A)[8]. The motor rate is independent of load, line voltage, and temperature variations. The principle of operation is shown in Fig. 16-10(B).

The motor is geared directly to a cam-operated lever extending between a pair of leaf-spring contacts and so arranged that its

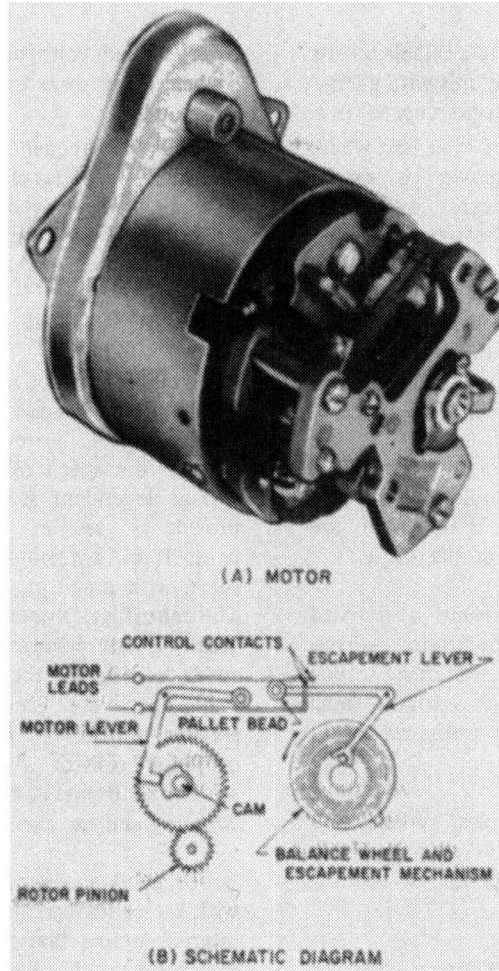

(A) MOTOR

(B) SCHEMATIC DIAGRAM

*Figure 16-10. Chronometric Governored DC Motor*

motion in either direction opens the contacts. Also located between the contacts is a pallet bead, insulated by mechanical attachment to the clock escapement lever in such a manner that the clock escapement merely releases these contacts allowing them to close while the motor-operated cam does the work of opening the contacts. The contacts are connected directly in series with the motor so that the faster the motor tries to go, the quicker the contacts are opened; yet they close at regular intervals as released by the escapement.

Thus pulses of full line voltage are applied to the motor at regular intervals, controlled by the escapement, and the duration of these pulses is determined by the travel of the motor. This results in a uniform travel of the motor during each time interval and, consequently, a constant rotor speed is obtained. The motor is designed to deliver maximum torque at slightly below minimum voltage under the least favorable temperature conditions. The motor, therefore, tends to overspeed but in doing so opens the governor contacts more quickly, thus shortening the duration of power pulses delivered to the motor. This results in a consequent reduction in motor speed.

Since the motor-driven cam transmits the same oscillating motion regardless of its direction of rotation, the motor is reversible. Further, when the clock balance wheel stops, it always tends to find its center position under the influence of the hairspring. Contacts always revert to a closed position when the current is turned off, and, consequently, the unit is reliably self-starting.

For power to keep the balance wheel oscillating, uniform mechanical impulses are delivered to the clock escapement through the contact springs that bear on the pallet bead, previously described, which is connected directly to the escapement lever. Uniformity of these mechanical pulses is assured since the contact springs are lifted a uniform distance by the oscillating motion of the cam-operated lever. Driving power for the balance wheel, therefore, is constant and independent of the motor speed or load and voltage variations applied to the motor.

The standard escapements are 300-beat and 900-beat types giving 150 or 450 complete oscillations per minute. The arrangement of the escapement and contact mechanism is such that the contacts are allowed to close at each half oscillation, and at this instant the escapement derives its energy from the contact springs. Thus the escapement mechanism is not called upon to perform any more work than when running free in an ordinary clock. Energy is stored in one contact spring just preceding the release of energy from the opposite spring, and the balance wheel thus gets its impulse at a time when the contact spring is not connected to the motor.

Some typical performance figures achieved by means of this governor are:

Voltage, ± 20%          Speed ± 0.1%
Temperature, −65° to + 165°F  Speed ± 0.3%
Vibration, 5-300 Hz, 10 g max.  Speed ± 0.5%

As in the case of AC motors, the power source for DC motors must be considered.

Often the source of DC power is a battery.

## 16-2.7 BATTERIES

The electrical battery has advantages by a large margin over a mainspring particularly in power density. This margin is claimed to be anywhere between 100 and 1500 times larger, depending upon the type of battery and the operating conditions. However, there is a restriction that the energy must be withdrawn at a relatively slow rate if all of it is to be made available for work. Other disadvantages of the electrical power supply include its change in output with temperature, its relatively short unattended storage life (1 or 2 yr for most batteries) and its inability to operate satisfactorily at temperatures below −40°F.

A battery that overcomes the last two disadvantages is the "reserve" battery that is inactive until ready for use. One method of making a reserve battery is to store the electrolyte in a glass ampule that is broken to activate the battery. Reserve zinc and silver oxide cells were in common use in World War II proximity fuzes.

Modern reserve batteries may be of the thermal type. A thermal battery uses electrolytes of various inorganic salts which are solid and nonconducting at normal military temperatures. Upon application of sufficient heat, the electrolyte melts and becomes conductive. A typical electrochemical system consists of calcium or magnesium anodes with various reducible cathodes, such as calcium chromate. The inorganic salts used as the electrolyte are generally potassium chloride and lithium chloride. Battery output from 2.5 to 450 V is available over the temperature range of −65° to 200°F.

Thermal cells typically are constructed either as a pellet or as a closed cup. The cells are then stacked into a cylindrical configuration with the diameter and the number of cells being varied to provide various capacities and voltages. The cell stacks then are insulated electrically and thermally, and encased in a metal container that is hermetically sealed. The container also includes, as heat source, an electric squib or mechanical primer. Thermal batteries range from 0.5 in. diameter by 0.5 in. length to 4 in. diameter and 5 in. length. Typical voltage tolerance is ± 10%.

Compared with other batteries then, the thermal battery has an acceptable shelf life. It has a versatile electrical output and is readily activated remotely. However, it shares the disadvantage with all batteries of a shortness of life, once activated. Also, thermal batteries are expensive, costing at least five times the amount of a good mainspring.

## 16-3 GEAR TRAINS

### 16-3.1 COMPUTING SYSTEMS

One of the basic components of a mechanical timer is the computing system. The computing system is essentially a counter that counts each of the periodic oscillations of the time base. For any given time interval, the system must be capable of registering the total number of oscillations which have occurred since time zero.

The most common type of computer is the gear train in which a set number of degrees of revolution of an output shaft represent a specific number of oscillations and, therefore, an interval of time. The output shaft is constrained to a constant rate of rotation by the oscillator or escapement.

### 16-3.2 TYPES OF GEAR TRAIN

There are several types of gear train which can be used to do the counting. These include straight spur gear and pinion sets (as found on ordinary clocks and watches), worm gear pairs, epicyclic trains, hypocyclic trains, harmonic drive, and synchronic index train.

### 16-3.2.1 Spur Gear Trains

Straight spur gear and pinion sets are used for the counting train in most timers. In small timing movements, the gears are about 0.3 to 0.4 in. in diameter, and pinions are about 0.06 in. in diameter and have as few as six teeth. The ratio per mesh is usually between 4:1 and 10:1 with the latter being a general upper limit.

There are two types of straight spur gear train: speed increasing and speed reducing. In the speed increasing type, a multiple stage train might be used to drive a fast moving escape wheel from a slow-moving mainspring barrel. This type is less efficient than the speed reducing train obtained when the input energy is supplied directly to the oscillator at the fast-moving end of the train. Torque variation and efficiency loss especially are pronounced when small pinions having six or seven teeth are used.

The overall efficiency of a typical five-stage timer gear train having a ratio of approximately 8600:1 is about 30 to 40 percent at best; and this can vary widely during operation due to gear eccentricities, tooth-to-tooth tolerance variations, engagement action, etc. Since the torques involved in driving the oscillator are very small, these gear trains can be a source of difficulty, especially when the speed increasing type is used, because the torque transmitted to the fast moving end becomes very irregular.

Angular errors in output shaft position as great as 1 or 2 percent can result from errors in gear concentricity, roundness, and tooth-to-tooth spacing since timer gears are not, in general, precision gears. A partial solution to these problems can be obtained if precision gear sets are used having at least 12 teeth on the pinions and using the involute form.

### 16-3.2.2 Worm Gearing

Worm gearing can provide high reduction

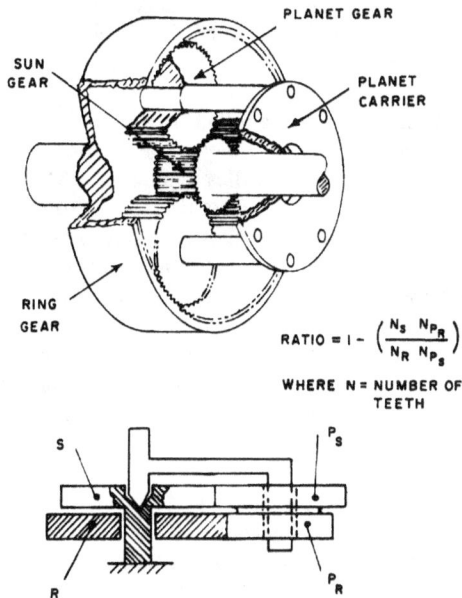

RATIO $= 1 - \left( \dfrac{N_S \; N_{P_R}}{N_R \; N_{P_S}} \right)$

WHERE N = NUMBER OF TEETH

*Figure 16-11. Epicyclic Gearing*

per stage (about 100:1) but the efficiency is lower because of the sliding contact. Worm gears have not been used to any extent in timers probably because of large effects on performance caused by frictional variations, greater cost to manufacture, and less positional accuracy than spur gear sets.

### 16-3.2.3 Epicyclic Gearing

Epicyclic gearing is a planetary arrangement of several gears in which one or more of the gears is constrained to move bodily around the circumference of coaxial gears that may be either fixed or rotating. The motion of the constrained gear is composed of a rotation about its own axis and a rotation about the axis of the coaxial gear. This form of gearing is used most often where a speed-changing device is required; it has not found application as a timer computer. For an epicyclic form that provides the highest ratios per stage obtainable, gears having a diametral pitch of approximately 600 would have to be used in order to fit a 100:1 stage with a 1 in.

diameter inclosure. Such gears are not obtainable, and the complexity of the epicyclic system rules it out for ratios that are much less than 100:1. Fig. 16-11[9] is an example of epicyclic gearing.

### 16-3.2.4 Hypocyclic Gear Trains

In this system, a relatively large-diameter gear rolls on or meshes with a slightly larger internal gear having two or three more teeth, Fig. 16-12[9]. As a result of this action, a wobbling motion is generated which can be converted to a rotary output. The advantages of this type of gearing are a large ratio per stage (50:1 to 150:1 for nominal 1-in. gears depending on diametral pitch) and the elimination of problems associated with very small pinions. Hypocyclic gear-train systems have not been used in timer-sized devices and, as a consequence, little can be said about limitations or merit in such an application.

### 16-3.2.5 Harmonic Gear Drive

The harmonic gear-drive mechanism is a relatively new invention introduced by The United Shoe Machinery Company in 1957 (see Fig. 16-13[9]). It consists of a flexible gear or spline having a large number of teeth which meshes with a rigid gear or spline having slightly fewer teeth. Very high ratios per stage are obtainable. They may be between 100:1 and 1000:1, depending on the size of the

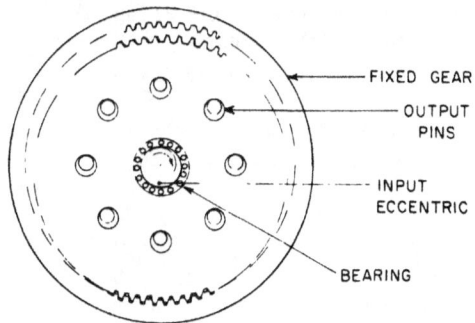

*Figure 16-12. Hypocyclic Gearing*

16-15

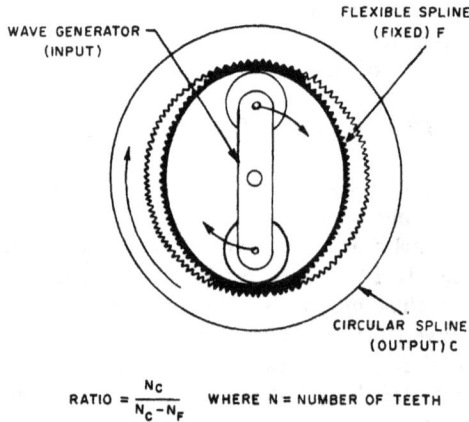

$$RATIO = \frac{N_C}{N_C - N_F} \quad WHERE \ N = NUMBER \ OF \ TEETH$$

*Figure 16-13. Harmonic Gear Drive*

mechanism and the number of teeth which can be used. It has a high positional accuracy and high torque loading capacity. This type of gearing has not been produced in miniature size suitable for timer applications.

### 16-3.2.6 Synchronic Index Gear Train

An accurate gear-train computing system can be designed by using the synchronic index of a gear train. For example, two concentric coplanar spur gears, one having 103 teeth and the other 100 teeth and both driven by a single pinion having 11 teeth (see Fig. 16-14[9]) will require 10,300 revolutions of the pinion to return the three gears simultaneously to their original positions. This coincidence of position could be used to close an electrical switch that would complete a firing circuit or release a mechanical latch. The counter would be set by rotating the pinion the desired number of revolutions (less than 10,300) in the reverse direction. Different setting methods and mechanical outputs could be devised. Various turn ratios can be achieved with relatively few gears and with good resolution of the readout or output function.

## 16-3.3 SPECIAL PROBLEMS ASSOCIATED WITH SMALL MECHANISM GEARING

When gears are applied to small mechanisms as clocks and timers, special problems arise. First, tooth profiles of small mechanism gears are widely different from those of power gearing. While almost all power gearing is used for a step-down in velocity ratio between the driver and the driven gear, in a clock or timer mechanism the prime mover— the mainspring barrel—may move a small number of revolutions per unit time and the escapement wheel several orders of magnitude faster.

The wheels are so tiny, that, in order to avoid using excessively small teeth, it is necessary to use driven gears having as few as 6 or 7 teeth. The tooth profiles of these gears must be those which permit operation over 60 deg of rotation of the driven gear. For this purpose, tradition has long dictated the use of the cycloidal based design for speed increasing trains. However, in the area of speed-increasing gear trains for timer mechanisms, much of the available data tends to disprove use of the cycloidal configuration and greatly favors the use of involute design.

The very small size and low power transmitted by timer gearing introduces special problems. It is inevitable that manufacturing errors are relatively large in tiny gearing, and rolling curves demanded by strict

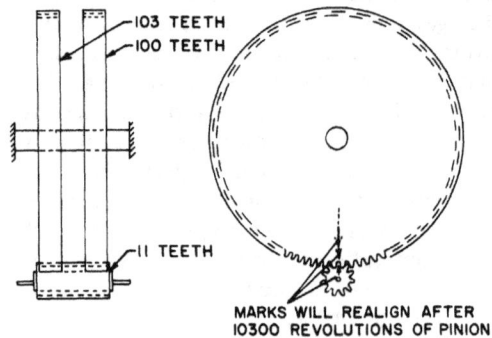

*Figure 16-14. Principle of the Synchronic Index of a Gear Train*

(A) GEARS WITHOUT TIP RELIEF

(B) GEARS WITH TIP RELIEF

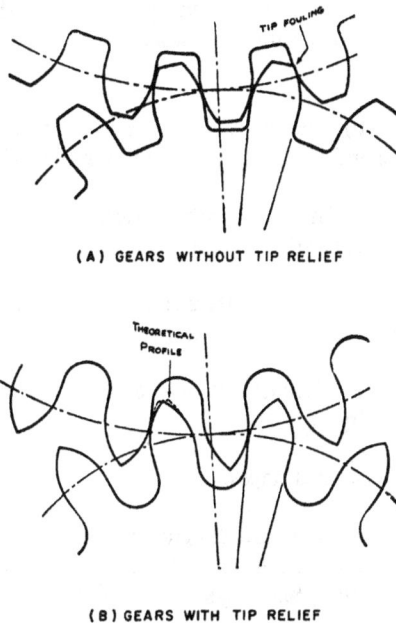

Figure 16-15. Comparison of Gear Tip Profiles[10]

theory need to be modified by a generous tip relief so that no contact can take place on the edges or tops in spite of errors of tooth form, spacing, or depth of engagement. Fig. 16-15(A)[10] shows the interference between two tooth surfaces caused by errors of form and spacing while Fig. 16-15(B) shows the relieved tip which avoids the effect. Generous clearance between flanks and between tips and roots are used on fine involute gears in all cases where the energy available to turn the gears is small.

The characteristics of gearing in timing mechanisms for ordnance are unique in comparison to most small gear applications. The more important ones are:

(1) The system is a step-up (or speed increasing) train, in which the pinions are driven by the gears.

(2) In order to conserve weight and space,

the gears are relatively small and the gear train employs large gear ratios per mesh, necessitating pinions with rather few teeth.

(3) Rotation usually is restricted to a single direction, thus a large backlash is not a particularly important disadvantage.

(4) The gears are subject to extreme shock, vibration, and acceleration and the teeth may be subject to high impulsive loads.

Any useful gear design should, therefore, meet the following requirements:

(1) The gear teeth, especially those of the pinions, must be rugged enough to withstand large shock loads, dictating thick teeth with little or no undercut.

(2) The mesh should have as high an efficiency as possible since two inherent flaws of a step up train—decreasing torque and tendency toward increased friction from driving a small pinion—combine to promote eventual self-locking for high ratios with several meshes. Although nothing can be done for torque loss, friction can be reduced by smooth, accurate mating surfaces; by continuous action and a high contact ratio to smooth transmission and distribute the load; and by placing the action largely in the arc of recess.

(3) The gears must achieve efficient action without extreme tolerance limitations on tooth profile or center distance.

(4) The design should be such that uniform accuracy in mass production is easily achievable.

(5) Adequate backlash is necessary to allow for center distance variations.

(6) Consideration of time, money, and equipment requires a degree of ease in manufacture of such designs, preferably in a variety of methods, without necessitating special tooling.

The design of the various gearing systems (involute, cycloidal, clock-tooth) are well documented in the literature (see Bibliography at the end of this chapter). However, it remains with the designer of military timers to make a choice as to which system or combination of systems will best approach or equal the useful gear-design requirements as outlined.

In order to facilitate the choice between gear systems, a comparison of the relevant features of each should be made. Advantages and disadvantages follow:

*a. Cycloidal Gears:*

(1) Advantages:

(a) Most contact tends to occur on the arc of recess.

(b) Train efficiency is high.

(c) They are in common use.

(2) Disadvantages:

(a) Teeth on small pinions are tapered to extremely thin roots.

(b) Tolerance requirements on tooth profile and center distance are rather small, and often critical.

(c) Cutter design for the cycloidal curve is complicated, being another cycloidal curve, and therefore prone to inaccuracy.

(d) Neither the cutters nor the gears are completely interchangeable.

(e) Contact ratios are usually less than unity.

*b. Involute Gears:*

(1) Advantages:

(a) Stronger tooth profiles are possible through versatility of design with standard equipment.

(b) Tolerance requirements on tooth profile and center distance are more liberal.

(c) Involute cutter design is quite simple, consisting of straight lines.

(d) Both cutters and gears are extremely versatile and interchangeable.

(e) Contact ratios are usually greater than unity.

(2) Disadvantages:

(a) Standard designs applied to pinions with few teeth result in rather large undercutting that weakens the teeth.

(b) Train efficiency is low.

(c) Standard designs tend toward greater approach action.

*c. Clock Tooth Gear:*

(1) Advantages:

(a) Minimum obliquity of thrust if largest possible generating circle is used.

(b) Minimum sensitivity to errors of diameter and center distance.

(c) Standard designs of teeth for certain numbers of teeth and gear ratios are available.

(2) Disadvantages:

(a) Highly specialized gearing for going trains.

(b) Standard design for the addendum of the driving gear varies with the number of teeth in driven pinion and the gear ratio. This system does not lead to easy standardization of tool stocks.

(c) It is desirable from friction considerations to have more recess contact than approach contact, so that special tooth forms must be employed rather than true epicycloidal curves.

From the standpoint of simple cutter design, interchangeability and higher allowance on tooth profile and center distance, a preference for involute gears can be justified with possible combination with clock tooth gear sets in the train.

## 16-3.4 FRICTION EFFECTS ON GEAR TRAIN EFFICIENCY

It is a well established fact that when two surfaces are in contact, and friction is completely absent, the line of thrust between the surfaces lies on the common normal. If sliding friction is present, the line of thrust is inclined to the common normal at an angle whose tangent equals the coefficient of friction.

Further, if two surfaces are shaped similar to a shaft and bearing, there exists an imaginary circle, called the friction circle, within the shaft whose radius is equal to the radius of the shaft multiplied by the sine of the angle of friction. If the line of thrust applied to the shaft passes inside the friction circle, the shaft will become friction locked and fail to rotate. If the line of thrust passes near to the friction circle, the shaft will rotate but at a low efficiency.

By using these facts, various phases of gear engagement can be studied, and variations of output torque and efficiencies can be determined. Fig. 16-16(A)[11] shows the locations of the normal force when two gears run together in the absence of friction. Fig. 16-16(B) indicates the location and direction of the sliding friction forces both before and

after the line of centers. The direction of the friction force will oppose the motion of the driver relative to the driven surface and be perpendicular to the common normal. Fig. 16-16(C) shows the gear mesh with the normal force applied to the drive in presence of frictson. The effective output under these conditions is then the vector sum of the normal force and the friction force. The angle of inclination of the resultant thrust to the normal thrust is the friction angle. Where brass and steel are in sliding contact, this angle is assumed to be $11°19'$. Fig. 16-17(A)[11] shows a gear and pinion in contact before the line of centers. When the torque is applied to the driver, a normal force $F_n$ will pass through point C in a direction normal to both tooth profiles. The moment arm of the normal force of the gear is $a$ and the moment arm of the pinion is $b$. Fig. 16-17(B) shows the same gear and pinion in contact before the line of centers with sliding friction present. The resultant force $F_t$ passes through the contact point C. The moment arm of $F_t$ for the gear is $c$ and the moment arm of the pinion is $d$. In order to evaluate gear mesh performance, the torque output with friction ($G_{driven}(F_R)$) is compared to torque output with no friction ($G_{driven}(F_o)$), where the same input torque is applied in both cases. Their ratio, termed instantaneous efficiency, is determined by

$$\text{Instantaneous Efficiency} = \frac{G_{driven}(F_R)}{G_{driven}(F_o)} = \frac{da}{cb}$$

(16-8)

Even though the friction force is reversed after the line of center, the equation for efficiency is the same. Eq. 16-8 is a general one which applies to both involute profiles and variations of the cycloidal form.

To determine the efficiency curve for a pair of gears in contact, the force diagram must be determined for each increment of the angle of action and the moment arms measured. By using Eq. 16-8 the efficiency can be

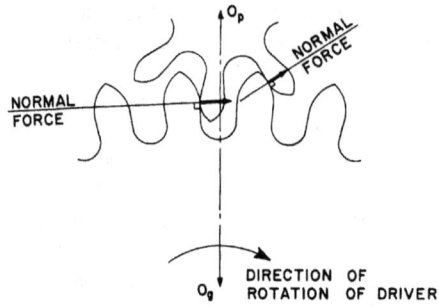

(A) DIRECTION OF NORMAL FORCE IN ABSENCE OF
FRICTION BEFORE AND AFTER CENTER

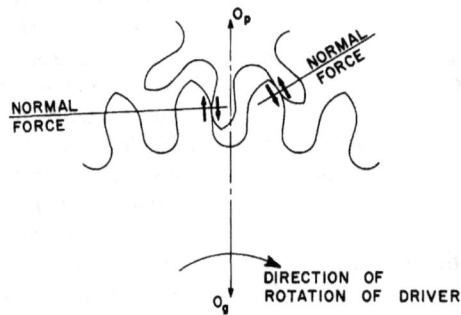

(B) FRICTION FORCES BEFORE AND AFTER CENTER

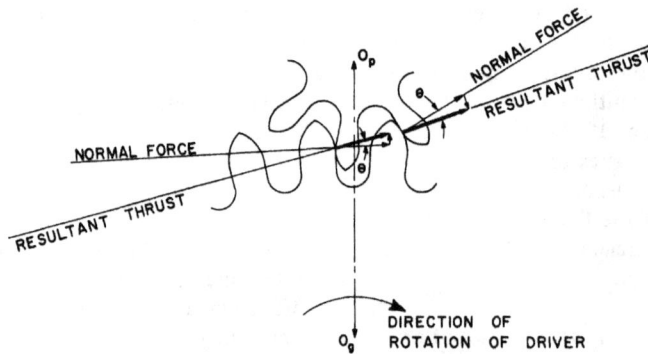

(C) NORMAL FORCE APPLIED IN DIRECTION
OF FRICTION

Figure 16-16. Direction of Resultant Effective Forces With Friction

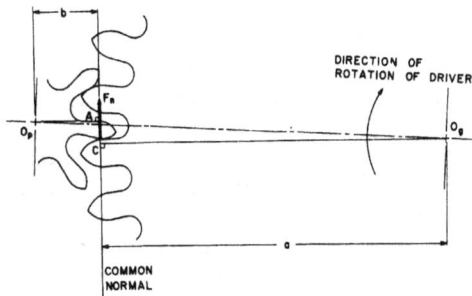

(A) NORMAL FORCE $F_n$ BEFORE CENTERS WITH MOMENT ARMS $a$ and $b$

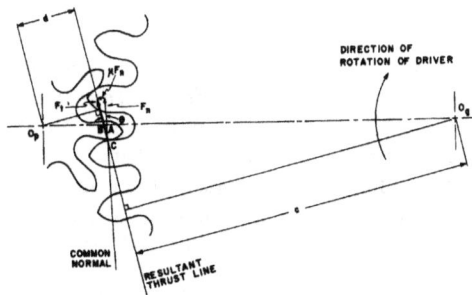

(B) EFFECTIVE RESULTANT THRUST BEFORE CENTERS WITH MOMENT ARMS $c$ and $d$

Figure 16-17. Forces on Teeth With and Without Friction

calculated. A graphic technique has been developed, for plotting torque fluctuations that are contributed by the tooth form of a gear and pinion set when going in and out of mesh[11]. Other effects such as gear eccentricities and mainspring torque are not considered.

Fig. 16-18[10] shows the cumulative effect on efficiency of three stages of gears with the individual efficiencies plotted on the same graph. It can be shown that step-down gear trains are more efficient than step-up trains and that incorrectly designed gear trains can be self-locking due to friction[10].

By referring to the direction of the friction

force in Fig. 16-16(B) it is evident why recess action is desirable. The transmitted forces and friction combine during approach action and tend to separate the gears while during recess action they oppose each other and exert less net radial force on the gears.

A study has been performed to justify the use of involute gearing over the traditional cycloidal form. The parameters studied include different pressure angles, different center distances, different gear blank sizes, different cutter feed depths, and combinations of these.

## 16-3.5 GEAR THEORY AND TOOTH DESIGN

The forms of the teeth of gears are based on either of two types of curves—the cycloid and the involute. A cycloid is the curve traced by a point on the circumference of a circle when the circle rolls without slipping along a straight line. Two variants of this curve are used in gearing: the epicycloid (used in clock gear trains) in which the circle rolls on the outside of another circle; and the hypocycloid, in which the circle rolls on the inside of another circle.

An involute is the curve traced by the end

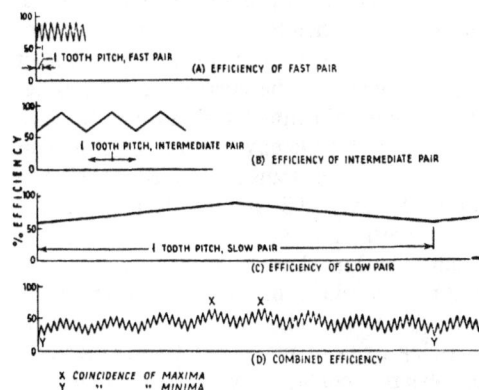

Figure 16-18. Variations of Torque Output of Several Stages[10]

16-21

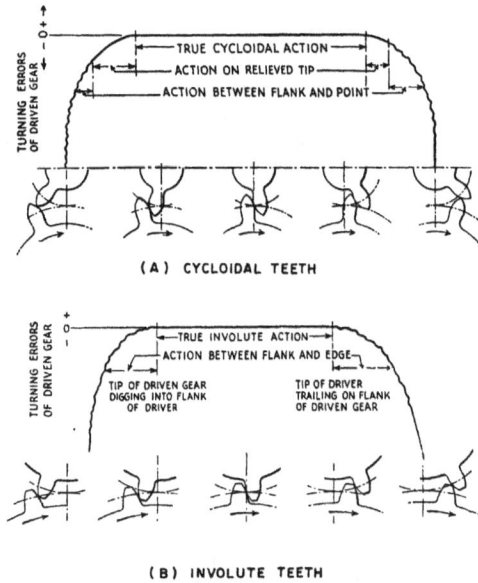

(A) CYCLOIDAL TEETH

(B) INVOLUTE TEETH

*Figure 16-19. Contact Action of Gear Teeth*[10]

of a string when it is held taut, as it is unwound from a drum. Involute gearing was developed more recently than cycloidal gearing. The theory of both types of gearing depends on geometrical principles, the fundamentals of which are available in a number of texts (see Bibliography at the end of this chapter).

These theories of involute and cycloidal curves have as their basis the requirement that the shape of the tooth profiles in contact must be such that the normals to the profiles at all points of contact shall all pass through the pitch point thereby insuring that the pair of engaging teeth transmit a constant velocity ratio. However, it is impossible to form the teeth with mathematical precision. As a result, within limits depending on the accuracy attained in practice, variations of velocity ratio take place, and successive engaging pairs of teeth sometimes enter into load-bearing contact more abruptly than they should. Fig. 16-19(A)[10] shows a cycloidal gear pair from which all of the teeth except

one on each wheel have been removed. Fig. 16-19(B) shows a similar pair, but with involute teeth. In each figure, a diagram of turning errors is shown which illustrates the sharp acceleration of the driven gear from rest, until true involute or cycloidal action is arrived at, with contact between the designed working faces of the teeth. Following this, there is a slowing down action until the driving tooth passes out of engagement, leaving the driven tooth stationary. The important fact to be noted here is the lead in and lead out on the turning error curves. If there is any kind of error of tooth form or spacing, part of the lead-in or lead-out curve has load bearing action.

In Fig. 16-19(B) the irregular rough lines, which represent contact upon an edge, are as a result of no tip relief, and there is an abrupt change from tip contact to true involute contact. Fig. 16-19(A) shows the lead-in and lead-out curves to be smooth, because the theoretical curves are blended into the relieved tips.

If several such curves, spaced at intervals of one circular pitch, are drawn in succession, the resulting diagram (see Fig. 16-20[10] will show the take-up of load bearing contact of the teeth. The parts of the curves which protrude above their preceding or succeeding curves represent load bearing contact. Since none of these curves can be exactly perfect in practice, it should be realized that contact on any given pair of teeth must in general always begin at a point of change over from retarded motion to accelerated motion. To minimize the reduction of output torque in step-up gear trains, special tooth profiles are designed. These variations cause a hump to occur in the turning error curve which indicates that most of the load bearing contact occurs after the point of contact has passed the line of centers (Fig. 16-20(A)). For gears in less particular applications than clocks or timers, the turning error is caused to be slightly humped (Fig. 16-20(B)). If by any combination of inaccura-

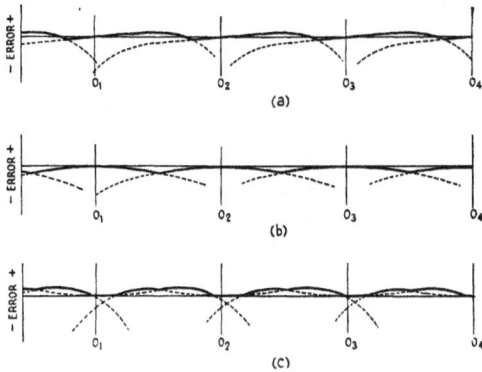

Figure 16-20. Curves of Turning Error[10]

cies the curve becomes double humped, (Fig. 16-20(C)), the load bearing contact would occur when the point of contact was remote from the line of centers and friction would be at its greatest value.

The single humped condition can be affected by arranging the tolerances on the tooth profile so that the curvatures are greater than dictated by rolling curve theory, never less.

## 16-3.6 GEAR RATIOS

In the design of gear trains, it is important to design the individual gear meshes so that a minimum of spur gear sets are used in order to avoid friction losses. In practice, gear ratio per gear mesh should not exceed 10/1 or go below 1/10, otherwise the large gear would have too many teeth (which is equivalent to having a large pitch diameter) or the small gear would have too few teeth.

One gear mesh is required for each decade of gear ratios, thus:

| Gear Ratio | No. Meshes |
|---|---|
| 10:1 or 1:10 | 1 |
| 100:1 or 1:100 | 2 |
| 1000:1 or 1:1000 | 3, etc. |

The ordinary gear velocity ratio $i$ is determined by

$$i = \frac{\omega_1}{\omega_2} = \frac{N_2}{N_1} \qquad (16\text{-}9)$$

where

$\omega_1$ = speed of first gear in mesh (driver), rpm

$\omega_2$ = speed of second gear in mesh (driven), rpm

$N_1$ = number of teeth in first gear (driver)

$N_2$ = number of teeth in second gear (driven)

For decreasing speeds, $i > 1$ and for step up gearing, $i < 1$. It must be remembered that the torque transmitted decreases as the speed is increased and vice versa.

If rotational speeds are to be changed significantly, particularly in a limited space, the transformation can be accomplished by introducing a second gear on the shaft of the driven gear of the first mesh, which in turn drives a gear on a third shaft, then introducing a second gear on the third shaft and driving a gear on a fourth shaft. This method can be carried on until the desired gear ratio is obtained. Fig. 16-21(A)[7] is an illustration of a speed increasing train ($i < 1$), and Fig. 16-21(B) shows a speed decreasing train ($i > 1$).

For a given gear ratio, the small wheels generally are determined first. Their selection usually is made according to desired degree of quality. Wheels with smaller numbers of teeth are usually cheaper to produce whereas wheels with higher teeth numbers provide a better degree of overlap and thus smoother running. By using the recommended number of meshes for a particular gear ratio, it remains the designer's job to provide a

16-23

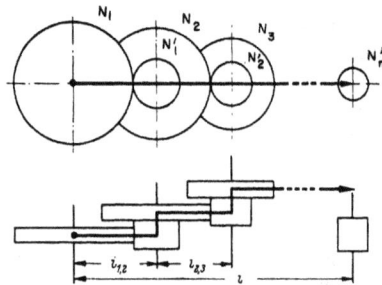

(A) SPEED INCREASING

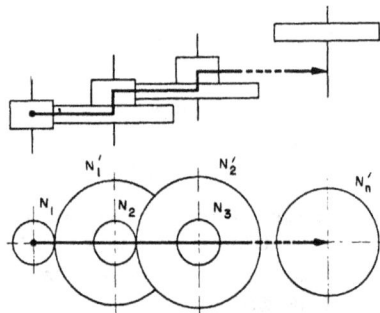

(B) SPEED DECREASING

*Figure 16-21. Gear Trains[7]*

combination that will fulfill the ratio requirements.

There are several methods for solving gear ratios whose numerators or denominators are higher-place prime numbers. Since it is not always necessary to maintain exactly the specified gear ratio, a good approximation to the solution can be made by the following two methods:

(1) Calculating with continued fractions. The originally specified gear ratio is reduced to a fraction with lower teeth numbers.

(2) Using Brocot's approximation method which yields considerably better results but which requires an auxiliary table[12].

### 16-3.6.1 Calculation With Continued Fractions

Theory will not be discussed here; it may be found in the appropriate mathematical literature. The procedure in the example that follows clearly shows the computation.

Given a gear ratio $i = 347/439 = N_2/N_1$.

The computation process is:

```
439:347 = 1 = a
-347
   92   347:92 = 3 = b
         -276
           71   92:71 = 1 = c
                 -71
                   21   71:21 = 3 = d
                         -63
                           8   21:8 = 2 = e
                               -16
                                 5   8:5 = 1 = f
                                     5:3 = 1 = g
                                     3:2 = 1 = h
                                     2:1 = 2 = i.
```

From the coefficients $a$ to $i$ thus found, associated teeth numbers are obtained which become increasingly greater and whose quotient continually approaches the true value:

$N_1 = A' = a\,1 + 0 = \quad 1 \qquad N_2 = A = a\,0 + 1 = \quad 1$

$\qquad B' = b\,A' + 1 = \quad 4 \qquad\qquad B = b\,A + 0 = \quad 3$

$\qquad C' = c\,B' + A' = \quad 5 \qquad\qquad C = c\,B + A = \quad 4$

$\qquad D' = d\,C' + B' = \quad 19 \qquad\qquad D = d\,C + B = \quad 15$

$\qquad E' = e\,D' + C' = \quad 43 \qquad\qquad E = e\,D + C = \quad 34$

$\qquad F' = f\,E' + D' = \quad 62 \qquad\qquad F = f\,E + D = \quad 49$

$\qquad G' = g\,F' + E' = \quad 105 \qquad\qquad G = g\,F + E = \quad 83$

$\qquad H' = h\,G' + F' = \quad 167 \qquad\qquad H = h\,G + F = \quad 132$

$\qquad I' = i\,H' + G' = \quad 439 \qquad\qquad I = i\,H + G = \quad 347$

Thus we obtain for the different approximation values $N_2/N_1$:

$$\frac{A}{A'} = \frac{1}{1} = 1$$

$$\frac{B}{B'} = \frac{3}{4} = 0.75$$

$$\frac{C}{C'} = \frac{4}{5} = 0.80$$

$$\frac{D}{D'} = \frac{15}{19} = 0.7894736$$

$$\frac{E}{E'} = \frac{34}{43} = 0.7906976$$

$$\frac{F}{F'} = \frac{49}{62} = 0.7903225$$

$$\frac{G}{G'} = \frac{83}{105} = 0.7904762$$

$$\frac{H}{H'} = \frac{132}{167} = 0.7904191$$

$$\frac{I}{I'} = \frac{347}{439} = 0.7904328$$

### 16-3.6.2 Brocot's Approximation Method

Example as above: $i = \frac{347}{439}$ 0.7904328.

From the table we obtain the two approximation values

$$\frac{34}{43} = 0.7906976 \quad \text{and} \quad \frac{49}{62} = 0.7903225$$

of which the one is greater and the other smaller than the exact value 0.7904328. It may be seen that these approximation values also can be obtained without the table. They are contained in the procedure using continued fractions (row E/E' and F/F'). Actually, Brocot's method represents an expansion of teeth number determination by continued fractions.

We let

$$\frac{34}{43} = \frac{a}{b} \quad \text{and} \quad \frac{49}{62} = \frac{c}{d} \, .$$

These two approximation values show a difference $\epsilon_i$ in comparison to the exact value of:

$$\frac{a}{b} - i = \epsilon_1 \quad \text{and} \quad i - \frac{c}{d} = \epsilon_2$$

From values $b$, $d$, $\epsilon_1$, and $\epsilon_2$ one forms the quotient

$$k = \frac{d \, \epsilon_2}{b \, \epsilon_1} \, ,$$

which at first yields a double fraction (numerator and denominator decimal fraction). This double fraction is expanded by 10, 100, etc. so that numerator and denominator become a low decimal number $> 1$, which after division by each other would yield the value $k_1$. By rounding off (or possibly by additional increase by one of both numerator or denominator) one finally obtains an ordinary fraction $x$, which is used to calculate the corrected value

$$i = \frac{a\,x + c}{b\,x + d}$$

which, upon suitable selection of $x$, approximates considerably closer the original $i$-value than the two original approximation values $a/b$ or $c/d$.

As applied to our example

$$\epsilon_1 = \frac{a}{b} - i = 0.7906976 \ - 0.7904328$$

$$= 0.0002648$$

$$\epsilon_2 = i - \frac{c'}{d} = 0.7904328 - 0.7903225$$

$$= 0.0001103$$

$$k = \frac{62(0.0001103)}{43(0.0002648)} \approx \frac{6.8}{11.4}$$

For $x = 7/12$ we obtain

$$i = \frac{34(7) + 49(12)}{43(7) + 62(12)} = \frac{14(59)}{19(55)} = 0.7904306.$$

Since Brocot's table is calculated only for true fractions whose numerators and denominators go no higher than 99 (two-place numbers), one can carry the approximation even farther by selection of even higher-place fractions as provided by the continued fraction method (e.g., F/F' and G/G' or G/G' and H/H' as limits).

The difference in comparison to the true value $i = 0.7904328$ when using continued fractions is for

$$F/F' = \frac{49}{62} \; (0.0001103)$$

$$G/G' = \frac{83}{105} \; (0.0000434)$$

$$HH' = \frac{132}{167} \; (0.0000237)$$

and according to Brocot's method is 0.0000022. In the last case two gear elements also are required.

### 16-3.7 GEAR TRAIN DESIGN

The design of gear train for a timer depends on the time interval to be measured. The following equation can be used to determine the gear train ratio

$$\chi = \frac{f_n \, t \, 360}{\theta \, N_w} \qquad (16\text{-}10)$$

where

$f_n$ = natural frequency of pallet operation, Hz

$t$ = functioning delay, sec

$\theta$ = required angle for last pinion, deg

$N_w$ = number of teeth on the escapement wheel

$\chi$ = gear train ratio, reciprocal of gear velocity ratio $i$

$N_w$ should be designed to place a maximum number of teeth on the escape wheel within the predetermined limits of the moment of inertia of the escape wheel. The natural frequency of the escapement system $f_n$ should be as high as possible for increased resolution; and $t$ is determined by the specifications placed on the timer.

After the gear train ratio is determined, the train can be designed so as to fit into the available space using a minimum of gears to reduce friction losses and inertia of the system.

### 16-4 ESCAPEMENTS

The escapement is the part of a mechanical clockwork which regulates the transmission of energy. The rate of energy release can be controlled by either a two-center untuned escapement or by a tuned escapement of two or three centers.

### 16-4.1 UNTUNED TWO-CENTER ESCAPE-MENT

Known variously as a runaway escapement, verge escapement, or inertial governor, the untuned two-center escapement (Fig. 16-22[13]) is an inertial system that lacks the resonance characteristics of the spring-mass type of timing escapement. In order to rotate,

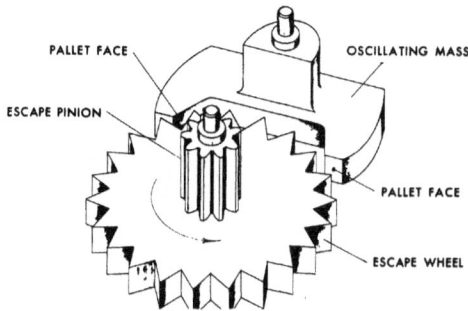

*Figure 16-22. Runaway Escapement, Flat Pallet Faces*

the escape wheel (sometimes referred to as the starwheel) must drive an inertial mass alternatively clockwise and counterclockwise. The direction of rotation of the inertial mass changes each time a tooth of the escape wheel passes by one of the two contact surfaces on the oscillating mass. These contact surfaces, called the "pallets" or "pallet faces", are contacted alternately by the escape wheel teeth, thereby producing the oscillatory motion of the mass. The mass itself often is referred to as the pallet. This continuous reversal of the rotational direction of the mass results in a drag that resists the input torque (energy is lost primarily by collision), and the average rotational velocity of the escape wheel becomes a function of the torque. The escapement, therefore, provides a useful timing function if the driving torque has acceptably small variations in its magnitude. The period $T$ for one rotation of the escape wheel is approximately

$$T = \frac{k}{\sqrt{G}} \text{ , sec} \qquad (16\text{-}11)$$

where

$G$ = input torque, lb-in.

$k$ = system constant, $(\text{in.-lb})^{1/2}$-sec

This escapement commonly is used in two

forms. One has an escape wheel and oscillating mass combination of the type shown in Fig. 16-22 where the teeth of the escape wheel engage the flat pallet faces formed by a special configuration of the mass. The other form, shown in Fig. 16-23[13], has a pin pallet. The escape wheel engages pallet faces formed by two pins that project from the oscillating mass perpendicular to its plane of oscillation. This second type is preferred by some designers who believe that the pin location is controlled more easily in production, providing better control of the geometrical position of the pallets. However, it is less reliable because a pin may be missing or bent.

### 16-4.1.1 Factors Affecting Escapement Operation

The following variables affect the running rate of a runaway escapement:

(1) Net driving torque (input torque less output load)

(2) Change in the geometrical relationship of the pallet and escape wheel (due to wear or manufacturing variations)

(3) Gear train and escapement tolerances

(4) Friction

All effects considered, typical lot-to-lot error for movements regulated with this type of escapement is usually within ± 15% when operating over the military temperature range, −70° to 125°F.

The timing error can be reduced to ± 5% with tight tolerances, good quality control, individual adjustment of the output angle of rotation to match a particular escapement rate, individual adjustment of balance inertia, and constant output load.

The frequency $f_n$ of pallet oscillation may be calculated from the net driving torque (Eq.

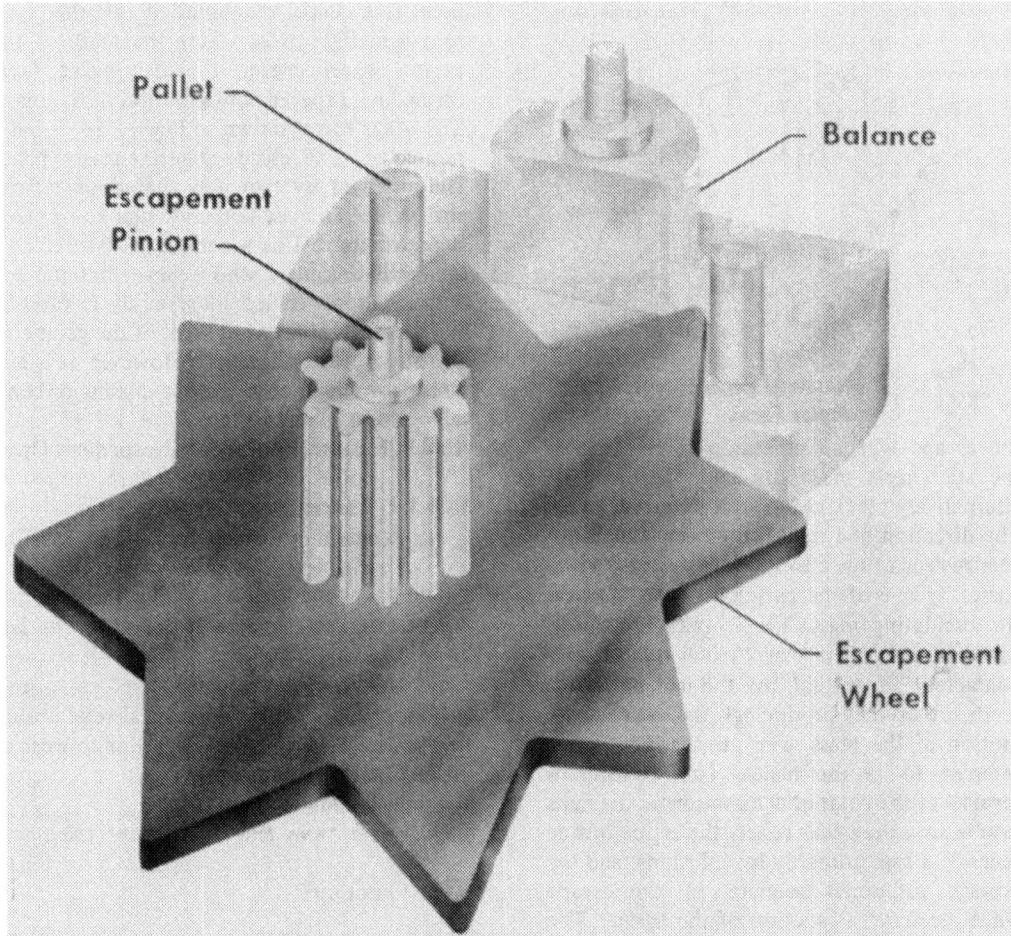

*Figure 16-23. Runaway Escapement, Pin Pallet Type*

16-12) on the escape wheel if the following assumptions are made[1] :

(1) Half cycles of the pallet are equal in time.

(2) Driving torque is constant.

(3) Impact is inelastic.

(4) Friction is negligible.

$$f_n = \frac{1}{2\pi} \sqrt{\frac{Gr_p/r_w}{2I_m \theta}} \text{ , Hz} \qquad (16\text{-}12)$$

where

$G$ = net driving torque, lb-ft

$r_p$ = radius of pallet pin, in.

$r_w$ = radius of escape wheel, in.

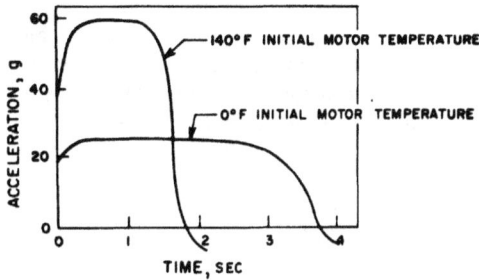

Figure 16-24. Typical Rocket Accelerations

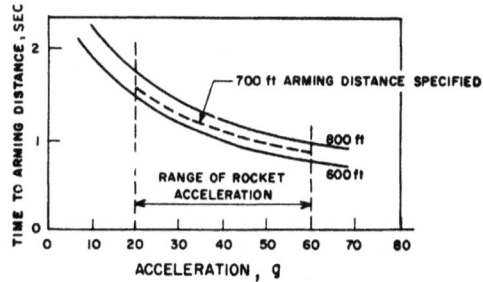

Figure 16-25. Variation in Rocket Arming Time

$\theta$ = angle between extreme positions of pallet, rad

$I_m$ = moment of inertia of oscillating mass of pallet, slug-ft$^2$

Thus the frequency varies as the square root of the escape wheel torque, or damping force is proportional to the square of the escape wheel angular velocity. When designing a gear train, the designer must remember that $G$ is the actual torque delivered rather than the theoretical torque. (Use 30% of the theoretical torque as a first approximation.)

### 16-4.1.2 Runaway Escapement Applied to Rocket Fuze

As an example, let us investigate the problem of arming a rocket fuze at $700 \pm 100$ ft from the launcher. For safety, the fuze must not become armed until the rocket has traveled a certain minimum distance.

Inasmuch as measuring this distance directly is a difficult if not impossible task, a timing interval is measured in a manner that is related directly to distance. If the speed of the rocket were constant, a timing device would suffice. The timing devices for arming artillery projectiles can be applied with reasonable confidence, but the behavior of rockets is too variable to measure arming distances with timing devices even if all rockets performed normally.

This variability in behavior of rockets can be illustrated by the influence of rocket motor temperature (at time of firing) upon the acceleration (see Fig. 16-24[1]). Other factors such as air density can have pronounced effects on the acceleration-time diagram.

Fig. 16-25[1] shows that the arming time must vary with the acceleration of the rocket if the arming distance is to be held within the specified tolerance.

A solution to this problem can be to use a runaway escapement. Using the acceleration of the rocket as a source of power to drive the escapement, the escapement can be designed to effect arming in the same distance even though the acceleration varies. Fig. 16-26[1] shows a device in which the torque applied to the escapement will be proportional to the setback acceleration.

The time to arm $t$ can be expressed as

$$t = \frac{1}{k'f_n} \quad \text{, sec} \tag{16-13}$$

because it depends upon the number of oscillations of the pallet and thence upon its frequency $f_n$. The distance $s$ that the rocket will travel during the arming time is

$$s = \frac{1}{2}at^2 \tag{16-14}$$

16-29

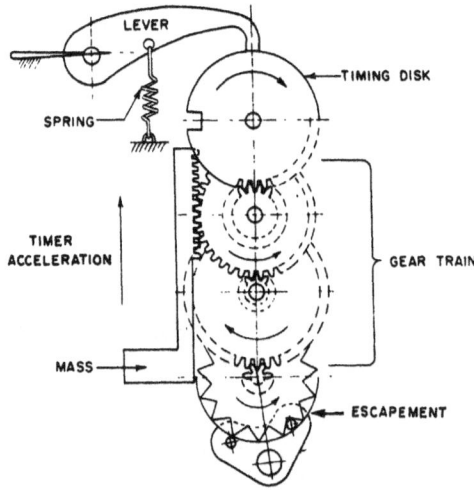

*Figure 16-26. Runaway Escapement*

The torque $G$ is given by

$$G = mar_w k'_1 \qquad (16\text{-}15)$$

where

$m$ = mass of the driving rack (Fig. 16-26), slug

$a$ = acceleration of the rocket, ft/sec²

$r_w$ = radius of the escape wheel, ft

$k'_1$ = ratio of number of teeth of driving gear and escape wheel

Combining Eqs. 16-12 to 16-15 results in

$$s = \frac{4\pi^2 I_m \theta}{k'^2 m \, k'_1 \, r_p} = k'_2 , \qquad (16\text{-}16)$$

in which all terms of definition are independent of the rocket ballistics.

The runaway can be employed to establish a constant arming distance in this circumstance. However, the analyses assumed that for any one rocket the acceleration during flight would be constant which is not necessarily true. Some rocket motors exhibit characteristics that make the rocket accelerations vary with time. Fortunately, the arming distance is only moderately affected as shown in Eq. 16-17.

$$s = k'_2 + \frac{1}{k'_3 a^3}, \text{ ft} \qquad (16\text{-}17)$$

Since both $k'_3$ and $a$ are large compared with $k'_2$, the second term becomes insignificant.

Eq. 16-12 describes an idealized device and cannot account for effects of friction or materials. For a particular one-second timer, the empirical equation for the average angular velocity $\omega$ of the escape wheel is given by[1]

$$\omega = \frac{0.231 \, I_w^{0.112} G^{0.5}}{I_m^{0.012}}, \text{ rad/sec} \qquad (16\text{-}18)$$

where

$I_w$ = moment of inertia of escape wheel, slug-ft²

$I_m$ = moment of inertia of oscillating pallet, slug-ft²

$G$ = torque, lb-ft

This is the same form as Eq. 16-12 because

$$f_n = N_w \, \omega/(2\pi), \text{ Hz} \qquad (16\text{-}19)$$

where $N_w$ is the number of teeth on the escape wheel. The constant coefficient (0.231) was found to depend upon various factors: center-to-center distance between escape wheel, friction in the gear train, and number of times that the mechanism had been "run down".

**16-4.1.3 Dynamics of Runaway Escapement**

There has been considerable analytical

work done in the dynamics of this type of escapement[14]. Three conditions of pallet wheel motion were studied. These include (1) cycle of no free motion, (2) cycle of no contact motion (all free motion), and (3) cycle of half free motion half contact motion.

The effect of geometrical factors upon the behavior and nonuniformity of running rates of this system was investigated[15]. Some of the results of this study were:

(1) Variations in apparently identical pallets may cause variations in period of as much as 0.13 sec.

(2) Variation in period introduced by different starwheels and drive springs is negligible.

(3) The factors in the pallet that caused radical changes of period are the shape and location of the leading pallet face.

(4) A change of 0.002 in. in the center-to-center distance of pallet and wheel results in a change of period of 8%.

(5) Friction is most important in the drive springs, and oiling the pallet surfaces strongly decreases the period.

(6) Physical smoothness of shafts, pallet faces, shoulders, or bearing holes is of negligible importance in affecting the period.

(7) Torque delivered at the escape wheel may vary greatly because of frictional effects from mechanism to mechanism.

(8) If a mechanism has been dormant for some hours or days, its period is large when first rerun and then decreases to the constant period.

(9) The nominal delay time may vary as much as 5% depending upon the operating positions.

## 16-4.2 TUNED TWO-CENTER ESCAPEMENTS

The tuned escapement is the part of a timing device that (1) counts the number of oscillations executed by the oscillating mass (pallet) and (2) feeds energy to its pallet. The pallet controls the rotation of the escape wheel while it receives energy that maintains its oscillations. Since the pallet teeth trap and release escape wheel teeth, the rotation of the escape wheel depends upon the frequency of oscillation of the pallet.

These escapements consist of a combination of a pivoted balance and a mass-restoring spring pulsed twice per cycle by an escape wheel. The axis of the escape wheel and balance mass pivot are usually parallel. The basic idea behind these, and all tuned escapements, is that the balance mass-restoring spring system has a natural frequency to which energy must be delivered so as to maintain the oscillation and alter the natural frequency as little as possible. These escapements might be classed in two groups, dead-beat and recoil. In dead-beat escapements, the escape wheel rests motionless against the locking face of the pallet. During the "free swing" phases of the cycle there is firm contact between escape wheel tooth and pallet except during the impulse, and catch-up phases. In recoil escapements almost all tooth action consists of impulse except for short angular displacements immediately after impacts where there may be a slight recoil of the escape wheel. Further discussion on the escapements follows:

(1) *Dead Beat Escapement.*
The original escapement of the dead beat type is the Graham dead-beat escapement used in pendulum clocks. To make this escapement suitable for military use, the number of teeth enclosed by the pallet was reduced to three and the impulse face was relocated to the escape wheel tooth. The resulting Junghans escapement is discussed in par. 16-4.2.1.

Another form of dead beat escapement is the cylinder escapement where the pallet is formed by removing a small section of a full cylinder and only one tooth is enclosed. Cylinder escapements are discussed in par. 16-4.2.4.

(2) *Recoil Escapement*. Recoil escapements in their classical form are not used in military mechanisms. There are some almost dead beat escapements that exhibit small amounts of recoil. The escape wheels—in another class of escapements that might be classed in this category—always drive, supplying too much torque to permit any recoil. They really are tuned runaways and are not very accurate, but are highly reliable and are suitable for some safing and arming mechanisms. A good example of this is the normal Junghans escapement operated backwards.

### 16-4.2.1 Junghans Escapement

The principal tuned two-center escapements used in mechanical time fuzes in this country are forms of an escapement originally designed by Dr. Ing. Junghans of Germany during World War I. The advantages of this escapement are its relatively simple construction, ability to withstand extreme environments (ruggedness), and its inherent high frequency of oscillation.

Basically, the Junghans system consists of an escapement lever (the oscillating mass) with a beam escapement spring and an escape wheel (see Fig. 16-27[16]). The lever is made of spring steel and formed as shown. The pallet is an integral part of the lever and is ground accurately to shape. Small weights sometimes are fastened to the lever arms in order to obtain the desired inertia for the spring-mass system. The arbor, to which the lever is staked, provides its pivots and also secures the escapement spring. While the escapement spring is fixed to the arbor near or at its midpoint, both of its ends are supported simply by guides. The position of the guides is

made adjustable so that the timing mechanism can be regulated to the design frequency thereby compensating for small material and manufacturing variations from unit to unit. The escape wheel is made of thin brass or beryllium copper and usually has 20 teeth on its periphery, three of which are enclosed by the lever pallets during operation. The wheel teeth are so shaped that, as the escapement wheel is rotated, the lever is caused to oscillate which in turn imparts intermittent rotation to the wheel. The normal mode of operation of the Junghans original escapement was dead beat, i.e., the escapement wheel stops and does not reverse its motion.

Fig. 16-28(A)[1] shows tooth A falling on pallet tooth A'. In Fig. 16-28(B) the lever is passing through the equilibrium point in its oscillation where tooth A is about to be released by the pallet. In Fig. 16-28(C) the escape wheel tooth C is the opposite part of the cycle. If the line of action of the impulse passes through the pivot of the lever, its motion will not be altered. As pallet tooth B' slides beneath tooth C, the escape wheel stops. In Fig. 16-28(D) the lever has returned to its equilibrium position and is being driven by the escape wheel as shown in Fig. 16-28(B). If the energy is added as the pallet passes through its equilibrium position, the frequency of the oscillating mass (regulator) is least affected.

Neglecting for the moment the interaction of the escapement wheel and pallets, the effects of variable driving torque, etc., the natural frequency $f_n$ —from elementary theory for a spring-mass vibrating system—neglecting friction, is

$$f_n = \frac{1}{2\pi} \sqrt{\frac{12k}{I_m}} = \frac{1}{\pi} \sqrt{\frac{3k}{I_m}}, \text{Hz} \qquad (16\text{-}20)$$

where

$k$ = torsional spring constant, lb-in./rad

$I_m$ = moment of inertia of the oscillating system, slug-in.[2]

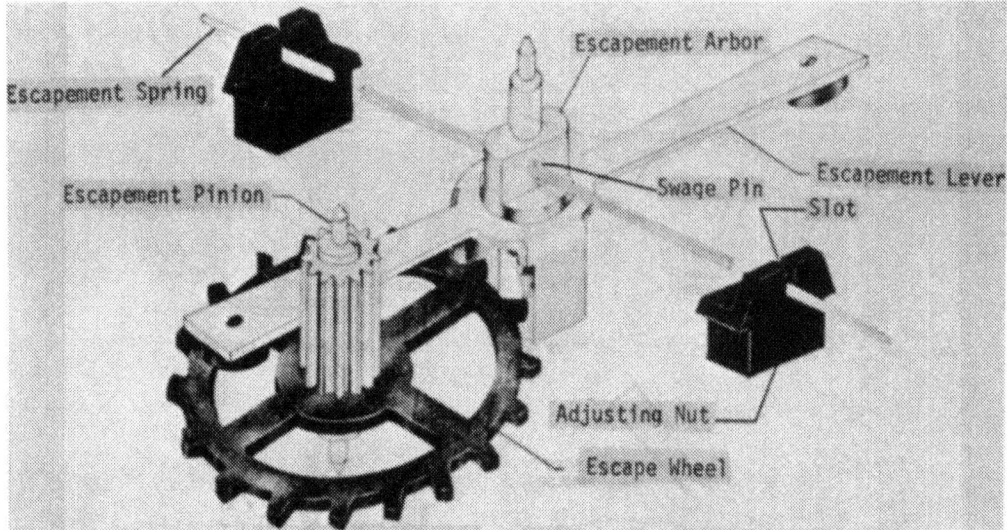

*Figure 16-27. Standard Junghans Type Escapement*

For the beam hairspring of a Junghans system with equal setting of guides (for small spring deflections)

$$k = \frac{12E I_A}{\ell}, \text{lb-in./rad} \qquad (16\text{-}21)$$

where

$E$ = Young's modulus of the spring, lb/in.$^2$

$I_A$ = moment of inertia of the cross-sectional area of the spring, in.$^4$

$\ell$ = total length of spring (twice length between arbor and guide), in.

thus

$$f_n = \frac{1}{2\pi} \sqrt{\frac{12E I_A \times 12}{\ell I_m}}$$

$$= \frac{6}{\pi} \sqrt{\frac{E I_A}{\ell I_m}}, \text{Hz} \qquad (16\text{-}22)$$

For large deflections, the value of $k$ is a function of escapement lever amplitude. A numerical solution for the nonlinearities is available[17].

If the beam spring is replaced with a torsion bar

$$f_n = \frac{d^2}{8} \sqrt{\frac{12G}{2\pi \ell I_m}} = \frac{d^2}{8} \sqrt{\frac{6G}{\pi \ell I_m}} \qquad (16\text{-}23)$$

where

$G$ = shear modulus of elasticity, lb/in.$^2$

$d$ = diameter of the torsion bar, in.

$\ell$ = free length of the torsion bar, in.

$I_m$ = moment of inertia of the cross-sectional area of the torsion bar, slug-in.$^2$

16-33

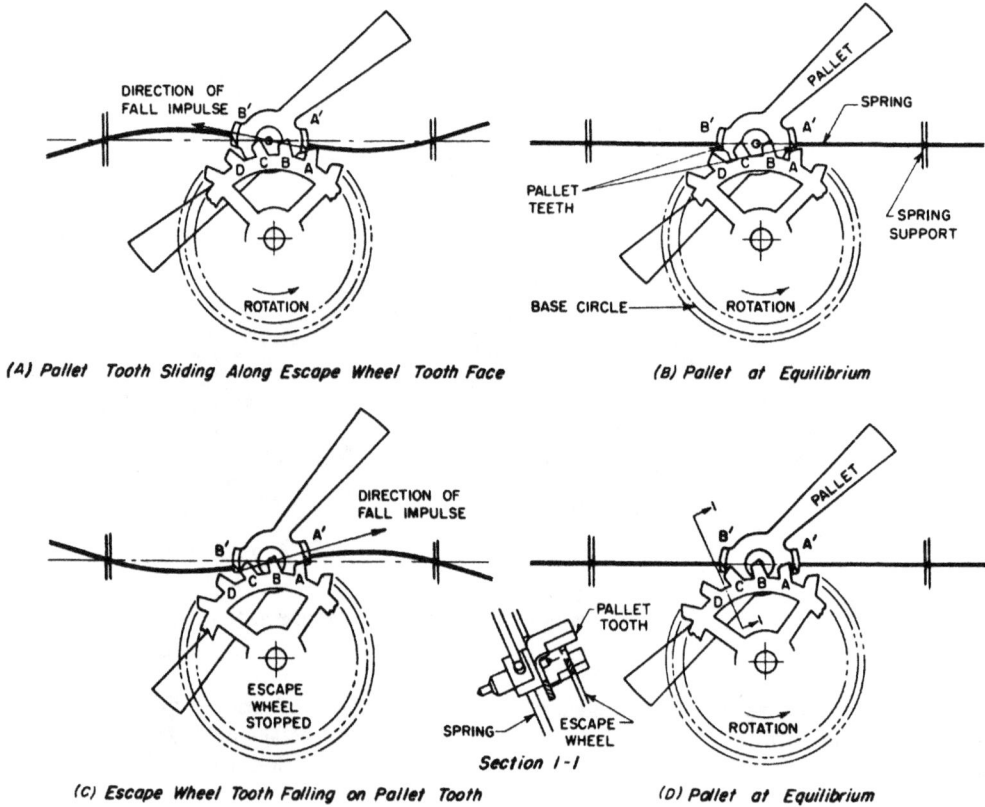

(A) Pallet Tooth Sliding Along Escape Wheel Tooth Face

(B) Pallet at Equilibrium

(C) Escape Wheel Tooth Falling on Pallet Tooth

(D) Pallet at Equilibrium

Section I-I

Figure 16-28. Action of Junghans Escapement

A typical experimentally determined frequency vs torque curve for a fuze movement containing a Junghans escapement is shown in Fig. 16-29[18]. Although this curve varies from movement to movement within a given lot of fuzes, the general shape of the curve is the same for all Junghans escapements.

Some preliminary analytical work indicates that the torque dependence on frequency can be expressed in the following form[18]

$$\frac{\Delta f}{f} = a \left[ c^2 - \frac{(bG + c)^3}{(1 + b)G} \right] \qquad (16\text{-}24)$$

where

$a$ = constant dependent on escapement geometry

$b$ = constant dependent on the friction coefficient between the tooth and pallet and on escapement geometry

$c$ = constant dependent on pivot friction of the escapement lever

By examining Eq. 16-24 one can readily see the reasons for variations of frequency shift from movement to movement. Small changes

16-34

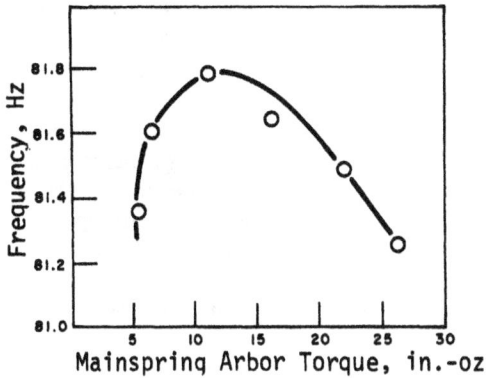

Figure 16-29. Junghans Escapement, Frequency-torque Curve

$A$ = amplitude, rad

$\omega$ = circular frequency, rad/sec

$\theta$ = phase angle, rad

$T$ = period, sec

($A$ and $\theta$ are constants determined by the initial conditions)

$$\omega = \sqrt{\frac{12k}{I_m}} = 2\sqrt{\frac{3k}{I_m}} \qquad (16\text{-}26)$$

$$T = \frac{2\pi}{\omega} = \pi\sqrt{\frac{I_m}{3k}}, \text{ sec} \qquad (16\text{-}27)$$

The Junghans escapement has undergone many changes of an evolutionary nature since its introduction approximately forty years ago into the United States for use in artillery fuzes. Materials have been improved and more efficient processes have been used. Further a mainspring has replaced centrifugal weights, gear train values have been increased, and safety features have been added. Two such improvements are the Popovitch modification and the Dock modification which are described in the paragraphs that follow. Two additional modifications, proposed specifically to reduce spin sensitivity (E-K and PA systems), are discussed in par. 16-4.4.

in $c$ and $b$ alter the value of $\Delta f/f$ appreciably. The relationship is especially sensitive to changes of $c$ at low torque and to changes of $b$ at high torques—i.e., at low torques, changes in pivot friction are critical; at high torques, changes in tooth friction are important. Since a larger random variation is expected of $c$ rather than of $b$, the escapement is less stable at low torques.

The fundamental equation for the motion of the balance wheel neglecting the effect of friction is

$$\frac{I_m}{12}\left(\frac{d^2\phi}{dt^2}\right) + k\phi = 0 \qquad (16\text{-}25)$$

where

$I_m$ = moment of inertia of balance wheel, slug-in.$^2$

$k$ = torsional spring constant, lb-in./rad

$\phi$ = angular displacement of balance wheel, rad, or

$\phi = A\cos(\omega T + \theta)$

### 16-4.2.2 Popovitch Escapement Modification

The Popovitch modification of the Junghans escapement is shown in Fig. 16-30[1]. It uses two outboard springs instead of a spring fastened to the arbor. The modification reduces the spin sensitivity of the mechanism.

A mathematical analysis has been made concerning the feasibility of the Popovitch modification in the design of escapements for artillery fuzes[19]. This report shows the behavior of the spring, in its horizontal and vertical positions, under the influence of

16-35

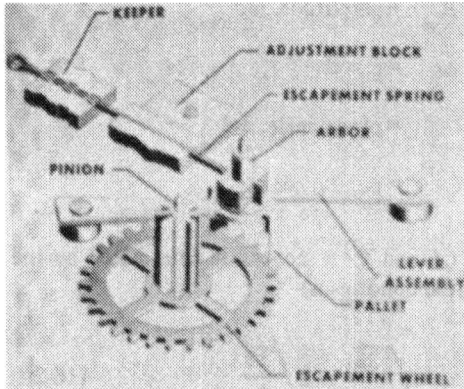

*Figure 16-30. Popovitch Modification of Junghans Escapement*

centrifugal and axial accelerations. Thermal expansion and gyroscopic effects also were considered in the analysis.

### 16-4.2.3 Dock Escapement Modification

The Dock modification is not directly concerned with the kinematical geometry of the escapement. It attacks the following problem: the natural frequency of the Junghans escapement changes in a spin environment because the spring must flex in a radial force field. The frequency change is proportional to the square of the spin speed. This is an appreciable effect. In addition to causing shifts in mean functioning time when the fuze experiences other than its regulation design spin, it also causes dispersion in functioning times. A large part of the timing variation of the Junghans escapement arises from this spin sensitivity (see par. 16-4.4).

In the Dock modification of the Junghans escapement the rectangular cross section spring is replaced by one of circular cross section (Fig. 16-31[16]). A reduction in spin sensitivity stems from this change alone. A

secondary benefit arises from the unique manner in which the round-wire escapement spring is fastened into the arbor. Fig. 16-32[16] shows in detail how the springs in the Junghans (standard) and Dock (new) versions are held in the arbor, and also indicates the attendant changes to the adjusting nuts. In the standard design, the escapement spring is distorted badly in the neighborhood of the arbor, necessitating a manual straightening operation after the spring is pressed into place.

### 16-4.2.4 Cylinder Escapements

Cylinder escapements are relatively simple and inexpensive. They are not used in applications requiring extreme timekeeping accuracy. The timekeeping errors are, in general, one order of magnitude higher than the errors in a detached lever escapement. These errors are due mostly to the torque sensitivity of cylinder escapements, i.e., the beat rate is somewhat dependent on the value of the torque that drives the escapement.

Fig. 16-33[20] shows the essential components of a typical cylinder escapement. The escape lever is a cylindrical tube with part of the wall area cut away. It spans one tooth. The inertia of this lever and the coupled hairspring constitutes an oscillatory system that meters the speed of rotation of the escape wheel. This is done by allowing the escape wheel to advance one tooth for each complete cycle of oscillation of the escape lever. Fig. 16-34[20] shows the action sequence for this escapement.

The motion of the various parts of the cylinder escapement were analyzed using Newtonian mechanics and, since the coupled equations of the impulse are nonlinear, solutions were obtained by computer methods[21]. Figs. 16-35 to 16-41[20] give the results of the variation of several escapement parameters. The definition of the symbols appearing on the graphs are:

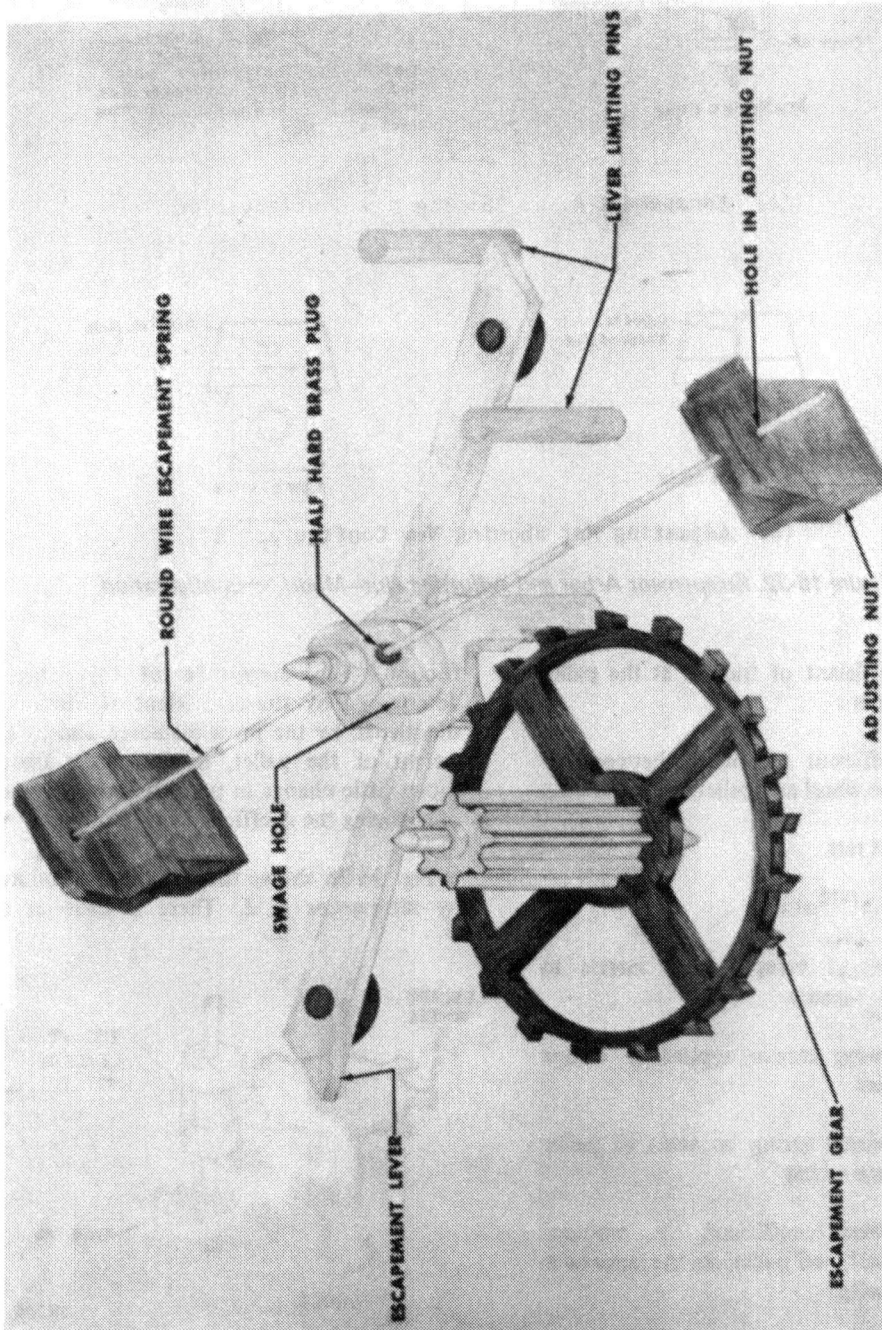

Figure 16-31. Dock Modification of Junghans Escapement

LEVER LIMITING PINS

HOLE IN ADJUSTING NUT

ROUND WIRE ESCAPEMENT SPRING

HALF HARD BRASS PLUG

ADJUSTING NUT

SWAGE HOLE

ESCAPEMENT LEVER

ESCAPEMENT GEAR

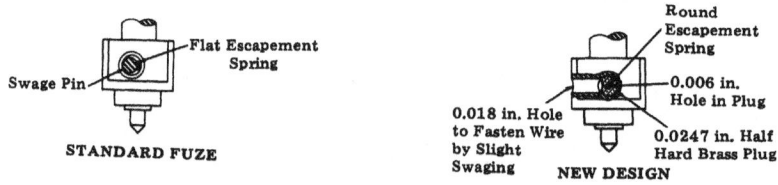

(A)   Escapement Arbor Showing New Configuration

(B)   Adjusting Nut Showing New Configuration

*Figure 16-32. Escapement Arbor and Adjusting Nut—Modified Configuration*

$\mu_c$ = coefficient of friction at the pallet pivots

$\mu_p$ = coefficient of friction between escape wheel and pallet surface

$BR$ = beat rate

$BR_n$ = natural beat rate

$J$ = ratio of escape wheel inertia to pallet inertia

$G_s$ = drawing torque applied to escape wheel

$k$ = torsional spring constant of pallet return spring

$E$ = rebound coefficient, i.e., rebound velocity of pallet to the approach velocity

Fig. 16-35 shows the effect of variation of torsional drag on the pallet due to pivot friction. The magnitude of this drag is determined by the coefficient of friction at the pivots, by the pivot diameter, and by the weight of the pallet. Study of the figures shows little change in performance as a result of changing the coefficient of friction.

Fig. 16-36 shows the variation introduced by differences in $E$. There is little or no

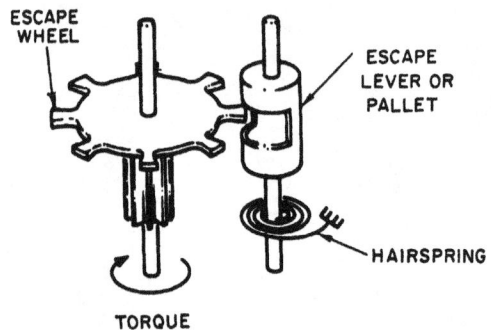

*Figure 16-33. Cylinder Escapement Assembly*

16-38

Figure 16-34. Action Sequence of Cylinder Escapement

sensitivity, for a change in $E$, in either timekeeping or amplitude properties.

Fig. 16-37 shows the results of varying the coefficient of friction $\mu_p$. Both timekeeping and amplitude are highly sensitive to this friction. The magnitude of the friction is sensitive to influences such as temperature and loading pressure. The drooping tendency of the timekeeping curves is probably too severe because most lubricants show a smaller coefficient of friction as the loading pressure is increased.

From these results, the choice of an ideal lubricant for the pallet surface should be one which is temperature insensitive, load insensitive, and has as small a friction coefficient as possible (say about 0.1). In addition, the lubricant should have wear resistance, chemical stability, and be immobile. The temperature sensitivity problem may be solved by selecting a lubricant whose coefficient is not sensitive over the required operating range.

Fig. 16-38 shows the effect of varying the ratio $J$. It is evident that a minimum of escape wheel inertia should lead to good timekeeping and at the expense of good timekeeping, increased pallet inertia could be used to control pallet amplitude.

A real problem in production is the variation of center distance. Fig. 16-39 shows the performance changes that occur as a result of center variations. Even though little sensitivity is indicated in the figure, experience says otherwise. A possible cause of the observed sensitivity may be rapid change of direction of the common normal to the contacting surface of escape wheel and pallet.

If the escape wheel tooth lift is increased, there tends to be a slight flattening of the timekeeping curves as well as some increase in amplitude as lift is increased (see Fig. 16-40).

The degree to which the escapement is "out of beat" is defined as the alignment error and is the parameter that is varied in

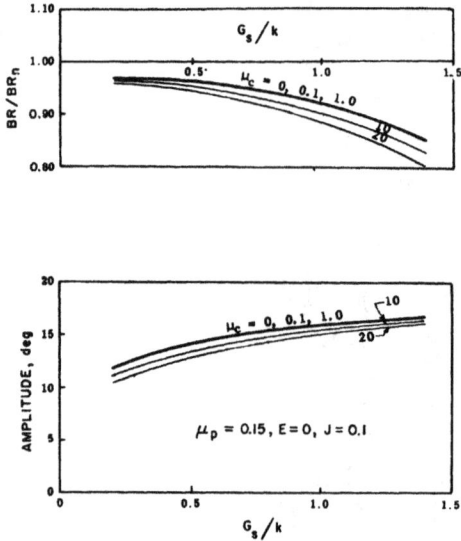

Figure 16-35. Pivot Friction Variation

Fig. 16-41. The figure shows that alignment errors—if they are small (2 or 3 deg)—have little effect but greater errors cause timekeeping degradation, increased torque sensitivity, and a lessening of the total range of torque over which the escapement will operate.

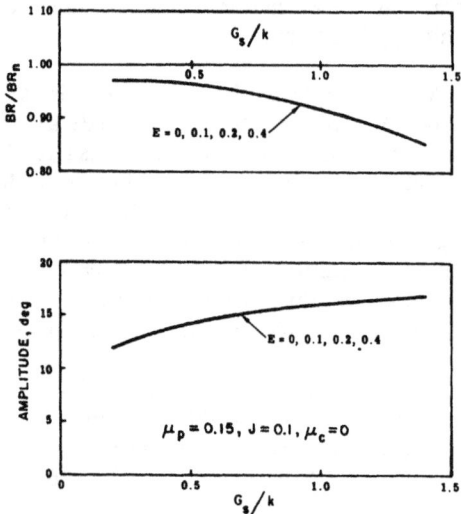

Figure 16-36. Rebound Coefficient Variation

Further work on the cylinder escapement indicates that the major contributor to torque sensitivity is the frictional drag between the pallet and escape wheel during the sliding phase of motion[20]. Possible improvement could be made by designing a recoil pallet that causes the escape wheel to back up slightly as the pallet moves into engagement. The energy that was extracted from the pallet due to recoil during engagement is restored to the mainspring and, during disengagement, it is returned to the pallet. This causes a negative drag that acts to cancel part of the positive drag. If these two effects could be proportioned properly, better timekeeping would result.

### 16-4.2.5 Escapement Self-starting Capability

This capability refers to the ability of the escapement to start the oscillator moving, without an external force, from any rest position and to sustain the oscillations as long as energy is supplied. Self-starting capability is

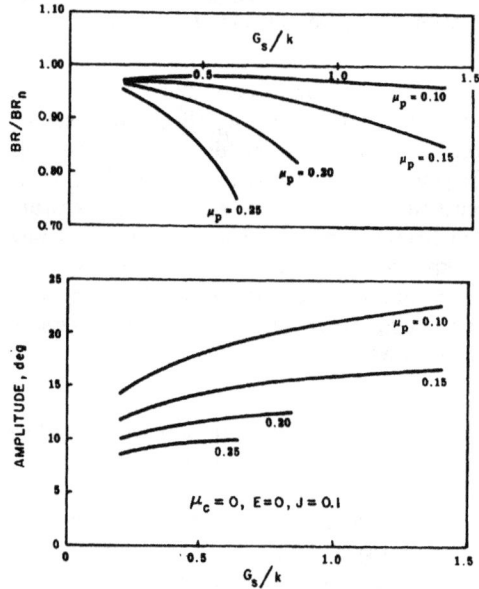

Figure 16-37. Pallet Friction Variation

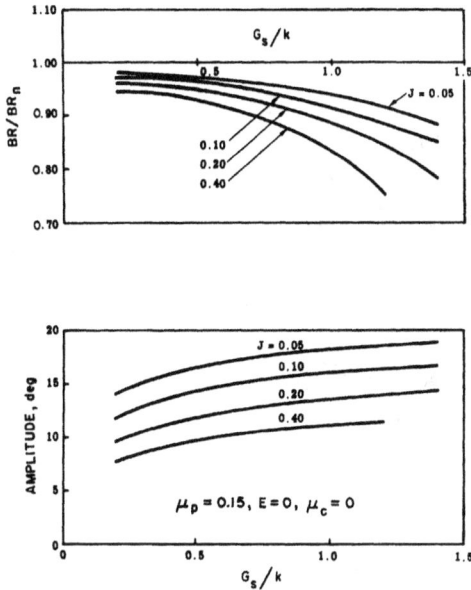

Figure 16-38. Inertia Variation

of great importance in many military applications.

Some escapements have dead spots or rest positions in which they will remain at rest, even when energy is being supplied, if they are allowed to stop in these positions. Restarting then requires an external distur-

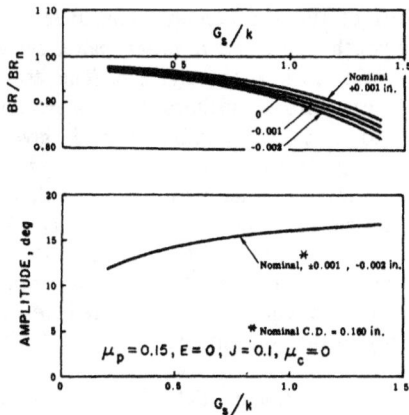

Figure 16-39. Center Distance Variation

bance. This is not a factor with the untuned two-center escapements that have no dead spots and are, therefore, inherently self-starting.

Some nonself-starting escapements can be made self-starting by putting them out of beat, i.e., making the relaxed position of the oscillator one other than that which is desirable for the best timekeeping accuracy. Another approach is to modify the tooth profile (see par. 16-3). However, care must be taken not to cause frictional drag. Gear trains can become self-locking due to friction (see par. 16-3.4).

An experimental and theoretical study was made of a variation of the Junghans system with a view to improve the self-starting capability[22]. The nonlinear equations of motion for the impulse phase are derived on the basis of energy considerations and are used to determine the relationships between escapement frequency and input drive torque. Experimental studies of frequency vs torque are described, and the test data are found to correlate reasonably the results of the theoretical investigation.

## 16-4.3 TUNED THREE-CENTER ESCAPE-MENTS

Tuned three-center escapements, which exist in several forms, have the following principal common characteristics that distinguish them from two-center escapements:

(1) There are three in lieu of two pivot centers.

(2) The oscillating balance is free (detached) of the escapement through as much as 85% of its swing, whereas in two-center escapements the balance is essentially never free from the escapement.

(3) Their timekeeping properties are improved significantly over two-center escapements.

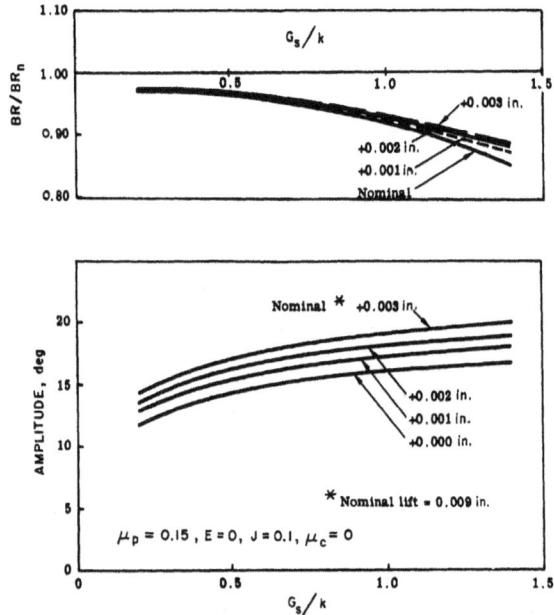

Figure 16-40. Variation of Lift

(4) The power requirements are minimal due to high operational efficiency.

(5) They are suited ideally to long runs because of their low natural frequency of oscillation.

The idea behind three-center or detached lever escapements is that an intermediate link between the escape wheel and the oscillating mass (balance) permits delivery of the impulses in a more precise manner, and also minimizes the drag torque imposed on the balance by the escape wheel.

The detached lever is capable of much greater accuracy and can be used for longer time delays than any other frictional movement. The poorest escapements are capable of accuracies of ± 0.1 to 0.2% while the best are capable of attaining accuracies of about ± 0.005%. The fundamental equations for the motion of the balance are the same as those

for the tuned two-center escapements (see Eqs. 16-25 to 16-27).

The detached lever escapement operates as follows: one end of a pivoted lever acts by means of two pallets in conjunction with the escape wheel; the other end acts on the balance. There are many slightly different versions of this escapement. The differences are in tooth shape and number, pallet design, lever ratio, fork geometry, etc. The escape wheel teeth are sometimes given a negative draw to improve reliability and give a self-starting feature but this destroys the detached lever characteristics because there is then a friction drag on the balance. For military applications, the amplitude may be reduced to permit higher frequencies. Both of these modifications tend to reduce the accuracy of the escapement.

Figs. 16-42[13] and 16-43[13] show the two main forms of the detached lever escapement,

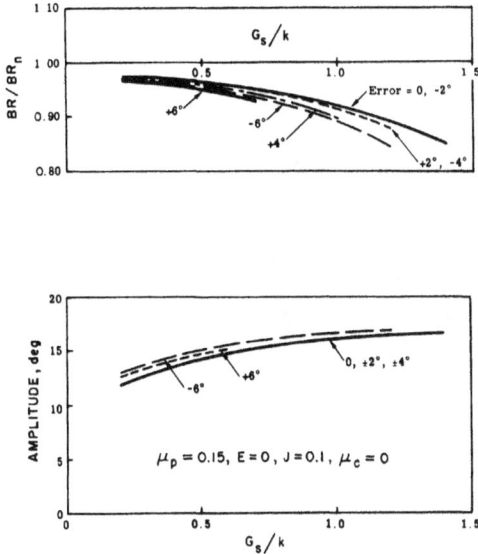

Figure 16-41. Alignment Error Variation

namely, the club-tooth type and the pin-pallet type. The pin-pallet escapement uses pins as pallets, and the impulse is entirely on the escape wheel tooth. This type is least expensive, rugged, and easiest to manufacture.

Detached lever escapements require a small

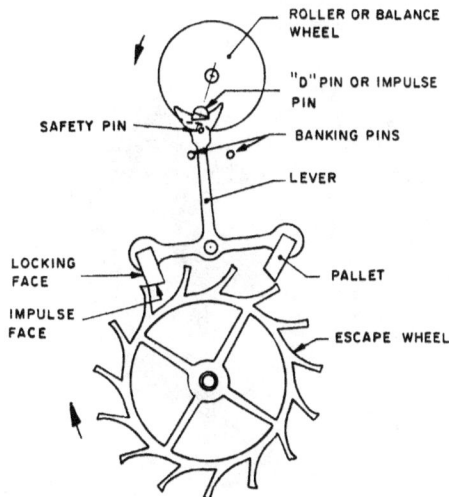

Figure 16-42. Detached Lever Escapement, Club Tooth Type

safety lever and roller on the balance staff (Fig. 16-43), and a pin on the lever to prevent unlocking the lever in the event the escapement receives a shock during the free motion of the balance wheel. In pin-pallet escapements, banking pins often are omitted and the pins are allowed to bank on the root diameter of the escape wheel.

The action of these escapements is as follows:

(1) The balance wheel travels in the direction indicated by the arrow (Fig. 16-42) with sufficient energy to cause the impulse pin to move the lever and pallet far enough to release the escape wheel tooth from the locking face, and allow it to enter on the impulse face of the pallet.

(2) The escape wheel, driven by the mainspring, moves in the direction of the arrow and pushes the pallet out of its path. After the wheel tooth has reached the end of the impulse face of the left pallet, its motion is arrested by the right pallet, the locking face of which has been brought into position to receive another tooth of the wheel.

(3) When the pallet is pushed aside by the wheel tooth, the lever imparts sufficient energy to the balance through the impulse pin to sustain oscillation of the balance wheel. The balance wheel proceeds on its excursion, deflecting the balance spring until it comes to rest. Then its motion is reversed by the uncoiling of the balance spring; the impulse pin again enters the notch of the lever but from the opposite direction, and the operation previously described is repeated.

### 16-4.3.1 Method of Increasing Torque Regulating Range of Detached Lever Escapements

An escapement has the twofold function of regulating the speed of the shaft and of dissipating the energy that is supplied through the shaft. While for watch making applications the speed regulation function is

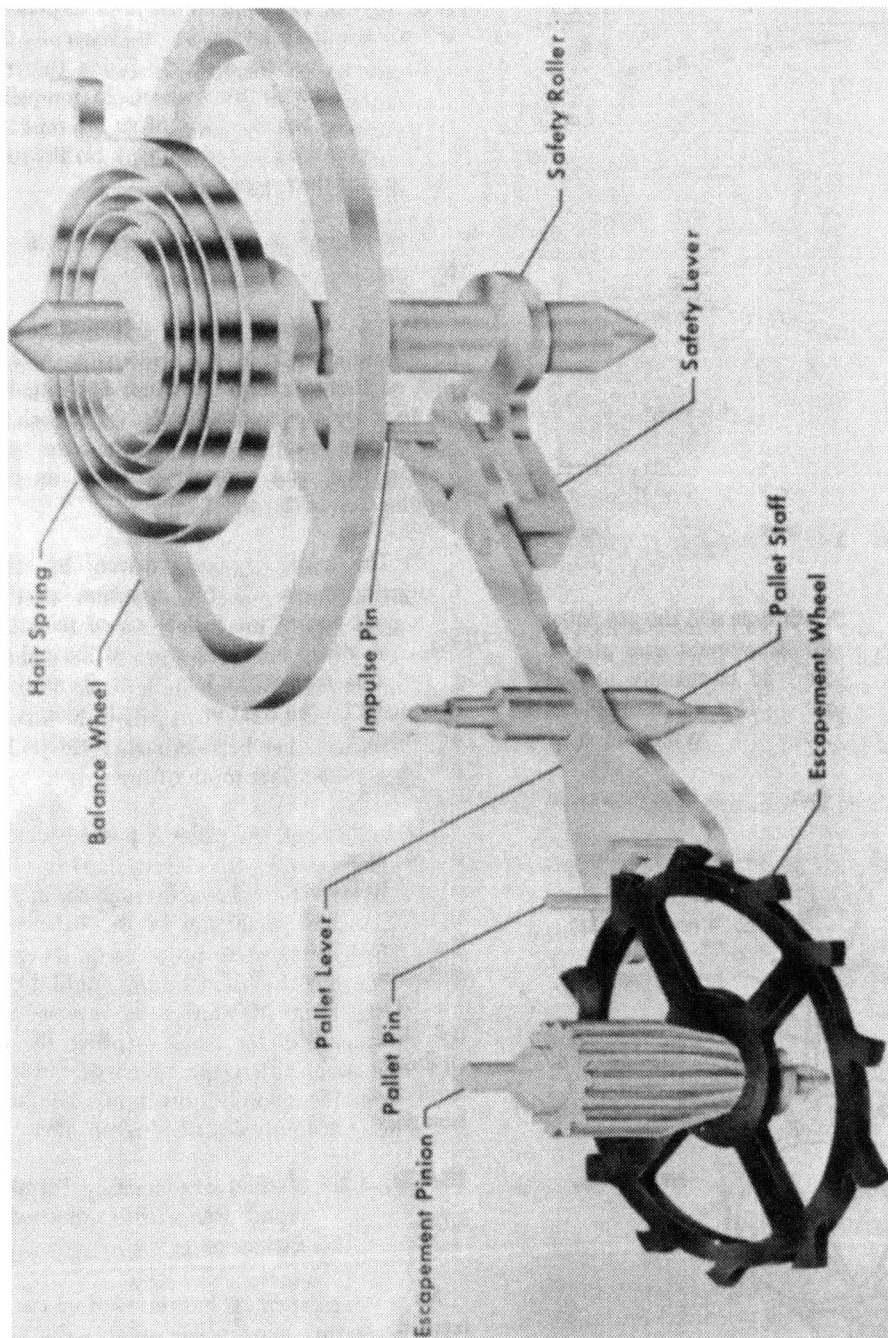

Figure 16-43. Detached Lever Escapement, Pin Pallet Type

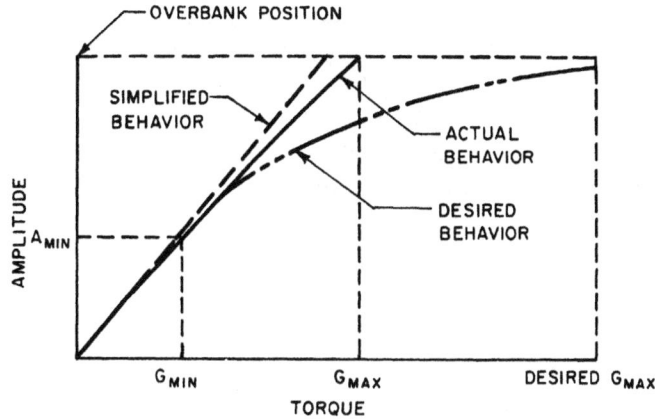

Figure 16-44. Amplitude vs Torque Behavior

paramount, in certain military applications the energy (or torque) regulation capacity is of primary concern.

The practically constant fraction of the energy that must be absorbed by the escapement is imparted to the balance during the impulse period; the remainder is dissipated as heat when the escape wheel impacts on the pallet. That energy that is imparted to the balance is dissipated primarily in coulomb friction at the balance pivots. It follows from an elementary energy balance that the amplitude of oscillation of the balance should increase in direct proportion to the applied torque. This is shown in Fig. 16-44[23] by the curve labeled "simplified behavior". At higher amplitudes, experimental curves deviate from linearity as is shown in the figure by the curve labeled "actual behavior" and indicate energy losses not directly proportional to input torque.

The permissible amplitude limits determine the range over which an escapement can operate. The maximum permissible amplitude is limited by the overbank position, i.e., the angular position of the balance wheel at which the pin strikes the outside of the lever fork. The minimum permissible amplitude is that amplitude for which the energy stored in the balance is just sufficient to provide

reliable unlocking action, accurate speed regulation, or both. These limiting amplitudes correspond to limiting torque values beyond which the escapement is said to be operating improperly.

The two methods available for increasing the torque regulation are either reducing the minimum permissible amplitude or introducing a nonlinearity into the amplitude-torque relationship so that higher torques can be accommodated without overbanking. This latter method is indicated in Fig. 16-44 by the curve labeled "desired behavior". Several methods for accomplishing nonlinearity are changes in windage drag, hairspring, escape wheel inertia, beat rate, and braking fork.

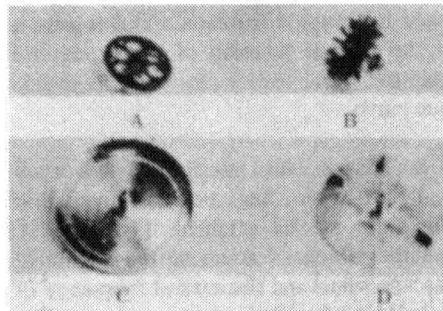

Figure 16-45. Balance Wheels

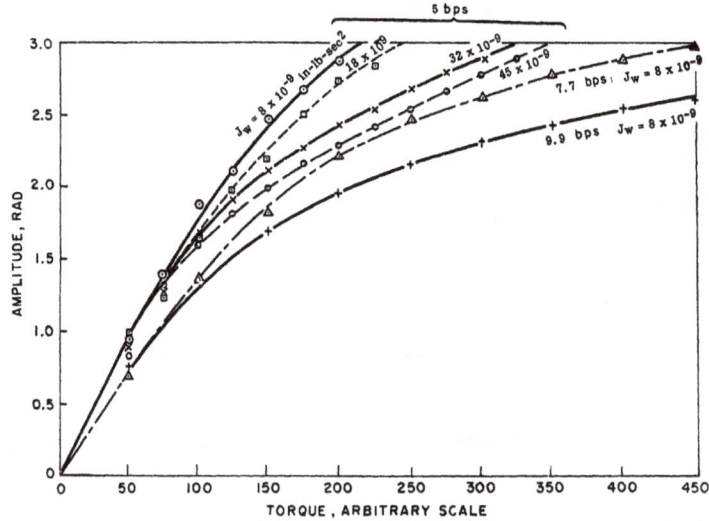

*Figure 16-46. Effect of Varying Escape Wheel Inertia and Frequency*

The windage drag, which is a function of velocity, can be amplified by altering balance geometry so that parameters bearing on windage can be made more prominent. Fig. 16-45[23] shows various arrangements of balances illustrating this principle. Although increases in windage effects are attainable, it was found that windage has little or no effect on nonlinearity.

Data and theory indicate that the hairspring generates a radial thrust comparable in magnitude to the weight of the balance wheel. Since this thrust increases with amplitude, nonlinearity like that shown in Fig. 16-44 can be demonstrated. The increased pivot loading at high balance amplitudes causes a rate of increase in energy dissipation which is greater than the rate of increase of amplitude and hence the nonlinearity in the amplitude torque curve.

If the escape wheel inertia and frequency are changed, it has been shown that nonlinearity can be attained. Fig. 16-46[23] shows the effects of changing the inertia of the escape wheel and the natural frequency of the balance-wheel/hairspring system. Fig. 16-47[23] shows the effect on frequency by changing the inertia of the escape wheel, as a function of amplitude.

Another way of accomplishing the desired amplitude behavior is by applying a drag torque to the balance which is proportional to the applied torque. Energy dissipation at the balance would then be dependent on the product of amplitude and applied torque. This can be accomplished by using a braking fork mechanism as shown in Fig. 16-48[23]. The effects of braking action which can be attained by this method are shown in Fig. 16-49[23]. Possible disadvantages are degradation of timekeeping and difficulties of manufacture. Fig. 16-50[23] shows the effect of several braking surface combinations.

## 16-4.3.2 Dynamic Analysis of Detached Lever Escapement

The dynamics of the detached lever escapement has been the subject of two analytical works[24,25]. The first is a Ph.D. dissertation on the dynamics of a detailed lever mechanism. The basic equation of an oscillating spring-mass system is augmented by adding terms that represent viscous damping, sliding friction, and a forcing function.

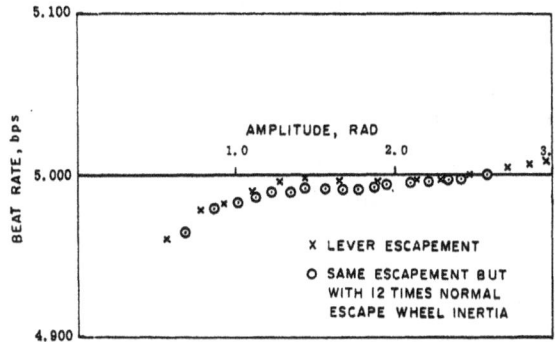

Figure 16-47. Beat Rate for a Detached Lever Escapement

The equation for the escapement is as follows

$$\ddot{\theta} + \omega_o^2 \, \theta + C\dot{\theta} + R\, \frac{\dot{\theta}}{|\dot{\theta}|} + X_\theta = 0 \qquad (16\text{-}28)$$

where

$\theta$ = angular displacement

$\omega_o^2$ = $k/I$

$k$ = torsional spring constant

$I$ = moment of inertia of balance

$C$ = (viscous damping coefficient)$/I$

$R$ = (sliding friction coefficient)$/I$

$X_\theta$ = forcing function

The equation of motion with certain values of $C$, $R$, $X$, was solved using computer programming and the results were plotted and compared with data taken from an actual operating unit in order to obtain the parameters $C$, $R$, $X$. Finally, tests were conducted on five experimental escapements. The experimental data gave excellent correlation to the theoretical data in both expected predictions and shape of the curves. A method of torque control is suggested to reduce the torque sensitivity of the detached escapement (see Fig. 16-51[24]). In this method a low-power escapement controls the

time rate of output of a high torque source while tapping a small fraction of the torque for its operation. With this arrangement, it is claimed that the torque is relatively constant and this provides better regulation.

The second report[25] points out that the high accuracy traditionally realized in watches and clocks using the detached lever escapement could not be maintained when the detached lever escapement is adapted for

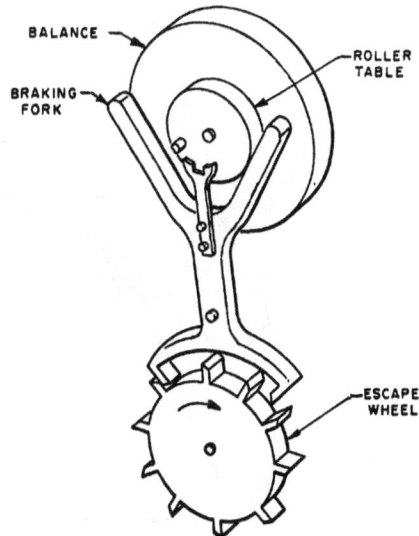

Figure 16-48. Braking Fork

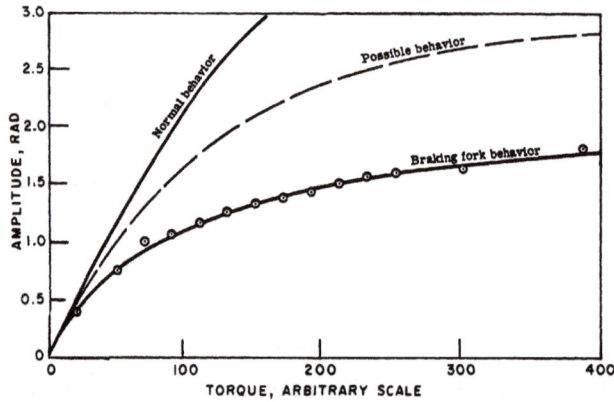

Figure 16-49. Effects of Braking Action

use in military timing mechanisms. Some of the causes believed responsible for the degradation in accuracy are:

(1) Elimination of costly jewelled bearings

(2) Strengthening of fragile, small diameter pivots

(3) Loosening of tolerances to decrease costs

(4) Increased variations in input torques and output loads resulting from military applications

(5) Increased escapement frequency necessary for high resolution in short timing intervals

(6) Operation over wide temperature extremes and after prolonged exposure to shock and vibration.

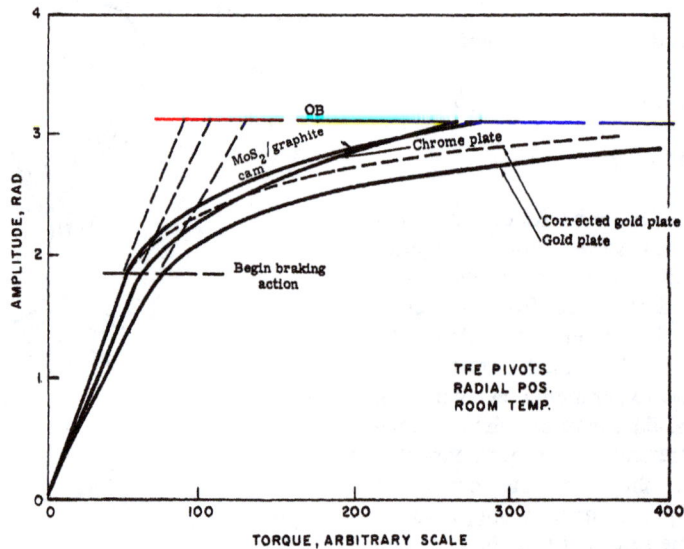

Figure 16-50. The Effects of Several Braking Surface Combinations

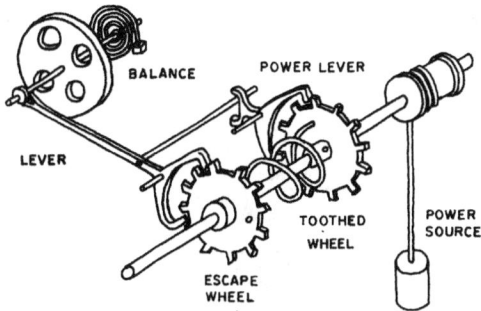

*Figure 16-51. Modified Escapement*

The gross effect of all these factors is to produce greater variations in the effective impulse delivered to the oscillating balance assembly.

Since an escapement that shows a large variation in timekeeping properties as the input torque is varied is said to be torque sensitive, the quality of a given escapement design can be assessed by determining its degree of torque sensitivity.

The main objective of the report[25] was to develop a mathematical model of the escapement which allows a more efficient investigation of parameter variations and the resulting performance changes not previously accomplished[26]. The differential equations governing the motion of the escapement system are derived, and the solution to these equations are obtained by computer methods.

The equations include most of the essential geometric and dynamic design parameters such as shapes, center distances, and inertias of parts. Energy loss mechanisms that are influenced by the input torque, collision processes, and escape wheel lag are also considered. One of the effects found to play a major role was that of hairspring radial sidethrust.

As a result of the investigation, several recommendations are made for an improved model:

(1) Provision for unlocking to end before zero balance displacement

(2) Addition of a draw angle in the escapement wheel tooth

(3) Addition of a variable geometry for the lever fork

(4) Provision for frictional losses during impulse

(5) Provision for elastic collisions during unlocking and impulse.

## 16-4.4 SPIN SENSITIVITY OF TUNED ESCAPEMENTS

It has been observed that the frequency of oscillation of the escapement lever (balance mass) changes when the escapement is spun in an artillery fuze. As previously indicated, this change arises from the fact that the escapement spring is now flexing in a radial force field that is proportional to the square of the spin speed. This frequency change can be either positive or negative depending on the conditions of fixity of the spring.

In addition to the frequency change caused by the force field, a change is also due to changes in output of the escapement wheel. In a typical movement, when the torque input to the arbor varies from 5 oz-in. (the minimum operating torque) to 30 oz-in. a frequency change of 1% is possible causing either random or periodic fluctuations. Random fluctuations are due, at least in part, to slippage between coils of the mainspring as it unwinds. Periodic fluctuations are due to eccentricities of the gears and to the variation of efficiency of gear tooth mesh on the different portions of the tooth profile. In addition, there is a gradual diminishing of the output torque of the mainspring as it

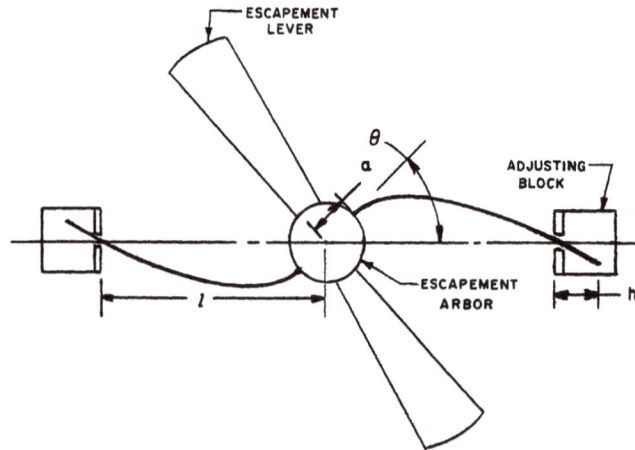

Figure 16-52. Escapement Spring Motion of Junghans Escapement

unwinds. Loss in torque due to this unwinding is on the order of 25%. (An exception to the decrease in output torque is the negator spring, see par. 16-2.1.)

Under spin conditions, the friction losses in the pivot of the off-center gears increase due to centrifugal force. If it is assumed that the spin has no effect on the mainspring, it can be shown that the torque output of the escape wheel can be expressed as a function of spin speed as follows[18]:

$$G = \eta G_a \left( 1 - \frac{\omega^2}{\omega_o^2} \right) \qquad (16\text{-}29)$$

where

$G$ = escape wheel torque under spin conditions, lb-in.

$G_a$ = arbor torque under nonspin conditions, lb-in.

$\omega$ = spin speed, rad/sec

$\omega_o$ = spin speed at which escape wheel gives no output torque, rad/sec

$\eta$ = gear train nonspin efficiency times train value (ratio of teeth, driver/driven)

Since $\omega_o$ is not measured easily, the minimum arbor torque $G_{min}$ and the spin speed $\omega_s$ at which the escapement stops running is found instead. Then

$$\eta G_{min} = \eta G_a \left( 1 - \frac{\omega_s^2}{\omega_o^2} \right) \qquad (16\text{-}30)$$

$$\left( \frac{\omega_s^2}{\omega_o^2} \right) = 1 - \frac{G_{min}}{G_a} \qquad (16\text{-}31)$$

but from Eq. 16-29

$$G = \eta G_a \left[ 1 - \frac{\omega^2}{\omega_s^2} \left( \frac{\omega_s^2}{\omega_o^2} \right) \right] \qquad (16\text{-}32)$$

Substituting

$$G = \eta G_a \left[ 1 - \frac{\omega^2}{\omega_s^2} \left( 1 - \frac{G_{min}}{G_a} \right) \right] \qquad (16\text{-}33)$$

The reduction in torque due to spin is appreciable. A typical timer operating at 80% of its stopping speed with $G_{min}/G_a = \frac{1}{4}$ will be receiving only slightly more than half of its nonspin torque.

The spin sensitivity of the escapement spring in a standard Junghans escapement (Fig. 16-52[18]) is given by

$$\frac{\Delta f}{f_n} = \frac{A\rho\ell^4\omega^2}{12 \times 112\,EI}\left[1 + \frac{7h^2}{\ell^2}\right]$$

$$\times (1 + 3\lambda - 12\lambda^2)* \qquad (16\text{-}34)$$

where

$f_n$ = natural frequency of escapement lever, Hz

$A$ = cross-sectional area of the spring, in.$^2$

$\rho$ = mass density of the spring, slug/in.$^3$

$\ell$ = distance from pivot center to the adjusting nut slot, in.

$E$ = modulus of elasticity of spring material, lb/in.$^2$

$I$ = moment of inertia of spring cross section, in.$^4$

$h$ = overhang of spring beyond adjusting nut slot, in.

$\lambda$ = ratio of 1/2 arbor width to $\ell$

Changes in certain parameters might reduce spin sensitivity, but care must be exercised. For instance, a change in $\lambda$ can introduce nonlinearities in the system.

The equation for spin sensitivity for the Dock escapement modification is similar to Eq. 16-34 with terms involving overhang ratio and $\lambda$ omitted as being too small to be of importance. Rewriting Eq. 16-34 we have

$$\frac{\Delta f}{f_n} = \frac{1}{12 \times 112}\left(\frac{\rho}{E}\right)\left(\frac{A}{I}\right)\ell^4\,\omega^2 \quad (16\text{-}35)$$

or

$$\frac{\Delta f}{f_n} = c_s\,\omega^2 \qquad (16\text{-}36)$$

*With restriction $\lambda \ll 1$. This expression is probably still good for $\lambda$ as high as 0.25 or 0.3.

where

$$c_s = \frac{1}{12 \times 112}\left(\frac{\rho}{E}\right)\left(\frac{A}{I}\right)\ell^4$$

To investigate what the change from a flat spring to the Dock round wire spring would do, form the ratio $c_s(Round)/c_s(Flat) = A/I(Round)/A/I\ (Flat)$ because the other parameters are unaffected. Substituting the approximate values of $A$ and $I$ for each type of spring, the ratio becomes

$$\frac{c_s\,(Round)}{c_s\,(Flat)} = \sqrt{\frac{\pi d}{3b}} \qquad (16\text{-}37)$$

By use of values for the movement of Fuze, XM565, spring diameter $d = 0.0033$ in. and spring width $b = 0.013$ in. (nominal) the ratio becomes

$$\frac{c_s\,(Round)}{c_s\,(Flat)} = 0.516 \qquad (16\text{-}38)$$

Therefore this change alone reduces spin sensitivity, and mean time shifts by one-half.

Two additional methods for reducing the spin sensitivity of the Junghans escapement are the E-K system and the PA system.

The E-K system (see Figure 16-53[18]) uses two springs with different end conditions and an off-center arbor. For this system, neglecting arbor width and overhang

$$\frac{\Delta f}{f_n} = \frac{A\rho\ell_1^4\,\omega^2}{12 \times 112\,EI}\left[\frac{1 - k\left(\dfrac{\ell_2}{\ell_1}\right)^3}{1 + \dfrac{\ell_1}{\ell_2}}\right](16\text{-}39)$$

where $k$ is a constant calculated to be 13/15. The advantage of this system is that one half, the left side, has positive spring spin sensitivity, while the other half, the right side, has negative sensitivity. This permits either positive, negative, or null spring sensitivity depending on the ratio of the spring length $\ell_1$

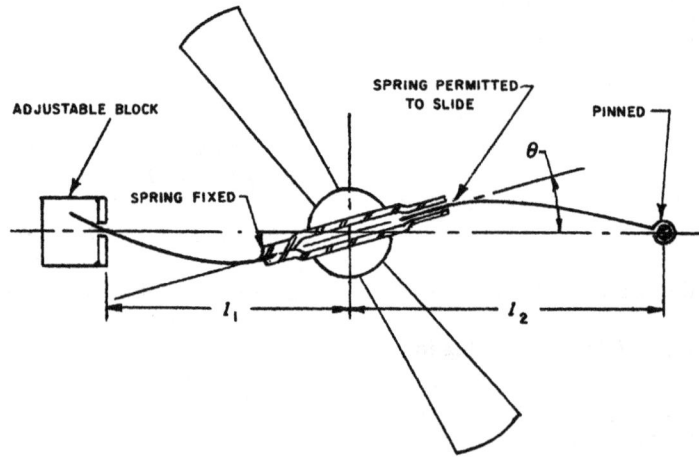

Figure 16-53. Escapement Spring Motion of E-K System

and $\ell_2$. Besides the apparent disadvantage of increased cost to manufacture, the main drawback might be the necessity of simultaneously juggling two adjustments to arrive at both the correct frequency and net null sensitivity.

The PA system is shown in Fig. 16-54[18]. The spring sensitivity of this system is given by

$$\frac{\Delta f}{f_n} = -\left(\frac{6}{7} + \frac{r}{\ell}\right)\frac{A\rho\ell^4\ \omega^2}{12 \times 16\ EI} \qquad (16\text{-}40)$$

where $r$ = radius of arbor that intercepts the spring.

Spring sensitivity is always negative. Much smaller $\ell$'s are required in this system than in the standard Junghans system for equal frequencies.

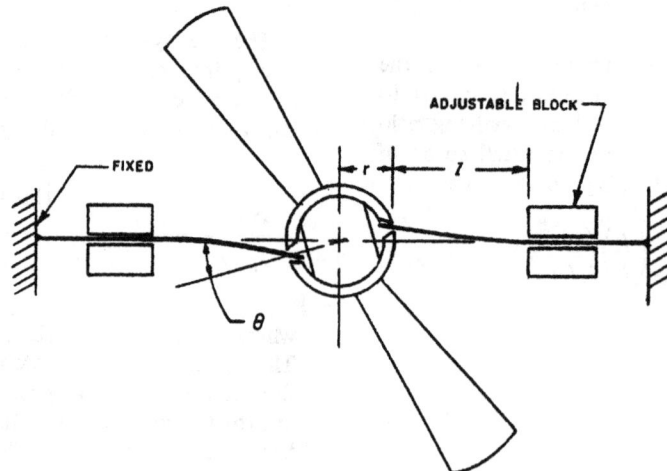

Figure 16-54. Escapement Spring Motion of PA System

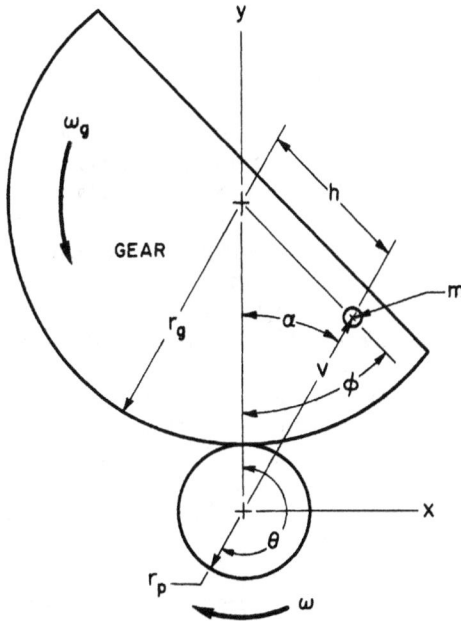

Figure 16-55. Segmental Gear Compensator

$r_a$ = radius of arbor, in.

$r_g$ = radius of gear, in.

$r_p$ = radius of pinion, in.

$\tau$ = tension in flexible connection, lb

$\omega$ = angular velocity, rad/sec

$\omega_s$ = angular velocity of gear, rad/sec

Analysis indicates that inertially generated torques can be used to compensate for friction, run-down, and spin decay energy losses in timing mechanisms. However, the analysis does not take into account two elements of the timing mechanism. These are torque gain due to spring spin, and torque loss

The problem of spin induced and run-down torque losses has been studied in order to determine the possibility of synthesizing an inertial (spin energized) compensator that will provide enough torque to balance out these effects[27]. Two methods were considered: (1) segmental gear compensator, and (2) connected mass compensator. The first type of compensator consists of two weighted segmental gears engaging a pinion. Such an arrangement (with both gears represented by a single gear) is shown schematically in Fig. 16-55[27], and the second type is shown schematically in Fig. 16-56[27]. The following symbols are used in the figures:

$F$ = force, lb

$h$ = distance from center of arbor or gear to center of mass, in.

$\ell$ = length of flexible connection, in.

$m$ = mass of gear, slug

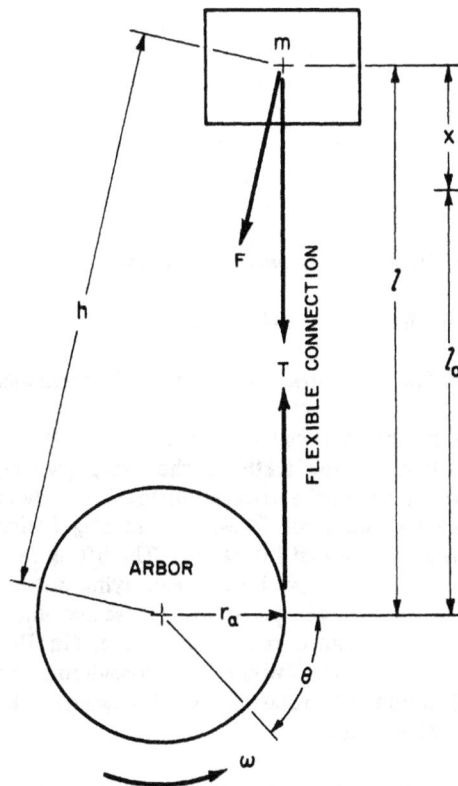

Figure 16-56. Connected Mass Compensator

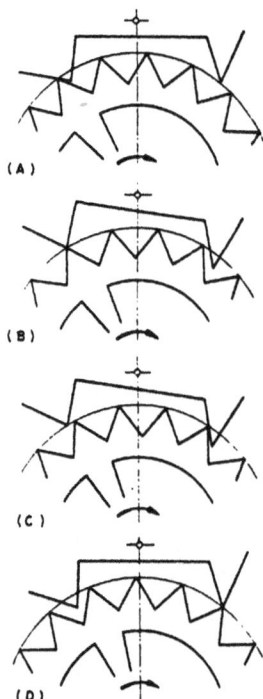

Figure 16-57. Operation of Runaway Escapement[7]

due to friction between spring leaves.

## 16-4.5 DESIGN FUNDAMENTALS

### 16-4.5.1 Untuned Two-center Escapements

In the runaway escapement the pallet is shifted by the teeth of the escape wheel, which act on the so-called lifting faces. There are two kinds of faces, the entering lifting face and the exit lifting face. The lifting face that is struck first by the teeth lying outside the span of the pallet as the escape wheel rotates is called the entering face. (In Fig. 16-57(A)[7] this would be, considering the direction of rotation of the wheel, the left-hand face.)

Assume that a tooth touching an entering face begins the movement (Fig. 16-57(A)).

The pallet reaches its maximum excursion when the tooth of the escape wheel reaches the tip of the pallet as in Fig. 16-57(B). Simultaneously, during rotation of the pallet, the exit face is moved far enough into the pitch circle of the approaching tooth tip that the pallet intercepts the increasing motion of the escape wheel. This brief free rotation of the escape wheel, sometimes called the drop, is necessary. To insure reliable movement, the rise of the pallet on the exit side cannot be initiated at the moment when the tooth tip on the entering side reaches the edge of the pallet. Manufacturing defects, such as out-of-round motion, could lead to jamming. Lifting begins on the exit side after the drop (Fig. 16-57(C)). The rotation of the pallet here also continues until the tip of the tooth reaches the extremity of the pallet on the exit side (Fig. 16-57(D)).

The movement described is a basic characteristic of all runaway escapements even though different escapements differ in design details.

For satisfactory functioning of every escapement, a number of quantities must be related in a very definite way. Only a limited range of choice is possible, namely:

(1) All points on a lifting face (shown in Fig. 16-58[7] in projection as a straight line) describe arcs during their motion which are referred to the center of motion of the pallet yoke (D) as their center. These circles are called the outer ($r_a$) and inner ($r_i$) pallet circles. The totality of these circles is limited by:

(a) The circle which runs through the intersection of the addendum circle with the lifting face on the entering side at the beginning of the lift (point R) and with the lifting face on the exit side at the end of the lift (point S'), see Fig. 16-58(A). (In this discussion, the escape wheel is assumed to rotate clockwise. The left-hand pallet, therefore, is the entering pallet.) and

16-54

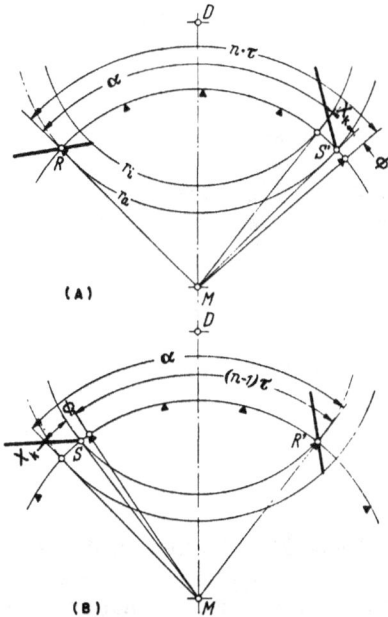

*Figure 16-58. Quantity Relationships for Pallet of Runaway Escapement*[7]

(b) A second circle running through the intersection of the addendum circle with the lifting face on the entering side at the end of the lift (point S) and the exit lifting face at the beginning of the lift (point R') (Fig. 16-58(B)).

(2) As shown later, the circle passing through S' need not join up with the circle through R. The same holds for the circles through R' and S. It is only necessary that the difference $r_a - r_i$ be constant.

(3) The two points R and R' (beginning of lift) form the escapement angle $\alpha$ with the center of motion of the escape wheel M, while points R and S, and R' and S', form the pallet lead $\chi_k$.

(4) The pallet lead and the drop cannot be chosen arbitrarily. Their magnitude is related to the escapement angle.

Let

$\alpha$ = escapement angle

$\chi_k$ = pallet lead

$\chi_z$ = tooth lead (also called the tooth tip width or tooth tip thickness)

$\phi$ = drop of the escape wheel

$n$ = span of pallet

$\tau$ = pitch

then we can formulate the following relationship from the one position in which the tooth at the entering pallet initiates the lifting process (Fig. 16-58(A)), in the first instance ignoring the lift on the tooth,

$$\alpha + \chi_k + \phi = n\tau \qquad (16\text{-}41)$$

Also on the exit side at the beginning of the lift

$$\chi_k + \phi + (n-1)\tau = \alpha \qquad (16\text{-}42)$$

Therefore

$$\chi_k = \frac{\tau}{2} - \phi \qquad (16\text{-}43)$$

If there is a run on the tooth, i.e., if the tooth has significant width at its tip, the pallet lead $\chi_k$ is reduced, if the drop interval of the wheel is to be preserved, to

$$\chi_k = \frac{\tau}{2} - \phi - \chi_z \qquad (16\text{-}44)$$

From Eqs. 16-41 and 16-43 we obtain for the escapement angle

$$\alpha = n\tau - \frac{\tau}{2} = \tau\left(n - \frac{1}{2}\right) \qquad (16\text{-}45)$$

The quantity $n$ which defines the span of the pallet takes on these values:

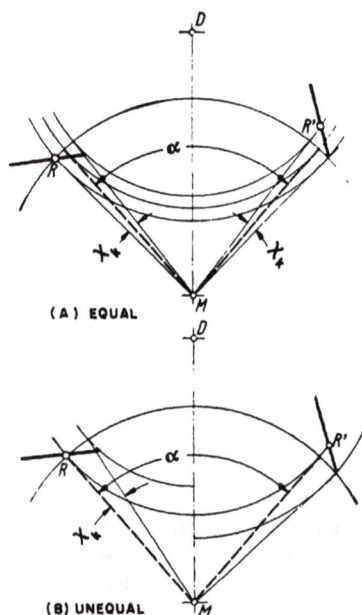

Figure 16-59. Runaway Escapement With
Equal and Unequal Arms[7]

For $n = 3$ the pallet spans 2.5 pitch
intervals

For $n = 4$ the pallet spans 3.5 pitch
intervals

For $n = 5$ the pallet spans 4.5 pitch
intervals, etc.

A discussion of pallet vs escapement arm
lengths follows:

(1) Pallet With Equal Arms. According to
Fig. 16-59(A)[7], which shows the same
situation as Fig. 16-58(A), the effective
sections of the two lifting faces lie on
concentric circles around the center of
motion of the pallet. Hence the points are
spaced unequal distances from the center of
motion of the pallet at the beginning of the
lift (R and R'). The two centers of the lifting
faces are seen to be the only corresponding

points on the two faces spaced equal distances
from the center of rotation of the escapement
yoke. These center points, of course, also are
spaced apart by the amount of the escapement angle. This design is called the equal-arm
escapement yoke.

(2) Pallet With Unequal Arms. Certain
detached lever escapements, for kinematic
reasons, make it desirable to rotate the
escapement angle in such a way that the two
points R and R' become equidistant from the
center of rotation of the escapement yoke
(Fig. 16-59(B)). In this way the outer pallet
circle of the entering pallet becomes the inner
pallet circle of the exit pallet. Escapements of
this type are called pallet escapements with
unequal arms.

(3) Pallet With Semi-unequal Arms. The
so-called semi-unequal-armed escapements
constitute an intermediate solution. These are
supposed to combine the advantages of the
escapements with equal and unequal arms and
avoid their disadvantages. They have not been
very widely used.

The pallet requires a certain time to
complete its swing. Since every pallet has a
definite inertia, we can consider it as a
physical pendulum in stable equilibrium. If
the impulse of the escape wheel tooth acting
on the entering and exit lifting faces is greater
than the directional moment of the pallet, the
pallet will execute forced oscillations. The
time in this case is a function of the mass of
the pallet, i.e., the moment of inertia of the
pallet and the moment of rotation acting on
the pallet.

The principles of construction of this
escapement, which is designed as an equal-
armed pallet, already have been shown in Fig.
16-58. Escapement angle and pallet lead can
be calculated using Eqs. 16-43 and 16-45. The
position of the center of rotation is arbitrary
but it is not without influence on the shape of
the pallet, especially the width of the lifting
faces.

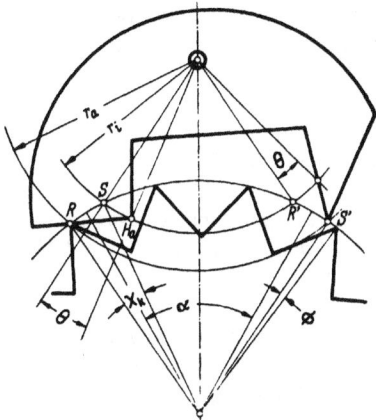

Figure 16-60. Untuned Two-center Escapement With Equal-arm Pallet[7]

Fig. 16-60[7] shows the complete design of an escapement with:

(1) An escape wheel of 15 teeth

(2) A pallet rotation or lift angle $\theta$ of 11 deg

(3) A drop $\phi$ of 2 deg

(4) A span $n$ of 3.

The position was chosen at a point when an escape wheel tooth had just left the edge of the lifting face (point S') on the exit side, and the tooth had just fallen on the lifting face (point R) on the entering side. If a pallet rotates through the lift angle $\theta$ on both the entering and exit sides, this angle is to extend inward from point S on the interior pallet circle and outward from point R'.

If the center of motion of the pallet is shifted toward the center of the escape wheel, the length of the lifting faces (distance R $H_a$) is shortened. Shorter lifting surfaces mean less space required for penetration of the pallet into the escape wheel.

In designing these escapements we find that several variables must be considered in choosing the dimensions, the effect of which in operation cannot be calculated. The choice of dimensions, therefore, is based entirely on experiment.

Especially important factors are:

(1) Size of the lift angle (depth of penetration)

(2) Coefficient of friction of lifting face and escape-wheel tooth

(3) Moment of inertia of the pallet

(4) Driving impulse.

The rapid swing of the escape wheel (untuned two-center escapements are used mainly in fast-running mechanisms) and the resulting strong impact of each tooth tip on the lifting faces of the pallet require that pallet and escape wheel be made of hard materials so that surface wear is not significant during service life.

Escapements can be adjusted to a limited extent by changing the lift angle through adjustment of the penetration depth. Experiment has demonstrated that the greater the depth of penetration, the greater the moment of inertia of the pallet and the smaller the lift angle. It follows conversely that errors due to

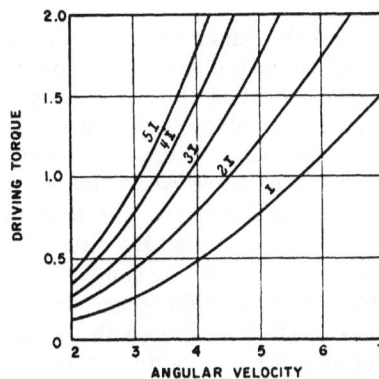

Figure 16-61. Escapement Curves for Different Moments of Inertia[7]

16-57

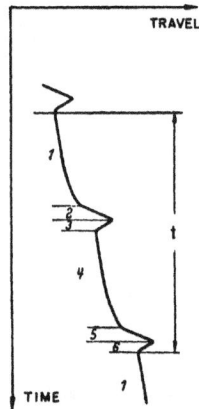

Figure 16-62. Rotating Drum Record of
Escapement Motion[7]

radial tolerances are smallest with small moment of inertia and large lift angle.

It is possible, for instance, to modify the running speed by changing the moment of inertia of the pallet by using adjustable weights. Fig. 16-61[7] shows the results of measurement on an escapement in which the moment of inertia of the pallet was changed in this fashion. It is apparent from the diagram that the regulation is simpler with increasing moment of inertia and that also, to maintain the desired rotational speed, the driving moment increases with increasing moment of inertia of the pallet.

Precise information about the individual phases of motion is obtained by recording the motion on a rotating drum. For this purpose both the motion of the escape wheel and that of the pallet can be used. To study the motion of the escape wheel, two or more small holes are bored near the teeth in the escape wheel spaced apart the distance of the pitch angle. These holes are illuminated with a condenser-equipped light source and projected by an optical system on the drum. The movement produces a curve like that shown in Fig. 16-62[7].

During one period (vibration time $t$) the escape wheel makes the following movements:

(1) Entering pallet face of the pallet touches the escape wheel:

(a) The escape wheel rotates forward and drives the pallet. Lift on the entering side (first interval)

(b) The escape wheel moves farther out of contact with the pallet. Drop on the entering side (second interval).

(2) The exit pallet face touches the escape wheel:

(a) The escape wheel is moved backward by the pallet. A recoil is present which is evident only with relatively high pallet mass (third interval).

(b) The escape wheel rotates forward and drives the pallet. Lift on the exit side (fourth interval).

(3) The entering pallet face touches the escape wheel. The escape wheel is rotated backward by the pallet. Exit recoil (sixth interval).

Additional design information is contained in Ref. 28.

16-4.5.2 Tuned Two-center Escapements

The simple untuned two-center escapement has a disadvantage—the oscillation period depends strongly on the driving impulse. This disadvantage can be eliminated by use of an oscillating mass. The inertia of the mass must be large enough so that the pallet is not subject initially to reversal of direction at the moment when the escape wheel teeth fall on the lifting faces of the pallet. The pallet, in addition to the intermittent locking action, returns to the oscillating mass the energy lost in friction, deformation of the escapement spring, etc. Without this resupply of energy the vibrations of the mass would be damped. This replenishment of energy is achieved by means of the escape wheel tooth during its run on the lifting surface.

The characteristic escapement values are as follows[7]:

(1) Period, Frequency, Number of Beats: The frequency, period, and number of beats in an escapement are related by Eq. 16-46

$$f_n = \frac{1}{T} = \frac{BR}{2} \qquad (16\text{-}46)$$

where

$f_n$ = natural frequency, Hz

$T$ = period, sec

$BR$ = beat rate (number of half vibrations of the oscillating mass)

For a flat escapement spring, the frequency is given in Eq. 16-22; for a torsion spring in Eq. 16-23. The period of a balance is given in Eq. 16-27.

(2) Escapement Constant: Since the rotational speeds of a timing mechanism with an escapement are determined entirely by the latter, a precise knowledge of the technical characteristics of the escapement and its effect on running of the mechanism is mandatory. Although in the case of air brakes, centrifugal governors, and the like, this effect can only be determined with difficulty, the relationship between the rotational speed of the escape wheel and the mechanical characteristics of the escapement—especially in the case of an escapement with a naturally oscillating mass—are much simpler to formulate.

When the oscillating mass (pallet) completes one vibration, the escape wheel has turned a distance of one tooth. If we designate the number of teeth on the escape wheel by $N_w$ the escape wheel will require for one rotation a time $t$ of

$$t = TN_w \qquad (16\text{-}47)$$

Thus the rotational speed $\omega$ of the escape wheel becomes

$$\omega = \frac{1}{TN_w} = \frac{f_n}{N_w} = \frac{BR}{2N_w} \qquad (16\text{-}48)$$

The far right-hand side of Eq. 16-48 is called the regulator constant. It is an important quantity and is used as an output equation in the design of clockworks. However, this equation is not valid for escapements without a naturally oscillating mass because the quantities $T$ and $BR$ are then not "inherent" values.

(3) Relationship Between Rotational Speed and Regulator Period: The relationships which follow compare unregulated and regulated timing movements

$$\omega_u = \frac{\chi}{T_u N_w} \text{ , rad/sec} \qquad (16\text{-}49)$$

and

$$\omega_r = \frac{\chi}{T_r N_w} \text{ , rad/sec} \qquad (16\text{-}50)$$

hence

$$\omega_u = \omega_r \left( \frac{T_r}{T_u} \right) \text{ , rad/sec} \qquad (16\text{-}51)$$

and

$\omega_u$ = unregulated angular velocity, rad/sec

$\omega_r$ = regulated angular velocity, rad/sec

$\chi$ = gear ratio between driver and escape wheel

$T_u$ = unregulated period, sec

$T_r$ = regulated period, sec

$N_w$ = number of teeth on escape wheel

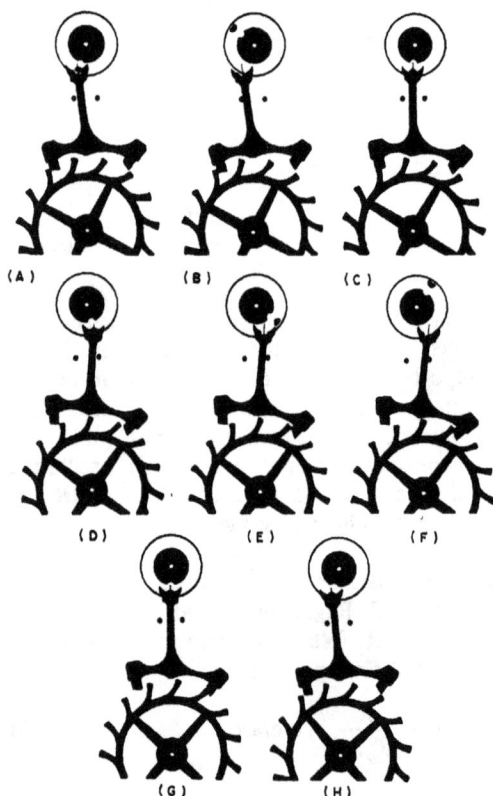

Figure 16-63. Operation Detached Lever
Escapement[7]

Thus the deviation $\Delta$ from normal speed is

$$\Delta = \omega_r - \omega_u = \omega_r \left( 1 - \frac{T_r}{T_u} \right) \qquad (16\text{-}52)$$

If $T_u > T_r$, the timer runs slow ($\Delta$ has a positive sign) and if $T_u < T_r$, the timer runs fast ($\Delta$ has a negative sign).

### 16-4.5.3 Tuned Three-center Escapements

Components, functioning, and operation are described in the paragraphs that follow:

(1) Component Parts and Functioning, General: In this escapement, an intermediate link (lever) is placed between the escape wheel and the oscillating mass (balance). This minimizes the drag torque imposed on the balance by the escapement. Important features of the tuned three-center escapement are (a) the energy is transferred over a relatively small angle during the rotary motion of the balance, (b) the balance must be free to rotate without any attachment, which is extremely important for a constant rate, and (c) this arrangement makes possible a greater speed of the escape wheel. The escapement is illustrated in Fig. 16-42 where the operating parts are labeled.

(2) Operation of Escapement: Fig. 16-63(A)[7] shows the starting position. The balance executes a swing to the right. The impulse pin leaves the fork notch in the lever. A tooth has fallen on the lock at the entering pallet. The lever is spaced the "lost distance" from the banking pin. The lost distance is the angle of rotation from the pallet position after the drop to the end of the swing.

Refer to Fig. 16-63(B). The pallet is drawn deeper into the escape wheel by the escape wheel itself, due to its run angle, until the lever touches the banking pin. Here the lever traverses the lost distance. In this way the pallet is protected at least partly against repulsion (e.g., due to vibrations). During this time the balance swings fully into the end swing and reverses (right swing).

Refer to Fig. 16-63(C). After return of the balance, the impulse pin again enters the fork notch, strikes its right wall, and transmits an impulse to the pallet by its impact. The pallet is withdrawn from the escape wheel and escape begins. The escape wheel tooth slides over the locking point on the lifting plane. Lifting begins. In this phase the direction of impulse between impulse pin and lever changes. Up to this time the impulse pin has driven the lever. Starting with the beginning of lifting, the lever drives the impulse pin.

Refer to Fig. 16-63(D). This is the end of lifting on the entrance side. The tooth is about to drop. The impulse pin leaves the

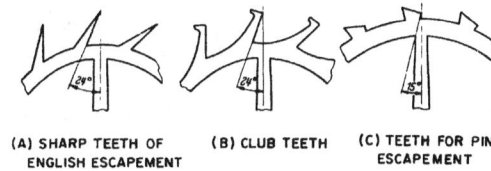

(A) SHARP TEETH OF ENGLISH ESCAPEMENT　(B) CLUB TEETH　(C) TEETH FOR PIN ESCAPEMENT

*Figure 16-64. Escape Wheel Tooth Shapes[7]*

fork notch; the balance moves free through the left swing.

Refer to Fig. 16-63(E). The tooth lying just in front of the exit pallet has fallen on the locking face. The balance swings farther into the end distance.

Refer to Fig. 16-63(F). The run angle on the exit side causes farther penetration of the pallet. The lever travels the lost distance on the exit side, until it meets the right banking pin. The balance reaches its extreme position and reverses its direction of rotation (it passes from a left swing to a right swing).

Refer to Fig. 16-63(G). The impulse pin strikes the fork wall and pulls the pallet out of the escape wheel. The escape on the exit side begins. After the escape, lifting begins and there is a transfer of energy through the lever and impulse pin to the balance.

Refer to Fig. 16-63(H). The tooth has reached the end of the lifting plane on the exit pallet and can fall away. The impulse pin leaves the fork notch. The tooth located in front of the entrance pallet drops on the lock.

(3) The Most Important Components and Their Function:

(a) Escape Wheel. There are three characteristic tooth shapes. Pointed teeth are used in the English lever escapement (Fig. 16-64(A)[7]). This escapement is less often used today because the tip of the tooth is very sensitive to shock, such as that following the drop. The club tooth predominates today (Fig. 16-64(B)). It has a greater mechanical

strength, ensures good oil retention, but tends to set more readily. The wheel of the pin escapement (Fig. 16-64(C)) is also very stable. It is relatively simple to cut with a milling machine, a fact of importance relative to cost.

(b) Pallet. A pallet with unequal arms mainly is used. The reason is to permit an equal distance of the locking planes from the center of motion and thus perform equal work on the entering and exit sides to release the pallet. Pallet insets are made of stone or steel.

(c) Draw Angle: The pallet locking planes, to do their job, must be arranged concentrically for minimum friction. Such a minimum, however, is not what is wanted here, because the lever, and therefore the pallet, would not be slowed down as the balance passes into the end of its swing. Thus the pallet could be dislodged from the escape wheel by vibration and shock. Only the escape wheel can supply the slowing action to the pallet during the time when the lever runs free of the impulse pin. For this reason, the circular shape of the pallet is abandoned, and its locking surfaces are replaced by planes. The position of these planes relative to the center of motion of the pallet is made such that the system tends to slide inward as a result of the pressure of the tooth until this "run" is halted by the banking pin.

The perpendicular force $N$ at point R (Fig. 16-65[7]) due to pressure of the tooth corresponds to the friction force $\mu N$ since there is bound to be friction. These forces act on the two lever arms $\ell_1$ and $\ell_2$, and tend to turn the pallet in a positive or negative

16-61

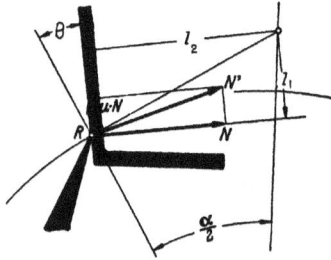

Figure 16-65. Determination of the Draw Angle[7]

direction (depending on the magnitude of the resultant moment or angle $\theta$). If we assume constant friction coefficients, $\theta$ may not exceed a definite value in order for the sum of the moments of $N$ and $\mu N$ (or in other words, the moment of the resultant $N'$) to be positive; and the pallet, therefore, to have a tendency to slide inward.

The length of the two lever arms is given by

$$\ell_1 = r \left( \tan \frac{\alpha}{2} \right) \sin \theta, \quad \ell_2 = r \left( \tan \frac{\alpha}{2} \right) \cos \theta \tag{16-53}$$

Torque of the perpendicular force is

$$N\ell_1 = Nr \left( \tan \frac{\alpha}{2} \right) \sin \theta \tag{16-54}$$

Torque of the force of friction is

$$\mu N \ell_2 = Nr \left( \tan \frac{\alpha}{2} \right) \cos \theta \tag{16-55}$$

If there is to be equilibrium, the two moments must be equal and opposite

$$Nr \left( \tan \frac{\alpha}{2} \right) \sin \theta = \mu Nr \left( \tan \frac{\alpha}{2} \right) \cos \theta \tag{16-56}$$

Hence

$$\mu = \tan \theta \text{ or } \theta = \text{Tan}^{-1} \mu \tag{16-57}$$

If we assume the coefficient of friction $\mu$ to be $\approx 0.17$ (steel and stone) for the most unfavorable case (dry surfaces and average

polish), then we obtain an angle of 10 deg for $\theta$. In order to ensure downward motion of the pallet, $\theta$ is made 12 deg. Further increase of the draw angle, however, must be avoided because, otherwise, the work of release and the return will be unnecessarily large.

(d) Lever. The lever serves to transmit the impulse from the pallet to the balance and vice versa. It is rigidly connected to the pallet and, therefore, moves together with it. The angle through which the lever turns is determined by the lifting angle, the lock, and the lost distance. Subtracting the lost distance, it amounts to about 10 deg. Shorter levers are preferred to longer ones. Longer levers mean greater mass and, therefore, heavier loading of the lever and pallet staff. Pallet and lever should be well balanced.

(e) Banking Pin. The purpose of the banking pin is to limit the run because, otherwise, the oscillating balance would have to expend too much work in release. Also the pallet should not touch the base of the tooth. The banking pin is positioned in such a way that the lost distance amounts to about 1 deg.

(f) Safety Pin. Only the extreme position of the lever can be delimited by the banking pin. It is conceivable, however, that the lever, during the phase when the impulse pin is outside the notch in the fork, might reverse as the result of a momentary shock in spite of the thrust due to the run angle. Premature reversal of the lever would cause the impulse pin to strike the outside of the fork on the way back; this would disrupt the system. In order to avoid this, a safety pin is attached to the lever that follows the center line of the lever and, if reversal should be induced, rides against the outer wall of the roller.

(g) Roller and Impulse Pin: The roller is in the form of a disk rigidly attached to the balance staff. Its purpose is twofold: (1) to act as a base for the impulse pin, and (2) to serve as a stop for the safety pin. The impulse

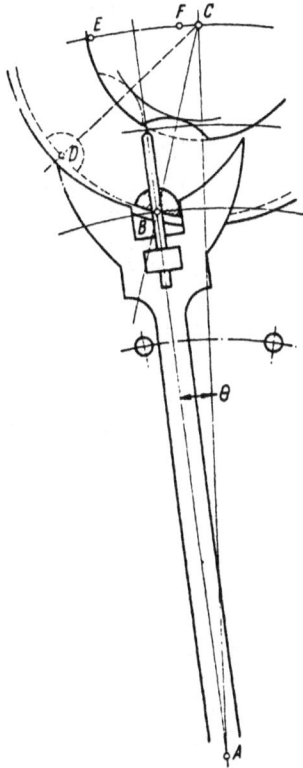

*Figure 16-66. Lever With Safety Pin*[7]

pin is set into the steel roller. It may be sapphire, but in military timers it usually is made of steel.

A single roller, which is seldom used, would carry the impulse pin and also serve as a stop for the safety pin. With increase in roller size, however, there is increased risk that the safety pin will jam on the edge of the roller. For this reason the safety roller is made as small as possible. A double roller meets the requirements of a larger roller diameter for the impulse pin and a smaller roller diameter for restraining the safety pin. The diameter of the safety roller is about half that of the impulse pin roller (see Fig. 16-43).

The safety roller is cut away on its outside

face in line between the center of rotation and the impulse pin to permit the safety pin to pass during reciprocation of the fork. This slot should be only as large as absolutely necessary because enlargement will not leave sufficient stop area for the safety pin after the impulse pin moves out.

(h) Lever. The lever transfers the impulse of the pallet to the balance and vice versa. It carries a number of safeguarding features, the satisfactory functioning of which must be ensured. Fig. 16-66[7] shows the lever at the end of lift on the exit side; information follows:

● Maximum lever swing angle

$$\theta = \frac{\text{lock} + \text{total lift}}{2} \qquad (16-58)$$

● Width of Impulse Pin: about 5 deg (relative to point A).

● Diameter of Small Roller: about 3/5 of the impulse pin circle diameter.

● Distance of Safety Pin Point From Small Roller: 0.5 deg.

● Distance From Side Face of Lever to Banking Pin (the so-called lost distance): about 1 deg.

● Slot to Pass Safety Pin: The slot must be situated in such a way that the safety pin has already entered the slot at the beginning of release. Its depth must be such that no contact is possible between the tip of the safety pin and the bottom of the slot.

● Arms of Fork: They must be long enough that the fork is prevented from a reverse stroke while the balance is traveling through the arc BD by their following the impulse pin. The safety pin cannot provide this safeguarding function in the sector BD. The inner surfaces of the fork arms are cylindrical surfaces. Their centers, for satisfactory movement of the impulse pin, should

(A) ENTRANCE SIDE

(B) EXIT SIDE

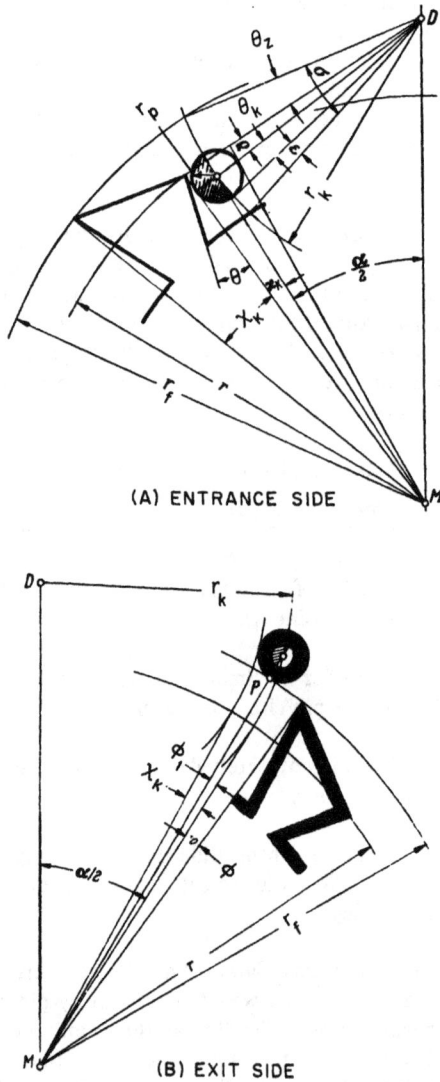

*Figure 16-67. Pin Pallet Escapement Geometry*[7]

not lie on the axis of the lever nor, in the center position, at the center C of the balance (points E and F).

● Lever Length: Length AB for short levers is 1.6 $r_a$ and for long levers 2.0 $r_a$; length BC is 0.5 $r_a$ for each, where $r_a$ is the outer pallet radius.

The short lever requires a heavy balance for dynamic reasons, while a balance with a smaller moment of inertia is preferred with a longer lever.

(i) Fork Arms. The fork arms on the lever provide added safety because the safety pin may still be in the roller slot when the impulse pin leaves the notch of the fork. In the single roller, the fork arms are not absolutely necessary because the safety pin can be restrained even after a slight rotation of the roller. In the double roller, the closer position of the safety pin slot to the center of rotation causes the angular distance over which the safety pin is in the slot, as measured from the center of rotation of the balance, to be considerably greater. During this time, the pallet is without adequate protection. The arms of the fork assume the protective function here. Incipient reversal of the lever causes the fork arms to ride against the impulse pin.

### 16-4.5.4 Detached Lever Escapements With Pin Pallet

In the pin pallet escapement (Fig. 16-43) the lift is effected almost entirely by the escape wheel. The pallet faces are replaced by round steel pins against which the escape wheel teeth act. On the average, the pin lift amounts to about one-fifth of the total lift.

While in the club tooth escapements the excursion of the lever is limited by the banking pin, in the pin pallet escapement the escape wheel itself performs this limiting function. As a result of the draw $\theta$ on the escape wheel tooth (Fig. 16-67(A)[7]), which is commonly 15 deg, the pallet is drawn so far in by the tooth draw (extra run $\theta$) that the pallet pin strikes the foot of the tooth. This simplifies manufacturing appreciably but introduces a risk with respect to operation of the lever, namely, that the pallet pin tends to stick at the base of the tooth because slightly thickened oil can lodge in the corners of the teeth. In spite of this disadvantage, the pin pallet escapement has been used widely, particularly in military timers.

Information on the design of this escapement follows:

(1) Pallet Pin Design:

The following notation, used for analysis, is illustrated in Fig. 16-67.

$r$ = tooth heel radius of escape wheel, in.

$r_f$ = tooth tip radius of escape wheel, in.

$r_k$ = radius about which pallet pin turns, in.

$r_p$ = radius of pallet pin, in.

$\alpha$ = escapement angle, rad

$\epsilon$ = run, rad

$\theta$ = total lift of pallet pin, rad

$\theta_k$ = pin lift, rad

$\theta_z$ = tooth lift, rad

$\rho$ = lock, rad

$\sigma$ = total pallet travel, rad

$\tau$ = pitch, rad

$\phi$ = drop of escape wheel, rad

$\phi_1$ = small increase to drop angle, rad

$\chi_k$ = pallet lead, rad

$\chi_z$ = tooth lead, rad

Starting with the total lift, we obtain the pin radius

$$r_p = r_k \theta_k, \text{ in.} \qquad (16\text{-}59)$$

Also

$$\chi_k = \frac{r_p}{r}, \text{ rad} \qquad (16\text{-}60)$$

so that

$$\chi_k = \frac{r_k \theta_k}{r}, \text{ rad} \qquad (16\text{-}61)$$

The tooth lead, obtained from Eq. 16-44, can be expressed as

$$\chi_z = \frac{\tau}{2} - \phi - \frac{r_k \theta_k}{r}, \text{ rad} \qquad (16\text{-}62)$$

Since from Fig. 16-67(B)

$$r_k = r \tan \frac{\alpha}{2} \qquad (16\text{-}63)$$

then

$$\chi_z = \frac{\tau}{2} - \phi - \theta_k \tan \frac{\alpha}{2} \qquad (16\text{-}64)$$

The total rotation $\sigma$ of the pallet and hence of the lever is

$$\sigma = \text{tooth lift} + \text{pin lift} + \text{lock} + \text{run}$$

or

$$\sigma = \theta_z + \theta_k + \rho + \epsilon \qquad (16\text{-}65)$$

$\theta_k = 2 \text{ deg} \qquad \rho = 2 \text{ deg}$

$\theta_z = 8 \text{ deg} \qquad \epsilon = 3 \text{ deg}$

(2) Draw of the Pallet:

If the point of intersection of the tangent to the tooth-tip circle (instead of half the escapement angle) with the center line (MD) is taken as the center of motion of the pallet, the draw distances on the entrance and exit sides are different. This disadvantage is clearly seen in Fig. 16-68(A)[7]. (The figure is not to scale to emphasize this fact.)

There is no cause, however, to put the center of motion at this point. If the center of motion D is shifted toward the center of the wheel (Fig. 16-68(B)), the entrance draw becomes smaller and the exit draw larger. If D is moved close enough, the relationships are

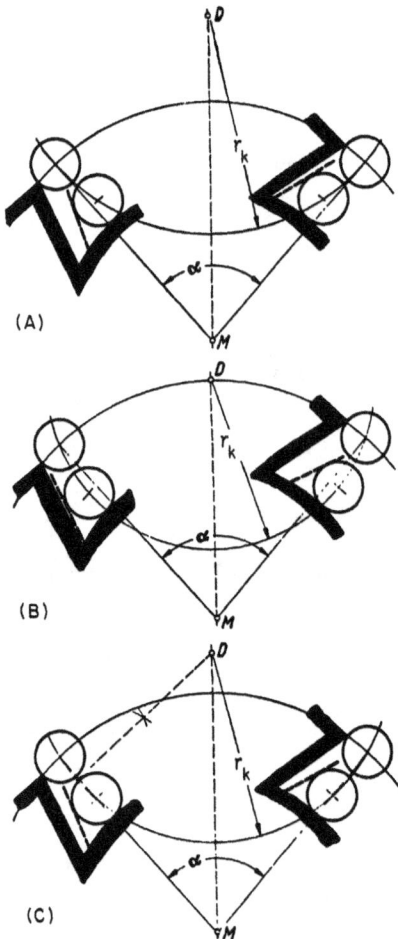

Figure 16-68. Entrance and Exit Draw Dependence on Pallet Center[7]

angle at the beginning of the lift when the pin rests at the base of the tooth (end of the run) (Fig. 16-68(C)). Then the draw on both sides is completely equal. The center of motion of the pallet then lies on the perpendicular to the midpoint of the line joining the centers of the pins.

Attempts have been made to improve the design of the pin pallet escapement by a variety of modifications. It has been suggested, for example, that the penetration of the pallet to the base of the tooth can be avoided and its swing limited by locating the centers of motion of the balance, the pallet, and the escape wheel in a straight line. Then the lever, elongated toward the escape wheel, can serve a protective function, with its split end suitably shaped to ride on the escape wheel spindle. Suggestions for changing the shape of pin and tooth (bending the back of the tooth) were not acceptable because of their great cost.

(3) Lever Design. The lever for pin pallets is simpler than for club-teeth pallets; safety pin and small roller are omitted (Fig. 16-69[7]). By reducing the radius of the impulse pin and by properly shaping the arms of the fork, the impulse pin is made to move over a greater range inside the arms of the fork, so that the latter assume the safeguarding function. Outside this range the fork tips take over the safeguarding function. These can ride on the part of the balance staff which was not cut away to pass them earlier. Lateral banking arms limit the movement of the balance.

## 16-5 SETTING

### 16-5.1 GENERAL

Setting is that action which fixes the time interval measured by the mechanical timer. In some mechanical timers, a clock mechanism is used. In operation it is similar to a mechanical alarm clock. The ordinary alarm clock can be set for a time delay up to 12 hr, while the

directly reversed. It should be kept in mind, however, that the center of motion of the pallet cannot be moved indefinitely close to the center of the wheel because this would require bending the pallet shaft. (The center of motion D in Fig. 16-68(B) already lies within the tooth heel circle.)

The center of motion of the pallet is positioned in such a way that the center of the pin lies on one leg of the escapement

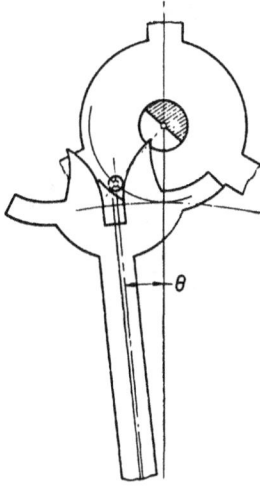

*Figure 16-69. Simplified Lever*[7]

clock mechanisms used in military timers usually can only be set for periods in the order of seconds or minutes. In order to set in a time delay, a knob on the face of the timer is turned so that the desired time is indicated on the dial. This action puts energy into the mainspring of the timer and corresponds to winding the alarm clock[29]. When the time has elapsed on the military timer, a switch is closed (or opened according to the application of the timer), or some mechanical motion may take place.

The time delay set into a clock mechanism may not be used until some other event takes place, hence some sort of triggering linkage is necessary. The triggering linkage usually consists of a catch that can be released by energizing a solenoid. This type of timing device can be used only once without some external operation being required. With timers driven by motors (AC or DC), clutching mechanisms are used to disengage the drive motor from the actuator to allow the timer to reset to its start position[30].

In one version of a mechanical timer[31], it is required to provide six specific times in the

interval of 40 to 120 sec by turning a single knob. The solution was to employ a "six-shot revolver" magazine arrangement of commercial odometer-type counters as shown in the exploded view of Fig. 12-4. This magazine permitted any counter to be coupled individually to the single timing movement by rotating the setting knob to any one of six positions. (A seventh "neutral" position also is provided.) The timer shown in the figure occupies about 6 in.[3]. The cylindrical magazine arrangement allows a simple, low cost, reliable mechanical setting mechanism while maintaining selectability for each of the six time intervals.

The readout is achieved by a mechanical switch contact on the hundreds or slowest moving wheel of the counter. The counter initially is set to the time desired (less 0.1 sec) and, when the timer is started, the counter counts backwards to zero. After the counter reaches zero on all wheels, the wheels turn to the 999.9 position in the succeeding tenth of a second; and this motion of the hundreds wheel closes the switch contact. Thus, all of the load to close the switch contact has been kept from the timing movement until a portion of the last 0.1 sec of timing. This digital system also increases the setting and readout resolution by at least 20 times over the traditional slow-moving cam and follower without having to resort to close tolerance components.

The power supply for this timer is a mainspring, and a detached lever escapement is used.

## 16-5.2 IMPROVED SETTING METHOD FOR MECHANICAL TIME FUZES

A mechanical time fuze to be adapted for tank-fired ammunition required a setting wrench to set the mechanism held in position by a high resisting torque (100 oz-in.). The fuze had a time range up to 100 sec, and each 0.1-sec setting represented a circumferential movement of only 0.007 in. Hence, a vernier

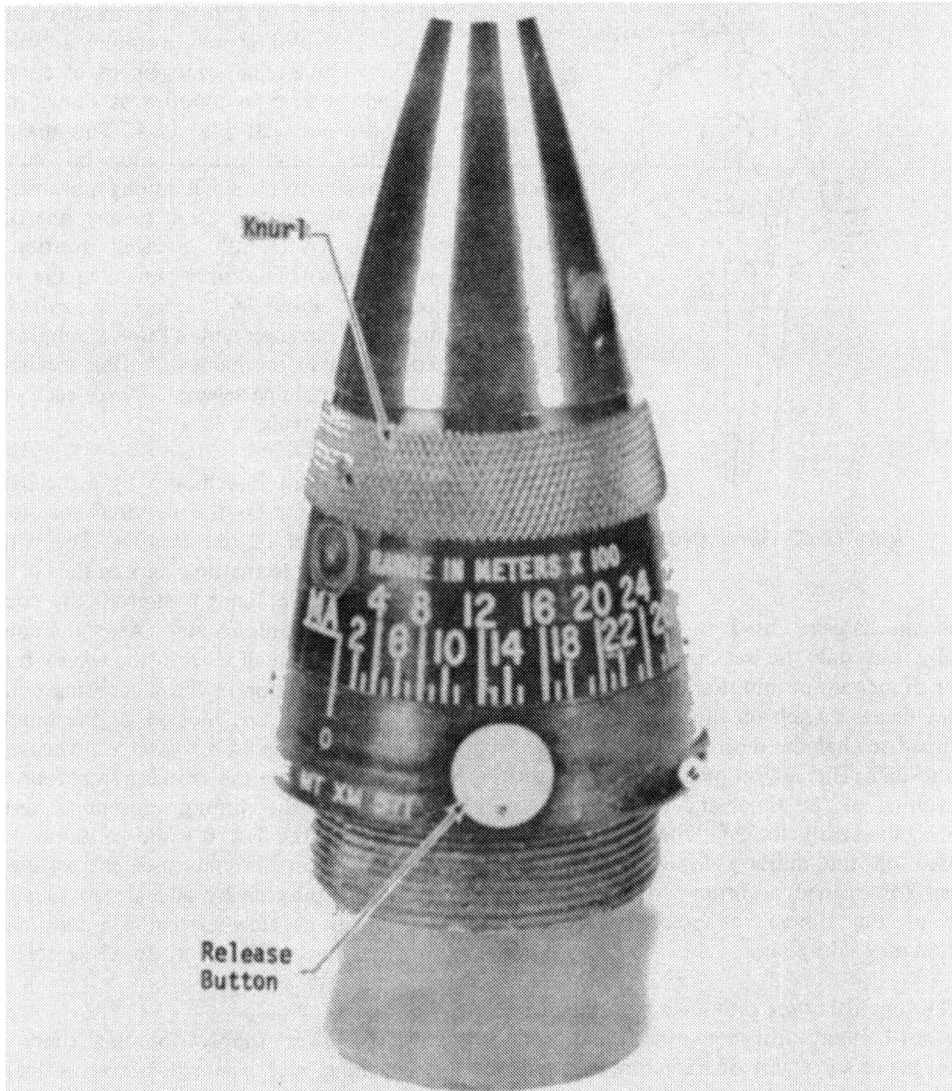

*Figure 16-70. Setting Mechanism for Fuze, MT, XM571*

scale had to be provided. This fuze obviously was not suitable for firing from a tank.

Fig. 16-70[1] shows Fuze, MT, XM571 with setting mechanism redesigned for tank firing. The design has the following features:

(1) There is no need for time settings

beyond 10 sec for tank-fired ammunition. The range of the setting was, therefore, reduced from 100 sec to 10 sec so that each 0.1-sec setting represents a circumferential movement of 0.07 in. This increase eliminated the need for a vernier.

(2) The setting torque was reduced so that

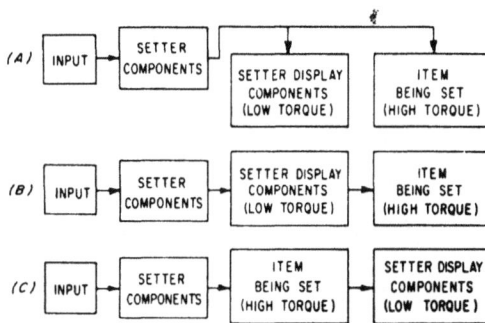

*Figure 16-71. Linkage of Setter Components*

the nose could be turned by hand. A wrench is thus no longer required. A knurl is provided on the nose to insure a good grip.

(3) The time setting is held by the release button. When the button is depressed, the nose turns freely. The button has five teeth that mate with an internal ring gear in the nose, whose pitch is such that each tooth represents 0.1 sec. When the button is released, it will lock the setting at any 0.1-sec increment.

(4) To eliminate the need for firing tables, the scale is calibrated directly in meters. Lines are numbered for every 200 m up to 4400 m. The intermediate 100-m settings have a tick mark. Incidentally, the scale is uneven because the even increments are on a time base.

(5) The size, shape, and thickness of the numbers and the numbered lines were selected experimentally so as to be readable under the red dome light during blackout conditions.

If one remembers the trying conditions under which the user must adjust a fuze, one can understand why this amount of attention is required for so simple a device as a time setting mechanism.

## 16-5.3 LINKAGE OF SETTER COMPONENTS

Designs of mechanical setter devices should include consideration of the linkage of the setter components and setter display components in conjunction with the device being set.

The parallel mechanical linkage (Fig. 16-71(A)) permits concurrent positioning of setter display components and the item to be set. This type of linkage could cause the display of a false reading because the setter display does not necessarily have to agree with the information actually set into item (the linkage to either the setter display components or item being set could be faulty). The series linkage (Fig. 16-71(B)) is little improvement over the first because the linkage to the item to be set could be deficient even though a setting is displayed. Deficiencies are more prone to occur in the high-torque gear trains of the item to be set rather than in the low-torque gear trains of the setter display assembly. The most reliable and safest linkage (at no increased cost) is a different series linkage (Fig. 16-71(C)) in which the setting actually positioned into item being set is displayed after the fact of setting (thus providing a true read-out).

## 16-5.4 SETTING MECHANISM FOR MECHANICAL TIME FUZE, XM577

The XM577 Fuze is shown on Fig. 16-72[32]. The mechanism of the fuze is housed in an upper body (fuze ogive) having a window area that indicates SAFE, "Point Detonating PD", or digital time sequences. A setting key (Fig. 16-73[32]) is used to rotate the digital setting module to the desired function. The digital setting module consists of: a setting shaft, three digital counter wheels (hundreds wheel, tens wheel, and seconds wheel), two counter wheel index

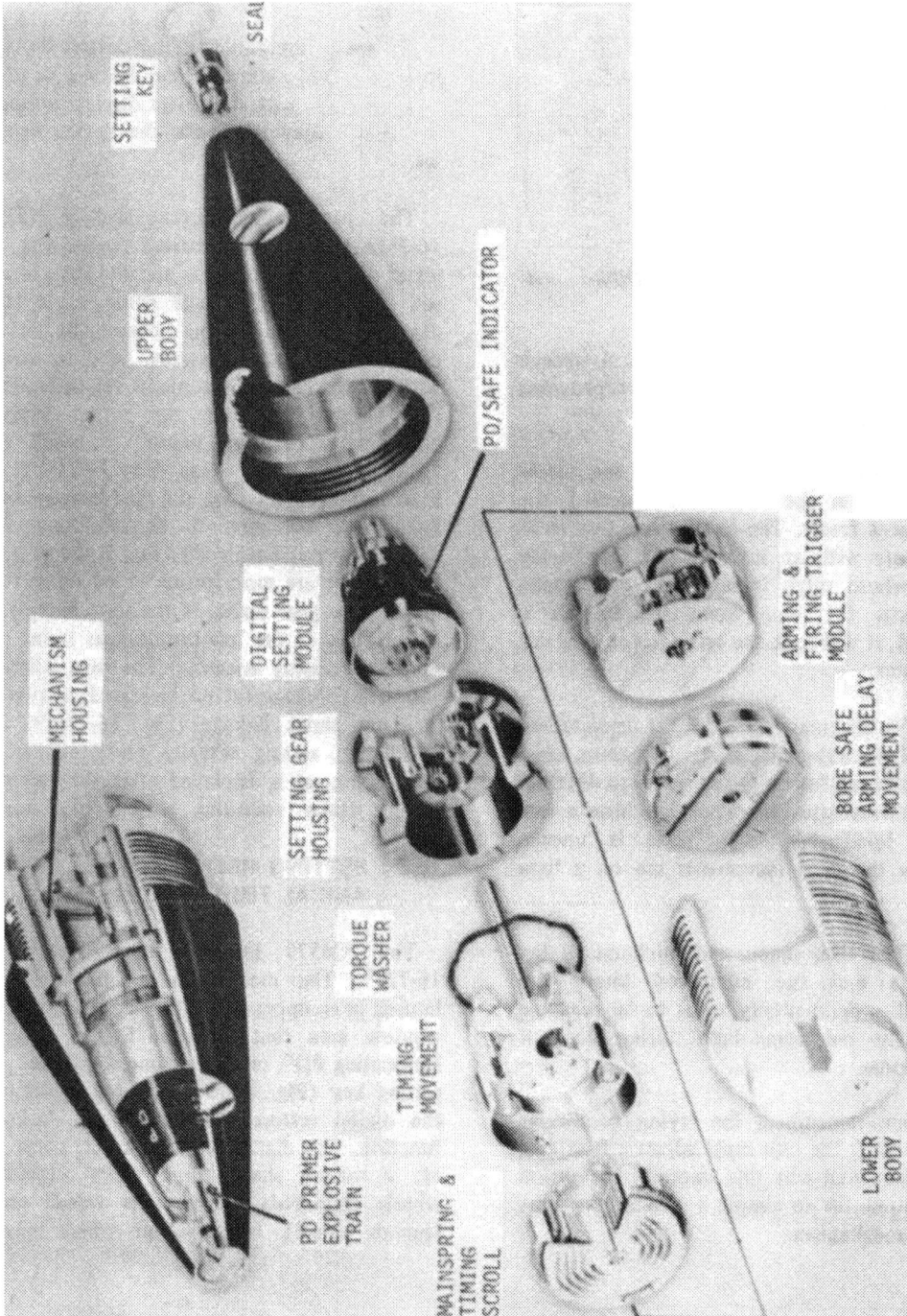

Figure 16-72. Mechanical Time Fuze, XM577

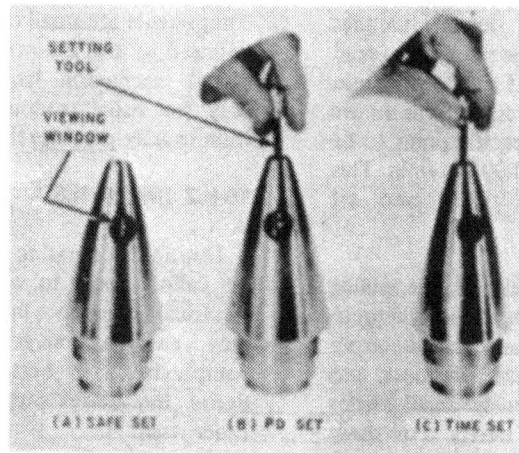

*Figure 16-73. Setting Positions for Fuze, XM577*

pinions, and a PD/Safe indicator flag. The setting shaft is coupled directly to the seconds wheel which in turn will index the tens and hundreds wheel, dependent upon the set time. Both the counter wheels and the indicator flag are observable through the fuze window when the fuze is set properly. Setting the fuze is accomplished by breaking the seal at the nose of the fuze and applying torque to the setting shaft with a setting key, as shown in Fig. 16-73. The applied torque rotates the timing mechanism module, displacing the scroll follower pin for the set time desired.

This fuze was developed to fill present requirements for artillery rounds. The fuze has demonstrated satisfactory performance under setback forces of 900 to 30,000 g, spin from 1600 to 30,000 rpm, and field settable time delays from less than 1 sec to more than 200 sec with better than 1% accuracy (0.06 standard deviation at a set time of 15 sec).

## 16-6 OUTPUT

### 16-6.1 ANALOG AND DIGITAL OUTPUTS

The overall objective of any timer output is to indicate the total number of oscillations of the time base occurring in a given interval of time. The output can be either visual (readout) or physical (actuation of cam or switch). The output function (visual or physical) can be either analog or digital. An example of an analog mechanism is a dial of a watch which has continuous motion with infinite resolution. The digital output is a stepping motion having no resolution between steps. Visual readouts of each of them are illustrated in Fig. 16-74[9].

An analog output may be obtained by placing a disk on the slowest moving gear (mainspring barrel) of the timer. With this setup, the angular position of a slow moving shaft (output shaft) must resolve the displace-

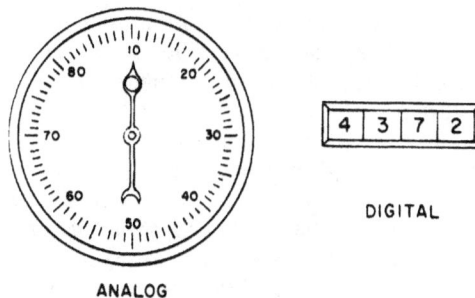

*Figure 16-74. Two Types of Visual Output*

16-71

ment of a fast moving shaft (input to gear train). For example, a 1-in. output disk that rotates once in 24 hr may be required to read out in intervals of 5 min of time. The 5-min interval on a disk 1-in. in diameter is an arc length of 0.01 in., and a specific point to be detected is moving at only 0.002 in./min. This makes it a problem to detect and set accurately.

Because of infinite resolution, the analog output is the most accurate device theoretically speaking. However, the practical considerations of precision manufacture, cost, and the need to detect extremely small angles limit its use for military timers. For these reasons a digital readout is better when high reading accuracy and incremental settings are desired.

The digital mechanism produces a discrete unit of output motion for each complete input pulse. A single pulse may represent a full 360-deg rotation of a shaft. The digital mechanism faithfully records each input pulse without error and is, therefore, accurate to the nearest pulse. For instance, again taking the 24-hr timer, suppose that the digital counter is arranged to record pulses that occur exactly every 30 sec. Then a time of 5 hr 10 min (18,600 sec) would be represented by a total of 620 pulses, and to an observer the figure 620 would be accurate to ± 30 sec when observed at any particular instant. This resolution is obtainable at any total time setting.

When the digital mechanism is used to provide a physical output at the end of the set time interval, it has an advantage of rapid transfer from one pulse to the next. Thus, the accuracy of the output at the set moment of transfer depends primarily on the accuracy of the time base. However, the digital device does not give the mechanical advantage of a gear train and, if a follower is applied to a cam surface on one of the counter wheels to sense a given time interval, it usually is found that performance is affected seriously because

the driving torques used to move these components are small compared with the load imposed by the follower. For this reason, the digital mechanism has been used in timers only for visual readout and the analog gear train usually provides the physical output.

## 16-6.2 SWITCHES AND RELAYS

During its operating interval, the timer may be called upon to operate a sequence of electrical contacts. In turn, these contacts may energize relays of any degree of complexity. It is better, where possible, to derive the useful output directly from the timer itself[31].

Contact ratings are important. The required number of cycles in the life of the unit, and the capacity and rating of the load are the principal factors affecting contact design. DC circuits are far more injurious to contacts than AC circuits. A frequent specification is to guarantee the same life (say, 100,000 operations) for 115 V AC and 28 V DC circuits drawing the same rms current. With loads drawing high in-rush currents (inductive loading), contacts may be derated to 25% to 10% of their resistive load ratings.

Because of the special importance of the contact problem, it is often advisable to incorporate spark suppression circuitry. This generally takes the form of series RC circuits across the contacts, or reverse energized diodes that break down when their inverse voltage rating is exceeded, thereby limiting peak voltages across the contacts. Design of these circuits is best carried out on a trial and error basis. Although spark suppression circuits introduce time delays in the completion of contact operation, they are *extremely important* where long life under difficult but uniform load conditions is required. Obviously load duty cycle, which is related to the frequency of timer operation and to the temperature rise occurring in the contacts (or in the entire timing apparatus for that matter), affects contact pressure. Spring

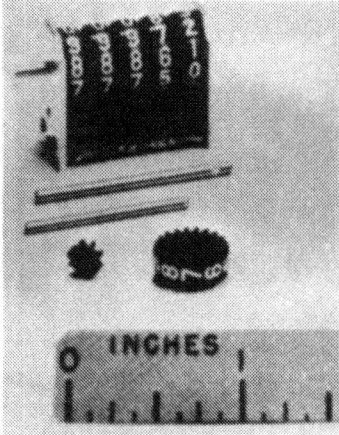

*Figure 16-75. Digital Counting Device*

constants establishing contact pressure are particularly sensitive to temperature.

In certain types of timers a constant speed motor may drive a set of brushes (contacts) over a printed circuit plate, where a pattern establishes the timing and contact closure sequences. At low brush speed, the contacts make and break so slowly that considerable arcing and pitting may occur. As a result contact rating may drop to 1% of the rating for contacts operating by snap action.

Other factors of importance include accuracy, repeatability, reliability, environmental conditions, and cost. Factors such as ease of adjustment, manner of setting, readout, and remote indication are more or less important depending on the specific application. Where a timer is to be used in a series of similar but not identical processes (where the timing will vary from one job to another), ease of setting up the timer is important. This set-up must include a check of accuracy which is best done by running through an approximate time cycle, and then making final adjustment. Where a timer is used over a long interval, readjustment may be necessary due to aging or wear of components.

It is important to be able to actuate the output circuit without taking too much energy from the escapement and associated components. For example, Ref. 31 discusses an experimental mechanical timer that uses commercially available digital counters as shown in Fig. 16-75[31]. The input shaft always is driven at the rate of 1 rps. The required time is preset into each digital counter which then is allowed to run down to zero. In the final 0.1 sec, all the wheels transfer as a single unit to read 999.9. At this instant a conductive knife blade attached to the hundreds wheel moves and completes an electrical circuit that provides an output signal. The advantage of this technique is that the switch closing load is not put on the timing movement until the final last 0.1 sec of timing. This contributes to the operational accuracy of the timer assembly.

## 16-6.3 TYPES OF SWITCH

Snap acting switches rather than those types where gradual contact is made are usually more desirable. A slowly rotating output shaft closing and opening gradually a cam operated switch carrying currents on the order of amperes can lead to contact destruction. Contact destruction also may be a consideration where the contacting mechanism consists of an arrangement of a printed circuit board and brushes. Current ratings in this type of circuit must be reduced substantially below those values that could be tolerated in snap-action type switching.

Snap action switches in a timer are best set by trial and error since the precise switching point is somewhat indefinite. Further limitations are inherent in the cam geometry, wear of cam surfaces, and switch surfaces.

Mercury switches are another good choice if fast make and break contacts are desired. Arcing caused when breaking inductive loads within the switch rating is nondestructive. Another advantage of mercury switches is low constant resistance at low voltage and absence of contact chatter. Since the contacts are

Figure 16-76. Adjustable Cam Timer[30]

Figure 16-77. Cam Cycle Timer[30]

isolated (contained within a sealed volume), contact corrosion from damaging atmospheres is avoided as well as possible ignition of combustible gases by the switching arc[33]. However, mercury switches are position and vibration sensitive and unable to operate under zero-g conditions.

A sealed dry reed switch may be actuated by a permanent magnet mounted on an output shaft or cam. It has advantages of small contact bounce, sealed contacts, and ability to operate in zero-g. Disadvantages are current limitations to a few amperes, glass-tube body, and susceptibility to vibration (but not at reed constant frequencies).

## 16-6.4 CAM OPERATED OUTPUTS

In these devices, rotating cams turn switches on and off according to a set program. The cams are driven by the output disk or shaft of the timer.

Cams fall into two classes, adjustable and nonadjustable. Nonadjustable cams are made of fiber, plastic, or metal disks into which a program is blanked or machined. The cams are attached to bushings, and the bushings are positioned in the drive shaft by set screws[4,30].

Adjustable cams basically consist of two disks whose diameter is reduced over approximately 180 deg. Fig. 16-76[30] shows an adjustable cam timer. The desired program is set by adjusting the disks relative to one another. The disks may be clamped together by a screw or locknut, or connected by friction. The cams are staked to a bushing, and are marked in a variety of ways to facilitate setting. Both snap-action and leaf switches are available. Two types of follower are used with cam outputs, roller and knife. The roller follower gives longer life and lower drive torque requirements, and the knife follower gives greater accuracy but shorter life and higher drive torque requirements. In the cam operated cycle timer of Fig. 16-77[30], switches are mounted separately so that they can be replaced easily. Also, set screws are used for fine adjustment of the operating point. There are many designs for timer outputs, and prices vary considerably. When selecting a timer, the designer should analyze his needs relative to ease of adjustment, frequency of adjustment, and accuracy of setting required. In specifying cam output timers, it is best to lay out a timing chart as shown in Fig. 16-78[30]. Remember, wide tolerances allow greater flexibility in the choice of actuators and switches, and reduce setting and testing costs.

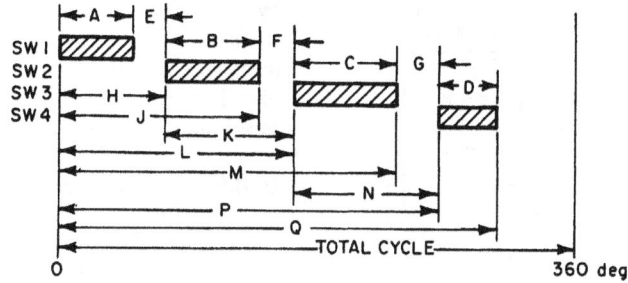

*Figure 16-78. Timing Chart for Cam-cycle Timer[30]*

## 16-6.5 READOUTS

Present-day mechanical time fuzes employ circumferential engraved scales and pointers for time settings and slotted screw buttons or segregated areas on the scale for optional function features. Usually, the rotary scales are limited to less than one full turn, with a secondary vernier scale to improve accuracy. This display is compatible with the internal setting portion of the fuzes and usually is achieved at low cost. Unfortunately, the display is difficult to read and can lead to gross setting errors by the operator. Significant improvement can be made by the design approach that follows.

### 16-6.5.1 Scale Layout

Fig. 16-79[34] shows the direction of increasing numbers for various scale types—i.e., for circular and curved, clockwise increasing; for vertical scales, bottom to top increasing; and for horizontal, left to right increasing. Except on multirevolution indicators such as clocks, a scale break should be included between the two ends of the scales as shown in Fig. 16-79(B). When pointer alignment is desired for check reading, the zero position should be at the nine o'clock position as shown in Fig. 16-79(C). For multirevolution indicators, the zero position should be at the top as shown in Fig. 16-79(D).

On circular scales, the numerals can be placed on the inside of the scale, or outside as shown in Fig. 16-79(E). In the former case, the indicating arm may blot out a number; however, in the latter case, the scale is somewhat constricted to allow for the numerals to be placed on the periphery. On

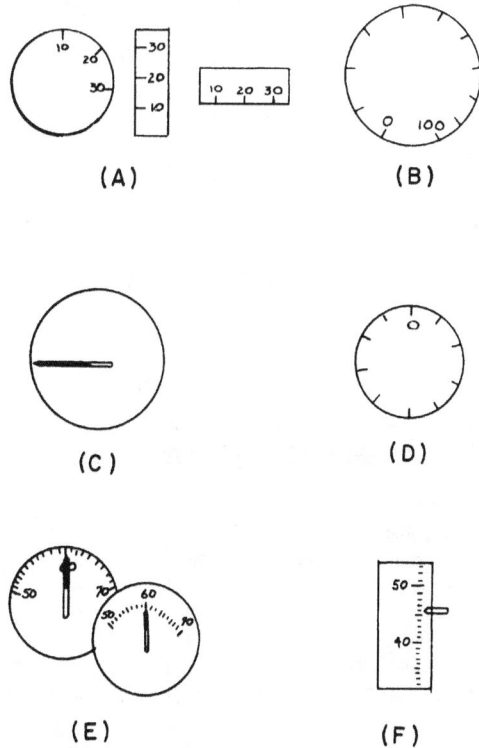

*Figure 16-79. Recommended Readout Display Scales*

16-75

"5" WHEN VIEWED FROM ABOVE

"5" WHEN VIEWED FROM BELOW

TIME

TIME

SEPARATION

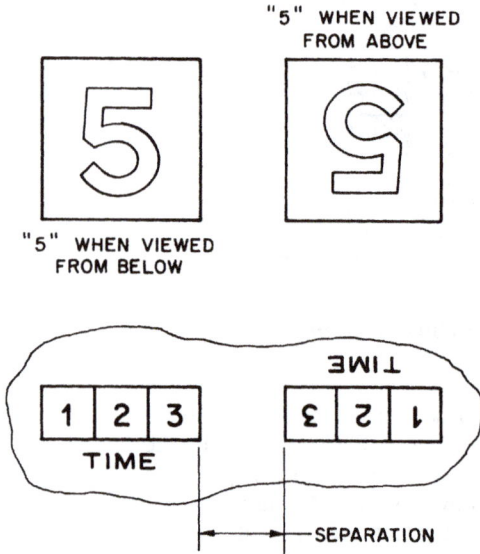

*Figure 16-80. Multioriented Displays*

vertical and horizontal straight scales, the numbers should be located on the side of the graduation as shown in Fig. 16-79(F).

The preceding general rules for scale layout may have to be modified in the use of fuzes to suit the particular circumstances imposed by the ogival shape and the random reading orientation between operator and round. If no definite orientation between operator and round is established, multioriented displays should be used as shown in Fig. 16-80[34].

### 16-6.5.2 Character Form

Table 16-3[34] shows a list of character type faces and their comparative legibility. This list should be used as reference when selecting a type face for display. From the chart, one can see that no one type face can be selected as being clearly superior to all others. For timer design, perceptibility and readability are more important than speed of reading because the amount of data is limited to several characters.

### 16-6.5.3 Grouping of Digits

The grouping of digits has a significant effect on the speed of perception. Thus the display arrangement should take full advantage of the most suitable grouping technique. The number of groups should be held to a minimum; and when minimized, should not be spread over wide areas.

**TABLE 16-3**

**EFFECT OF CHARACTERISTICS OF PRINTED MATTER ON LEGIBILITY**

| Type Face Comparison: Scotch Roman Versus | Visibility | | Perceptibility | | Speed of Reading | | Reader Opinions of Relative Legibility | |
|---|---|---|---|---|---|---|---|---|
| | Percent Difference | Rank* | Percent Difference | Rank* | Percent Difference | Rank* | Mean | Rank* |
| Antique | +56.3 | 1 | −14.8 | 3 | − 0.2 | 3 | 2.4 | 2 |
| Cheltenham | +38.6 | 2 | −22.2 | 2 | − 2.5 | 8 | 2.3 | 1 |
| American Typewriter | +36.5 | 3 | −37.7 | 1 | − 5.1 | 9 | 5.5 | 6 |
| Cloister Black | +25.0 | 4 | + 2.3 | 10 | −16.5 | 10 | 9.8 | 10 |
| Bodoni | +19.5 | 5 | − 4.6 | 7 | − 1.1 | 4.5 | 4.2 | 3 |
| Garamond | +16.9 | 6 | − 6.8 | 6 | + 0.5 | 1 | 5.4 | 5 |
| Old Style | +15.8 | 7 | −11.4 | 4 | − 1.1 | 4.5 | 4.6 | 4 |
| Caslon Old Style | +11.1 | 8 | − 7.9 | 5 | − 1.3 | 6 | 6.4 | 8 |
| Kabel Light | + 8.2 | 9 | + 0.1 | 9 | − 2.3 | 7 | 8.2 | 9 |
| Scotch Roman | 0.0 | 10 | 0.0 | 8 | 0.0 | 2 | 6.2 | 7 |

.*Increasing order of desirability

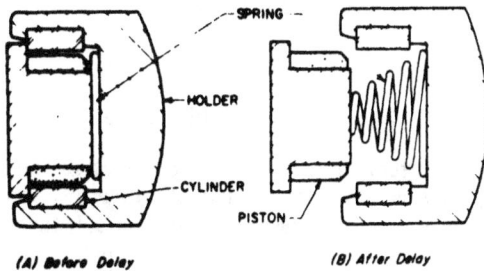

Figure 16-81. Pneumatic Dashpot for Arming Delay

### 16-6.5.4 Background

The background of the numerals or characters has an effect on reading speed and legibility. Distance is of secondary concern because in almost all situations the reading will be done at no greater than arm's length from the display. Black on white is superior to white on black from the standpoint of reading speed, distance, peripheral legibility, and eye movements. With respect to achromatic vs chromatic backgrounds, eye movements for black and white are far superior to color on color, a red on say green display. High contrast seems to be the most important factor with respect to reading speed.

Surface texture has little to no effect on reading speed or distance, but shiny surfaces or any highly reflective background should be avoided to reduce glare. If possible, windows or viewing ports should be nondistorting in nature.

### 16-6.5.5 Illumination

Since the nature of the process involved in setting a timer usually requires that the operator be in an outdoor environment, there is little the designer can do to reduce over-illumination when in bright sun conditions other than reduce obscurity due to glare. When nighttime operations are dictated by the battlefield situation, the use of synthetic illumination should be considered either by self-illumination or by light provided by auxiliary equipment. Red light should be used when possible to preserve night vision.

## 16-7 FLUID TIMERS

In a fluid timer, a gas or liquid regulates the motion of mechanical parts. Fluid timers should not be confused with flueric timers (Chapters 22-25) in which mechanical parts do not move. Rather, the fluid serves much the same purpose as the escapement, viz., to slow down the motion and to achieve a uniform rate. Fluid timers are not as accurate as clockworks but have found application in arming delays of fuzes. The two greatest problems with fluids are leakage and viscosity change over the military temperature range which lead to inaccuracies. However, both of these problems can be overcome through proper design. The application of two currently used fluids is described in the paragraphs that follow. Additional devices are covered in Ref. 1.

### 16-7.1 PNEUMATIC DASHPOT

An annular orifice dashpot is shown in Fig. 16-81[35]. The "orifice" is the minute clearance between piston and cylinder. By selecting materials for piston and cylinder having different thermal coefficients of expansion, the orifice will change with temperature, thus affording a means of approaching a constant flow in spite of air viscosity changes with temperature. A glass cylinder can be produced accurately and the piston can be ground from ceramic or metallic materials.

The piston is pushed by a spring. The holder, of silicone rubber or polyethylene, holds and seals the parts. Models have been made from 1/8 in. diameter and 1/3 in. long to 1-1/2 in. diameter and 6 in. long. Time delay varies between 0.1 sec and 1 hr. The

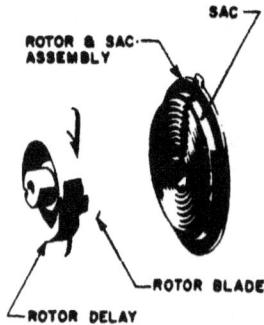

Figure 16-82. Delay Assembly of Fuze, XM218

dashpot has been used in experimental fuzes and is incorporated in Fuze, XM431, for the 2.75-in. rocket.

Theory of operation is based on the compressible flow of air through the orifice. The annular orifice, in this case, acts the same as a rectangular duct of the same dimensions; i.e., width equal to clearance, and length equal to circumference. The clearance is

$$h = \left[ \frac{6\eta\, r\, \ell^2 P_1}{t\,(P_1{}^2 - P_2{}^2)} \right]^{1/3} \text{, in.} \qquad 16\text{-}66)$$

where

$h$ = clearance from piston to cylinder, in.

$\eta$ = viscosity of the air, slug/ft-sec

$r$ = radius, in.

$\ell$ = length of air travel, in.

$P_1$ = pressure in the cylinder, psi

$P_2$ = ambient pressure, psi

$t$ = desired time delay, sec

## 16-7.2 SILICONE GREASE CAPSULE

The temperature-viscosity curve of silicone grease is flatter than that of other oils and greases. In the past, use of this substance therefore has been attempted to provide time delay. However, the leakage problem was severe, and the grease gummed up the arming mechanism so as to render it useless. This problem was overcome by sealing a silicone gum in a plastic sac made up of heat sealable Mylar tape.

Fig. 16-82 shows the sac and rotor delay mechanism of Grenade Fuze, XM218. The sac assembly consists of a metal backing disk and a plastic capsule, about 3/4 in. in diameter and 1/8 in. thick, containing silicone grease. The periphery and a segment of the plastic disk are heat sealed to the metal disk to form a pocket for the delay fluid. The sac assembly is placed against the delay rotor assembly (the space between the two assemblies in the illustration was introduced solely to show the sac assembly clearly). In operation, the delay is obtained when the four blades of the delay rotor, by virtue of a torsion spring, slide over the surface of the fluid sac, thus displacing and metering the fluid from one side of each blade to the other.

The design described was obtained by empirical means. The analysis is complex because the flow in the fluid sac passages varies as a function of rotor radius. Analytical techniques relating to the interactions of timer geometry, silicone fluid properties, and friction levels must be employed.

# REFERENCES

1. AMCP 706-210, Engineering Design Handbook, *Fuzes*.

2. AR 70-38, *Research, Development, Test, and Evaluation of Materiel for Extreme Climatic Conditions*, Dept. Army, July 1969.

3. C.M. Saltsman, Jr., "Timer Types and Features", Instruments and Control Systems, 38, 86-9 (October 1965).

4. R.S. Walton, "Clock-Driven Timers", Instruments and Control Systems, 35, 87-9 (March 1962).

5. A.M. Wahl, *Mechanical Springs*, McGraw-Hill Book Co., N.Y., 1963.

6. *Spring Design Data*, Section 7: "Spring Motors", Hunter Spring, Hatfield, Pa., 1965.

7. Fr. Assmus, *Fundamentals Concerning Technical Mechanisms*, Springer Verlag, Berlin, 1958, p. 168 (translated).

8. "Governed D.C. Motor", The A.W. Haydon Co., Waterbury, Conn., Form MM 802-R1, 1969.

9. D.S. Bettwy, D.L. Overman, and H. Rosenberg, *Long-Delay Timer for Air Force Bomb Fuze*, US Army Harry Diamond Laboratories, Washington, D.C., Report TR-1126, 29 April 1963.

10. W.O. Davis, *Gears for Small Mechanisms*, N.A.G. Press Ltd., London, 1953.

11. V.A. Rogers and R.A. Shaffer, *A Graphic Technique Applied to the Analysis of the Torque Fluctuations of a Speed Increasing Gear Train*, Frankford Arsenal, Philadelphia, Pa., Report M64-28-1, June 1964 (AD-608 483).

12. Hütte, *Des Ingenieurs Taschenbuch, Part A, Maschinenbau*, Ernst & Sohn, Berlin, 1954, Chapter XIV.

13. *A Compendium of Mechanisms Used in Missile Safety and Arming Devices (U)*, Journal Article 27.0 of the JANAF Fuze Committee, Part 2, Section 1, June 1967 (AD-384 530) (Confidential Report).

14. *A Study of the Dynamics of an Untuned Clock Mechanism*, Virginia Military Institute, Lexington, Virginia, September 1953.

15. *A Study of the Effect of Geometrical Factors Upon the Behavior of an Untimed Clock Mechanism*, Virginia Military Institute, Lexington, Virginia, June 1954.

16. K. Schulgasser and C. Dock, "Development of the 'Dock' Escapement", *Proceedings of the Timers for Ordnance Symposium*, Volume I, US Army Harry Diamond Laboratories, Washington, D.C., November 1966, pp. 15-34.

17. E. Pearson, *Analysis of the Junghans Escapement Spring; Restoring Torque vs. Angular Deflection for Static Conditions*, Frankford Arsenal, Philadelphia, Pa., Report M61-35-1, June 1961.

18. K. Schulgosser, *A Comparison of Three Methods for Supplying the Restoring Torque to a Junghans Escapement When That Escapement is Subjected to Spin*, Frankford Arsenal, Philadelphia, Pa., Report M63-17-1, October 1962.

19. M.M. Stonisic, *Mathematical Analysis Concerning the Feasibility of the Popovitch Escapement in the Design of Artillery Fuzes*, Purdue University, Indiana, September 1964 (AD-453 012).

20. J.N. Shinkle, "Cylinder Escapement Performance", *Proceedings of the Timers for Ordnance Symposium*, Volume III, U S Army Harry Diamond Laboratories, Washington, D.C., November 1966, pp. 145-170.

21. J.N. Shinkle, *Cylinder Type Escapement Study*, Sandia Laboratory, Albuquerque, New Mexico, Report SC-TM-65-399, October 1965.

22. N. Czajkowski, *A Theoretical and Experimental Analysis of the Dynamics of Junghans Escapement*, M.S. Thesis, University of Maryland, 1955.

23. J.N. Shinkle, *Detached Lever Escapement Study*, Sandia Laboratory, Albuquerque, New Mexico, Report SC-RR-65-57, May 1965.

24. W.C. Monday, *Detached Lever Clock Mechanism Dynamics*, Oklahoma State University, Ph.D. Thesis, May 1965.

25. R.B. Minnix, *The Development of a Mathematical Model of the Detached Lever Escapement*, Virginia Military Institute, Lexington, Virginia, July 1968.

26. J.N. Shinkle, *Environmental Sensitivity of Detached Lever Escapements*, American Society of Mechanical Engineers, New York, Paper WA/MD-11, November 1966.

27. R.C. Geldmacher and H.H. Pon, *Inertial Compensation of Energy Losses in Timing Mechanisms*, Technical Sciences Corporation, Elgin, Illinois, Report TR1-2, February 1966 (AD-486 080).

28. M.E. Anderson and S.L. Redmond, *Runaway (Verge) Escapement Analysis and Guide for Designing Fuze Escapements"*, Naval Weapons Center, Corona Laboratories, Report NWC CL TP 860, December 1969.

29. Dept. of the Air Force, *Guided Missiles — Operations, Design and Theory*, McGraw-Hill Book Co., N.Y., 1958.

30. "Timers", Machine Design, 38, 70-4, December 15, 1966.

31. D.L. Overman and D.S. Bettwy, *Experimental Mechanical Timer With Detached Lever Escapement and Digital Readout System*, U S Army Harry Diamond Laboratories, Washington, D.C., Report TM-65-44, August 1965.

32. D. Popovitch, S. Alpert, and M. Eneman, "XM577 MTSQ Fuze", *Proceedings of the Timers for Ordnance Symposium*, Volume I, U S Army Harry Diamond Laboratories, Washington, D.C., November 1966, pp. 131-194.

33. K.H. Olson, "Timers", Electromechanical Design, 10, 143-56, January 1966.

34. David Goldstein, "Setting Mechanisms for Analog Electronic Projectile Timers", *Proceedings of the Timers for Ordnance Symposium*, Volume II, U S Army Harry Diamond Laboratories, November 1966, pp. 55-112.

# BIBLIOGRAPHY
## GEAR AND TOOTH DESIGN (PAR. 16-3)

*20-Degree Involute Fine-Pitch Systems,* American Gear Manufacturers Assoc., Washington, D.C., Report 207.04.

*Standard Systems,* American Gear Manufacturers Assoc., Washington, D.C., Report 207.05A, January 1965.

E.K. Buckingham, "Recess Action Gears", Product Engineering, 35, 82-9, June 1964.

A.H. Candee, "A Simple Method for Determining the Thickness of Involute Gear Teeth", Machine Design, 28, 90-100, May 1956.

D.W. Dudley, *Gear Handbook,* McGraw-Hill Book Co., N.Y., 1962.

H.M. Durham, "Curves Simplify Involute Gear-Tooth Layout for Strength Calculations", Machine Design, 28, 117-8, May 1956.

D.A. Goldstein, *Involute and Cycloidal Tooth Forms for Use in Mechanical Escapement Systems,* Fairchild Space and Defense Systems, N.Y., n.d.

A.P. Isayer, *Experimental Investigation of the Effect of Impact Overloads on Clockworks,* Foreign Science and Technology Center, Army Materiel Command, Report FSTC-HT-23-35-66, February 1966.

L. Martin, *The Case for Involute Gearing in Fuze Mechanism,* Eastman Kodak Co., Rochester, N.Y., n.d.

G.W. Micholec, *Precision Gearing Theory and Practice,* John Wiley and Sons, N.Y., 1966.

Miles and Samuelson, *Horological Literature Survey (Gear Trains),* Frankford Arsenal, Philadelphia, Pa., Report R-01735, August 1964.

R.H. Pearson, "Gear Overdesign and How to Avoid It", Machine Design, 40, 153-7, May 1968.

A.L. Rawlings, *The Science of Clocks and Watches,* Pittman, N.Y., 1948.

M.E. Spotts, "How to Predict Effects of Undercutting Hobbed Spur Gear Teeth", Machine Design, 28, 123-7, April 1956.

R.L. Thoen, "Precision Gears", Machine Design, 28, 93-6, April 1956.

C.H. Wickenberg, *Research and Development Work on Gear Tooth Forms for Fuze Timing Devices,* General Sintering Corp., Ill., n.d.

PART FOUR – PYROTECHNIC TIMERS

## CHAPTER 17

## INTRODUCTION

### 17-1 GENERAL SYSTEMS

A pyrotechnic timer is a self-contained device in which a low explosive mixture burns linearly with respect to time to provide a controlled time delay. It is commonly called a delay element or just a delay. Its burning action often is likened to that of a cigarette, which also burns linearly. However, the delay housing sleeve is not consumed and air is not required for combustion. Low explosives are divided into two groups: (1) gas-producing low explosives that include propellants, certain primer mixtures, igniter mixtures, black powder, photoflash powders, and certain delay compositions; and (2) nongas-producing low explosives that yield the gasless type delay composition. Additional information on pyrotechnic timers is contained in other engineering design handbooks[1,2].

Pyrotechnic delays are the oldest military timers. Early descriptions of artillery tell of firing cannon balls with flaming wicks that served as pyrotechnic delays to provide for detonation after the ball arrived at the target. Whereas modern delays have been improved significantly, it is curious to note that black powder, the explosive employed originally, is still in current use for some applications.

Compared with electronic, mechanical, and flueric timers, pyrotechnic delays are smaller, simpler, and less expensive, but they are also less accurate. Accuracy over the military temperature range is only about ± 10%.

In its simplest form, the delay element consists of a cylindrical housing containing the pyrotechnic composition. Sometimes a primer is added at the input end to provide the initiating flame. The delay has a flame output that can be augmented into a detonation wave, if desired, by use of a relay.

Delay compositions are either gaseous or gasless depending on the quantity of gas generated when they burn. Combustion products of less than 10 or 15 $cm^3$ of gas per gram are considered gasless; true gasless compositions do not exist. Provision must be made for the gas generated in gaseous delays by either venting or, in obturated delays, designing a free internal volume. For highest accuracy, gasless delays must have one of these two features. However, addition of these features makes the housing more complex. The current trend is toward the use of gasless delays. While delays adjustable in time have been used in the past, they also add complexity. Hence, most modern delays have a fixed delay time.

Advantages and disadvantages of pyrotechnic delays are summarized as follows:

(1) Advantages:

(a) Normally, delays contain no moving parts.

(b) Time delays are available ranging from microseconds to several minutes.

(c) By comparison, they are smaller and lighter than equivalent mechanical, electrical, and flueric delays.

(d) The devices are rugged and can pass the standard military rough-handling tests. (Gasless delays operate over a wide temperature range and withstand high impact acceleration.)

(e) Large amounts of energy are available in small weights and volumes.

(f) Usually, the fabrication cost is lower than that of other timing systems.

(2) Disadvantages:

(a) Since they contain pyrotechnic and explosive material, delays require that explosive safety precautions be taken.

(b) Pyrotechnic materials are sensitive to humidity and temperature, and careful control is necessary in processing, storage, and application.

(c) Time reproducibility at a given temperature is about 3% within one batch under the most controlled laboratory conditions. Batch-to-batch variation in production is considerably larger. This variation in burning time can be due to insufficient mixing, incomplete drying, difference in ingredient composition, variation in packing density, purity, particle size, and particle size distribution of the ingredients. Variations in column height also can introduce lack of reproducibility.

## 17-2 MILITARY APPLICATIONS

Pyrotechnic delays are used extensively in military devices, primarily in fuzes and demolition devices, to provide functioning or self-destruction delays. The *Fuze Catalog*[3] describes many fuzes containing pyrotechnic delays and lists, in Volume 3, the delays themselves. Delays as elements of explosive trains are discussed in Refs. 4 and 5.

The classic use of these delays is in the time train of pyrotechnic fuzes. Fuze, M37A1[3] is typical of the artillery fuzes employed during and before WW II. (Almost all of these fuzes are now obsolete for another reason, namely,

outdated arming safety.) The fuze contains a black powder delay column in the shape of almost a complete ring. One end of the ring is blind, the other has a relay leading to lead, booster, and main charge. A primer is mounted in the ogive, at right angles to the ring. The ogive turns so that the primer can be positioned anywhere around the ring. When it is at the relay, the delay is nil; when it is at the blind end, the entire column must burn before setting off the relay. A scale on the outside of the ogive permits setting the time from zero to 25 sec. Of course, the black powder in the column burns both ways when the primer is in the middle but the flame is extinguished at the blind end.

Fig. 17-1[3] shows Detonator, Delay, T65, which is more typical of a simple modern application. The delay itself is the small center charge of a barium chromate/boron composition. The input end of the detonator consists of the electric contacts (pin and case), a carbon bridge, a spotting charge of lead styphnate, and an igniter mix. At the output end, the lead azide and RDX charges produce a detonation wave. It takes 50 msec for the flame to burn through the length of the delay composition.

## 17-3 REQUIREMENTS

The basic requirements for a military timer include reliability, ruggedness, and simplicity. The timer must function as intended when subjected to the correct stimuli. On the other hand, it must withstand the rigors of handling, transportation, and storage over the military temperature range. A pyrotechnic delay can meet these requirements.

To these general requirements must be added the important factor of safety. Of course, safety is a general requirement, which is certainly implied in ruggedness. Some of the delay compositions, like black powder, can burn very rapidly under certain conditions. Safety in pyrotechnic timers is therefore of primary importance. Safety aspects are discussed in more detail in par. 20-2 and in Ref. 6.

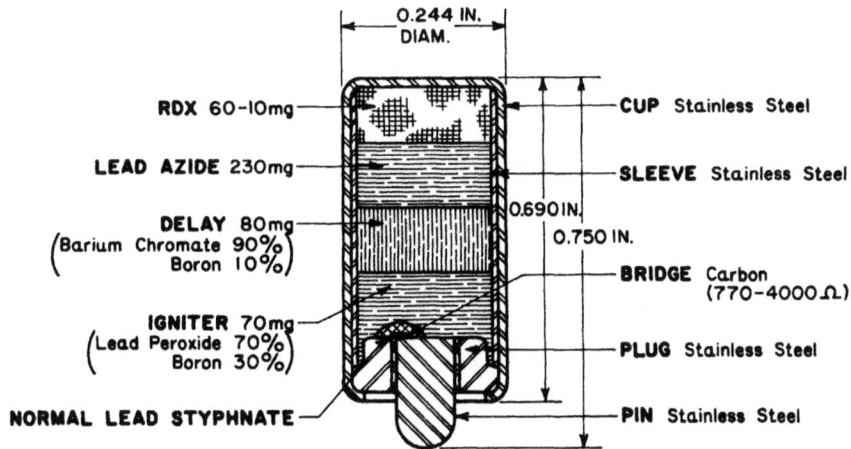

*Figure 17-1. Detonator, Delay, T65*

## REFERENCES

1. AMCP 706-188, Engineering Design Handbook, *Military Pyrotechnics, Part Four, Design of Ammunition for Pyrotechnic Effects.*

2. AMCP 706-185, Engineering Design Handbook, *Military Pyrotechnics, Part One, Theory and Application.*

3. MIL-HDBK-137, *Fuze Catalog*, Department of Defense, 20 February 1970.

    Volume 1, *Current Fuzes (U)*, (Confidential Report).

    Volume 2, *Obsolete and Terminated Fuzes.*

    Volume 3, *Fuze Explosive Components (U)*, (Confidential Report).

4. AMCP 706-179, Engineering Design Handbook, *Explosive Trains.*

5. *A Compendium of Pyrotechnic Delay Devices*, Journal Article 30.0 of the JANAF Fuze Committee, 23 October 1963 (AD-474 833).

6. AMCR 385-100, *Safety Manual*, Army Materiel Command, April 1970.

# CHAPTER 18

# DELAY SYSTEM CHARACTERISTICS

No single pyrotechnic delay mechanism is suitable for all applications. Hence, the selection of a delay device must be based on the overall requirements of the particular military application in which it will be used. Considerations differ depending on whether the delays are vented or obturated and must take into account the space limitations imposed.

## 18-1 SPACE LIMITATIONS

The designer of a pyrotechnic timer always will be faced with space limitations when trying to fit the device into a military item. Factors affecting the amount of space required are the length of the delay column (a function of the time delay and the delay mix), the diameter of the column (each mix having a particular failure diameter below which propagation is not reliable), internal volume, the need for baffles and retainers, and the method of initiation (i.e., mechanical or electrical).

All of these factors must be taken into consideration and each factor balanced against the other so that an inexpensive, reliable, and rugged item results.

## 18-1.1 LENGTH

Length is a linear function of delay time: the longer the delay column, the longer it burns. Thus the burning rate of a particular delay composition and the total delay interval desired define column length. Whereas total delay time usually is specified, the determination of burning time is complex.

The delay time in a particular delay device is affected by numerous design parameters. These include delay composition, free volume, heat sink, igniter material, loading conditions, and space limitations. The user must specify the delay time required, the environment under which ignition must take place, and the severity of the environmental conditions from factory to field. Having this information, the designer is in a position to know approximately what delay compositions are required by referring to data on various delay mixes (see par. 21-2.1). Knowing the delay time required and the delay mix and its burning rate, the designer can now determine the approximate column height. If the resulting length exceeds the available space, substitution of another delay mix or the formulation of a new mix may be necessary.

Loading pressure also must be borne in mind in the design of a delay column. Data relating burning rate to density for barium chromate/boron compositions are given in Table 18-1[1]. Almost all compositions demonstrate a slow but systematic change of burning rate with loading pressure, although not necessarily to the same extent. While this change suggests using an adjustment of pressure to compensate for lot-to-lot variation in burning rate, the practice of loading gasless delays at pressures between 30,000 and 40,000 psi (and at least 60,000 psi for gas producing compositions) has been established for considerations of ruggedness. It should be remembered that a crack in the delay column can result in a "blow-by" and hence instantaneous functioning. For best results, delay columns should be loaded in increments not over one-half diameter long.

TABLE 18-1

EFFECT OF LOADING PRESSURE ON BaCrO$_4$-B COMPOSITIONS

Loading Pressure, $10^3$ psi

| 95/5 BaCrO$_4$-B | 36 | 18 | 9 | 3.6 | 1.3 | 0.5 |
|---|---|---|---|---|---|---|
| Mean burning rate (sec/in.) | 1.69 | 1.60 | 1.49 | 1.39 | 1.29 | 1.21 |
| Mean burning rate (sec/g) | 0.648 | 0.655 | 0.645 | 0.642 | 0.646 | 0.693 |
| % Coefficient of Variation | 1.2 | 0.6 | 0.7 | 0.7 | 0.8 | 0.8 |
| **90/10 BaCrO$_4$-B** | | | | | | |
| Mean burning rate (sec/in.) | 0.670 | 0.653 | 0.619 | 0.586 | 0.558 | 0.544 |
| Mean burning rate (sec/g) | 0.272 | 0.276 | 0.280 | 0.287 | 0.297 | 0.309 |
| % Coefficient of Variation | 1.5 | 0.9 | 1.1 | 1.6 | 2.0 | 1.8 |

## 18-1.2 DIAMETER

The burning of a delay column can be retarded or even extinguished by radial heat losses. These losses become more serious as the column diameter, burning rate, and ambient temperature are reduced. The effects combine to result in a failure diameter associated with delay mix and temperature. For a manganese, barium chromate, lead chromate composition at −60°F, the failure diameter increases from less than 0.109 in. for a 3-sec/in. mix to between 0.156 and 0.203 in. for a 12.5-sec/in. mix[1]. As shown in Table 18-2[2] the effect of column diameter proved to be significant for all 95/5 barium chromate-boron delay systems and for obturated 90/10 barium chromate-boron delays. Present usage indicates that for practical delay mixtures at −65°F, a 0.25 in. diameter is well above the limiting failure diameter[1].

## 18-1.3 WALL THICKNESS

The wall thickness of the body into which the delay is loaded is another important factor in the design of a delay column. This body serves as a heat sink because the metal body tends to have higher coefficients of heat transfer than the delay composition. It is desirable to have a body that acts as an infinite heat sink—above a minimum wall thickness the body acts as an infinite heat sink—of sufficient mass to transfer at a constant rate enough of the heat generated so that the burning of the delay column remains approximately constant. If the surrounding wall thickness is increased, delay columns close to their low-temperature failure diameters tend to have larger thermal coefficients. However, the effects of wall thickness becomes less important for materials well above their failure diameters. Fig. 18-1 illustrates the change in delay time with the mass of the metal body[3].

## 18-2 PHYSICAL CHARACTERISTICS OF BODY MATERIAL HOUSING

Another factor to be considered is the strength of the body material. Erratic delay times have been known to result due to yielding under the loading pressure. A conservative way to design the body is to consider it a tube stressed hydraulically and provide adequate strength to prevent yielding.

The quality of materials for the body or subassembly immediately surrounding the delay column is critical. Porosity of the metal envelope or contamination at the interface of body and delay column occasionally can

## TABLE 18-2

### EFFECT OF FUZE HOUSING MATERIAL AND DIMENSIONS ON BURNING TIME OF BARIUM CHROMATE-BORON COMPOSITIONS

| Parameters | Effect |
|---|---|
| 1. Metal Housings<br> a. Aluminum, Brass and Stainless Steel | a. No effect on burning times over test temperatures, internal diameter, wall thickness, and for either vented or obturated columns.<br> b. No interaction between metal and composition. |
| 2. Internal Diameter, in.<br> a. 0.250 95/5 vented vs obturated | a. Results for vented columns significantly different at 95% confidence level. |
| b. 0.375 95/5 vented vs obturated | b. Results for obturated columns significantly different at 95% confidence level. |
| c. 0.250 90/10 vented vs obturated | c. No significant difference for either column at this diameter. |
| d. 0.375 90/10 vented vs obturated | d. Results for obturated column significantly different at 95% confidence level. |
| 3. Temperature, °C<br> a. − 54, Room Temperature, and 76 | a. The effect of temperature was significantly different at the 95% confidence level for all metals, internal diameters, wall thickness, and for vented and obturated columns. |
| 4. Wall Thickness, in. (0.05, 0.15, 0.30, 0.50, 0.75, and 1.00)<br> a. 95/5 vented and obturated columns | a. Results for different wall thickness were significantly different, although no apparent trend was observed. |
| b. 90/10 vented and obturated columns | b. No significant differences or trends in results due to wall thickness for either type of column. |

cause erratic performance due to impurities formed by traces of gas.

Many metals as well as plastics and ceramics have been investigated as delay element housings. However, aluminum and brass are the choice of most designers because of fabrication ease and availability, and also because of compatibility of these materials with most delay compositions. It has been shown that the burning times of delay compositions are not affected by the material used for the housing or its thickness above that necessary to resist melting provided that the inside diameter of the housing is well above the failure diameter of the delay composition.

## 18-3 VENTED DELAYS

Vented delay elements have openings to permit the escape of gases produced by their functioning. They are used when large quantities of gas are produced by the burning of the delay powder. They may even be necessary for gasless compositions when long delay times are required in order to eliminate the pressure build-up within the delay element and subsequent unpredictable burning rates. Venting exposes the burning delay composition to ambient pressure and, as a result, the burning rate is sensitive to changes in altitude except that manganese delay compositions show no significant effect. In addition, these vents require sealing up to the

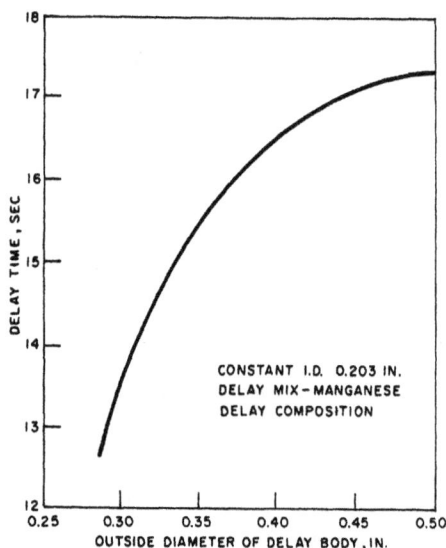

*Figure 18-1. Variation in Delay Time With Body Diameter*

time of functioning in order to protect the delay element from humidity. Two methods of sealing vented delays—by a disk, Method (A), and by a solder plug, Method (B)—are shown in Fig. 18-2[2].

## 18-3.1 INITIATION MEANS

The purpose of the delay initiator assembly, whether it is mechanically or electrically activated, is to produce hot gases and particles which will impinge on the delay column. Delays are easily ignited by the flame of a primer; however, slow-burning delays require an igniter. For ignition charges, see par. 21-1.1; for sensitivity characteristics, par. 21-4.

The initiator assemblies can take several forms. In some instances the assembly contains a charge holder that sits directly on top of the delay column (conventional),

allowing very little or no free volume, as shown in Fig. 18-3(A)[3]. Others use a chimney-type charge holder that may or may not sit on top of the delay column but provides for a free volume in which to vent gases (Fig. 18-3(B)). It also directs and concentrates the hot gases and particles into a definite area.

Another type of pyrotechnic delay assembly frequently uses a configuration employing a primer holder subassembly in which only the primer is held securely. This subassembly then is screwed into, staked into, or otherwise held rigidly in the main delay assembly. The igniter charge is made a part of the main pyrotechnic delay column. Fig. 18-6 illustrates the use of such a primer subassembly.

## 18-3.2 OUTPUT MECHANISM

The burning rate of a gas producing material is, in general, nearly proportional to pressure. For instance, a 95.4% barium chromate—4.6% boron composition was tested at various external pressures and the results are shown in Fig. 18-4[2]. Examination of the figure shows that an increase in external pressure resulted in a decrease in burning times (an increase in burning rate). The relationship has been shown to be hyperbolic and can be represented by an equation of the form[2]

$$t = aP^n \qquad (18\text{-}1)$$

where

$t$ = burning time, sec

$P$ = pressure, lb/in.$^2$

$a$ = factor depending on the mixture, $(1/\text{sec}) (\text{in.}^2/\text{lb})^n$

$n$ = factor depending on the mixture, dimensionless

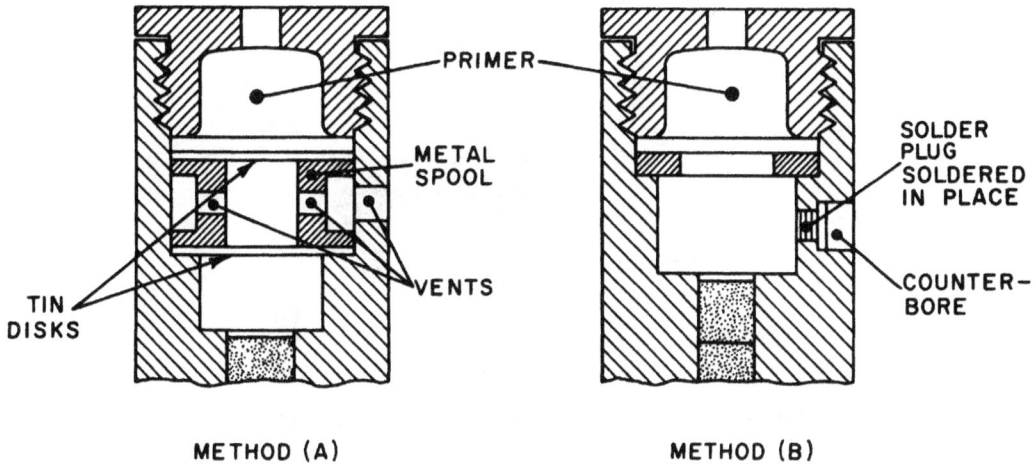

METHOD (A)                    METHOD (B)

*Figure 18-2. Sealing of Vented Delay Element*

The numerical values of the factors are $n =$ 0.13 and $a = 2.52$ (1/sec) $(in^2/lb)^{0.13}$ for the 95.4/4.6 barium chromate-boron composition.

Two other barium chromate-boron compositions, 90/10 and 81/19, were investigated for external pressure effects and these results, also shown in Fig. 18-4, indicate no significant change.

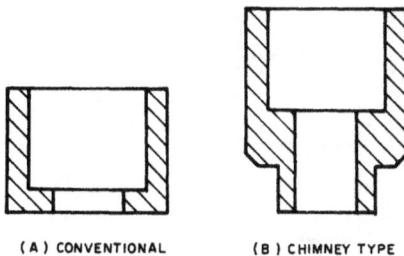

(A) CONVENTIONAL        (B) CHIMNEY TYPE

*Figure 18-3. Charge Holders*

As the delay column burns to the end, the base or output charge is ignited. This charge, called the relay charge, must be initiated by the delay mix and have the desired output characteristics as discussed in par. 21-3.1. While the delay column is burning, heat is being transferred through the metal body. In cases of long delay times (5 sec or greater) it may be necessary to insulate the output charge from the delay body if the output charge material is heat sensitive. This can be done by enclosing the output charge in a cup or ring made of insulating material.

### 18-3.3 BAFFLES AND RETAINERS

The design of a delay column must include means to resist forces such as spin, setback, shock, and internal gas pressure if they are present in the specific application. These forces can cause separation of the delay column and resulting failure, either in the form of no-fire or instantaneous functioning.

This separation usually occurs while the delay column is burning and at the burning front. Here, the material is usually molten and

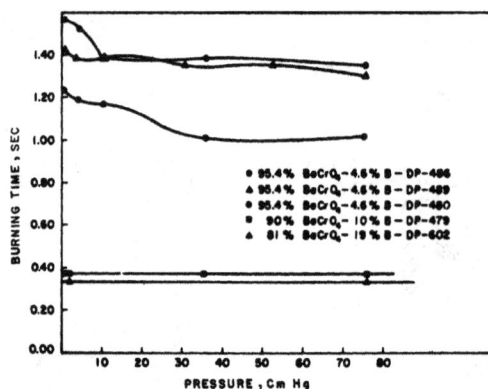

Figure 18-4. Burning Time of Barium Delay Compositions (M112 Fuze)

any of the forces can cause the column to separate on either side of the burning front. Means of alleviating separation include threading the inner diameter wall, and using retainer rings and disks. A more successful solution is to baffle both ends of the delay column. These baffles can be in the form of slotted disks, washers, or porous metal disks. A typical baffled delay column is shown in Fig. 18-5[3]. The baffles are always force fitted into the delay body with approximately 0.002 to 0.005 in. interference fit.

Baffles also are employed between a delay and its primer to reduce blast effects and particle impingement. In general, increasing the free volume between the two will make initiation more difficult. Decreasing confinement of the delay column will have the same effect.

A retainer or delay holder is a separate member into which the delay column and relay element are pressed. When one of these elements is used, the designer may take advantage of a different material from that of the body, thus providing a better heat sink or other advantages such as good mechanical strength, while the body material can be selected for lighter weight, if desired.

## 18-4 OBTURATED DELAYS

An obturated delay element is constructed to retain all the gases produced by the functioning of the initiator and delay composition before the base or terminal

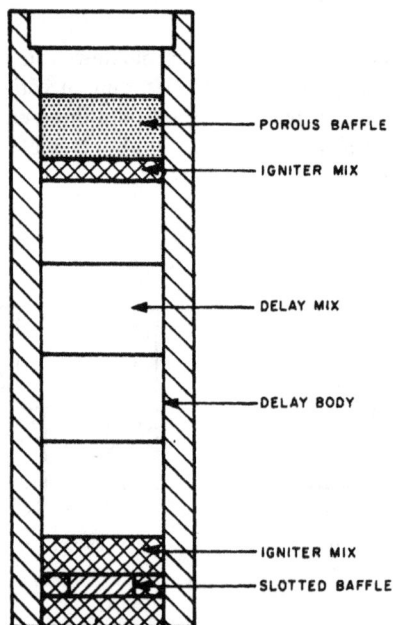

Figure 18-5. Typical Baffled Delay Column

charge is ignited, see Fig. 18-6[2]. See par. 21-2.1 for delay charges.

Delays are considered to be obturated if the gases produced are vented internally into a closed volume in the pyrotechnic device. The effects of ambient pressure and humidity are eliminated in obturated delays because they are sealed from the external environment. Possible harm to other components of the system is prevented because the combustion products are contained. If a short time delay is required, an obturated delay often is used because obturation tends to increase the average burning rate of the delay composition.

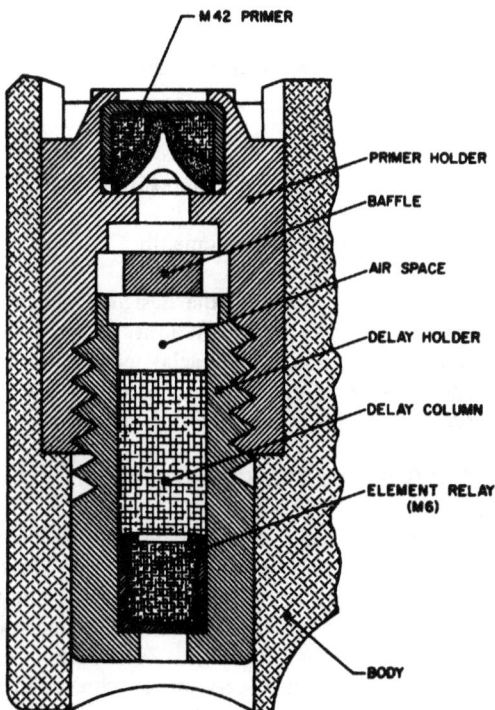

*Figure 18-6. Delay Element, Obturated, M9*

## 18-4.1 INITIATION MEANS

When the primer or flash charge is ignited in an obturated system, the pressure in the enclosed free volume is increased. At first, this happens very quickly and then the pressure is increased progressively by gas liberated by the burning delay column.

As a result, the burning rate accelerates continuously and is usually nearly proportional to pressure. Unless the free volume is increased along with the delay column length, the delay time does not increase directly. This requirement for a volume that is nearly proportional to the delay time limits obturated gas producing delays to about 0.4 sec with a column diameter commonly in the range of 0.1 to 0.125 in.[1]. In addition to its direct relationship to the free volume, the delay time of an obturated delay element is related inversely to the gas volume and heat of explosion of the primer, see Fig. 18-7[1]. See par. 21-1.1 for ignition charges.

## 18-4.2 OUTPUT MECHANISM

The need for a volume proportional to delay time is discussed in par. 18-4.1. After the delay column burns to the end, it ignites the relay. See par. 21-3.2 for output charges.

## 18-4.3 BAFFLES AND RETAINERS

Baffles and retainers are included in an obturated design for the same reasons that they are used in the vented design. Design considerations are the same in either instance, see par. 18-3.3.

## 18-4.4 INTERNAL FREE VOLUME

The internal free volume is that volume formed by a cavity in the delay housing which is designed to contain the gases produced by the chemical reaction. Containing the gases makes the delay independent of the effects of pressure or humidity of the ambient atmo-

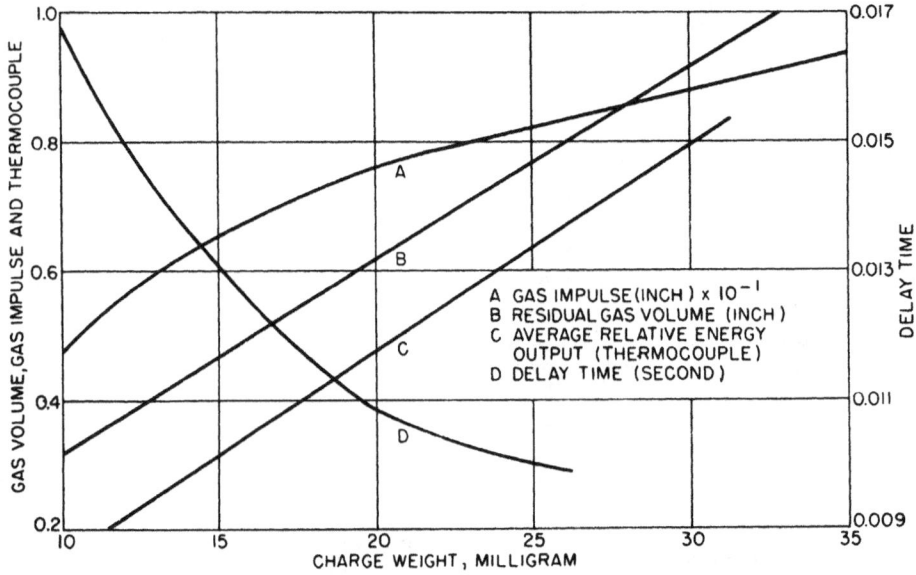

*Figure 18-7. Characteristics of an Obturated Black Powder Delay Element*

sphere and the fumes which might have harmful effects on other components of the system.

In an obturated system, the time will be greatly increased or the item may not burn through if the pressure rise is sufficient to cause bursting or significant leakage. The pressure can be calculated by defining the thermodynamic relationship between the heat and gas volume liberated by the primer, and delay column and the enclosed free volume in which the gases are confined. A reasonable estimate, for design purposes can be derived from the empirical equation[1]

$$P = \frac{30\,(W_p + W_d)}{V} \qquad (18\text{-}2)$$

where

$P$ = pressure, lb/in.$^2$

$W_p$ = weight of priming composition, mg

$W_d$ = weight of delay composition, mg

$V$ = enclosed free volume, in.$^3$

Using this equation, the designer can run through a series of calculations, varying the weights of priming and delay compositions and the free volume in order to obtain a safe pressure.

## 18-4.5 SEALS AND CLOSURES

The three major elements of a pyrotechnic timer (initiator, delay column, and base charge) usually are integrated into one body or housing. It is good practice to design for hermetic sealing. Sealing is desirable even when venting is necessary and is achieved in a number of ways. Stab-initiated delay devices are sealed and obturated until initiation. The puncture caused by the firing pin allows venting. In percussion-initiated delays, venting, if required, may be obtained by use of blow-out disks or plugs (Fig. 18-2). A system

18-8

of double cups has been employed in electrically initiated devices which allows them to be hermetically sealed until, at initiation, the thin-walled cup is fractured by the initiator output allowing the gases to be vented. In some instances, hermetic sealing is required in addition to venting. It is possible to seal the delay device into a tube or container of sufficient volume to provide an adequate vent volume. Figs. 18-8[4] and 18-9[4] are typical examples of hermetically sealed delay actuator devices. An experimental percussion-initiated delay detonator is shown in Fig. 18-10[5].

## 18-5 SYSTEM DESIGN AND PERFORMANCE

In delay system design, the delay compositions being the critical component of the delay element, should ideally have the following characteristics:

(1) They should be stable and nonhygroscopic; should have the highest purity consistent with requirements; should be readily available and inexpensive; and should be compatible with each other.

(2) They should be as insensitive as possible, meaning they should be capable of being blended, loaded, and assembled into an item with minimum risk from impact, friction, moisture, heat, and electrical discharge.

(3) They should be readily ignitible, and should change little in performance characteristics with small changes in percentages of ingredients. Their burning rates should be reproducible within each batch and from batch to batch with a minimum of variation.

(4) They should be compatible with their container as well as with other contacting compositions. Performance characteristics should not change appreciably with long term storage.

(5) They should be relatively insensitive to changes in pressure and temperature.

(6) They should be capable of withstanding the vibration and shock of transportation, setback, rotation, and impact, and should be resistant to physical abuse inherent in the loading and firing of ammunition.

Pyrotechnic designers must continue to explore new compositions and investigate the effects of parameter changes when existing or new compositions are rearranged in their

All dimensions in inches

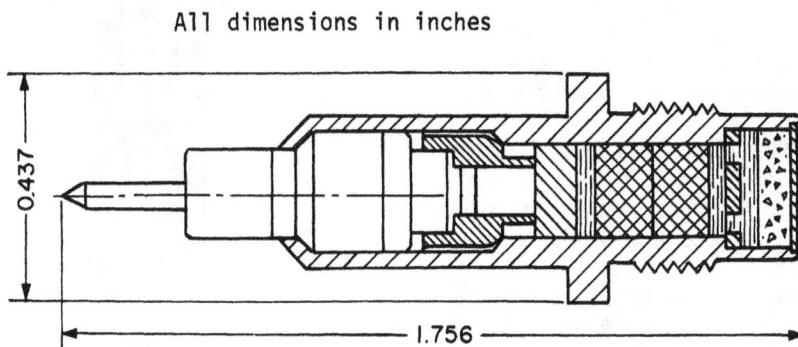

*Figure 18-8. Stab Initiated Delay Actuator (MARK 14 MOD 0)*

All dimensions in inches

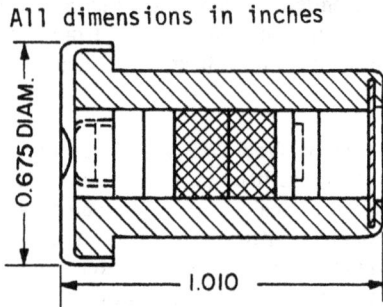

Figure 18-9. Electrically Initiated Delay
Actuator (MARK 18 MOD 0)

metal parts, when metal parts themselves are
varied in their geometry and metallurgy, and
when combinations of these may yield a new
end item.

The intended end-item requirement of the
delay element will guide the choice of the
most suitable type. If timing is to be
controlled carefully, selection of obturated
elements is called for; if operational physical
abuse is severe, the inclusion of a binder
seems the correct approach. Extreme climatic
diversification or use outside the earth's
appreciable atmosphere would call for prime
consideration of obturated designs. Other
considerations to be considered include the
method of ignition, the delineation and
tolerancing of housings, their physical properties, their costs or the costs of the assembly.

## 18-5.1 SYSTEM ASSEMBLY PROCEDURES

In the assembly of pyrotechnic timers the
procedures vary according to the type of
delay (i.e., vented or obturated). For instance,
in an experimental obturated percussion delay
(Fig. 18-10), the delay body is loaded before
the detonator is crimped in place or the
primer assembly is screwed into the body.
The detonator cup is loaded before assembly.
The seal is provided at one end by the

shoulder of the primer holder bearing on the
primer flange and the soft metal primer
washer. Sufficient igniter is loaded into each
end of the delay space to transfer ignition to
and from the delay.

In a vented delay, the vents often are sealed
or plugged before the delay element is
assembled to protect the element from
adverse atmospheric conditions, to facilitate
the ignition of the composition, or both (see
par. 18-4.5). If the vent is not sealed, the
requirement of surety of ignition of the delay
column by the primer will limit the size of the
vent, which in turn limits the time that the
burning rate can be slowed down to conserve
space. Any sealing of the vent should be
broken as the primer is initiated, therefore,
the seal cannot be too firm.

The delay body must be loaded before
assembly. It is important that the delay
composition be loaded flush with the end of
the delay body. If this is done, the delay

All dimensions in inches

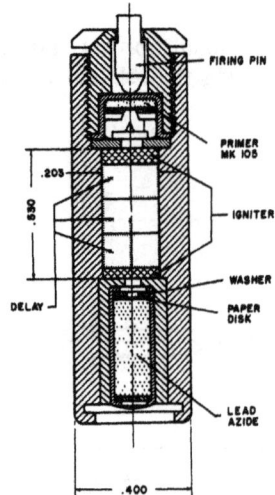

Figure 18-10. Experimental Obturated Percussion Delay Detonator

mixture is held firmly in place when the delay body is screwed tight into the holder. In cases where loaded black powder is pressed against metal, it is considered good practice to insert a thin paper disk between the powder and metal to serve as a cushion and thus to minimize the possibility of initiating the powder by friction.

Visual inspection of the bore surface of the delay body with a magnifying glass should disclose no tool marks or scratches running in an axial direction because these are conducive to blow-bys. Sharp edges (no burrs) are desirable at the ends of the bore. Concentricity between the bore and the expansion chamber, and smoothness of the bore are necessary to facilitate the use of a close-fitting loading ram with the bore[5].

Information relating to the packaging of the assembled item either individually or mounted in the military end item are covered in par. 20-1.

## 18-5.2 PERFORMANCE REQUIREMENTS

There are a number of factors that influence the performance of a pyrotechnic delay timer. These include:

(1) Quantity of charge (delay time)

(2) Vented vs obturated design

(3) Pressure and temperature environment

(4) Output charge (relay)

(5) Column diameter

(6) Loading and production techniques

(7) Storing and handling

(8) Delay compositions.

The important aspects of factors (1) through (5) have been discussed while factors (6), (7), and (8) are the subjects of Chapters 19, 20, and 21, respectively.

Because delay compositions contain all ingredients necessary for a self-propagating reaction, their burning is metastable. The effect of any factor that tends to cause an increase or decrease in burning rate is magnified. For this reason, satisfactory performance requires accurate control of all such factors. Control must be maintained from the procurement of raw materials until the end item, in which the delay is a component, accomplishes its intended use. The designer should be governed, therefore, by the following rules:

(1) Use delay compositions prepared by a well-established procedure from ingredients of known and controlled characteristics.

(2) Use obturated or externally vented construction when practical.

(3) Where obturated construction is impractical, use a seal that opens upon ignition.

(4) If a sealed unit is not practical, use delay compositions of demonstrated resistance to conditions of high humidity.

(5) Calculate the effect of cumulative tolerances upon such pertinent factors as external free volume.

(6) Provide for adequate free volume in obturated units.

(7) Analyze stresses induced by both internal and external forces that may be anticipated during loading, shipping, launching, and operation.

(8) Make sure that all components will survive these stresses, taking into account the elevated temperatures that result from burning of the delay columns.

(9) Specify adequate loading pressures (at least 60,000 psi for gas producing compositions and at least 30,000 psi for gasless delay powders), and short enough increments (one-half diameter or less).

(10) Provide for proper support of the delay column.

(11) Use diameters well above the failure diameter at $-65°$F. (Usual practice is 0.2 or 0.25 in. for gasless mixtures; 0.1 or 0.125 in. for black powder.)

## REFERENCES

1. AMCP 706-179, Engineering Design Handbook, *Explosive Trains.*

2. AMCP 706-185, Engineering Design Handbook, *Military Pyrotechnics, Part One, Theory and Application.*

3. *Some Aspects of Pyrotechnic Delays,* Journal Article No. 22.0 of the JANAF Fuze Committee, 5 December 1961 (AD-270 444).

4. MIL-HDBK-137, *Fuze Catalog, Vol. 3, Fuze Explosive Components,* Dept. of Defense, 20 February 1970.

5. *Ordnance Explosive Train Designers' Handbook,* U S Naval Ordnance Laboratory, Silver Spring, Md., Report NOLR 1111, April 1952 (AD-029 151).

# CHAPTER 19

## PRODUCTION TECHNIQUES

### 19-1 GENERAL

The production of pyrotechnic timers involves a series of steps starting with selection and processing of the pyrotechnic ingredients, production of metal parts, and ending with final assembly. It is important in the selection and processing of the ingredients that their chemical and physical properties are known in advance, especially particular hazards to personnel and property caused by the reaction of these materials to various stimuli. For detailed information on the properties of pyrotechnic materials, see Ref. 1. The laboratory and plant procedures that must be followed for the safe production of pyrotechnic compositions are described in the *Safety Manual*[2], and processing procedures are listed in Refs. 3 and 4. The production of the metal parts that house the pyrotechnic compositions involves techniques that are covered in detail in pars. 14-1 and 14-2.

### 19-2 DESIGNING FOR MASS PRODUCTION

As a rule of thumb, small explosive components (initiators, delays, relays) are pressed. The most common procedure for pressing powdered explosives is pouring the powder into a mold and pressing it with a ram that fits snugly. The pressure most frequently specified for charges used in military items is 10,000 psi. Charges may be pressed directly into their containers or pressed into molds and ejected as pellets. When a container of length greater than its diameter is used, the explosive usually is loaded in increments that are one diameter long.

Explosive charges are first weighed or measured and then pressed either directly into place or into pellets. If the explosive is pelletized, the pellets usually are reconsolidated by being pressed into place.

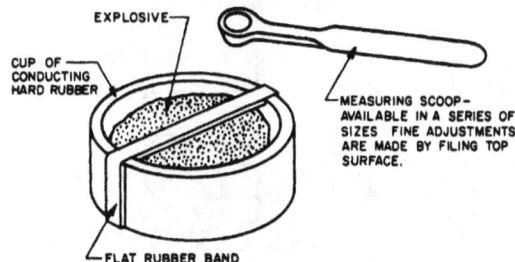

Figure 19-1. Scoop Loading

## 19-2.1 MEASUREMENT OF EXPLOSIVE CHARGES

The desire is always to load a specific weight of explosive. For small test quantities or for some premium quality production, direct reading, one-pan balances are used that provide an accuracy within one percent[5]. The two most common volumetric measuring devices are scoops and charging plates. To obtain the desired weight, the loading plant must adjust the volume to account for bulk density. Scoops (Fig. 19-1[5]) are filled and leveled against a rubber band. Careful scooping is accurate within 4 percent[5]. Charging plates (Fig. 19-2[5]) lend themselves to production rates. After filling the holes in the top plate and scraping off the excess, plates are aligned with the cup holes.

## 19-2.2 DIRECT PRESSING

Fits and tolerances of explosive charge cases and loading tools must be determined by consideration of (1) production costs, (2) clearances between run and case, and (3) binding resulting from interference between run and case. The cost of a set of loading tools is, of course, distributed over a large number of items.

Production loading tools should be hardened (60 Rockwell C is common), and the die should be lapped and polished. Cups are supported by close fitting loading tools while the charge is being pressed. Standard dimensions and tolerances of cups are listed in MIL-STD-320[6]. When loading explosives directly into an assembly, a pin is used to hold the component in alignment while loading. Fig. 19-3[5] shows equipment for hand loading of leads.

A quantity of explosive can be pressed either to a controlled height (stop loading) or to the limit of an applied load for a given diameter (pressure loading). Stop loading is faster but not as accurate as pressure loading. In normal production, a reasonable weighing tolerance for initiator charges is three or four

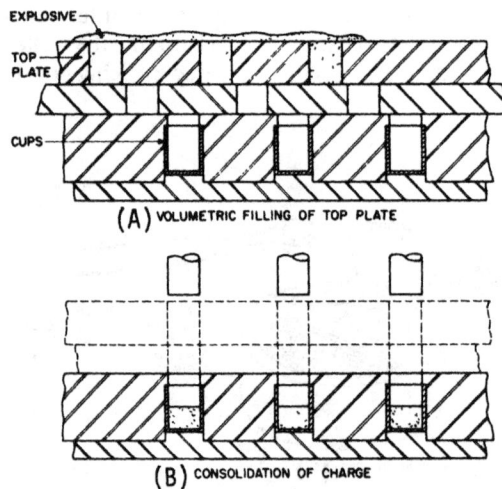

*Figure 19-2. Charging Plate Loading*

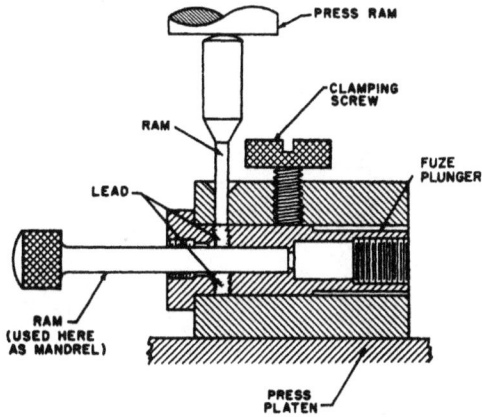

*Figure 19-3. Tool for Direct Loading of Component*

## 19-2.3 PELLETIZING

Many powdered explosives are prepressed into pellets. Better finishes are obtained than for cases loaded by direct pressing. Quantity production of pellets is accomplished on automatic pelleting machinery, in which the explosive is metered volumetrically by controlled movement of punches. Single-stroke presses produce about 90 pellets per minute while rotary presses have rates of about 700 pellets per minute[5]. Immediate expansion upon ejection is about 0.3 percent. Pellet-to-pellet variations are usually less than 0.1 percent[5]. With frequent pellet density checks and adjustment of the pelleting press, pellets can be made reproducibly to one percent in an automatic pelleting press.

## 19-3 REQUIREMENTS FOR TECHNICAL DOCUMENTATION

Production requires complete documentation that is assembled into a technical data package. This documentation is to ensure that military items are capable of reliably performing the functions for which they were designed. See par. 14-4 for detailed information on the contents of a technical data package.

percent. In stop loading, if the height of an increment is exactly reproduced, the density may vary as much as seven percent[5]. In either type of loading, the density should be checked for each production lot.

## REFERENCES

1. AMCP 706-187, Engineering Design Handbook, *Military Pyrotechnics, Part Three, Properties of Materials Used in Pyrotechnic Compositions.*

2. AMCR 385-100, *Safety Manual,* Army Materiel Command, April 1970.

3. F. B. Pollard and J. H. Arnold, Eds., *Aerospace Ordnance Handbook,* Prentice-Hall, Inc., Englewood Cliffs, N. J., 1966.

4. AMCP 706-186, Engineering Design Handbook, *Military Pyrotechnics, Part Two, Safety, Procedures and Glossary.*

5. AMCP 706-179, Engineering Design Handbook, *Explosive Trains.*

6. MIL-STD-320, *Terminology, Dimensions, and Materials of Explosive Components for Use in Fuzes,* 3 August 1959.

# CHAPTER 20

## PACKING, STORING, AND SHIPPING PROCEDURES

### 20-1 GENERAL

Military materiel is packaged to protect it from harmful environments, including transportation. Packaging will protect the item after production, during storage, during transport, and until delivery to its ultimate user[1]. During the transportation phase, which includes both handling and carriage, the Department of Transportation (DoT) regulations must be strictly observed for movement within the U.S. Since items must be stored for long periods, they must be protected against physical damage and deterioration; it may also be necessary to conduct periodic inspections of stored materiel.

From the military standpoint, good packaging methods protect the item through all phases and environments with minimum cost. Items must be protected for one of three required levels: overseas shipment (Level A), long-term storage (Level B), or interplant shipment (Level C).

The most damaging environments during transportation by truck, rail, ship, or aircraft are usually shock vibration. However, temperature extremes and other potentially harmful environmental factors should be considered as relevant to damage assessment or prevention.

These subjects are treated in more detail in Chapter 15; see par. 15-1 for packing, par. 15-2 for storing, and par. 15-3 for shipping.

### 20-2 SAFETY

Pyrotechnic timers like other military materiel also must be suitably packaged at minimum cost. These timers, as well as the munitions of which they are a part, are subject to another important requirement: they must be safe during packing, storing, and shipping.

Explosives and pyrotechnics can be shipped and stored safely if they are handled correctly and carefully, and with all of the necessary precautions. The safety record of both the military and the explosives industry is a result of careful preparation, not chance. Pyrotechnics are set off by energy concentrations such as sparks, friction, impact, hot objects, flame, chemical reactions, and excessive pressure. Established safety practices will avoid these conditions in order to minimize hazards.

Probably the most important aspect of safety is a correct individual attitude. The *Navy Transportation Safety Handbook*[2] puts it this way:

"It is imperative that anyone who is engaged in the handling, transportation, or storage of ammunition, explosives, and other dangerous articles think safety, act safety, and live safety until it becomes a habit. Accidents usually result from failure to observe regulations, failure to understand hazards, or failure to take necessary precautions. In each case, the 'failure' denotes human error, and human error denotes ignorance, specific carelessness, or poor judgment. The individual must realize that when he does not comply with safety regulations, he not only endangers his own life, but also puts his fellow worker's life in jeopardy. In the military, safety responsibility follows the line organization.

The commanding officer is responsible for the accident-free performance of his supervisors and their individual employees. A superior is not exonerated for an employee's poor judgment, carelessness, or failure to comply with safety regulations. The supervisor must advise and train his men so that they are constantly aware of the hazardous nature of their work. Further, he must indoctrinate his men with the philosophy that each individual is responsible for his own and his fellow worker's safety. And he must promote safety-mindedness by persuasion and by authoritative force when necessary."

The basic reference for safety is the *Safety Manual*[3]. It contains detailed discussions of the established safety practices for packing, storing, and shipping. These topics are discussed in the paragraphs that follow. The reader is referred to the *Safety Manual* for the following related safety topics that are outside of the scope of this Handbook:

(1) Building construction features

(2) Material handling equipment

(3) Hand and machine tools

(4) Electrical equipment and wiring

(5) Lightning and static electricity protection

(6) Fire fighting

(7) Protective clothing and equipment

(8) Good housekeeping practices.

The safety regulations have been slightly abbreviated and rearranged in two volumes for Department of Defense agencies[4] and their contractors[5]. These separate publications, available from the Superintendent of Documents, U S Government Printing Office, are convenient references.

## 20-3 HAZARD CLASSIFICATION

Dangerous materials are arranged into eight classes according to their level of hazard. Pyrotechnics are divided into classes 1 and 7, depending on whether they merely burn or whether they can detonate. Class 1 items are those which have a high fire hazard but no blast hazard and for which virtually no fragmentation or toxic hazard exists beyond the fire hazard clearance distance ordinarily specified for high-risk materials. In contrast, class 7 items are those for which most items of a lot will explode virtually instantaneously when a small portion is subjected to fire, severe concussion, impact, the impulse of an initiating agent, or considerable discharge of energy from an external source[5].

Pyrotechnic timers usually are packed, shipped, or stored as components of military ammunition. The ammunition assembly defines the hazard class. It is not feasible to specify here the hazard classes of all munitions that contain pyrotechnic timers. However, this information is tabulated in the AMC *Safety Manual*[3], which lists ammunition classes, and in the Navy *Safety Handbook*[2], which lists every munition by Federal Stock Number.

The hazard and shipping class must be established before an item containing pyrotechnics can be packed, shipped, or stored. If the hazard class of a particular pyrotechnic delay has not been established, it must be obtained by means of standard tests devised for this purpose[6].

## 20-4 PACKING

Pyrotechnic timers and ammunition containing them must be packed properly. Packing drawings and specifications have been prepared for essentially all military items containing hazardous materials. The drawings and specifications cover all applicable details of wrapping, boxing, bracing, palleting, and handling. If such drawings and specifications are not available for a particular item, Department of Transportation regulations apply and they specify minimum requirements. Packing for different levels of protection are discussed in par. 15-1.

Federal regulations require all containers of hazardous materials be conspicuously marked and labeled.

## 20-5 STORING

Hazardous materials are stored in accordance with quantity-distance requirements. These requirements are defined as "the quantity of explosives material and distance separation relationships which provide defined types of protection. These relationships are based on levels of risk considered acceptable for the stipulated exposures and are tabulated in the appropriate quantity-distance tables. Separation distances are not absolute safe distances but are relative protective or safe distances"[4].

Quantity-distance tables are contained in the safety manuals[3-5]; a typical excerpt is shown in Table 20-1[3]. The largest minimum distances are required where a hazard exists to personnel, i.e., inhabited buildings. Intraline refers to the minimum distance between any two buildings within one operating line or assembly operation. The magazine distances given in the excerpt are for above-ground storage, which is the least desirable. Earth-covered, arch type magazines are preferred because they are safer; their required separation distances are much less than those of above-ground magazines. Note that separation distance is roughly proportional to the quantity of explosive, and that a barricade of proper construction cuts in half the distance used for unbarricaded storage.

To determine distances between different types of magazines, Ref. 4 is the easiest to

use. It contains a group of diagrams like the following:

It states that the minimum distance from a barricaded above-ground magazine to the door end of an earth-covered, arch type magazine is found in Table 5-6.3, column 5 in the reference.

In addition to quantity-distance, compatibility also must be considered in storage. Only compatible hazardous items may be stored together in one magazine. Compatibility is established by consideration of the following factors[3]:

(1) Effects of explosion of the item

(2) Rate of deterioration

(3) Sensitivity of initiation

(4) Type of packing

(5) Effects of fire involving the item

(6) Quantity of explosive per unit.

## 20-6 SHIPPING

The safe transport of hazardous materials is the responsibility of the shipper. It has become expedient to pack and label hazardous cargo to meet requirements for all kinds of transportation. The Navy is the largest shipper of military cargo because most of it ultimately ends up aboard ship. If a commercial shipper is used, he should be properly licensed in all states and countries involved. Shipping regulations are complex and a qualified shipper is needed to cope with them.

All safety regulations are enforced in the

## TABLE 20-1

### EXCERPT FROM QUANTITY-DISTANCE TABLES

| Pounds of Explosive | Inhabited Building | | Highway & or Railway | | Intraline | | Above Ground Magazine | |
|---|---|---|---|---|---|---|---|---|
| | bar. | unbar. | bar. | unbar. | bar. | unbar. | bar. | unbar. |
| Class 1 | | | | | | | | |
| No limit | 100 | | 100 | | 100 | | 80 | |
| Class 7 | | | | | | | | |
| 1 | 40 | 80 | 25 | 50 | | | | |
| 10 | 90 | 180 | 55 | 110 | 30 | 40 | | |
| 100 | 190 | 380 | 115 | 230 | 40 | 80 | 28 | 51 |
| 1,000 | 400 | 800 | 240 | 480 | 95 | 190 | 60 | 110 |
| 10,000 | 865 | 1730 | 520 | 1040 | 200 | 400 | 130 | 235 |
| 100,000 | 1855 | 3630 | 1115 | 2180 | 415 | 830 | 280 | 510 |
| 500,000 | 4510 | 4510 | 3245 | 3245 | 715 | 1430 | 475 | 875 |

shipment of hazardous materials to protect life, property, and the cargo itself. All cargo must be properly blocked and braced during shipment. For some hazard classes, the vehicle must be placarded and inspected. Mixed shipments in the same vehicle must be compatible. In case of an accident on any mode of shipping, Form F5800 must be filed with the Department of Transportation when the incident involves death or serious injury, $50,000 property damage, or continuing danger.

Considerations for specific modes of shipping follow:

(1) Rail Transport. Railroad shipment of hazardous materials is covered in Department of Transportation Tariff No. 25[7].

(2) Truck Transport. Motor Vehicle shipment of hazardous materials is covered in Department of Transportation Tariff No. 11[8]. Motor vehicle shipment is more complex than rail shipment. A train is made up of many cars watched over by the engineer in front and caboose personnel behind. The engineer is in voice communication with the tower. The railroad controls traffic flow over its route. It provides trained inspectors. In contrast, each truck solos. It has no control over traffic on the public highway, and the driver must cope with any situation that may arise.

For these reasons, drivers of hazardous materials are given careful training and detailed instructions (e.g., Ref. 9), and the vehicle is carefully inspected for safety (e.g., lights and brakes) and compliance with local laws (e.g., weight limit).

(3) Ship Transport. All water shipment is regulated by the Coast Guard[10]. There are many restrictions to the transport of hazardous materials by ship which must be taken into account. Some dangerous articles are not permitted on passenger carrying vessels. Also, many ports do not permit the anchorage of

vessels carrying dangerous articles. The Army Corps of Engineers therefore has established suitably isolated explosives anchorages at various ports.

(4) *Air Transport.* Aircraft shipment of

hazardous materials is covered by Department of Transportation Tariff No. 6-D[11]. As in ship transportation, dangerous cargo is prohibited on passenger carrying craft.

## REFERENCES

1. AMCP 706-121, Engineering Design Handbook, *Packaging and Pack Engineering.*

2. OP 2165, *Navy Transportation Safety Handbook,* Naval Ordnance Systems Command.

3. AMCR 385-100, *Safety Manual,* Army Materiel Command, April 1970.

4. DOD 4145.27M, *DOD Ammunition and Explosives Safety Standard,* Dept. of Defense, March 1969.

5. DOD 4145.26M, *DOD Contractors' Safety Manual for Ammunition, Explosives and Related Dangerous Material,* Dept. of Defense, October 1968.

6. TB 700-2, *Explosives Hazard Classification Procedures,* Dept. of Army, 19 May 1967.

7. R. M. Graziano, Agent, *Hazardous Materials Regulations of the Department of Transportation,* Tariff No. 25, Association of American Railroads, Washington, D.C., 1972.

8. Tariff No. 11, *Regulations for Transportation of Explosives and Other Dangerous Articles by Motor, Rails, and Water, Including Specifications for Shipping Containers,* published by Agent F. G. Freund, American Trucking Assoc., Inc., 1616 P St., N.W., Washington, D.C. 20036.

9. OP 2239, *Drivers Handbook – Ammunition, Explosives, and Other Dangerous Articles,* Naval Ordnance Systems Command.

10. CG 108, *Rules and Regulations for Military Explosives and Hazardous Munitions,* U S Coast Guard.

11. Tariff No. 6-D, *Official Air Transport Restricted Articles Tariff,* published by C. C. Squire, Airline Tariff Publishers, Inc., Agent, 1825 K Street N.W., Washington, D. C. 20006.

# CHAPTER 21

## PYROTECHNIC DELAY COMPONENTS

The delay element is the heart of the pyrotechnic timer. The timer also includes an initiator and a relay or output charge. A knowledge of (1) the various ways of applying input stimulus, (2) the delay compositions currently in use and their characteristics, (3) the relay or output charge, and (4) the sensitivity of the delay compositions to external stimuli will aid the designer in meeting the system specifications.

## 21-1 INITIATORS

The first element of the explosive train of a pyrotechnic timer is the initiator or primer. It is categorized according to its input stimulus as (1) stab, (2) percussion, or (3) electric and according to its output as (1) pressure or (2) flame. Stab and percussion primers are known as mechanical primers.

The primer is a transducer that converts mechanical or electrical energy into explosive energy. It contains materials sensitive to stimuli such as heat or impact, has a relatively small explosive output, and is not designed to initiate secondary high explosive charges.

### 21-1.1 STAB PRIMER

#### 21-1.1.1 Construction

The stab primer is a rather simple item consisting of a cup made of aluminum, stainless steel, copper, or gilding metal. It is loaded with explosives and covered with a closing disk that is crimped in place. MIL-STD-320[1] describes design practices and specifies standard dimensions, tolerances, finishes, and materials for initiator cups. Two examples of stab primers are shown in Fig. 21-1[2]. For more detailed information on the

design of initiators, see Ref. 3.

#### 21-1.1.2 Ignition Charge and Loading

Stab primers are initiated by a pointed firing pin that punctures the cup. The energy required to fire stab primers increases nearly linearly with the thickness of the metal which the firing pin penetrates. It is also a function of the firing pin dimensions and tolerances, and the composition and density of the priming mix. Table 21-1[3] lists the common primer mixtures.

Most explosive charges are loaded by pressing them directly into cups at pressure between 10,000 and 20,000 psi. It is the usual practice, when the charge of an explosive is longer than its diameter, to load in increments

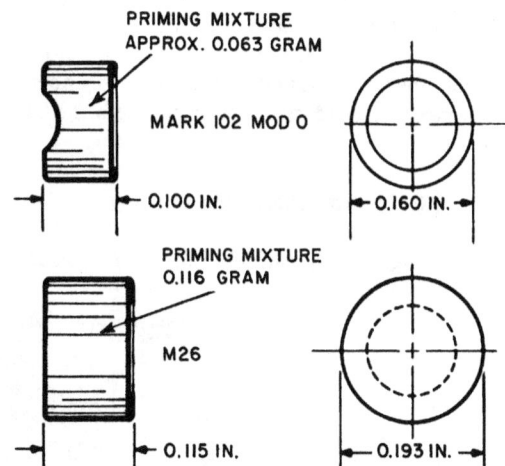

PRIMING MIXTURE
APPROX. 0.063 GRAM

MARK 102 MOD 0

0.100 IN.          0.160 IN.

PRIMING MIXTURE
0.116 GRAM

M26

0.115 IN.          0.193 IN.

*Figure 21-1. Typical Stab Primers[2]*
*Reprinted by permission of Prentice-Hall, Inc., Englewood Cliffs, N.J.*

TABLE 21-1

COMMON PRIMING COMPOSITIONS

Composition, (percent by weight)

| Ingredients | FA956 | FA982 | PA100 | PA101 | NOL60 | NOL130 |
|---|---|---|---|---|---|---|
| Lead Styphnate, basic | – | – | – | 53 | 60 | 40 |
| Lead Styphnate, normal | 37 | 36 | 38 | – | – | – |
| Barium Nitrate | 32 | 22 | 39 | 22 | 25 | 20 |
| Lead Azide | – | – | – | – | – | 20 |
| Tetracene | 4 | 12 | 2 | 5 | 5 | 5 |
| Lead Dioxide | – | 9 | 5 | – | – | – |
| Calcium Silicide | – | – | 11 | – | – | – |
| Aluminum Powder | 7 | – | – | 10 | – | – |
| Antimony Sulfide | 15 | 7 | 5 | 10 | 10 | 15 |
| PETN | 5 | 5 | – | – | – | – |
| Zirconium | – | 9 | – | – | – | – |

not over one diameter long (see par. 18.1.1).

## 21-1.2 PERCUSSION PRIMER

### 21-1.2.1 Construction

The percussion primer differs from the stab primer; it is fired by a firing pin that does not puncture the container. The essential components of a percussion primer are a cup—commonly brass—a thin layer of priming mix, a sealing disk, and an anvil. Percussion Primer, M42 is shown in Fig. 18-6 and Percussion Primer, M39A1 is shown in Fig. 21-2[3]. In general, percussion primers require more input energy to function than stab primers. For more detailed information see Refs. 1 and 3.

### 21-1.2.2 Ignition Charge and Loading

Priming compositions for percussion primers are shown in Table 21-1. The loading pressures of percussion primers are similar to those of stab primers. Loading pressure has a negligible effect upon sensitivity of percussion primers in the range from 10,000 to 60,000 psi. Firing energy requirements can be expected to increase with thickness and hardness of the primer cup, and with thickness of the primer mix between anvil and cup.

## 21-1.3 ELECTRIC INITIATOR

### 21-1.3.1 Construction

Electric initiators contain the initiation mechanism as an integral part. A plastic plug holding the initiation mechanism makes up one end of the cylindrical housing. Electrical connection is by means of lead wires or pins. Electric Squib, M2 is shown in Fig. 21-3[3] As a group, electric initiators are more

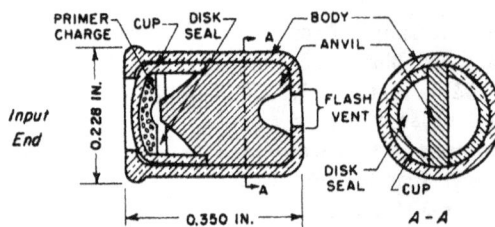

Figure 21-2. Percussion Primer, M39A1

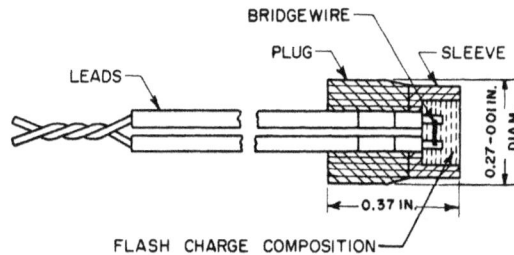

*Figure 21-3. Electric Squib, M2*

sensitive than mechanical primers. While several types of transducers have been employed—viz., hot wire bridge, exploding bridgewire, carbon bridge, conductive mix, and spark gap—the hot wire bridge is the most common initiation mechanism. See Ref. 3 for detailed design information.

### 21-1.3.2 Ignition Charge and Loading

The input characteristics of electric initiators are subject to precise control over quite remarkable ranges. Firing energies can be selected to range from less than one erg to thousands of ergs, current requirements from hundredths to hundreds of amperes, and resistance from a few hundredths of an ohm to tens of megohms. The input sensitivity varies with the type of transducer and each type must be considered separately.

The explosive in direct contact with the bridgewire is known as the spot charge. Whereas normal lead styphnate has the broadest general use, lead azide has been used for applications where extremely rapid response is needed. Information is given in MIL-L-17186[4] for normal lead styphnate and MIL-L-3055[5] for lead azide. Basic lead styphnate, as procured under specification MIL-L-16355[6], has particle sizes in the range between 5 and 95 micron which is highly satisfactory for spot charges.

Loading pressure may either increase sensitivity by improving contact between wire and explosive or decrease it by increasing the rate of heat dissipation through the explosive.

In lead styphnate, loaded at pressures between 1000 and 4000 psi, the latter trend apparently predominates. On the other hand, lead azide loaded at pressures between 3000 and 90,000 psi becomes more sensitive as loading pressure is increased.

The response times of hot bridgewire initiators fired by capacitor discharge vary with spot charge material as well as bridgewire characteristics. Information that is needed to design hot bridgewire initiators is available, including design data on bridgewire material, size, diameter, and resistance as well as firing energy and power[3].

### 21-2 DELAY ELEMENTS

The delay element consists of a metal tube, usually aluminum or brass, loaded with a delay composition. It is placed between the initiator and the relay or other output charge. Sometimes all three are combined into one unit. See pars. 18-2 and 18-3 for design details of vented and obturated delays. Fig. 18-2 shows an example of a vented delay element while Fig. 18-6 shows an example of an obturated delay element. Data on representative delays covering various time ranges have been compiled. In the paragraphs that follow there is a discussion of the basic chemistry of delays and the characteristics of various compositions that are used.

### 21-2.1 BASIC CHEMISTRY OF DELAY COMPOSITIONS

The basic ingredients of a delay composi-

**TABLE 21-2**
**FUELS FOR DELAY COMPOSITIONS[2]**

| Ingredients | Percent of Ingredients | | | Avg. burning rate at 70°F, sec/in. | Coefficient of | | Variation |
|---|---|---|---|---|---|---|---|
| | | | | | Temp msec/ deg C/sec | Pressure msec/ mm/sec | |
| $BaCrO_4$-$KClO_4$-W | 40 | 10 | 50 | 8.53 | 1.5 | 0.73 | |
| | 70 | 10 | 20 | 31.31 | 1.0 | 0.40 | |
| CuO-B | 90 | 10 | — | 1.74 | 2.56 | 0.52 | 1.72 |
| $MoO_3$-B | 88 | 12 | — | 2.75 | 1.77 | 1.14 | 8.5 |
| | 92 | 8 | — | 5.21 | 0.58 | | 2.11 |
| $Co_2O_3$-B | 90 | 10 | — | 4.17 | 1.47 | 1.89 | 1.92 |
| $MnO_2$-B | 87 | 13 | — | 4.35 | 3.88 | 1.44 | 1.88 |
| | 90 | 10 | — | 6.41 | 6.02 | — | 2.18 |
| $V_2O_5$-B | 91 | 9 | — | 4.50 | — | 0.57 | 1.11 |
| $WO_3$-B | 91 | 9 | — | 5.40 | 3.68 | 0.15 | 3.4 |
| | 93 | 7 | — | 9.90 | 2.12 | 0.02 | 1.2 |
| $Fe_2O_3$-Ti | 69 | 31 | — | 0.84 | 2.9 | — | 3.0 |
| | 73 | 27 | — | 1.41 | 2.4 | — | 1.9 |
| $Fe_2O_3$-Ti-$SiO_2$ | 69 | 31 | 1 | 1.39 | 1.0 | — | 3.4 |
| $BaCrO_4$-Zr | 79 | 21 | — | 5.11 | 2.1 | — | 5.6 |
| $BaCrO_4$-Ti | 82 | 18 | — | 1.39 | 2.6 | — | 0.9 |
| | 87 | 13 | — | 3.35 | 2.0 | — | 0.9 |
| | 89 | 11 | — | 10.30 | 1.7 | — | 2.7 |
| Mo-$BaCrO_4$-$KClO_4$ | 30 | 60 | 10 | 9.0 | 1.25 | 0.54 | |
| | 20 | 70 | 10 | 23.0 | 0.33 | 1.4 | |
| | 40 | 50 | 10 | 5.5 | | | |
| Zr-$MoO_3$ | 49 | 51 | — | 0.18 | 0.66 | | |

tion are a fuel, an oxidant, a binder, and a lubricant. Delay compositions react when oxidant and fuel in the proper ratio are mixed intimately, and then ignited. The rate of burning is dependent on the proportions of the ingredients and their particle size. For a general discussion of the chemistry and technology of primary explosives, see Ref. 8.

### 21-2.1.1 Fuels

Principal fuels for delays are finely powdered metallic elements, nonmetallic elements, and their alloys. Ideally the fuel should have a high heat of reaction, yield reaction products stable at high temperature,

be resistant to moisture, and be nonpyrophoric. Fuels finding the greatest use are boron, silicon, zirconium, manganese, tungsten, zirconium-nickel alloys, and sulphur. Table 21-2[2] lists common fuel ingredients as well as their burning rate and the coefficients of temperature and pressure.

### 21-2.1.2 Oxidants

To be suitable for delay compositions, an oxidant must be stable at relatively high temperatures, reactive, nonhygroscopic, and yield solid reaction products. These criteria limit oxidants to those shown in Table 21-3[2].

**TABLE 21.3**

**OXIDANTS FOR DELAY COMPOSITIONS[2]**

| Chromates | Perchlorates | Oxides | Peroxides | Nitrates |
|---|---|---|---|---|
| Barium | Potassium | Ferric | Barium | Potassium |
| Lead | | Cuprous | | |
| Strontium | | Tungsten | | |
| | | Lead | | |
| | | Manganese | | |
| | | Bismuth | | |

Reprinted by permission of Prentice-Hall, Inc., Englewood Cliffs, N.J.

### 21-2.1.3 Binders

For a delay composition to resist setback forces or acceleration, a small quantity of binder may be added to reduce the changes of the charge breaking. A minimum quantity of binder is added to minimize the amount of gas generated and when it becomes necessary to granulate the mixture to provide a free-flowing powder for pelleting. Binders used include linseed oil, glycerine, and vinylalcohol-acetate copolymer[2]. Delay compositions adequately supported have resisted over 100,000 g setback without a binder. It is necessary to minimize additives, such as binders, to minimize evolution of gas.

### 21-2.1.4 Lubricants

A composition may be difficult to pelletize or press if it contains a hard material or if it sticks to the punch. A minimum quantity of a lubricant may be added to alleviate these difficulties. Those used include graphite, stearic acid, and stearates[2].

### 21-2.2 DELAY COMPOSITIONS

#### 21-2.2.1 Early Compositions

A discussion of early delay compositions follows:

(1) Black Powder:

Black powder is not favored by most engineers for use in new designs. Still there are instances where the unique ballistic properties of black powder are difficult or impractical to duplicate. Formed into compressed pellets, columns, or ring segments, black powder has been used to obtain delay times from milliseconds to a minute.

The advantages of black powder are great sensitivity to ignition even at low temperature, economy, multiplicity of uses, and relative safety in handling. The disadvantages are hygroscopicity and limited stability, excessive flash and smoke, undesirable solid residue, difficulty in controlling burning rate (high pressure increases the rate), poor burning qualities under diminished pressure, and finally, limited supply[9].

Perhaps the main reason that black powder is considered for new designs is its relatively high burning rate at low pressure and small-pressure burn-rate exponent. An empirical relation that has proven satisfactory for predicting performance with an interior-ballistic computer program is[9]

$$r = 0.133P^{0.325} \qquad (21\text{-}1)$$

where

$r$ = burning rate, in./sec

$P$ = pressure, psia

Experimental data indicate that a transition

21-5

in burning rate occurs in the region of 100 psia. It should be noted that powders produced by different manufacturers exhibit different ballistic properties.

(2) Studies of Gasless Compositions:

To overcome the disadvantages of black powder as a delay composition, research was initiated to develop nongaseous delay powders, making use of inorganic exothermic reactions similar to those used for thermite mixtures. Reactions which were studied are summarized in Table 21-4[10].

A large number of these compositions have been considered and many subjected to experimental investigations[11,12]. However, most of them have been discarded for one or another of the following reasons[3]:

(1) Erratic burning rates

(2) Too large a column diameter necessary for reliable propagation

(3) Failure at low temperatures

(4) Hygroscopicity

(5) Rapid deterioration

(6) Unavailability of reproducible supply of raw materials

(7) Large pressure coefficient of burning rate

(8) Failure at low pressure

(9) Reaction products that are liquid or otherwise subject to movement by projectile acceleration during burning of the delay column.

It was found that for each oxidizing agent used, under the standard test conditions, very fast burning times were obtained with magnesium, aluminum, zirconium, and titanium. Slower burning times were obtained with silicon, manganese, and chromium; while still slower ones occurred with iron, tungsten, and others. Metals that gave fast burning times with silver oxide, silver chromate, barium peroxide, and lead chromate gave slower burning times with cuprous oxide, barium chromate, and iron oxide.

### 21-2.2.2 Current Gasless Compositions

Table 21-5[10] lists the gasless delay combinations in current use. The range of compositions given for some of the combinations allows for adjustment of the burning rates over wide ranges. The powdered mixtures are pressed into tubes at high loading pressures.

Two types of fast delay mixture have been used at US Army Harry Diamond Laboratories for producing delays in the millisecond range[14]. One of these is a mixture of zirconium/ferric oxide/Superfloss and the other a mixture of zirconium and barium chromate. (Superfloss is a pure form of diatomaceous earth produced by Johns-Manville Corp.) A gasless igniter study has shown that the addition of about 10% Superfloss improves the loading characteristics of zirconium/barium chromate mixtures and its use in these delay mixtures is recommended[13]. Although both of these compositions have been tested in the M9 Delay Element (Fig. 18-6), they may be readily adapted to other designs.

Two slow burning mixtures have been used or investigated at Harry Diamond Laboratories[14]. These include (1) manganese/lead chromate/barium chromate and (2) tungsten/barium chromate/potassium perchlorate/Superfloss. Details of these compositions as well as a barium chromate/barium composition follow:

(1) Manganese Delay Compositions. Mixtures of manganese, lead chromate, and barium chromate can be produced with controlled burning times varying from about 2 to 13.5 sec/in. Generally, a standard

## TABLE 21-4

### HEATS OF REACTION OF INORGANIC MIXTURES CONSIDERED FOR DELAYS

| Metals | Silver Oxide −ΔH° | Silver Oxide −ΔH°/n | Silver Chromate −ΔH° | Silver Chromate −ΔH°/n | Barium Peroxide −ΔH° | Barium Peroxide −ΔH°/n | Lead Chromate −ΔH° | Lead Chromate −ΔH°/n | Cuprous Oxide −ΔH° | Cuprous Oxide −ΔH°/n | Barium Chromate −ΔH° | Barium Chromate −ΔH°/n | Iron Oxide −ΔH° | Iron Oxide −ΔH°/n |
|---|---|---|---|---|---|---|---|---|---|---|---|---|---|---|
| Magnesium | 139.1 | 69.6(I) | 688.5 | 68.9(I) | 126.7 | 63.4(I) | 560.7 | 56.1(I) | 103.6 | 51.8(5) | 292.9 | 48.8(5) | 239.8 | 40.0 |
| Aluminum | 378.0 | 63.0 | 1869.0 | 62.3 | 340.8 | 56.8(I) | 1485.6 | 49.5 | 271.5 | 45.3 | 253.6 | 42.3(N) | 200.5 | 33.4 |
| Zirconium | 244.1 | 61.0 | 1206.5 | 60.3 | 219.7 | 54.9(I) | 950.9 | 47.5 | 173.1 | 43.3 | 483.5 | 40.3 | 377.3 | 31.4 |
| Titanium | 211.0 | 52.8 | 1041.0 | 52.1 | 186.2 | 46.6(I) | 785.4 | 39.3(5) | 140.0 | 35.0 | 384.2 | 32.0 | 278.0 | 23.2 |
| Silicon | 187.0 | 46.8 | 921 | 46.1(N) | 164.8 | 41.2(I) | 665.4 | 33.3 | 116.0 | 29.0(N) | 312.2 | 26.0(12) | 206.0 | 17.2(N) |
| Manganese | 89.5 | 44.8 | 440.5 | 44.1 | 77.1 | 38.6(4) | 312.7 | 31.3 | 54.0 | 27.0(10) | 144.1 | 24.0(23) | 91.0 | 15.2(N) |
| Chromium | 252 | 42.0(N) | 1239 | 41.3(N) | 214.8 | 35.8(5) | 855.6 | 28.5 | 145.5 | 24.3 | 63.8 | 21.3 | 74.5 | 12.4 |
| Zinc | 76.5 | 38.3 | 374.5 | 37.5 | 63.9 | 32.0 | 247.7 | 24.8 | 41.0 | 20.5 | 104.6 | 17.4 | 52.0 | 8.7 |
| Tin | 124.1 | 31.0 | 606.5 | 30.3 | 99.3 | 24.8 | 350.9 | 17.5 | 53.1 | 13.3 | 123.5 | 10.3 | 17.3 | 1.4 |
| Iron | 177.5 | 29.6(8) | 866.5 | 28.9(8) | 140.3 | 23.4 | 483.1 | 16.1(11) | 71.0 | 11.8(N) | 53.1 | 8.9(N) | 0 | 0 |
| Cadmium | 58.2 | 29.1 | 284 | 28.4 | 45.8 | 22.9 | 156.2 | 15.6 | 22.7 | 11.4 | 50.2 | 8.4 | −2.9 | −0.5 |
| Tungsten | 174.7 | 29.1 | 852.5 | 28.4 | 137.5 | 22.9 | 469.1 | 15.6 | 68.2 | 11.4 | 50.3 | 8.4 | | |
| Molybdenum | 155.5 | 25.9 | 759 | 25.2 | 118.3 | 19.7 | 373.1 | 12.4(17) | 49.0 | 8.2 | 31.1 | 5.2 | | |
| Nickel | 50.5 | 25.3 | 250.0 | 25.0 | 39.0 | 19.5 | 122.2 | 12.2 | 15.9 | 8.0 | 29.8 | 5.0 | | |
| Cobalt | 51.4 | 25.7 | 245.5 | 24.6 | 38.1 | 19.1 | 117.7 | 11.8 | 15.0 | 7.5 | 27.1 | 4.5 | | |
| Antimony | 145 | 24.2 | 704.0 | 23.5(10) | 107.8 | 18.0(6) | 320.6 | 10.7 | 38.5 | 6.4 | 20.6 | 3.4 | | |
| Bismuth | 116.1 | 19.4 | 544.0 | 18.1 | 78.9 | 13.2 | 176.1 | 5.9 | 9.6 | 1.6 | −8.3 | −1.4 | | |
| Copper | 31.5 | 15.8 | 150.5 | 15.1 | 16.4 | 8.2 | 22.7 | 2.3 | | | −29.9 | −5.0 | | |
| **Nonmetals** | | | | | | | | | | | | | | |
| Phosphorus | 195 | 39(I) | 192.4 | 38.5(I) | 487.6 | 48.8(I) | 128.4 | 25.7(2) | 213 | 21.3 | 552.8 | 18.4(6) | 35.3 | 7.1(15) |
| Sulfur | 142.1 | 23.7 | 682.5 | 22.8(4) | 158.2 | 26.4(4) | 101.2 | 16.9(9) | 95.9 | 4.8 | 28.8 | 4.8 | −15.1 | −0.5 |
| Selenium | 67.9 | 11.3 | 311.5 | 10.4 | 86.6 | 14.4 | 31.3 | 5.2 | −6.2 | −0.3 | | | | |

−ΔH° heat of reaction.
−ΔH°/n equivalent heat of reaction (heat of reaction per electron charge).
(N) no reaction observed.
(I) very fast burning rate.
( ) burning time, sec.

## TABLE 21-5

### GASLESS DELAY COMPOSITIONS IN CURRENT USE

| Fuel, % | | Oxidants, % | | Inert, % |
|---|---|---|---|---|
| Manganese 30 to 45 | Barium Chromate 0 to 40 | Lead Chromate 26 to 55 | | None |
| Boron 4 to 11 13 to 15 | Barium Chromate 89 to 96 40 to 44 | Chromic Oxide — 41 to 46 | | None |
| Nickel-Zirconium Alloy 26 | Barium Chromate 60 | Potassium Perchlorate 14 | | None |
| Nickel-Zirconium Mix 5/31 5/17 | Barium Chromate 22 70 | Potassium Perchlorate 42 8 | | None |
| Tungsten 27 to 39 39 to 87 20 to 50 | Barium Chromate 59 to 46 46 to 5 70 to 40 | Potassium Perchlorate 9.6 4.8 10 | | Diatomaceous earth 5 to 12 3 to 10 |
| Molybdenum 20 to 30 | Barium Chromate 70 to 60 | Potassium Perchlorate 10 | | |
| Silicon 20 | Red Lead 80 | | | Diatomaceous earth Max 8 parts by weight |
| Zirconium 28 | Lead Dioxide 72 | | | |

pressure of 30,000 psi is used to load the dry mixtures in a column of 0.203 in. diameter[14]. Burning rates are affected markedly by the ratio of barium chromate to lead chromate. Increasing the barium chromate content decreases the burning rate. Experimental data on this effect are shown in Table 21-6[2]. Burning rates also are affected by particle size; the finer sizes burn faster.

Table 21-7[2] shows the effect of the housing material on the burning rate while Table 21-8[14] shows the effect of dry storage on burning time of manganese delay mixtures.

(2) Tungsten Delay Compositions. Tungsten delay compositions were developed to satisfy the requirements for reliable, long burning times (40 sec/in.). Potassium perchlorate was added to the binary mixture of tungsten and barium chromate to increase the heat of reaction, increase ignitibility, and to insure propagation at low temperatures[2]. Burning times, gas volumes, and heats of reaction for various mixtures of these ingredients are shown in Table 21-9[2]. Variation in the burning rate of a 55/10/35 barium chromate/tungsten/potassium perchlorate composition, loaded at 36,000 psi in vented bodies—as a function of tungsten particle size—is given in Table 21-10[2].

The effects of the potassium perchlorate level on the burning characteristics of tungsten delays modified with a ceric oxide flame sustainer have been studied[15]. The results indicated that there is an optimum level of potassium perchlorate (9%) at which minimum delay time variability occurs. Increasing potassium chloride sublimation introduces delay failures and higher burning times.

**TABLE 21-6**
**EFFECT OF INGREDIENTS ON BURNING RATE OF MANGANESE DELAY COMPOSITIONS[2]**

| Comp. | Ingredients | % | Nominal burning rate, sec/in. | Heat of reaction, cal/g | Gas volume, cm$^3$/g | Gas analysis, % H$_2$ | % CO | 5-sec value | Ignition temp, °C |
|---|---|---|---|---|---|---|---|---|---|
| A | Manganese* | 45.0 | 2.1 | 260 | 15.4 | 550.0 | 45.0 | 608 | 382 |
| | Lead chromate | 55.0 | | | | | | | |
| | Barium chromate | .... | | | | | | | |
| B | Manganese* | 33.0 | 8.4 | 256 | 18.3 | 77.7 | 22.3 | 660 | 522 |
| | Lead chromate | 37.0 | | | | | | | |
| | Barium chromate | 30.0 | | | | | | | |
| C | Manganese* | 32.8 | 13.5 | 262 | 11.4 | 65.4 | 34.6 | 702 | 478 |
| | Lead chromate | 30.2 | | | | | | | |
| | Barium chromate | 37.0 | | | | | | | |

*Micron size 5.7
Reprinted by permission of Prentice-Hall, Inc., Englewood Cliffs, N.J.

(3) Barium Chromate/Boron Compositions. A series of mixtures based on barium chromate and amorphous boron has been found to be easy to manufacture, readily ignitible, and capable of withstanding storage under adverse conditions with a high degree of reliability. Burning time values, heats of reaction, gas volume, and ignition temperature with increasing percentages of boron are shown in Table 21-11[2]. Increasing the loading pressure results in a small decrease in burning rate. Typical data for 95/5 and 90/10 compositions loaded in M112 Fuze Housings are shown in Table 21-12[2].

## 21-3 RELAYS

The output charge of a pyrotechnic delay timer is usually a relay. The relay is either the last charge increment in the delay element, or a separate component inserted into the delay assembly. Typical relays are shown in Fig. 21-4[3].

The usual relay consists of a cup into which lead azide is pressed at 10,000 psi. In some relays, a sealing disk is crimped over the open end, while in others, the end is left open; but the skirt left by partial filling is crimped at an angle. When such relays are inserted into delay elements and crimped in place, the crimp is compressed just enough to obtain a firm and snug fit.

## 21-4 SENSITIVITY

### 21-4.1 NEED FOR CHARACTERIZATION

To be able to minimize the hazards of pyrotechnic delay compositions, it is necessary to determine their reaction and sensitivity characteristics. This information forms a basis for the establishment of optimum safety procedures in processing pyrotechnic devices. The goal looked for is minimum risk to personnel who prepare, handle, test, and transport the compositions.

Sensitivity is established by tests that simulate such external stimuli as impact, friction, heat, moisture, and electrostatic discharge[3]. The tests also are used to (1)

**TABLE 21-7**
**EFFECT OF HOUSING MATERIAL ON BURNING RATE OF MANGANESE DELAY COMPOSITIONS**

| | Aluminum bodies | Brass bodies |
|---|---|---|
| Nominal avg burning time, sec | 8.15 | 10.3 |
| Column length, in. | 0.676 | 0.845 |
| Nominal avg burning rate, sec/in. | 12.1 | 12.2 |

TABLE 21-8

EFFECT OF DRY STORAGE ON BURNING TIME OF MANGANESE DELAY COMPOSITIONS

Batch[a]  Burning Time[b] (sec) After Indicated Exposure in Sealed Flasks at $+165°F$

| | Before Storage | | 1 Wk Storage | | 2 Wk Storage | | 4 Wk Storage | | 8 Wk Storage | |
|---|---|---|---|---|---|---|---|---|---|---|
| | Avg.[c] | Std. Dev. | Avg.[c] | Std. Dev. | Avg.[c] | Std. Dev. | Avg.[c] | Std. Dev. | Avg.[c] | Std. Dev. |
| 1 | 6.56 | 0.123 | 6.73 | 0.089 | 6.73 | 0.095 | 6.85 | 0.127 | 6.86 | 0.117 |
| 2 | 5.35 | 0.122 | 5.46 | 0.028 | 5.50 | 0.036 | 5.56 | 0.077 | 5.62 | 0.064 |

[a]33% Mn, 30% $BaCrO_4$, 37% $PbCrO_4$.
[b]Mixtures loaded at 30,000 psi with 2-100 mg increments of DM-3 igniter in delay bodies 0.203 in. I.D., 0.500 in. O.D., 0.750 in. long.
[c]Average of ten tests.

determine the energy required to initiate a delay composition, (2) serve as a quality assurance gage, and (3) to indicate the effect of storage under adverse conditions.

## 21-4.2 SENSITIVITY TO INITIATION

There are a number of tests that measure how easily explosive materials are initiated. They use various stimuli that are capable of setting off the explosive. The stimulus most widely used is impact by dropping, others used are friction and rifle bullet impact.

TABLE 21-9

BURNING RATES OF VARIOUS TUNGSTEN DELAY COMPOSITIONS[2]

| Ingredient, % | | | Nominal avg burning rate, sec/in. | Gas volume, cm³/g | Heats of reaction, cal/g |
|---|---|---|---|---|---|
| W | BaCrO₄ | KClO₄ | | | |
| 85 | .. | 15 | 1.6 | .... | .... |
| 80 | .. | 20 | 2.5 | .... | 304 |
| 28 | 62 | 10 | 48.6 | 8.5 | .... |
| 30 | 60 | 10 | 31.2 | 10.4 | .... |
| 32 | 63 | 5 | 28.7 | 3.3 | .... |
| 35 | 30 | 35 | 24.1 | .... | .... |
| 40 | 47 | 13 | 5.9 | .... | 346 |
| 50 | 40 | 10 | 7.8 | 4.3 | 305 |
| 65 | 20 | 15 | 3.0 | 2.1 | .... |
| 70 | 10 | 20 | 7.3 | 1.6 | .... |

Reprinted by permission of Prentice-Hall, Inc., Englewood Cliffs, N.J.

### 21-4.2.1 Impact

The impact test consists of dropping a weight on a sample of explosive. In practice, two types of tests are used, each requiring a special type of equipment. The two most prevalent impact tests are the Picatinny Arsenal and the Bureau of Mines tests. The method of the Bureau of Mines consists of dropping on the sample a known weight in guided free fall from a preselected height. The method developed by Picatinny Arsenal (PA) uses the same procedure except that the sample fills a hardened steel cup that is covered by a brass cap. A vented steel plug that is centered on the brass cap is hit by the falling weight. Fig. 21-5[3] shows the PA test apparatus.

The main difference between the two tests is that the PA test involves greater confine-

TABLE 21-10

VARIATIONS OF BURNING RATE DUE TO CHANGE IN PARTICLE SIZE OF TUNGSTEN DELAY COMPOSITION[2]

| Tungsten, Micron Size | Average Burning Rate, sec/in. |
|---|---|
| 4 | 9.0 |
| 8 | 17.2 |
| 15 | 28.1 |

Reprinted by permission of Prentice-Hall, Inc., Englewood Cliffs, N.J.

TABLE 21-11

DELAY CHARACTERISTICS FOR VARIOUS PERCENTAGES OF BORON[2]

| Boron, % | Nominal* avg burning time, sec | Heat of reaction, cal/g | Volume of gas, cm³/g | Impact test PA, in. | Ignition temp, °C 5-sec value |
|---|---|---|---|---|---|
| 3.0 | 7.56 | Incomplete reaction | | 40+ | . . . . |
| 4.0 | 1.72 | 354 | 5.0 | 40+ | . . . . |
| 5.0 | 1.09 | 420 | 8.0 | 40+ | app. 700 |
| 6.0 | 0.77 | 431 | 8.4 | 37 | . . . . |
| 8.0 | 0.56 | 462 | 7.9 | 28 | . . . . |
| 10.0 | 0.47 | 515 | 7.3 | 21 | 680 |
| 13.0 | 0.40 | 556 | 8.9 | 20 | . . . . |
| 15.0 | 0.39 | 551 | 7.0 | 16 | . . . . |
| 17.0 | 0.38 | 543 | 11.6 | 13 | . . . . |
| 19.0 | 0.37 | 535 | 8.8 | 16 | . . . . |
| 21.0 | 0.38 | 526 | 8.6 | 34 | . . . . |
| 23.0 | 0.41 | 503 | 4.2 | 40+ | . . . . |
| 25.0 | 0.43 | 497 | 10.2 | 40+ | . . . . |
| 30.0 | 0.57 | 473 | 10.4 | 40+ | . . . . |
| 35.0 | 0.97 | 446 | 12.7 | 40+ | . . . . |
| 40.0 | 2.19 | 399 | 14.1 | 40+ | . . . . |
| 45.0 | 5.25 | 364 | 15.0 | 40+ | . . . . |
| 50.0 | 14.5 | Incomplete reaction | | 40+ | . . . . |

*Loaded in M112 Fuze Housings at 36,000 psi
Reprinted by permission of Prentice-Hall, Inc., Englewood Cliffs, N.J.

ment of the sample, distributes the translational impulse over a smaller area, and involves a frictional component (between cap and sample). Hence, PA test values are affected greatly by sample density.

## 21-4.2.2 Friction

To measure sensitivity to friction, a 7-g sample (50-100 mesh) is rubbed by a steel or fiber shoe swinging at the end of a long rod. The behavior of the sample is described qualitatively to indicate its reaction to this stimulus. The most energetic reaction is explosion; it decreases in the order of severity to no effect[3].

## 21-4.2.3 Rifle Bullet Impact

The traditional bullet sensitivity test consists of firing a cal .30 bullet into the side of a 3-in. pipe nipple that contains approxi-

mately 0.5 lb of the explosive being tested and is capped at both ends. Because of the curved target surface, test results can be affected greatly by the condition of the weapon, the characteristics of the ammunition and the impact angle[3]. An improved test with a flat target plate was devised at PA[16].

## 21-4.3 SENSITIVITY TO THE ENVIRONMENT

Pyrotechnic compositions are subject to degradation or initiation by several environmental conditions. Burning rate can be adversely affected by absorption of moisture. Initiation may be caused by heat or electrostatic discharge.

## 21-4.3.1 Heat

A heat test is used to determine a threshold temperature below which there is no readily

TABLE 21-12

VARIATION OF BURNING RATE WITH
LOADING PRESSURE OF BARIUM
CHROMATE/BORON COMPOSITIONS[2]

| | Composition | | | |
|---|---|---|---|---|
| | 95/5 | | 90/10 | |
| Loading pressure, psi | Nominal burning rate, sec/in. | Density, g/cm$^3$ | Nominal burning rate, sec/in. | Density, g/cm$^3$ |
| 200 | 1.21 | 1.88 | 0.53 | 1.89 |
| 500 | 1.21 | 1.91 | 0.54 | 1.95 |
| 1,300 | 1.29 | 2.22 | 0.56 | 2.09 |
| 3,600 | 1.39 | 2.41 | 0.59 | 2.26 |
| 9,000 | 1.49 | 2.55 | 0.62 | 2.43 |
| 18,000 | 1.60 | 2.70 | 0.65 | 2.57 |
| 36,000 | 1.69 | 2.89 | 0.70 | 2.73 |

Reprinted by permission of Prentice-Hall, Inc., Englewood Cliffs, N.J.

detectable reaction and above which a reaction takes place within some short time interval. Several ways of doing this are[2]:

(1) Measuring time to explosion or time to ignition

(2) Isothermal heating

(3) Adiabatic heating

(4) Differential thermal analysis

(5) Thermogravimetric analysis.

## 21-4.3.2 Moisture (Hygroscopicity)

In this test, the absorption of, or reaction with, external moisture by an ingredient or composition is determined because variations in moisture can cause changes in sensitivity. Basically, a known weight of sample is placed in a vessel prepared to have a known relative humidity at a specific temperature. At predetermined intervals, the percent gain or loss of weight of the sample is calculated[2].

## 21-4.3.3 Electrostatic Discharge

The relative ease of initiation of a pyrotechnic composition by an electric spark is termed its sensitivity to electrostatic discharge[2]. In this procedure, the sample is loaded in a cylindrical depression in a heavy metal plate that serves as an electrode. A second electrode, whose height is adjustable, is placed above the sample which is then subjected to a high-voltage spark. The maximum, 50%, and minimum energy levels at which initiation occurs then are determined. Humidity levels must be controlled in this test.

## 21-4.4 INPUT AND OUTPUT SENSITIVITY

### 21-4.4.1 Input

A discussion of the input sensitivity of initiators follows:

(1) Mechanical Initiators. The input sensitivity of stab and percussion primers is determined by the impact test (see par. 21-4.2.1).

(2) Electric Initiators. Depending upon the application, the sensitivity of electric initiators should be characterized in terms of the

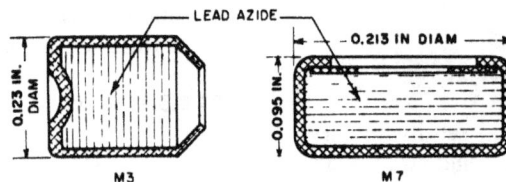

Figure 21-4. Typical Relays

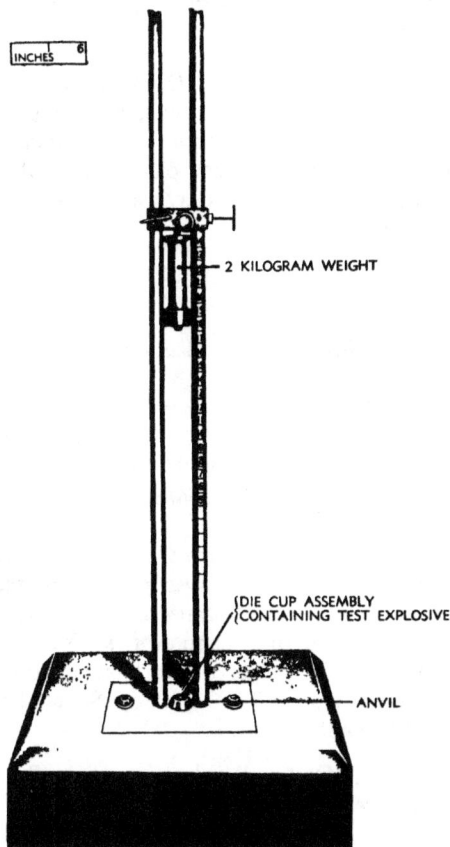

*Figure 21-5. Picatinny Arsenal Impact Test Apparatus*

threshold current, voltage, power, energy, or some combination of these. Tests used for sensitivity of electric initiators include[3]:

(1) Capacitor discharge test

(2) Voltage sensitivity

(3) Steady current functioning.

## 21-4.4.2 Output

The output of a delay is a flame or hot slag. Tests for flame output include (1) light output, as measured by a photocell, (2) a

modification of the lead disk test (amount of lead displaced by output of device), (3) a test to measure flame temperature and length, and (4) a closed chamber test using a manometer to measure flame output momentum[3]. A stop watch is often satisfactory for measurement of relatively long burning times. However, the electronic chronometer is suitable for all ranges of burning time[2]. In this method, the timing instrument starts when the initiator functions and stops when the charge flashes. Correction must be made for the burning times of the initiating source and the relay charge.

## 21-4.5 ANALYSIS OF DATA

The sensitivity of an explosive charge is the magnitude of the minimum stimulus that will result in its initiation. Stimuli too weak to initiate charges can alter them, sometimes in a readily noticeable way and sometimes by changing the sensitivity of the charge to the stimulus. In recognition of this variability a number of statistical plans have been devised for sensitivity studies. Some of the following plans are designed to characterize sensitivity in terms of an assumed normal distribution and still others determine some point in the distribution of particular interest:

(1) Staircase Method, the Bruceton Test

In the Bruceton test[17], the magnitude of the stimulus used in each trial is determined by the result obtained in the immediately preceding trial. If the preceding trial was a misfire, the stimulus for the next trial is made one step higher, and if it fired, the stimulus should be one step lower. The step used is a difference between stimulus levels, such as 5 V or 2 in. of drop height. Misfires are discarded because their sensitivity could have been affected by the previous test.

The validity of this procedure depends on whether the assumptions are valid that (a) the steps are of a uniform size, and (b) the frequency of explosions is distributed normally.

(2) Frankford Run-down Test Method. A run-down test has been developed by Frankford Arsenal[18] that, with the expenditure of a much larger sample, makes possible a greatly improved assessment of the distribution of the underlying population. The number of trials is constant at each energy level. The maximum level for all samples activated, the minimum level for no samples

activated, and several points in between are determined.

(3) Other Procedures. Other procedures include the Probit[19], Normit, and Logit[20] which are analytical methods of determining the underlying statistical distribution of a measured parameter. They are not data collecting schemes.

## REFERENCES

1. MIL-STD-320, *Terminology, Dimensions and Materials of Explosive Components for Use in Fuzes,* 3 August 1959.

2. F. B. Pollard and J. H. Arnold Jr., Eds., *Aerospace Ordnance Handbook,* Prentice-Hall, Inc., Englewood Cliffs, N.J., 1966.

3. AMCP 706-179, Engineering Design Handbook, *Explosive Trains.*

4. MIL-L-17186, *Lead Styphnate, Normal Commercial Grade,* Department of Defense.

5. MIL-L-3055A, *Lead Azide,* Department of Defense.

6. MIL-L-16355A, *Lead Styphnate, Basic,* Department of Defense.

7. *A Compendium of Pyrotechnic Delay Devices,* Journal Article 30.0 of the JANAF Fuze Committee, 23 October 1963 (AD-474 833).

8. Tadeusz Urbanski, *Chemistry and Technology of Explosives,* Pergamon Press, London, Volume III, 1967.

9. Richard A. Whiting, "The Chemical and Ballistic Properties of Black Powder", Explosives and Pyrotechnics **4,** Nos. 1 and 2, January and February, 1971.

10. AMCP 706-185, Engineering Design Handbook, *Military Pyrotechnics, Part*

*One, Theory and Application.*

11. AMCP 706-187, Engineering Design Handbook, *Military Pyrotechnics, Part Three, Properties of Materials Used in Pyrotechnic Compositions.*

12. Raymond H. Comyn, *Summary of Pyrotechnic Delay Investigations for the AEC and Sandia Corporation,* U S Army Harry Diamond Laboratories, Washington, D.C., Prepared for Sandia Corp., Order No. ASB 14-9986, September 1963 (AD-683 807).

13. Ira R. Marcus, *Development of a Split Igniter for Initiating Gasless Delays,* U S Army Harry Diamond Laboratories, Washington, D.C., Report TR-875, 14 November 1960.

14. Raymond H. Comyn, *Pyrotechnic Research at DOFL, Part II, Pyrotechnic Delays,* U S Army Harry Diamond Laboratories, Washington, D.C., Report TR-1015, 15 February 1962.

15. James E. Rose, *Effect of Potassium Perchlorate on Flame Propagation in Modified Tungsten Delay Composition,* Naval Ordnance Station, Indian Head, Md., Report IHTR 322, 30 July 1970.

16. S. D. Stein, *Quantitative Study of Parameters Affecting Bullet Sensitivity of Explosives,* Picatinny Arsenal, Dover, N.J., Report TR2636, September 1959.

17. *Statistical Analysis for a New Procedure*

*in Sensitivity Experiment,* Statistical Research Group, Princeton, N.J., AMP Report 101.1R SRG-P No. 40, July 1944 (ATI-34 558).

18. C. W. Churchman, *Manual for Proposed Acceptance Test for Sensitivity of Percussion Primers,* Frankford Arsenal, Phila., Pa., Report R-259A, January 1943.

19. D. J. Finney, *Probit Analysis,* Cambridge University Press, London, England, 1947.

20. J. Berkson, "A Statistically Precise and Relatively Simple Method of Estimating Bio-Assay with Quantal Response", J. Am. Statistical Assoc., **48,** 565-99 (1953).

PART FIVE — FLUERIC TIMERS

CHAPTER 22

INTRODUCTION

Flueric devices represent those elements that perform control or logic functions by means of a fluid (liquid or gas) in motion without the use of moving or mechanical parts. Flueric timers are timers fabricated from fluid elements to provide an "output" pulse after some known time interval has elapsed following an "input" pulse.

A flueric timer can be constructed by series interconnection of binary stages that are driven by a fluid oscillator. In order to construct a useful timer, e.g., one which indicates the elapse of a pre-programmed time interval, additional auxiliary components are required. A typical digital interval timer is shown in block diagram form in Fig. 22-1[1]. As indicated in the figure, such a system may include a power supply, a time base oscillator, a multistage counter, a setting mechanism, a decoder to determine when the set time has elapsed, and an amplifier stage which feeds the output transducer. At present the feasibility of flueric timing has been demonstrated by several independent investigators.

Standard flueric terms and symbols are covered in Ref. 2 and a bibliography appears in Ref. 3.

## 22-1 ADVANTAGES

The use of flueric devices instead of the more conventional electromechanical types is attractive for the following reasons:

(1) High Reliability. Due to the lack of moving parts, flueric elements in general can operate for almost indefinite time periods. No parts are subjected to flexure, bending, wear, or sticking. If the supply fluid is filtered, erosion practically is eliminated and fluid flow in the element itself produces a self-cleaning effect. Even with the addition of some contaminants (dirty air), tests have indicated that an oscillator performed satisfactorily for two weeks before cleaning was necessary[4]. Of course, the degree to which a circuit can tolerate contaminants will vary with size and circuit requirements. A sound design principle is always to employ contaminant free supply fluid.

(2) Large Temperature Range. Although temperature will play a most significant role in system fluid whose dynamic properties are inherently temperature dependent (and this fact will alter element function), the element itself (fabricated from metals, glass, ceramics, etc.) can be made to withstand extreme temperature ranges extending from −100° to 700°F. At high temperatures, the problems associated with sealing, connecting, or joining various elements can be more important than the fact that the system fluid is at a high temperature.

(3) Not Affected by Electromagnetic and Nuclear Radiation. The components of a flueric element, if properly made, are in general not affected by RF fields or nuclear radiation. Flueric elements can be constructed to represent a complete Faraday shield that will minimize effects of electromagnetic waves. Extending this line of reasoning, a nuclear radiation shield also can be applied over the final design to minimize penetration of gamma and X rays. Effects of local

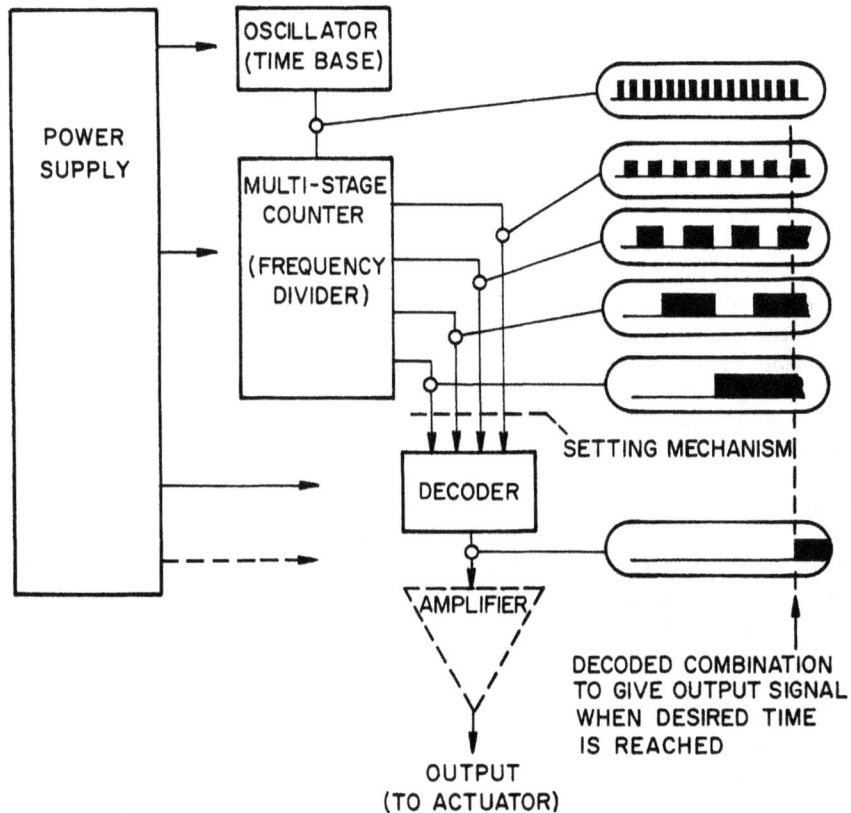

Figure 22-1. Digital Interval Timer Block Diagram

environment can be eliminated by proper design and fabrication.

(4) Shock and Vibration. The effect of shock and vibration on a stored element or a standby element can be reduced to practically zero, by initial choice of fabricated materials. The shock effect on an operating unit, however, is another matter. Generally it is agreed that the relation between the density of the working jet compared with the density of the fluid surrounding the jet is of prime importance. Many devices, however, require shocks in excess of 1000 g's to affect their performance. Other devices, such as a properly adjusted turbulence amplifier, can be "upset" by acoustic shock signals at one discrete frequency and remain relatively unaffected by other frequencies. One could conceivably capitalize on this phenomenon in designing an acoustic switching device.

(5) Reduction of Interfaces. Flueric devices can simplify or reduce the number of interfaces in a system because the flueric circuit can operate using the medium of liquid or gas being controlled, thus reducing interfaces and the need for expensive transducers. This not only adds to the simplicity of the system but increases reliability[5].

(6) Low Cost. As the technology develops, undoubtedly the system cost will decrease.

Potential savings often overlooked are the lowering of

(a) Cost of maintenance

(b) Servicing expenses

(c) Down time.

(7) Not Affected by Corrosive Fluids. If a proper choice of fabrication material has been made, flueric circuits can function with corrosive fluids, and within corrosive atmospheres.

(8) Miniaturization. The three major advantages of miniaturization are:

(a) Speed of response can be increased.

(b) Smaller components require less power.

(c) Smaller dimensions and less mass are desirable so that payload can be increased.

## 22-2 DISADVANTAGES

Some of the most common disadvantages associated with the use of flueric circuits and elements are:

(1) Limit of Response Time or Switching Speed. Response time or switching speed is very important in digital application[5]. Changes in design will affect the switching speed of the element, but for various wall-attachment devices having a nozzle width of 0.010 in. and using air at pressures below 50 psi, an average range is 100 to 1000 Hz, or response times of 0.005 to 0.0005 sec (switching occurs twice per cycle). Higher speeds have been achieved at higher supply pressures but at the expense of unacceptable power consumption. Due to the present limits of output transducing devices, it is questionable whether speeds in excess of several hundred hertz can be used effectively. Units (fluid oscillators) have operated at 100,000 to 200,000 Hz, but the outputs were not used in a meaningful manner.

(2) Limit of Signal Propagation. The limit of signal propagation for a gas operated system is equal to the local speed of sound through the gas. For dry air at ambient temperature, the propagation speed is about one foot per millisecond.

(3) Large Power Consumption. Flueric elements require a continuous supply of fluid while in operation, and considerable amounts of power may be required[5]. This can be offset to some extent by micro-miniaturization. The power consumption of an element generally depends upon the flowing fluid, jet velocity, and nozzle size. In wall-attachment devices, these variables can be related to two nondimensional parameters: the Reynolds number and the ratio of nozzle depth to width (aspect ratio). Bistable operation of the device requires a minimum Reynolds number, below which attachment no longer occurs. However, this occurs only at very low Reynolds numbers.

Pressure recovery in the output receivers is also important in determining the power consumption. Efficient geometries will require less power.

In fluid timer applications, power consumption is an important consideration. If the timer is being supplied from a bottled source or pyrotechnic gas generator, the power consumption of each stage must be known more critically than if ram air is used. This is because bottled supplies have a limited useful life, but ram air is available in almost unlimited amounts once the projectile or missile is air-borne.

(4) Need For Contaminant Free Supply. Micro-miniaturization of circuits has created a need to employ pure bottled fluids to prevent clogging and malfunctioning, etc. Ram-air power supplies will require a closure to exclude water and other contaminants prior

to firing. The closure will be opened or removed prior to or during firing. This can be done automatically by the firing cycle, such as setback opening the ram tube or by having a plastic cap that is blown off by blow-by as the projectile leaves the muzzle. A small separator such as the cyclone type will be included in the ram-air power supply to eliminate rain, snow, and dust in flight. The use of a ram-air power supply increases safety since the system is completely inert until it achieves a minimum velocity.

(5) Dependence on Temperature and Pressure:

Although a fair degree of temperature and pressure insensitivity has been accomplished, especially with compensated flueric oscillators, perfect insensitivity has not been achieved. This is discussed fully in Refs. 6 and 7.

Theory and experiments indicate that temperature insensitivity in the resistance-inductance type of oscillator is obtained at the expense of pressure sensitivity, and a compromise is necessary in the actual oscillator. With fluid oscillators of the RCR feedback type, claims of ± 1% have been reported over a pressure range of 6 to 30 psig and 77° to 175°F[6]. At the present time, RC oscillators have been operated from −60° to +170°F, using power jet pressures from 6 to 30 psig with only a ± 1% change in frequency. The oscillator must be miniaturized for practical use in a fuze. Hybrid systems (such as the torsion bar oscillator) that use vibrating mass systems are virtually temperature and pressure insensitive, but are not strictly flueric devices. All uncompensated flueric oscillators are temperature sensitive because they depend on the velocity of sound of the operating fluid which is, in turn, temperature sensitive. The sonic velocity is proportional to the square root of the absolute temperature.

Since the precision of the basic oscillator governs the overall precision of the flueric timer, its temperature and pressure sensitivity is a major consideration.

## REFERENCES

1. "The Application of Flueric Devices to Ordnance Timers", Journal Article 51.0 of the JANAF Fuze Committee, 3 May 1967 (AD-834 083).

2. MIL-STD-1306, *Fluerics: Terminology and Symbols,* Dept. of Defense, 17 July 1968.

3. *Fluerics. 23. A Bibliography,* U S Army Harry Diamond Laboratories, Washington, D.C., Report TR-1495, April 1970.

4. *Fluidics System Design Guide,* Fluidonics Division of the Imperial Eastman Corp., Chicago, Ill. 1966, p. 99.

5. *Fluidics,* Fluid Amplifier Associates Inc., Boston, Mass., 1965.

6. C. J. Compagnuolo and S. E. Gehman, *Flueric Pressure and Temperature Insensitive Oscillator for Timer Application,* U S Army Harry Diamond Laboratories, Washington, D.C., Report TR-1381, February 1968.

7. C. J. Campagnuolo and H. C. Lee, *Review of Some Fluid Oscillators,* U S Army Harry Diamond Laboratories, Washington, D.C., Report TR-1438, April 1969.

# CHAPTER 23

## FLUERIC SYSTEMS DESIGN

### 23-1 FACTORS

Among the factors to consider before developing a fluid system are gain, signal-to-noise ratio, frequency response, ease of interconnection of individual components, output pressure, flow and power. A discussion of each factor follows:

(1) Gain:

The three types of gain are pressure, flow and power. Gain is defined differently for digital and proportional devices.

Digital-amplifier *flow gain* is the ratio of change of flow output to change in control flow necessary for switching to occur. *Pressure gain* is the ratio of change in output pressure to change in control pressure required for switching. *Power gain* is the ratio of change in output power to change in control power required to initiate switching.

(2) Signal-to-noise Ratio:

This is the maximum output-signal amplitude divided by the maximum noise amplitude. Good signal-to-noise ratio is not as important for the devices used in timers as it is for proportional amplifiers.

In timing devices, output is at one of two discrete levels, each of which has a tolerance band. Provided the noise riding on the output signal does not exceed limits of the tolerance band, the flueric device can be relatively noisy and still function properly. However, care must be taken to avoid spurious switching by a noisy input signal.

(3) Frequency Response:

Response for flip-flops is given by the switching time; the time interval that begins when the changing input signal reaches 50% of its final value and ends when the subsequent output signal reaches 50% of its final value. In each instance the output signal can be flow, pressure, or power.

(4) Interconnectability:

A good measure of the ease with which a digital flueric device can be staged is its *fanout* capability—how many downstream devices it can control. In determining fanout capability, all of the devices should be identical in size, input impedance, gain, and power-supply conditions.

Fanout applies primarily to digital systems where the flueric devices are used as logic elements to handle information. Gain between successive stages is not a primary objective.

Another approach to interconnection is to match input and output impedances. If output impedance of the driving element is matched to input impedance of the driven element, maximum power transfer occurs. The impedance of the elements is adjusted by such techniques as providing output flow dividers, vents, bleeds, and varying duct cross sections.

Table 23-1 compares types of digital devices with respect to power range, switching time, geometry and ease of interconnection.

## TABLE 23-1

### CHARACTERISTICS OF DIGITAL FLUERIC DEVICES

| Type of Device | Power Range | Switching Time | Geometry of Ducts | Ease of Interconnection |
|---|---|---|---|---|
| Wall-attachment | Low to high | Medium | Rectangular | Difficult |
| Edgetone | Low to high | Fast | Rectangular | Moderate |
| High-pressure low-flow | Low to medium | Medium | Circular | Moderate |
| Turbulence | Low | Slow | Rectangular or circular | Easy |

## 23-2 MODULAR COMPONENTS

In order for a flueric timer to replace other types in fuze applications, the overall size must be reduced to a minimum. The use of photoetching and other recent processes has made it possible to fabricate microminiature flueric elements. These are usually thin wafers of rectangular or circular shape made from metal (phosphor-bronze). The external shape of these wafers is held constant and various geometries are etched through the entire wafer thickness. These elements can then be stacked together readily to form a final package. This idea typifies the modular component concept.

With modular components, the aspect ratio of nozzles readily can be changed simply by stacking two or more elements of similar shape together. Also an entire stage can be removed and replaced quite readily. Fig. 23-1 shows two typical element wafers, a relaxation oscillator (A) and its binary amplifier (B). These measure about 0.5 in. by 0.75 in. and are 0.003 in. thick. The corner holes on each wafer are used to clamp the wafers together (by use of long screws) and the hole spacings are standardized. The top and bottom of the center holes are supply pressure manifold holes (power supply) which may be used at a particular stage or may carry flow through to another stage. Output arms feed through other holes to become inputs to subsequent stages. The small, centrally located holes are bleed-off ports as is the etched section open to the side of the element. It is quite obvious that the proper order of stacking must be maintained throughout.

Experiments have shown that if the proper torque is applied to each mounting screw and the wafers are etched properly, the use of gasket sealing material between each wafer is unnecessary. In fact, the use of a sealant is to be discouraged because the tiny nozzle openings can easily clog.

Mechanical strength of the wafer also must be considered since high pressures permanently can deform overhanging sections of the etched element.

At both ends of the stacked elements (as well as at any intermediate location) one may use thicker plates of similar external size which serve to introduce or remove fluid signals, and also give mechanical strength for clamping by the through screws.

A complete timer package employing modular components is shown in Fig. 23-2.

## 23-3 INTEGRATED CIRCUITS

Integrated circuitry as applied to fluerics is

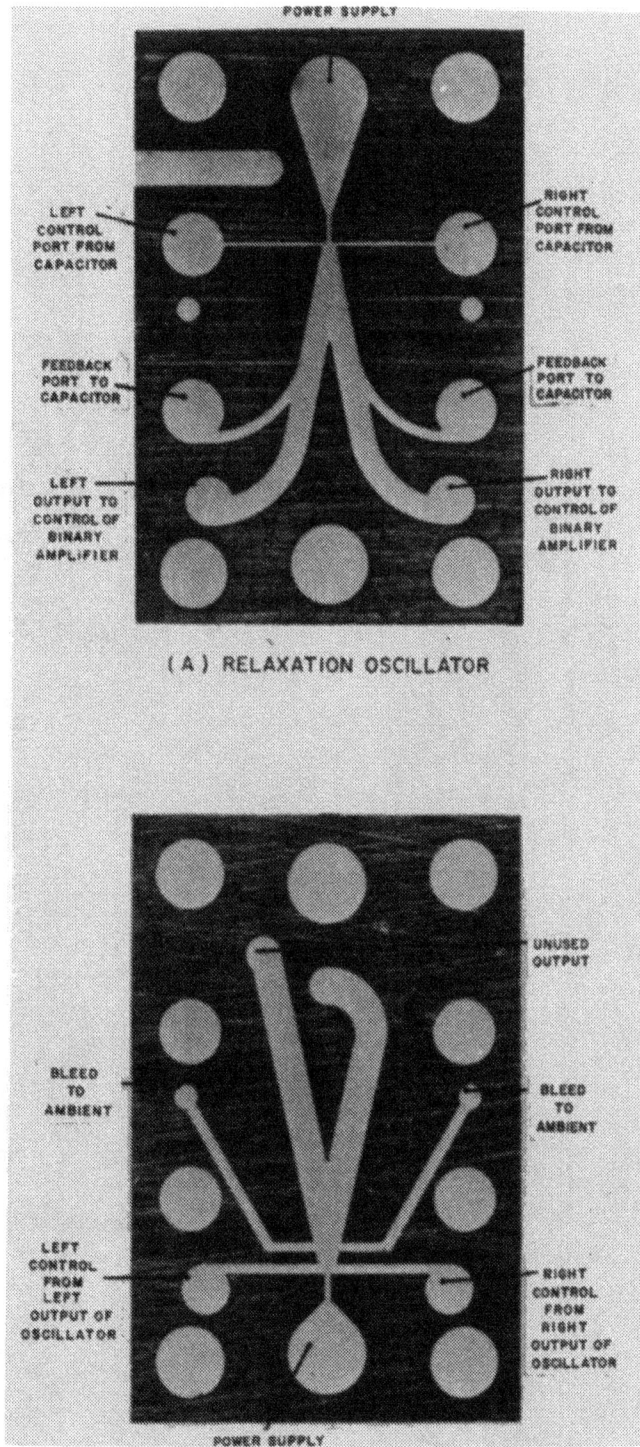

Figure 23-1. Miniaturized Modular Circuit Elements

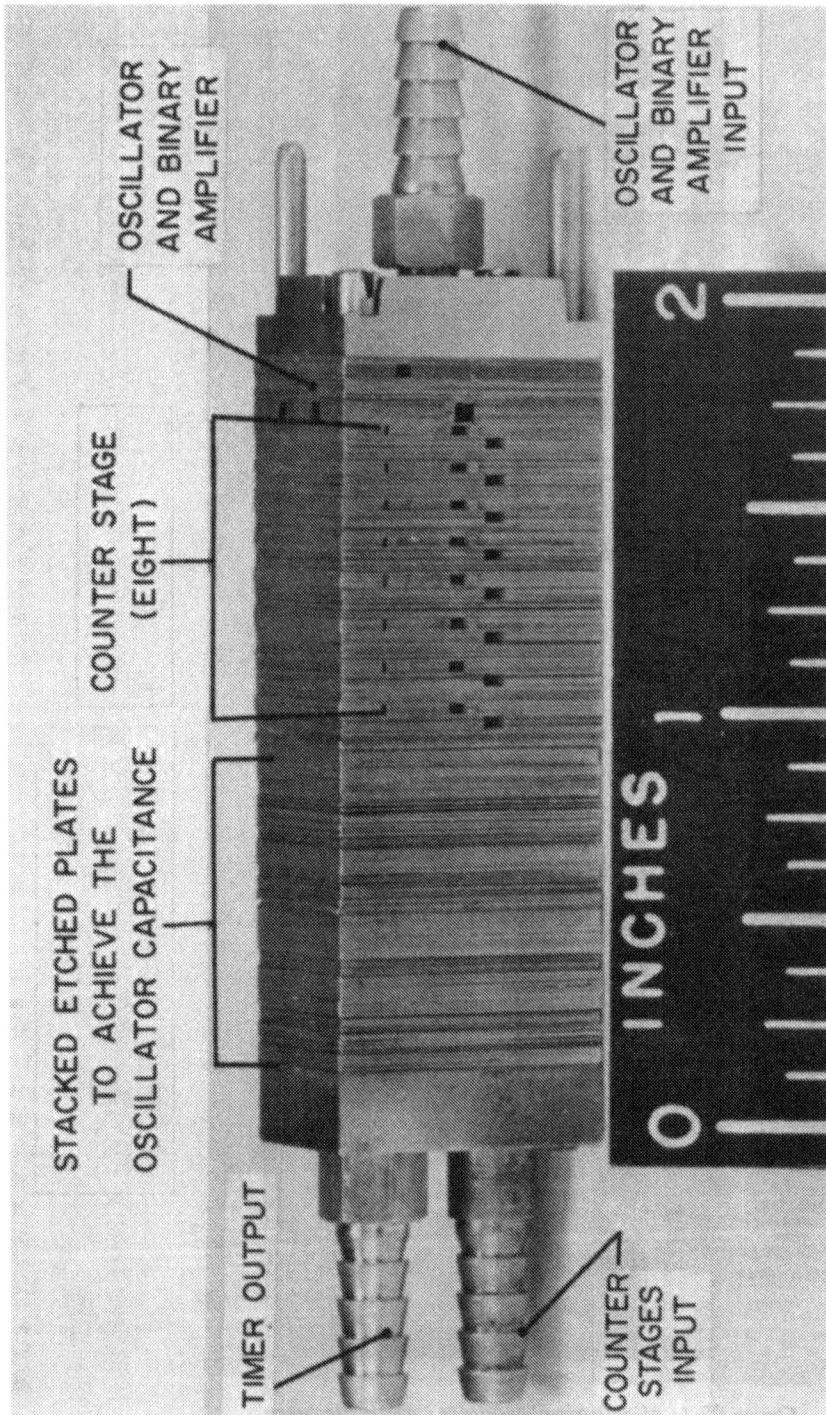

Figure 23-2. Fluid Timer Assembly Illustrating Modular Design

OSCILLATOR AND BINARY AMPLIFIER

OSCILLATOR AND BINARY AMPLIFIER INPUT

COUNTER STAGE (EIGHT)

STACKED ETCHED PLATES TO ACHIEVE THE OSCILLATOR CAPACITANCE

TIMER OUTPUT

COUNTER STAGES INPUT

INCHES

the incorporation of two or more functional geometries on a single planar sheet. This concept has several desirable features: (1) space can be used fully on any given wafer, (2) overall size of the final unit can be reduced, and (3) more efficient interstage matching can be accomplished. The desirability of (1) and (2) cannot be overemphasized. In par. 23-4 the importance and significance of source and load impedance matching are discussed. Integrated circuitry offers a way for better inter-stage impedance matching since the input and output channels can be controlled more critically and fluid flow need not confront any discontinuities that may be present from wafer to wafer.

Since the geometries of the etched wafer cannot be altered after fabrication, the designer must know beforehand the effect of each part of the element. At present a design problem cannot be tackled straightforward from a purely analytical technique, and usually a large-scale model initially is fabricated, and development proceeds from this model by empirical methods. Once the geometries have been defined and the elements function as desired, the model is reduced proportionately to its final size.

Integrated circuit concepts can produce elements of high reliability. Some integrated circuit devices have been fabricated which include 10 or more individual elements on a single planar sheet.

## 23-4 MATCHING TECHNIQUES

The coupling of fluid elements, stage to stage, may introduce a matching problem. Matching may not be necessary with bistable flip-flops having bleed ports and in devices where each stage receives a supply pressure from a common manifold. However, in staged amplifiers, matching must be considered[1]. The main difficulty is that the performance of a stage depends not only on its own configuration but also on the geometries of the previous and the following stages. A

means is needed to predict the performance of a stage in a system without actually assembling the components. Since system conditions vary widely and generally are not known in advance, it is necessary to simulate a large number of operating conditions when testing a single stage. In addition, the system may be required to operate over a range of frequencies and power supplies. Thus units may have to be matched dynamically as well as statically.

*In general, the matching of fluid components is analogous to matching in the design of vacuum-tube and transistor circuits in electronics, which usually is accomplished by using static characteristic curves. It is also possible to approach the interconnection of fluid amplifiers in this way. Indeed, in view of the mathematical difficulties involved in describing fluid motion in complex geometries, it may provide the only feasible approach.*

Before applying characteristic curves to fluid amplifiers it is necessary to understand the general case of a fluid flowing from one black box to another. The first box may be called the *source* and the second box the *load*. Under steady flow conditions, the flow supplied from the source must equal the flow delivered to the load.

With any source of energy there is associated an impedance, limiting the flow rate at which a quantity of energy can be removed from the source. This is called the *source impedance*. Similarly, every load has an impedance that determines the flow rate at which energy can be supplied to the load. This is called the *load impedance*. The most useful parameter to describe the performance of a source or load is, therefore, its impedance. In steady-state flow, the impedance is considered as purely resistive. Resistive impedance, for incompressible flow, is defined as the total pressure drop across the source or load divided by the volume flow through it. Very often the impedance of a

(A) SOURCE SCHEMATIC

(B) TYPICAL SOURCE CHARACTERISTIC

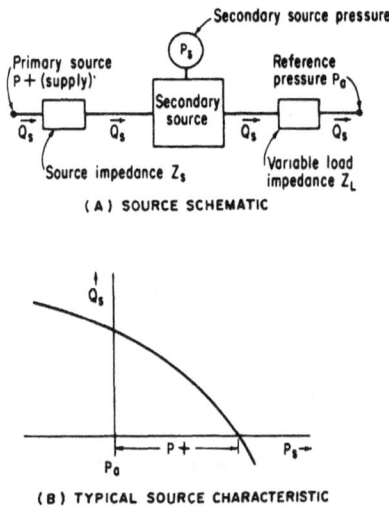

**Figure 23-3. Characteristics of the Source[1]**
*Reprinted with permission of McGraw-Hill Book Company*

fluid component is nonlinear, i.e., it is a function of the volume flow. For this and other reasons associated with the complexities of fluid motion, it is very difficult to derive an analytical expression for the resistance of a fluid component. Resistance, however, can be determined readily from the experimental characteristic curves. In discussing fluid devices it is convenient to use primary and secondary sources, as shown schematically in Fig. 23-3[1]. A primary source is the constant-pressure source or manifold which supplies fluid to the secondary source. The impedance of the primary source is zero, i.e., it can supply an unlimited amount of flow at constant pressure. The secondary source is the active fluid interaction device on which characteristics are desired. To obtain these characteristics the secondary source is connected to a variable load impedance $Z_L$. The primary source supply pressure $P+$ is connected to the secondary source through the secondary-source impedance $Z_s$. In steady flow the mass flow in the circuit is constant. When density changes are small, the volume flow $Q_s$ may also be considered constant. A typical source characteristic is shown in Fig.

23-3(B). This curve is obtained experimentally by measuring the volume flow and secondary source pressure over a range of load impedances. Each load impedance provides one point on the source characteristic curve. The complete source characteristic is the locus of these individual points over the entire range of load impedances that could possibly occur in practice. Points on this characteristic curve which fall in the first quadrant represent passive loads (resistances). The curve can be continued into the second and fourth quadrants by using active loads (pressure sources) that lower or raise the reference pressure $P_a$.

A schematic drawing of the method of measuring load impedance is shown in Fig. 23-4(A)[1]. To obtain the characteristics of the load, one must connect it to a variable source that might occur in practice. Again, for small density changes, the volume flow $Q_L$ passes from the source through the load impedance to the reference pressure $P_b$. Experimentally, the total pressure $P_L$ is measured upstream of the constant load impedance $Z_L$. A typical load characteristic is shown in Fig. 23-4(B).

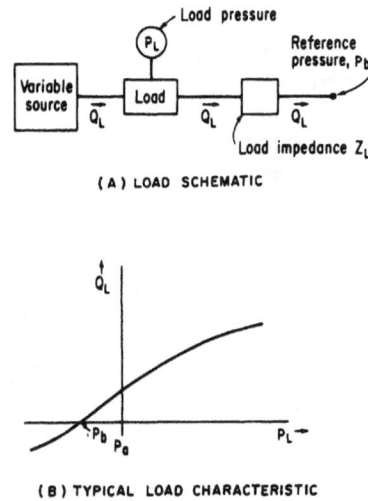

(A) LOAD SCHEMATIC

(B) TYPICAL LOAD CHARACTERISTIC

**Figure 23-4. Characteristics of the Load[1]**
*Reprinted with permission of McGraw-Hill Book Company*

23-6

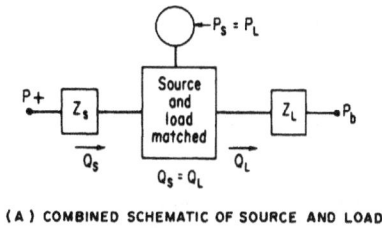

(A) COMBINED SCHEMATIC OF SOURCE AND LOAD

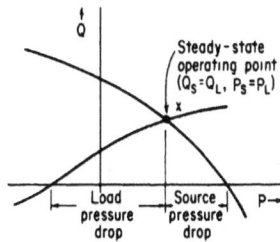

(B) MATCHING SOURCE AND LOAD CHARACTERISTIC CURVES

Figure 23-5. Matching Source and Load
Characteristics[1]
*Reprinted with permission of McGraw-Hill Book
Company*

To obtain each point on the load characteristic, it is only necessary to set the load pressure at some level by varying the source and to measure the resulting flow. When the load pressure is equal to the reference pressure, the load flow is zero. Increasing the load pressure increases the flow. If the load pressure is maintained below the reference pressure, a reverse flow will occur. Thus the curve represents the locus of possible operating points for this one particular load subjected to a wide variety of source conditions.

Fig. 23-5(A)[1] is a schematic drawing of a specific load connected to a specific source. As previously mentioned, under steady flow conditions the source must equal the load flow to maintain continuity. It is also necessary for the source pressure to equal the load pressure since they are measured at the same point in the circuit. The connection of a specific source to a specific load, therefore, results in

one particular value of flow and one particular value of pressure. Graphically this means that the source and load curves intersect at one point when they are superimposed on each other as shown in Fig. 23-5(B). This point is called the operating point. Very often, in fluid amplifiers, the load must be adjusted for a particular source, so that the operating point falls in a desirable operating region. The coordinates of the operating point shown in Fig. 23-5(B) are

$$P_x = \frac{Z_L}{Z_L + Z_s}\left(P+ + \frac{Z_s}{Z_L}P_b\right), \text{psi} \quad (23\text{-}1)$$

$$Q_x = \frac{P+ - P_o}{Z_L + Z_s}, \text{in.}^3/\text{sec} \quad (23\text{-}2)$$

where

$P_x$ = pressure at the operating point, psi

$Z_L$ = load impedance, $\dfrac{\text{lb}}{\text{in.}^2}\Big/\dfrac{\text{in.}^3}{\text{sec}}$

$Z_s$ = source impedance, $\dfrac{\text{lb}}{\text{in.}^2}\Big/\dfrac{\text{in.}^3}{\text{sec}}$

$P+$ = primary source pressure, psi

$P_b$ = reference pressure, psi

$Q_x$ = flow at the operating point, in.$^3$/sec

Thus the operating point depends upon the source and load impedances as well as the supply and reference pressures.

## 23-5 DC AND AC SYSTEMS

The concept of direct-current (DC) in electrical phenomena implies flow of current (electrons) in one direction, the amount of flow being controlled by the driving force (voltage). This concept applies equally well to the field of DC fluerics wherein the fluid flow is the current and the driving force is the pressure applied to the fluid. Since the control of air flow is in many respects similar to the control of unidirectional electric

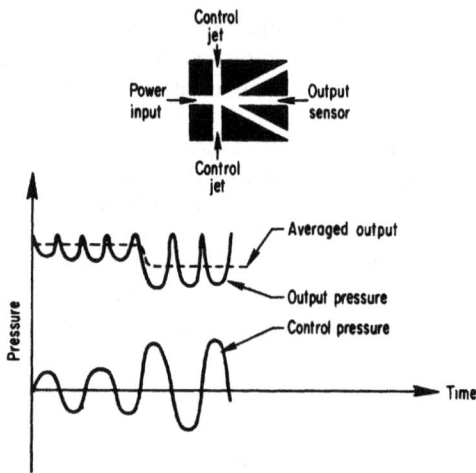

*Figure 23-6. AC Rectification Flueric Device*[3]

current, the design principles of DC fluerics have grown to parallel closely those of DC electronics. Unfortunately, some types of flueric circuits also have limitations similar to those of DC electronics. Weak signals are difficult to transmit and detect, and noise from turbulent flow in fluidic systems is even more troublesome than thermionic noise in electrical systems. To help alleviate this problem, AC flueric systems are being developed[2]. The use of the term AC applied to flueric signals may be misleading to those who carry the analogy between electrical and flueric circuits too far. A more precise description might be *undulating* or *oscillatory flow superimposed on a steady flow*. The important difference between electric and flueric AC signals is that the flueric signal always has a DC component in addition to the AC component. In some ways, however the features of the new AC fluerics often parallel those of AC electronics, and the terminology is also similar. Designers of AC fluerics systems speak of inductance, frequency modulation, phase shift, and pulse width in the same manner as do designers of electronic systems.

In AC fluerics the term signal refers to the pressure changes occurring in the circuit. These pressure variations can be either referenced to local atmospheric pressure, or to some other point in the circuit. The signal is therefore a differential pressure. Signal amplitude in an AC circuit generally ranges from 0.01 to 10 psi. More typical values are between 0.1 and 1 psi. Speculation on the ultimate frequency limit for flueric devices ranges from a conservative 10 kHz up to a rather optimistic 100 kHz.

The transmission speed for an AC flueric device is about equal to the velocity of sound in the fluid medium, or about 1000 ft/sec in air which is commonly used. The corresponding time delay is about 1 msec/ft.

The specific advantages of AC over DC flueric systems include faster response, greater accuracy, and less vulnerability to system noise. Present indications are that AC fluerics is especially well suited to the measurement and control of speed, temperature, and pressure in environments that are too extreme for conventional sensors or control devices. The present types of flueric timers which incorporate oscillators and counters are a form of AC fluerics. A widely used component that well illustrates AC flueric operation is the rectifier or absolute-value amplifier. This device has an output port centered downstream from the supply nozzle as shown in Fig. 23-6[3]. Two opposing control nozzles impinge on the power stream and direct it away from the outlet port to a degree proportional to the imbalance in the opposing control streams. When the two control pressures are equal, the power stream is centered on the outlet and the pressure output is maximum, as shown in the figure. When the differential between control pressure is highest, the beam is deflected away from the outlet by a maximum amount, and the resulting power output is a minimum. Sinusoidally varying control pressures thus flip the power stream back and forth across

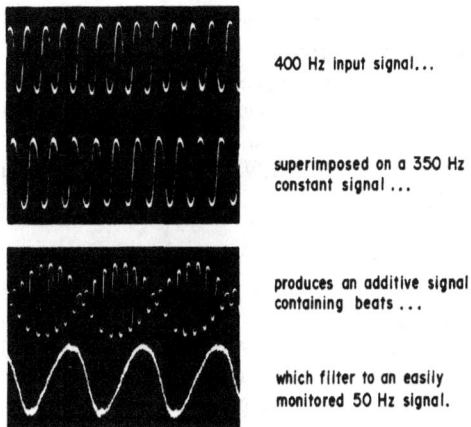

400 Hz input signal...

superimposed on a 350 Hz constant signal ...

produces an additive signal containing beats ...

which filter to an easily monitored 50 Hz signal.

*Figure 23-7. Heterodyning a Flueric Signal*[3]

the outlet port to produce output pulsations of the type shown in the figure. These pressure pulses are then usually filtered to produce an average constant DC signal having an intensity inversely proportional to the amplitude of the AC signal.

Another example of AC fluerics which has its counterpart in AC electronics is in heterodyning a flueric signal. In the wave forms shown in Fig. 23-7[3], a 400-Hz input signal is superimposed on a constant carrier frequency of 350 Hz to produce an amplitude-modulated wave that can then be filtered to produce a beat frequency of 50 Hz. This principle has been used to produce a flueric/acoustic transceiver, each unit powered by inflated balloons. In the transmitter unit the power supply air is passed through an ultrasonic whistle, and the tone generated is amplitude modulated by acoustical interaction with a human voice. At the receiver, which receives the line-of-sight signal, demodulation occurs back to audible frequencies by AC flueric devices also powered by an inflated balloon. A successful transceiver developed by H. H. Unfried of Genge Industries Inc. transmitted voice information over several hundred feet with fidelity approaching a pocket radio. The acoustic

radio, however, is very limited. It is generally known that to get high directivity, one must use high frequency flueric signals. These signals are severely attenuated by atmospheric conditions, and air currents can deflect the beam direction. One possible use, however, could be in short range communications in explosive environments where the use of electrical energy would be prohibited.

One of the major difficulties facing developers of AC flueric systems is the need to strengthen the analytical basis of the present design technology. As a result, a considerable amount of work is now done by empirical methods. The lack of analytical tools makes it extremely difficult to optimize a device, or in fact to analytically design a system to serve the particular intended function. Continued effort and work with flueric systems will alleviate this problem.

An analysis of an AC flueric circuit follows very closely the approach used in electrical-network analysis. The circuit is first subdivided into active and passive components. Active elements include amplifiers, power supplies, and auxiliary sources. Passive elements are components composed of idealized resistive, capacitive, and inductive elements. A simplified approach then assumes that the characteristics of the individual components combine in a linear manner. This simplified analysis does not completely define the real situation over all operating conditions, but does provide an adequate bench mark for initial circuit design.

For a more detailed comparison of flueric and electronic logic, see Appendix A.

## 23-6 HYBRID SYSTEMS

Hybrid systems use flueric elements together with mechanical elements (diaphragms, etc.) to make an efficient final unit. For example, output from a flueric circuit can be used to deflect a diaphragm that is connected to mechanical linkage to perform switching,

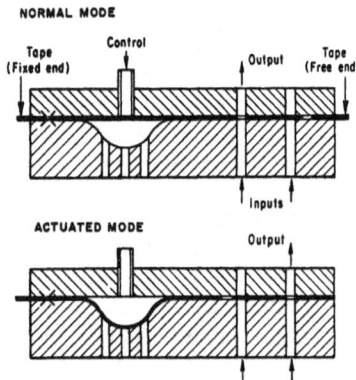

Figure 23-8. Monostable Coded-tape Valve[4]

etc. Some typical hybrid devices are:

(1) Coded tape valves

(2) Tape actuators

(3) Vibrating-reed frequency sensors

(4) Moving part detectors

(5) Flow control devices

(6) Pneumatic switches.

## 23-6.1 CODED TAPE VALVES

A coded, flexible strip can be used as a moving element in a valve, Fig. 23-8[4]. The strip is actuated by pneumatically deflecting the strip into a profiled cavity. When pressure is applied to the control port, the tape

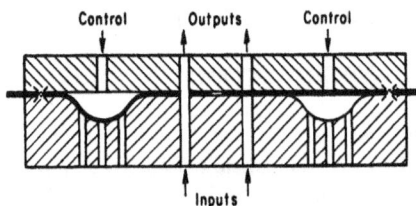

Figure 23-9. Bistable Coded-tape Valve[4]

deflects into the cavity and the free end moves in its groove. Holes in the tape are arranged so that, when the tape is deflected into the actuating cavity, the formerly open port is blocked and the opposite port is opened. Consequently, the device acts as a switch.

The vent holes in the bottom of the actuating cavity allow the fluid to escape as the tape is deflected by control pressure. Since the tape is of an elastic material, when the control signal is removed, the tape snaps back to the original position. Consequently, this device is monostable.

Between application of control pressure and completion of switching to the deflected position, some of the control signal leaks around the tape and out of the venting ports. When the tape is completely deflected, the tape seals the venting port and blocks the leakage.

The same principles also can be the basis for a bistable device, Fig. 23-9[4]. Essentially the same as the device shown in Fig. 23-8, it incorporates two actuating cavities, and both ends of the tape are clamped. The tape is just long enough to lie flat across one cavity when it is completely depressed into the other. The configuration is stable and the tape remains deflected into one cavity until a control signal is applied to the other cavity. Thus, this device can be used both as a switching device and a storage element.

The coded-tape device has two important characteristics:

(1) The element is capable of controlling many inputs. In the sketches shown, only two ports are indicated. In practice, however, many ports could be controlled by a single tape.

(2) The output is isolated completely from the input.

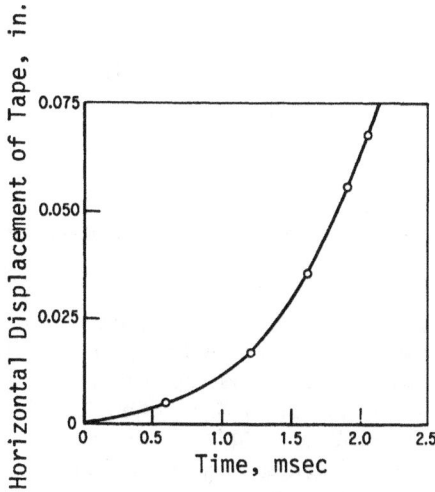

*Figure 23-10. Displacement-Time Relationship for Tape-element Switching[4]*

Although the coded-tape device has moving parts, the moving part is of light weight. Furthermore, the device is simple and inexpensive. In practice, Mylar tape has been found to be a suitable material.

Switching time of the device depends on the signal pressure applied and the mass of the flexible member. In working devices, switching times of 2 or 3 msec can be obtained. A typical time-displacement relationship for a control pressure of 20 in. water is shown in Fig. 23-10[4].

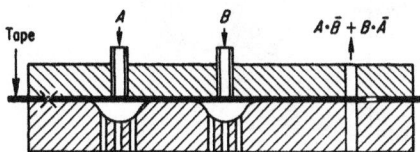

*Figure 23-11. Exclusive OR Coded-tape Element[4]*

In digital-circuitry applications, coded tape devices can generate logic functions, Fig. 23-11[4]. Here a flexible tape with two cavities is actuated by control ports, A and B. This is an exclusive OR circuit; i.e., the signal port is open when the pressure is applied to either control port A or B, but not both. A variety of other logic circuits also can be generated with devices of this type.

Digital code-translators have been made by placing several of these logic units in parallel. When so arranged, all the inputs are in parallel. As they switch, they produce parallel outputs, but in a different code.

## 23-6.2 TAPE ACTUATORS

The coded-tape device also can be used as an actuator; i.e., when the tape is deflected, the moving end of the tape can be used to apply a force rather than to switch signals, Fig. 23-12[4]. Here, the actuating cavity does not limit tape deflection.

A sketch of the forces $F$ acting on such a tape is shown in Fig. 23-13[4]. Output force is given by[4]

$$F = Pb \left( \frac{\ell^2}{8d} - \frac{\Delta}{2} - \frac{\ell\mu}{2} \right), \text{lb} \qquad (23-3)$$

where

$P$ = supply fluid pressure, psi

$b$ = width of the tape, in.

$\ell$ = span of the cavity, in.

$\mu$ = coefficient of friction between tape and cavity, dimensionless

$\Delta$ = tape deflection, in.

From this equation, as the deflection increases, the force should decrease. An experimental measurement of force vs displacement for a typical device is shown in Fig. 23-14[4].

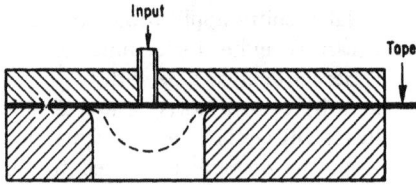

Figure 23-12. Tape Actuator[4]

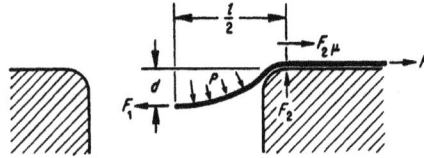

Figure 23-13. Forces Acting on Tape Segment[4]

## 23-6.3 VIBRATING REED FREQUENCY SENSORS

Vibrating reed frequency sensors may be used in applications where there is a requirement for the measure of a time-variable, repetitive, mechanical motion. A typical application of this type of sensor is shown in Fig. 23-15[4] where it is used with a wobble-plate signal generator. This signal generator has two nozzles 180 deg from each other (with reference to the shaft) which can be used as shown to provide pressure pulse outputs 180 deg out-of-phase with each other.

The outputs from the wobble-plate signal generator are directed against the sides of a vibrating reed, Fig. 23-16(A)[4], and act as out-of-phase, sinusoidal driving forces. When the reed is at rest, the paddle deflects all the flow from the supply nozzle and none of the pressure is recaptured in the output tube. As the reed vibrates, the intermittent opening between the supply nozzle and the output tube permits a portion of the supply pressure to be recaptured and used as an output signal.

When the frequency of the driving jets is not close to the resonant frequency of the reed, the reed is not deflected appreciably. However, as the driving frequency goes through the resonant frequency of the reed, the reed deflections increase and the pressure in the output tube goes through a maximum, Fig. 23-16(B).

Frequency error can be sensed by use of two reeds tuned to slightly different frequencies, Fig. 23-17(A)[4]. Pressures $P_1$ and $P_2$ are the signals from a wobble-plate generator.

Output pressures as a function of input frequency are shown in Fig. 23-17(B). The two signal pressures are applied to opposite sides of a stream interaction or other proportional amplifier. A curve of the pressure differential between the two output ports is shown in Fig. 23-17(C).

Test results on such a device, with air as the working fluid, are shown in Fig. 23-18[4].

## 23-6.4 MOVING PART DETECTORS

Mechanical motion or flow can be detected by a tightly covered foam-rubber filled cavity, Fig. 23-19[4]. An impregnated rubber-fabric diaphragm is placed across the surface of the cavity. When moving parts hit the diaphragm of the foam-filled cavity, they create an

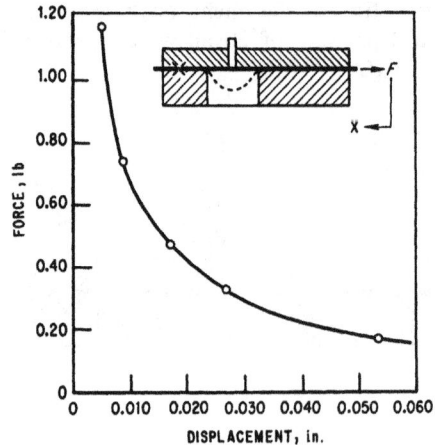

Figure 23-14. Force-Displacement Relationship for Actuator[4]

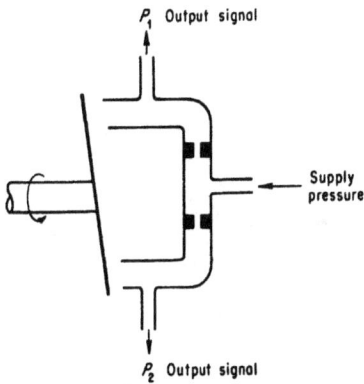

Figure 23-15. Wobble-plate Signal Generator
With Out-of-phase Signal Outputs[4]

output fluid pulse. This output pulse can be
used to actuate a pneumatically operated
switch; it can be amplified by flueric
amplifiers for other uses; it can operate flueric
counter circuits; or it can be incorporated

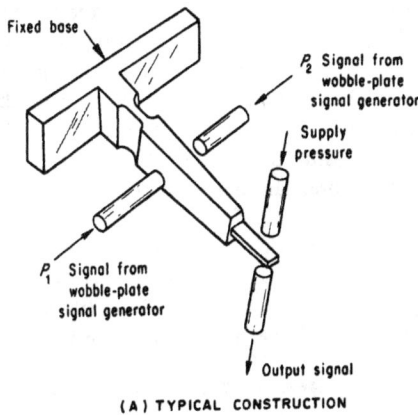

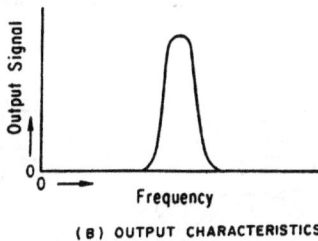

(A) TYPICAL CONSTRUCTION

(B) OUTPUT CHARACTERISTICS

Figure 23-16. Vibrating-reed Frequency
Sensor[4]

with an oscillator to determine time between
events or rate of production.

A large variation in the sensitivity of the
detector can be obtained by varying the
density of the foam. Sensitivity varies
inversely with density. Sensitivity can be
made so high that a few grams of material
striking the surface give usable output-pres-
sure pulses, or the sensitivity can be so low
that several pounds are required to give a
usable output pulse.

## 23-6.5 FLOW CONTROL DEVICES

In principle, most flow controllers are
usually variable fluid resistors whose resistance depends on some external parameters
such as pressure or temperature. These
controllers often involve moving parts and are
thus subject to wear. As a general rule of
thumb, an element that employs deformable
parts is generally more reliable and longer
lasting than an element that uses moving
parts.

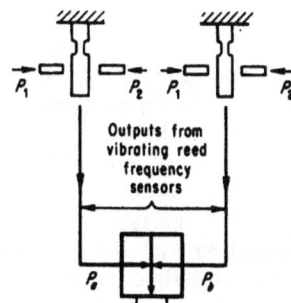

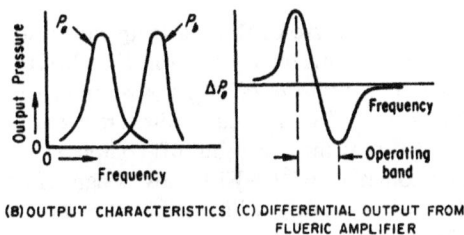

(A) SCHEMATIC ARRANGEMENT

(B) OUTPUT CHARACTERISTICS     (C) DIFFERENTIAL OUTPUT FROM
FLUERIC AMPLIFIER

Figure 23-17. Double-reed Frequency Sensor[4]

23-13

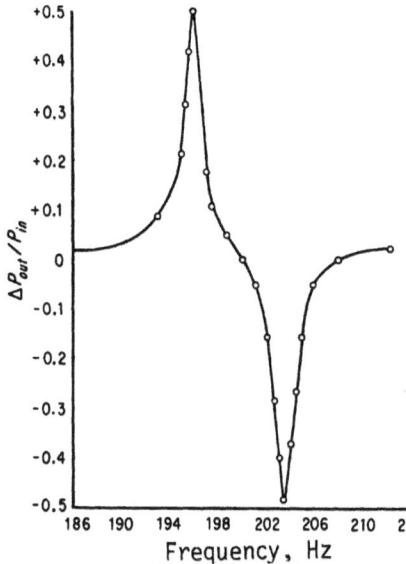

Figure 23-18. Test Results for Double-reed Frequency Sensor[4]

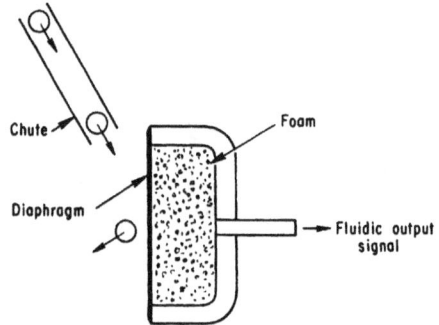

Figure 23-19. Flueric Moving-part Detector[4]

One type of temperature-controlled variable fluid resistor is shown in Fig. 23-20[4]. Silicon rubber has an unusually high coefficient of volumetric thermal expansion. As the valve body and the O-ring increase in temperature, the O-ring expands and decreases the clearance between itself and the tapered needle screw. As the clearance decreases, the fluid resistance increases. The lateral position of the tapered needle screw can be adjusted so that the valve shuts off completely at any desired temperature. Thus, as temperature is increased, resistance increases until, at some predetermined temperature, the resistance becomes infinite.

A typical curve of flow vs temperature for this type of resistor is shown in Fig. 23-21[4]. A constant pressure of 20 psi was applied to the element. The tapered needle screw was so adjusted that the valve shut off completely at approximately 65°C. With this design, fluid flow varies linearly with temperature over a relatively wide range.

A patented adjustable, laminar-flow vari-

able fluid resistor consists of two flat plates sealed by O-rings, Fig. 23-22[4]. Resistance is varied by varying the spacing between the plates.

Spacing of the plates can be controlled by temperature-responsive elements, Fig. 23-23(A)[4], when used in conjunction with the variable fluid-resistor. Here the dimensions of the inner rod and the casing vary with the temperature changes, varying the spacing between the two plates. An example of this resistor circuit is shown in Fig. 23-23(B). Output pressure depends on the resistance of the adjustable, laminar-flow restrictor. Typical relationships of output pressure and temperature for this device are shown in Fig. 23-24[4]. Both plots show significant ranges of temperature variation over which the output pressure varies linearly with temperature.

## 23-6.6 PNEUMATIC SWITCHES

Historically, pneumatically operated

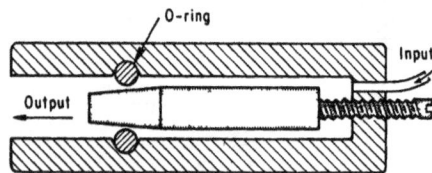

Figure 23-20. Temperature-sensitive Flueric Resistor[4]

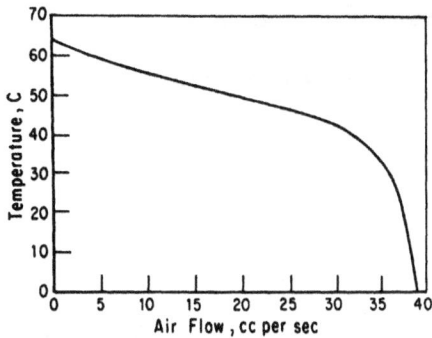

Figure 23-21. Flow-Temperature Characteristics for Resistor Shown in Fig. 23-20[4]

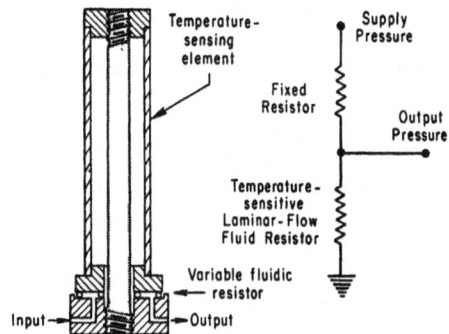

Figure 23-23. Temperature-sensitive Flueric Resistor[4]

switches have been used extensively throughout industry and have the potential for wide applications in hybrid flueric circuits. In principle, most of these switches include a membrane or other flexible part which is deflected by applied pressure to actuate electrical contacts. Some of these switches, even though insensitive to external shock and vibration, sense pressures of 0.5 in. water or less[4].

## 23-7 TEST AND EVALUATION

An initial effort in any new fleuric design problem is usually made with a breadboard setup in which various parameters of the circuit easily can be changed. At this time the use of pressure and flow meters and special test equipment will aid in the development. Once a circuit design is developed, it most likely will be scaled down to a size that will not permit further measurement. Because of this, the designer must have a thorough

knowledge of the effects of scaling down a large model to one of greatly reduced size. Furthermore, in his evaluation of the breadboard, he will most likely conduct static as well as dynamic tests. He should be aware in the dynamic tests that even a short tube, such as a pressure connector, can add a significant shunt capacity as the frequency is increased and thus alter circuit response. Due

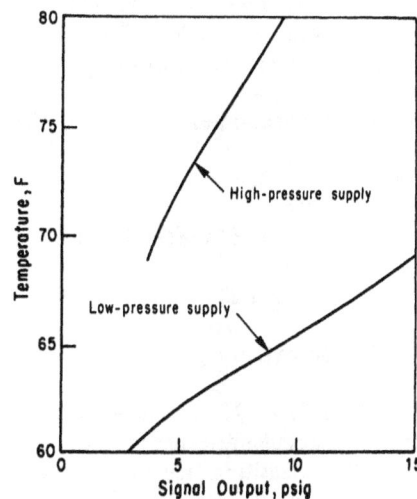

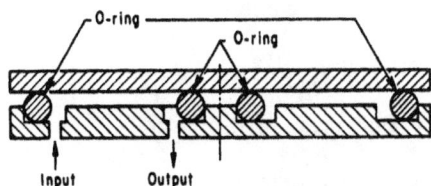

Figure 23-22. Adjustable Laminar-flow Flueric Resistor[4]

Figure 23-24. Output Pressure-Temperature Characteristics for Circuit Shown in Fig. 23-23[4]

to the nonlinear behavior of most flueric circuits, an analytical solution to system design, at the present time, has not been as practical as the cut-and-try method. As fluid technology develops, however, it is hoped that much of the work being done will be documented in the form of handbooks and textbooks which might then facilitate the design process.

## 23-7.1 FLOW METERS

Fluid flow can be recorded as volume flow per unit time, or mass flow per unit time. If the fluid is compressible, the flow is usually corrected to SCFM units (standard cubic feet per minute), generally at a reference level of 14.7 psia and 59°F[5].

There are two types of fluid flow of interest to the flueric designer: laminar flow and turbulent flow. Laminar flow is characterized by fluid particles in a tube moving in straight parallel lines down the tube. Turbulent flow is characterized by an unpredictable tumbling, churning flow as the fluid progresses down the tube. If the Reynolds number of the flow in a tube is below 2000, the flow will be laminar; if the Reynolds number of the flow is above 4000, very probably it will be turbulent. The Reynolds number $R_e$ may be determined by

$$R_e = \frac{vd}{\nu}, \text{ dimensionless} \qquad (23\text{-}4)$$

where

$v$ = mean velocity of the fluid, ft/sec

$d$ = tube diameter, ft

$\nu$ = kinematic viscosity, ft²/sec

For air, $\nu$ is $1.57 \times 10^{-4}$. The flow laws for the laminar and turbulent regions are well known, but in the transition region the flow is complex and no known law describes the fluid behavior.

An expression for Reynolds number which is more useful for flueric circuitry is

$$R_e = \frac{Q}{Q_r}, \text{ dimensionless} \qquad (23\text{-}5)$$

where

$Q$ = actual weight flow, lb/sec

$Q_r = \dfrac{\pi g \mu d}{4}$ = reference weight flow, lb/sec

$g$ = sea level acceleration due to gravity, 386.4 in./sec²

$\mu$ = fluid (dynamic) viscosity, lb-sec/in.²

$d$ = tube diameter, in.

Thus for each size of tube and for a given fluid, there is a definite reference current. If the actual fluid current through the tube is less than 2000 times this reference current, the flow is in the laminar region. If the actual current is greater than 4000 times the reference current it will be turbulent flow. Between these values is the transition region. Fig. 23-25[6] shows a plot of tube diameter versus flow for air and water constructed from Eq. 23-5. The fluid viscosities used were $\mu_{air} = 2.6 \times 10^{-9}$ and $\mu_{water} = 1.46 \times 10^{-7}$. Similar figures can be easily constructed for any working fluid if the viscosity is known.

The product of fluid pressure and flow is power

$$H = kPQ \qquad (23\text{-}6)$$

where

$H$ = power (see note)

$k$ = a factor

$P$ = fluid pressure, lb/ft²

$Q$ = flow, ft³/min

Note: When $k = 1$, power is in ft-lb/min. For

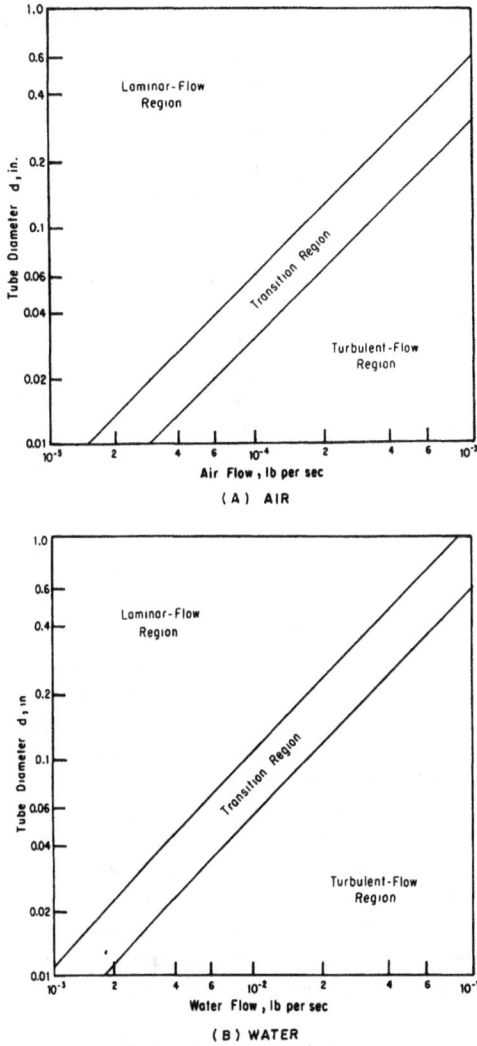

Figure 23-25. Flow Regions for Air and Water in Round Tubes[6]

Figure 23-26. Air Power Expended in Typical Rectangular Nozzles

power in watt, use $k = 0.5645$ when air is the fluid.

Fig. 23-26[5] shows the air power, in watts, of various size nozzles with rectangular cross section as the pressure and flow are varied at constant temperature. Exit is to the atmosphere. Fluid power need not be converted to

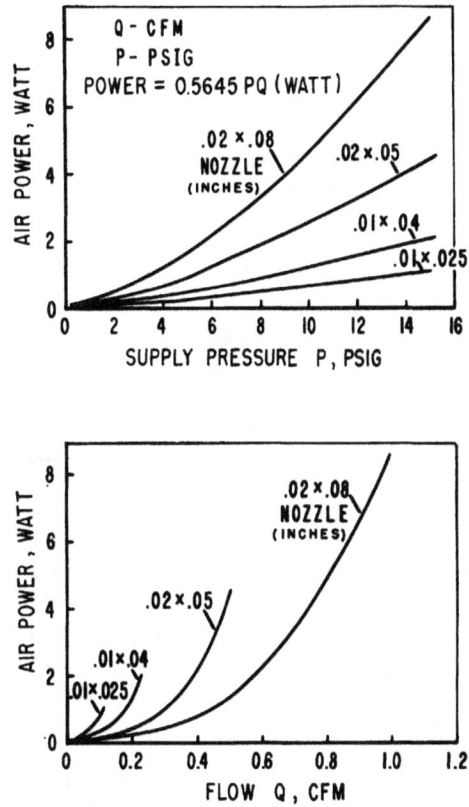

watts unless some form of electrical analogy is desired. In some types of applications the product $P_A Q$ ($P_A$ is absolute pressure), is used as a measure of fluid *current*. This term commonly is called *throughput*.

A wide variety of flow meters is commercially available. Many of them place a known fluid resistance in the flow stream and monitor the pressure drop across it. Others use rotating cups and vanes whose rotational velocity can be related to flow rate. Some of these latter types are in reality volume meters from which a $Q$ can be calculated.

Another type is an electrical flow meter

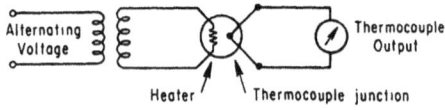

Figure 23-27. Indirectly Heated, Single-thermocouple Flow Meter[7]

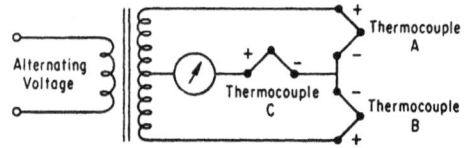

Figure 23-28. Three-thermocouple Compensated Type Flow Meter[7]

shown in Fig. 23-27[7]. In this type of a thermocouple with an adjacent heater is placed in the flow path. Output of the thermocouple can be related to flow rate; i.e., for a fixed heater current, voltage output will decrease for an increase in flow since the fluid will capture more heat from the heater-thermocouple bulb. This type must be calibrated with a given fluid and a similar fixed geometry. For a given flow, output readings will change if the fluid temperature changes. To eliminate this, a compensation thermocouple can be integrated into the circuit as shown in Fig. 23-28[7]. Thermocouple C is the compensating thermocouple. A and B are heated, C is not.

Changes in the ambient temperature of the flowing fluid affect all three thermocouples. Since the output of thermocouple C opposes that of the other two, temperature-fluctuation effects of the flowing fluid are cancelled. The compensation is so effective that no correction need be made for fluid temperatures from 40° to 200°F when measuring the flow of air. Instruments of this sort are commercially available. The circuit of Fig. 23-28, although normally used as a steady-state flow meter, can measure flow fluctuations up to 10 or 12 Hz.

## 23-7.2 VELOCITY METERS

When fluid flows in a tube, the velocity is not constant along any radial line, but usually varies from zero at the tube walls to a maximum at the centerline. The shape of the velocity profile is a function of the surface roughness, fluid viscosity, and diameter of the

tube. This applies to laminar flow. With turbulent flow, a different velocity profile exists. In many cases, with laminar or turbulent flow, the designer is interested in an average velocity that can be assumed uniform over the cross-sectional area of interest, so that flow rate can be related to this velocity and the area through which it acts.

The basic principle of the flow meters just described can be used to measure fluid velocity, but most velocity meters employ a light weight metallic film or thin wire. These are known as hot-wire or hot-film anemometers. They work on the following principle: A hot wire (or film) in a moving fluid stream loses heat energy to that stream, if the stream is at lower temperature than the wire. The amount of heat loss and resulting temperature drop of the wire depend on the temperature and velocity of the moving fluid stream. Resistance of the wire changes with the temperature. Energy loss or temperature of the wire can be monitored, and is a direct measure of the velocity of the fluid.

Due to the small mass of the sensing element (5-micron tungsten wire, 1 mm long), response is fairly fast and velocity fluctuations to 60 kHz are possible. This is far above the frequency range of interest in most flueric circuit designs. To eliminate the fragile nature of tiny wires of brittle materials, metallic film deposits on cylinders of quartz or glass are the heart of the hot-film anemometers. With the introduction of these types, fluid velocity measurements from 150 m/sec (a maximum for hot-wire types) to above 500 m/sec were realized.

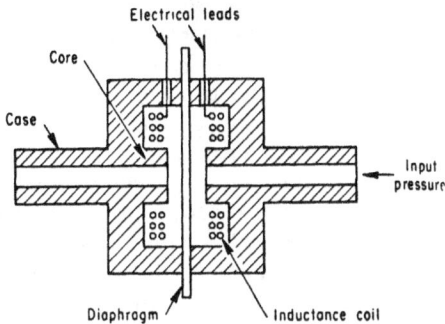

Figure 23-29. Stretched-metal Diaphragm
Pressure Meter[7]

Under conditions of thermal equilibrium, the rate of heat loss from the wire must be equal to the electrical power supplied to the wire. A useful empirical equation has been developed as follows

$$I = A + B(v)^n, \text{ A} \qquad (23-7)$$

where

$I$ = current supplied to wire or film, A

$A, B,$ and $n$ = constants

$v$ = velocity, m/sec

The values of $A$, $B$, and $n$ are determined empirically to give the best fit to the experimental data within the desired velocity range. This formula is known as King's Law. The value of $n = 0.5$ fits the hot-wire case well when the probe is operated in atmospheric air and the velocities lie between 1 and 100 m/sec. For a hot-film probe in the same velocity range, a value of $n = 0.33$ gives a better fit to the experimental data[7].

The anemometers can be instrumented such that the current through the element is held constant, then the voltage across the element is related to fluid velocity; or the current is varied such that the element always remains at a fixed temperature, thus heater

current is related to flow velocity. For a thorough discussion of the instrumentation employed with anemometers, consult Ref. 7.

## 23-7.3 PRESSURE METERS

No problem exists in the measurement of steady-state or static pressure because many gages are commercially available. These measurements can be made on extremely small devices such as those typifying flueric elements. It is in the area of measuring fluctuating pressures within small orifices and channels that problems arise, primarily because the gage must be physically small with low inertia. Since flueric devices have frequency characteristics that could extend to 1000 Hz or more, an adequate gage must have a frequency response to at least several thousand hertz. The gage should not have an acoustical resonant point within the range of interest, and its mechanical resonance frequency should be high.

Excellent piezoelectric gages with the required frequency response are available for transient pressure measurements. They are available in very small size and often can be mounted directly in a flueric element, flush with some internal wall. These gages however are high-impedance devices, and are vibration sensitive. Furthermore, they have a low frequency cut-off, and cannot be used to measure either a steady-state or a slowly changing pressure. Since the gage is a high-impedance device, it requires a preamplifier reasonably close to the gage.

In applications where a preamplifier is prohibited, an electrical pressure gage with low internal impedance is desirable. Such a gage, using a stretched metal diaphragm as the pressure-sensing element, has been developed, as shown in Fig. 23-29[7]. This gage is small and can be mounted in a very small volume. However, it must be connected by a tube to the region in which the pressure is to be sensed. In high-frequency measurements, correction must be made for capacitive effects of the tube. A typical calibration curve is

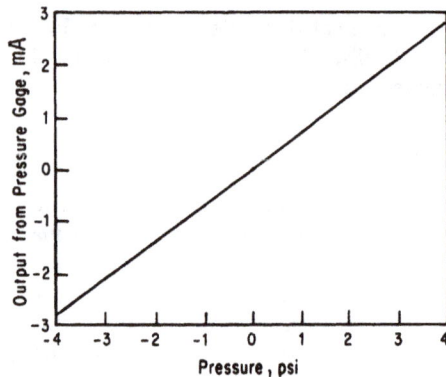

Figure 23-30. Calibration Curve of a Typical
Pressure Gage[7]

shown in Fig. 23-30[7]. This plot shows that the linearity over the indicated pressure range is very good.

A recently developed pressure meter uses a heated filament mounted inside a very small ceramic tube. One end of the tube is attached to a reference pressure while the other end is connected to the pressure to be monitored. The ceramic tube is sized so that the flow velocity through it is a direct function of the difference between the pressure being monitored and the reference pressure. The hot filament measures the velocity, which in turn, gives a measure of the pressure difference.

The reference pressure is set slightly lower or slightly higher than the extreme of pressure fluctuation expected in the monitored source. One advantage of holding the reference pressure higher than the pressure being measured is that gas is bled into the system being monitored rather than out of it. Since the reference pressure can be filtered, dirt is kept out of the pickup tube. Flow from the reference also can be kept at a constant temperature. If gas is bled into the system being measured, pressure measurements will not be affected by fluctuations in the temperature of the measured source.

A typical inside diameter for the ceramic tube is 0.002 in. This diameter is so small that very little disturbance is introduced into the region where pressure is being measured. In applications where this small amount is important, another tube and a valve can be installed adjacent to the pressure-measuring tube, and adjusted so that as much fluid is bled off by the second as is introduced by the first. Frequency response of this type pressure gage, which is comparable with that of the hot-wire anemometer, extends into the kilohertz range.

## 23-7.4 FLOW VISUALIZATION

In device development, knowledge of the actual flow paths of the fluid through the devices is often useful. These paths can be determined by water-flow tables, smoke utilization, and optical techniques.

### 23-7.4.1 Water Table

In this technique, the device to be studied is built to a scale, usually many times its actual size, and water is pumped through it. The flow paths of the water give an indication of the flow paths through the actual device. Dye usually is added to the water supply, and the flow paths may be photographed for permanent record as the dyed water proceeds through the element under test. Since the fluid behavior of the free surface of water and air are similar, a knowledge of the behavior of the typical flueric device in the water table may be used as the basis for the scaled-down design of a device in another fluid. Water tables are useful in new element design because the geometry of the element can be changed readily. For example, the shape of the channel walls can be changed easily since they usually are made from a soft malleable material such as lead.

Water tables are expensive to construct. A less expensive method is to pass smoke through a transparent plastic model of the device under study.

### 23-7.4.2 Smoke

When smoke is introduced in the input nozzles of a transparent model, the flow patterns of the smoke, and hence the flow pattern of the fluid within the device, readily are seen. Smoke can be made by vaporizing kerosene in a smoke generator producing a smoke white in color. With time, however, the kerosene droplets condense and collect on the sides of the device under test. In most instances, this collection is not enough to disturb the experiment. Commercially available smoke-generating machines produce a pure white smoke. In some of these, the smoke is generated by oxidizing mineral oil and mixing it with carbon dioxide.

### 23-7.4.3 Optical Techniques

One of the most common optical techniques is the shadowgraph. It is the simplest and cheapest to employ for flow visualization. With this method, light is passed through a transparent model normal to the fluid flow impinging on a photographic plate on the opposite side. The resultant pictures represent the rate of change of the density gradient in space normal to the transmitted light. Since this method represents the second derivative of density, it is insensitive to uniform density gradients. Consequently, its usefulness for visualization is restricted to suddenly changing conditions (such as in turbulence or shock fronts in supersonic flow); therefore, quantitative analysis seldom is made with this technique.

Another optical technique, the Schlieren method (based on Toepler's theory), requires more costly and precise equipment than the shadowgraph. With this method, the resultant picture represents the density gradient and furnishes detail in more "gentle" flow than the shadowgraph. It generally is considered the most sensitive of all optical methods, and thereby offers much use for visualization. Care must be taken, however, in interpreting the flow characteristics, in that a single picture of a density gradient can be misleading unless proper analysis be made. Quantitative analysis becomes more practical for Schlieren methods than the shadowgraph, especially if the flow is two dimensional.

Another optical technique, the interferometer (Mach-Zehnder type), is the most expensive, most precise, and most difficult to adjust. This method is considered favorably because of its ability to record density directly; therefore, quantitative analysis becomes relatively simple for two-dimensional problems.

In consideration of all factors involved, the Schlieren method has received the greatest application. A relatively simple design of this system can be constructed, capable of permitting (1) wide-range sensitivity, (2) the use of color to enhance contrast, and (3) simple adaptation to a shadowgraph or an interferometer of the polarizing or diffraction grating type. In addition, the Schlieren system is more compatible for the restrictions imposed upon the field of coverage and the mechanical requirements of the test section than the interferometer. Also, a preliminary system can be set up using readily available optics such as photographic lenses. A thorough description of these methods is given in Refs. 8 and 9.

### 23-7.5 FREQUENCY METERS

Measurement of fluid oscillations can be accomplished with a hot-wire anemometer or a sensitive pressure transducer. The hot-wire anemometer has the advantages of very small size, and thus minimum obstruction to flow, and has excellent response. Unfortunately, considerable care must be used in handling these anemometers. Calibration is time-consuming, and the output is a fourth power function.

### 23-7.6 SIGNAL GENERATORS

A low-frequency signal generator can be

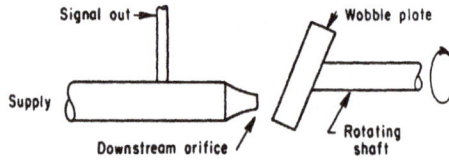

Figure 23-31. Wobble-plate Type Signal Generator[7]

obtained from a variation of the flapper valve; a wobble-plate attached to a rotating shaft, as shown in Fig. 23-31[7]. Shaft rotation varies the distance between the surface of the flapper, or wobble-plate, and the downstream orifice. This changes the resistance of the orifice and the pressure within the volume.

The lower frequency limit of this type of signal generator is zero, or a steady-state pressure. The upper limit depends on wobble-plate speed and the frequency characteristics of the flapper valve itself. Response of the flapper valve drops off with frequency due to the reactive capacitance of the volume and the resistances of the upstream and downstream orifices.

Since the simple wobble plate is dynamically unbalanced, it produces undesirable mechanical vibrations. This fact limits the upper frequency available from this type signal generator. This limitation can be overcome partially by use of a circular cam Fig. 23-32[7]. Since the cam can be balanced

dynamically, higher speeds can be obtained. A variety of output waveforms can be obtained by varying the cam shape.

An electrically-driven flapper provides for easily changeable output wave shapes and higher frequencies. The flapper part of the valve is attached to the voice cone of a loudspeaker, Fig. 23-33[7]. Since the speaker cone can be driven from an electrical-signal generator, a wide variety of output-wave shapes can be obtained by varying the shape of the electrical-input signal. This type of signal generator is very versatile and useful in laboratory development work.

Pressure oscillations can also be provided by piezoelectric elements. In this type of signal generator, the elements are mounted in a cavity. When excited electrically, the elements distort and create a pumping action. This pumping action excites the chamber, and a tube connected to the chamber produces a varying pressure output. These generators can produce pressure oscillations from 20 Hz to between 5,000 and 10,000 Hz. The upper end of this spectrum is well beyond the requirements of most flueric work. The pressure variations available from piezoelectric generators are small, at least in the pressure ranges of interest in flueric work. The variation in output is a function of the electrical power applied, the bias-pressure level (the supply pressure applied to the load), and the volume of the attached load.

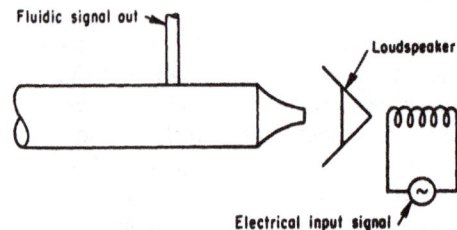

Figure 23-32. Circular Cam Signal Generator[7]

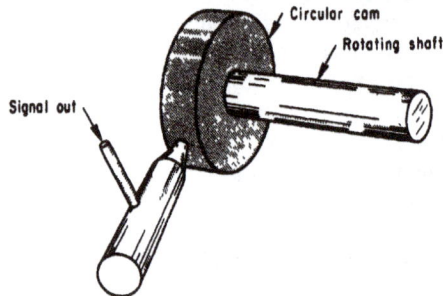

Figure 23-33. Loudspeaker-actuated Signal Generator[7]

# REFERENCES

1. J.M. Kirshner, Ed., *Fluid Amplifiers,* McGraw-Hill, Inc., N.Y., 1966, pp. 258-272.

2. "Conference Digest, A-C Fluidics", Machine Design, **40**, 163-74 (November 21, 1968).

3. Ronald Khol, "A-C Fluidics", Machine Design, **41**, 126-30 (February 6, 1969).

4. D.L. Letham, "Fluidic System Design, 18. Hybrid Devices", Machine Design, **39**, 231-5 (February 16, 1967).

5. *Fluidics System Design Guide,* Fluidonics Division of the Imperial Eastman Corp., Chicago, Ill., 1966.

6. D.L. Letham, "Fluidic System Design, 2. Analysis of Fluid Flow", Machine Design, **38**, 128-35 (March 3, 1966).

7. D.L. Letham, "Fluidic System Design, 17. Test Equipment", Machine Design, **39**, 142-8 (February 2, 1967).

8. J.R. Keto, *Fluid Amplification 2. Flow Vizualization-Compressible Fluids,* U S Army Harry Diamond Laboratories, Washington, D.C., Report TR-1041, 20 August 1962 (AD-286 666).

9. A.S. Dubovik, *Photographic Recording of High-Speed Processes,* NASA Technical Translation, NASA, Washington, D.C., TT-F-377, November 1965 (N66-11258).

# CHAPTER 24

## COMPONENT DESIGN

The timer package is composed of a number of individual elements connected together in a modular fashion to fit within a specified volume. For example, present flueric timers have as components a power source, a stable oscillator, a counter circuit, and an output transducer. The operation of each of these as an entity must be thoroughly understood as well as their operation when integrated in the timer assembly. The paragraphs that follow discuss some of these elements with the emphasis on their use in a military timers or timing circuits. Standard terminology and symbols are used[1].

### 24-1 POWER SUPPLIES

All flueric elements require a continuous flow of fluid through them during the period of time that they are functional. Thus the source of power for a flueric timer becomes a most basic component and should be considered carefully before a supply source is decided upon. Also present devices operate mainly on "flow" rather than "pressure". To achieve isolation between elements, "venting" or "bleeding" is employed. These conditions necessitate an excess of supply fluid.

It is completely feasible to use almost any type fluid as a source supply for a flueric device. Some examples of tested fluids are air, gases, water, oil, and process fluids. At present, air is used more widely than any of the others.

There are primarily three distinct ways of obtaining a power source:

(1) Use of environmental fluids (ram air, water, etc.)

(2) Packaged supplies

(3) Evaporation of liquids, sublimation of solids, and gas liberated from chemical reactions.

Ram supply appears to be a logical choice for timer circuits for missiles and projectiles, and for objects which have a relative motion with respect to their environment. In military projectile applications, the utilization of ram air has certain advantages and disadvantages, namely:

(1) Advantages:

(a) Inherent safety because the system cannot be activated until the projectile is airborne.

(b) No internal power supplies needed; therefore, a savings in weight, volume, and cost.

(c) Unlimited supply of source fluid.

(2) Disadvantages:

(a) Contaminants in ram air (dirt, water, ice) can make the system inoperative.

(b) Filtering to eliminate contaminants causes an increase of weight, volume, and cost.

Packaged supplies, on the other hand, offer several desirable features, and are being used in missile timers, primarily because the timer can now be a complete package, independent of environment and motion with environment. One serious drawback is its limited supply. If we consider bottled supply fluids, we can cite the following disadvantages:

(1) Additional weight, volume, and cost

(2) Limited fluid supply

(3) Pressure variation

(4) Supply leakage intolerable.

A summary of the advantages and disadvantages of environmental and bottled gas types of power supply is shown in Table 24-1.

Power supplies in the third category — evaporation of liquids, etc. — are primarily potential ways of generating a fluid but at present have not been employed in any military timer application.

### 24-1.1 RAM AIR POWER

The fluid streaming past a missile, projectile, or vehicle which moves through the atmosphere or water provides a direct source of power without transduction for flueric systems. The movement must be fast enough so that there is sufficient ram pressure to

operate a flueric system. The stagnation pressure is obtained from Fig. 24-1 from altitude, air speed, and air temperature data as follows:

(1) Read the stagnation pressure in standard atmospheres from the graph for a certain air speed and air temperature (use the standard temperature from the table above 35,000 ft).

(2) Multiply the stagnation pressure in standard atmospheres by the standard atmospheric pressure in pounds per square inch corresponding to the proper altitude from the table.

Effective velocities can be quite low for specifically designed systems. US Army Harry Diamond Laboratories in cooperation with Army Aviation Laboratories has designed a flueric flow separation indicator which is a stall sensor for a wing. The device indicates flow separation down to 35 kt, with effective pressures at the instrument as low as 2 in. of water.

Ram air power supplies have many advantages for fluerics:

**TABLE 24-1**

**COMPARISON OF FLUID POWER SUPPLIES**

| Environmental Supply | | Bottled Gas Supply | |
|---|---|---|---|
| **Advantages** | **Disadvantages** | **Advantages** | **Disadvantages** |
| Less weight, size, and cost | Contaminants from environment can clog system | Pure source fluid | Additional weight, volume, and cost |
| Inherent fail safe | Pump required for stationary systems | Operate in any environment | Limited fluid supply |
| Unlimited supply | Filters required | Tailored for precision circuits | Supply leakage intolerable |
| Any fluid, liquid or gas | | | Pressure regulation required |
| | | | Limited to gases |

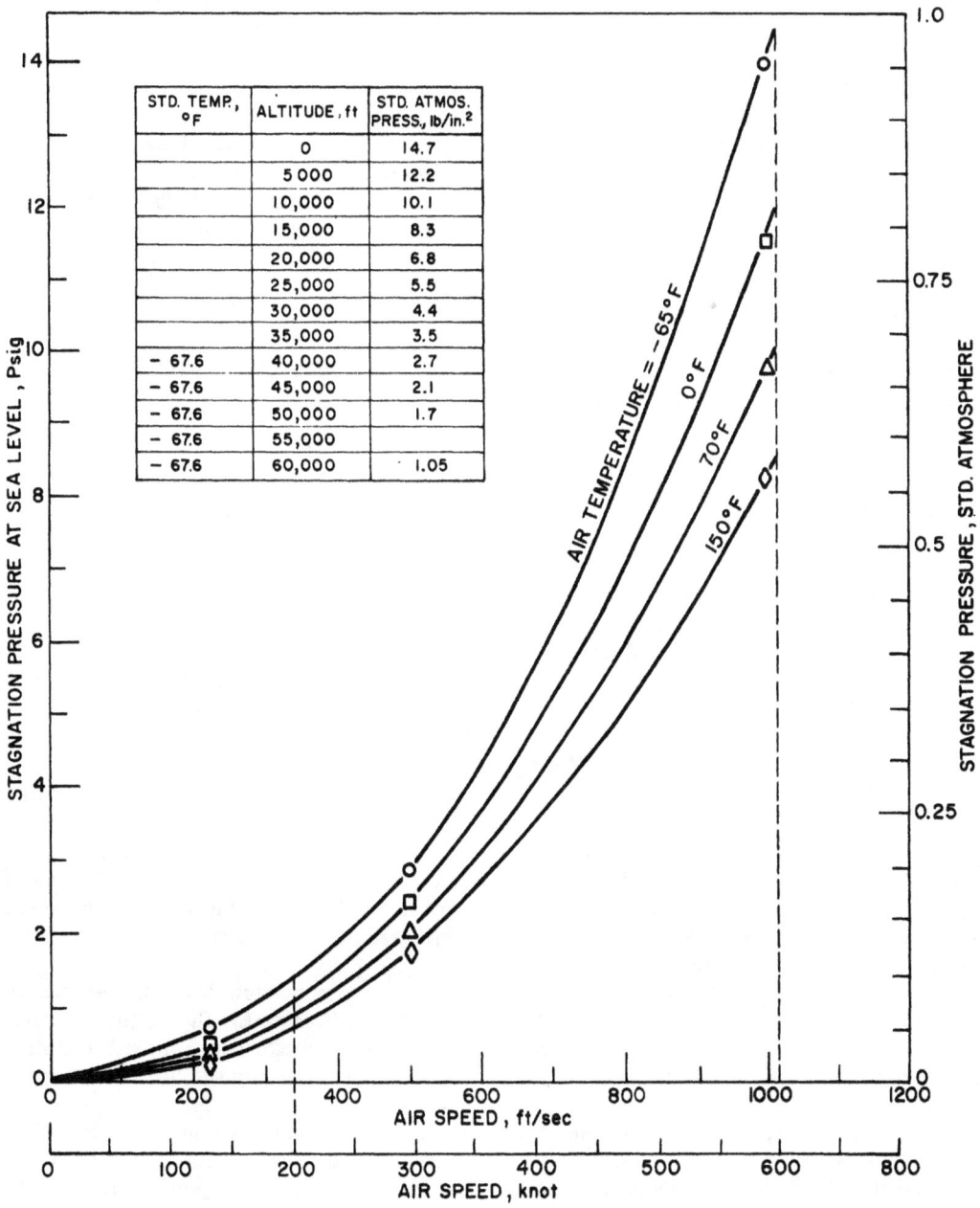

*Figure 24-1. Relation of Altitude and Air Speed to Pressure*

(1) Safety. There is no power until the projectile achieves a predetermined velocity. It is completely inert during storage.

(2) Reliability. For example, the power supply unit can be sealed until it is fired in a projectile. When the projectile is fired, the seal is automatically opened. Before ram air is used, rain drops and particles can be removed by a centrifugal separator.

(3) Less weight and space. No fuel supply or high-pressure tanks are needed.

(4) Less pressure regulation. At high velocities, the flow can be choked to limit pressure recovery.

(5) Long term storage. Completely inert in storage.

Electrical power also can be supplied by a frueric generator which is driven by ram air. The generator has no rotating parts. However, it does have a diaphragm which flexes at a very low stress level. Ram air also can be used to drive a pump[2].

## 24-1.2 PACKAGED PNEUMATIC POWER SUPPLIES

Packaged supplies are usually metallic containers with auxiliary equipment—such as relief valves, filters, and regulators—from which a gas emerges to operate downstream flueric elements. The gas within the package may be generated from the combustion of solid or liquid propellants, or a bottle may have been charged with an inert gas such as air, nitrogen, or argon at a high pressure.

The types using the combustion of propellants usually are classified as steady-state supplies. With these, ideally, one wishes to supply a quantity of gas at a fixed pressure (no change with time) and temperature. By virtue of a chemical reaction, the gas temperatures of steady-state supplies are high.

This fact necessitates flame resistant containers, and heat and erosion resistant downstream hardware.

Containers that are charged initially with an inert gas under pressure are classified as transient supplies. With these a fall-off of pressure with time occurs. However, with transient supplies the designer can almost guarantee a particle-free fluid at ambient temperature, and filtering may be unnecessary.

## 24-1.3 STEADY-STATE SUPPLIES

### 24-1.3.1 General

The steady-state supplies (gas generators) are devices producing gas in a controlled manner[3]. The supply may be considered as either a constant-rate or constant-pressure source.

The prime advantage of this type of gas generator lies in its ability to deliver a maximum of power for a given weight and volume. The gas in most cases is generated by the reaction of a pyrotechnic mixture. Efficiency is maximized when the gases can be used at high temperatures. This advantage in the gas generator may be a disadvantage so far as the flueric timer is concerned because it makes it more expensive.

In some applications lower temperature gas is available because the flueric circuits can be operated by waste gas from other systems such as a servo control system.

Of prime importance in the design of gas generators is provision for furnishing the gas at the time, rate, and pressure required for operation of the flueric system. This provision is met in most cases by control of the shape, size, and weight of the propellant grain used to supply the gas.

## 24-1.3.2 Solid-propellant Gas Generators

Solid-propellant gas generators occupy little volume, are light in weight, can be controlled easily for quick initiation or termination of the generating action, and can supply gas at steady-state conditions. The properties of the propellant to be considered are:

(1) Burning rate of the propellant

(2) Equilibrium combustion pressure

(3) Temperature dependency of burning rate and equilibrium combustion pressures

(4) Density

(5) Characteristic flame temperature.

The most commonly used types of propellants are double-base or composite. Nitrocellulose and nitroglycerin are components of the double-base propellants. The composite types contain an oxidizer and a reducing agent or fuel. Ammonium nitrate in a butyl or acetate binder is an example of a composite propellant.

Both double-base and composite propellants are presently used as gas generators, but are not considered ideal materials. Composite type propellants provide the lowest burning rates and are cooler burning. As disadvantages, composite propellants require pressure regulation, burning rates are dependent on temperature, ignition is difficult at low pressure, and gases may contain solid material.

Nitrocellulose and nitroglycerin have higher burning rates, and also higher burning temperatures. Unlike the composite materials, the burning rates of double-base propellants are less dependent upon grain temperature.

Fig. 24-2[4] shows a gas generator in its simplest form. It consists of a section of

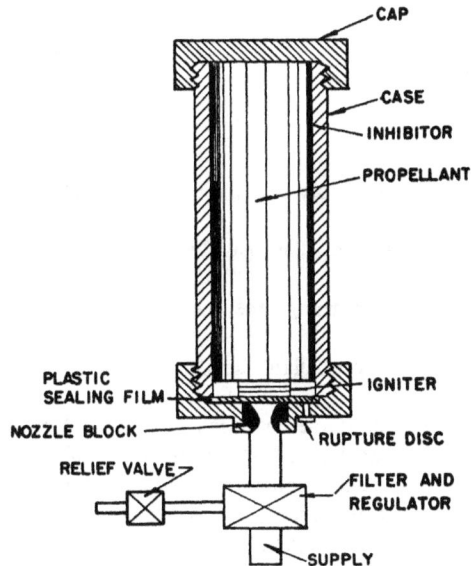

Figure 24-2. Basic Gas Generator

cylindrical casing, capped at both ends, and contains a single grain of propellant which completely fills the case. Gas is allowed to emerge through a nozzle in one end after the grain is initiated by an igniter. Various auxiliary equipment such as a filter, regulator, and relief valve are needed to operate the unit successfully.

With restricted, end-burning grains the supply time is a direct function of grain length and burning rate. The amount of gas generated is directly proportional to burning rate, burning surface area, and density of the propellant. The mass flow rate of the unit is a function of the exit orifice, chamber pressure, and characteristic flame temperature. In addition, the chamber pressure is a function of the ratio of burning area to exit area. In designing a gas generator, the interdependence of these variables should be considered to provide a satisfactory unit.

The volume for steady-state supplies can be expressed in terms of the mass flow rate, propellant density, and operating time as

$$V = \frac{w_t t}{\rho_p}, \text{in.}^3 \qquad (24\text{-}1)$$

where

$V$ = supply container volume, in.$^3$

$w_t$ = propellant gas flow rate, lb/sec

$t$ = burning time, sec

$\rho_p$ = propellant density, lb/in.$^3$

The container volume in cubic feet is plotted in Fig. 24-3[4], using a $\rho_p$ of 0.053 lb/in.$^3$, against required flow $w_t t$ in pounds. From this figure it is seen that the steady-state supplies occupy smaller volumes than do the transient or bottled gas supplies when the flow rate is equal.

### 24-1.3.3 Chemical Gas Generators

Gases may be produced from chemical reactions or chemical decomposition of liquid oxidizers and reductants. The gaseous products may be complex mixtures of CO, $H_2O$, $N_2$, $CO_2$, and many other gases. Some examples of combination fuels are nitromethane, nitric acid-aniline, and hydrazine hydrate-hydrogen peroxide. Nitromethane and other monopropellants can be decomposed into hot gases by increasing the temperature and pressure of the reaction. Nitric acid-aniline and hydrazine hydrate-hydrogen perioxide are self-igniting (hypergolic). Other generators depend upon chemical decomposition of compounds through the use of catalysts, such as sodium permanganate, calcium permanganate, and potassium chromate to produce hot gases.

Gas generation by chemical decomposition presents several problems. Fuels and oxidizers usually are highly reactive and extreme care must be taken in storage, transfer, and reaction initiation. The chemical compounds have rather short shelf lives and therefore will not conform to the usual military requirements.

### 24-1.4 TRANSIENT SUPPLIES (PRESSURIZED BOTTLES)

Bottled supplies are the most reliable and simplest of packaged power supplies. Some of the factors affecting the selection of a gas for these supplies are:

(1) Safety in handling

(2) Toxicity

(3) Critical temperatures and pressures

(4) Purity

(5) Corrosive properties

(6) Noncondensible content

(7) Dew point

(8) Availability of the gas.

The most common gases selected are air, argon, and nitrogen. Because of leakage problems, compressed gas bottles should be charged shortly before use. Leakage rates of bottles produced to commercial tolerances may be too high for long-term storage. For a given minimum mass flow it is known that compressed gas systems require greater volume and weight for the tanks, as compared with other packaged systems. Both mass flow rates and storage-tank pressures decrease with time after initial use. Therefore, the bottle must carry a greater supply than is necessary to provide the minimum mass flow.

With transient supplies, the volume required to supply gas for a given time, pressure ratio, and flow rate is given as[5].

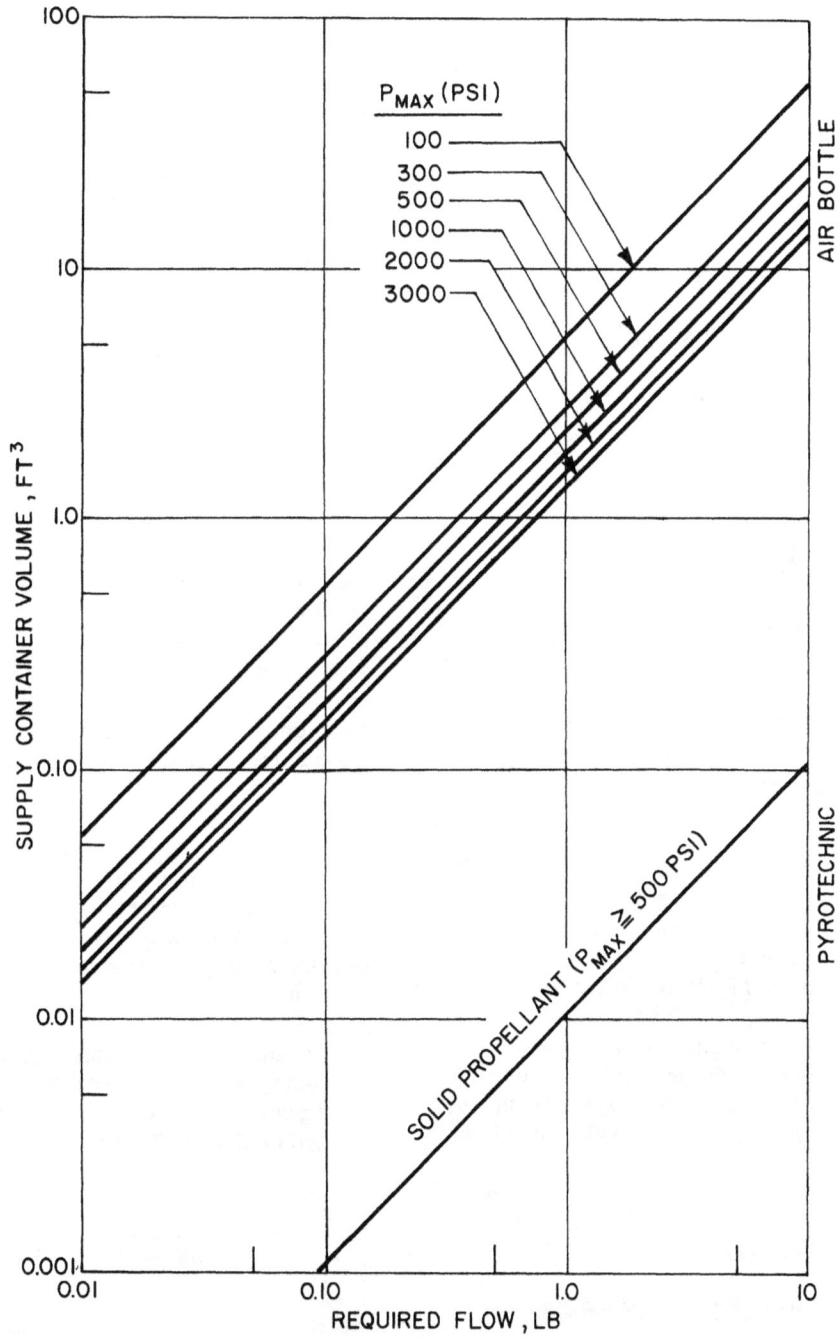

Figure 24-3. Container Volume vs Required Flow for Air Bottle
and Gas Generator[16]

$$V = \frac{12 \, R T_o \, w_t t}{P_t \ln \left( \dfrac{P_{max}}{P_t} \right)} \; , \text{in.}^3 \qquad (24\text{-}2)$$

where

$V$ = supply container volume, in.$^3$

$R$ = universal gas constant, ft-lb/lb-°R

$T_o$ = gas temperature, °R

$t$ = time required to supply the gas, sec

$w_t$ = required flow rate, lb/sec

$P_t$ = minimum pressure to supply a flow of $w_t$, psi

$P_{max}$ = initial tank pressure, psi

Maximum flow will occur when the velocity at the throat is Mach 1. This will take place when the pressure ratio is equal to or less than critical.

If air at 70°F is the fluid considered and $P_t$ is taken as 30 psi (to guarantee supersonic flow since vent is to the atmosphere), the container volumes for various $P_{max}$ versus required flow $w_t t$ is plotted in Fig. 24-3.

Usually the minimum mass flow rate $w_t$ is specified along with the time duration $t$ to supply this flow. Since space is at a premium, the smallest practical volume should be considered. This will necessitate using tanks charged to high pressures, and usually supersonic rather than subsonic flow will occur in the supply nozzle.

## 24-1.5 STORAGE CONTAINER DESIGN

In the interest of safety, foremost in a designer's mind should be the ability of the container to withstand the maximum pressure without rupture and the incorporation of a relief valve in case the pressure inadvertently is exceeded. Safety factors between 1.5 and 4 usually are incorporated, and a minimum wall thickness of 0.1 in. is specified. Steady-state supply containers should be fabricated from a good grade of stainless steel that is flame resistant, but transient containers have been made successfully from mild carbon steel. Typical design shapes for containers are shown in Fig. 24-4[4]. This container is somewhat idealized to aid analysis, and is composed of a cylindrical section with two end caps of hemispherical shape. The ratio $\ell/d$ (length-to-diameter) is a useful quantity and is denoted by $S$. When $S$ equals 1 the container reduces to a sphere. The container may be analyzed in terms of a uniform maximum principal stress (tension in hoop or hemisphere) or on the basis of a uniform wall thickness. If the uniform stress is used, theory shows that the cylindrical section thickness always should be twice as great as the spherical section thickness. Junction stresses are neglected, assuming smooth transition from one thickness to the other.

The container external volume, thickness, and weight as a function of $S$ for each design type can be expressed as shown on Table 24-2[4]. If a wall thickness less than 0.1 in. is calculated, the designer should use a thickness of 0.1 in.

For the uniform stress condition, by combining and rearranging the volume and weight equations, a weight to volume ratio as a function of $S$ can be expressed as

$$\frac{W_c}{V} = k \left( \frac{6S - 3}{3S - 1} \right) \qquad (24\text{-}3)$$

where

$$k = \frac{P_{max} \, \rho_c}{\sigma}$$

and other terms are defined in Table 24-2.

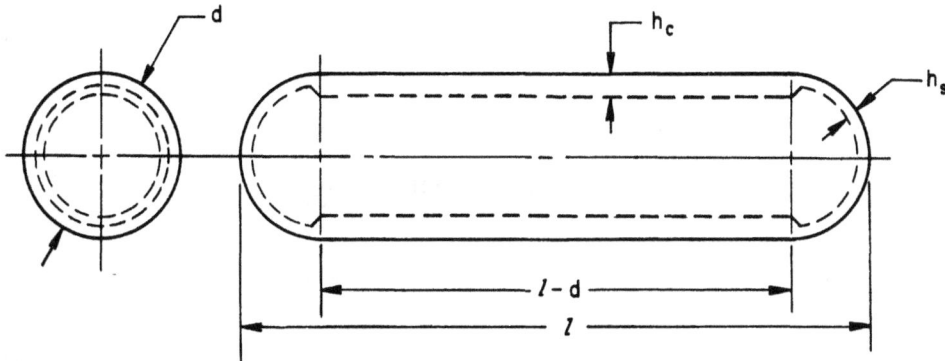

(A) UNIFORM STRESS DESIGN

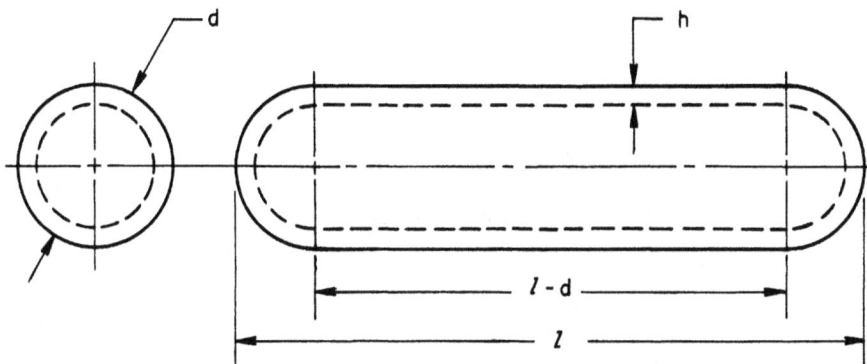

(B) UNIFORM THICKNESS DESIGN

*Figure 24-4. Two Examples of Idealized Container Designs[1][6]*

Eq. 24-3 shows that the best shape is a sphere for which $S = 1$ (since $S$ can never be less than 1), and $W_c/V = 1.5\ k$. This gives the minimum weight to volume ratio. As $S$ is increased from 1 to $\infty$, $W_c/V$ increases from $1.5\ k$ to a maximum of $2.0\ k$.

For the uniform thickness design

$$\frac{W_c}{V} = k\ \left(\frac{6S}{3S - 1}\right) \qquad (24\text{-}4)$$

It is seen that this expression approaches 2 $k$ as $S$ approaches infinity, but remains greater than $2\ k$ for practical values of $S$. Since the container thickness $h_c$ was used in the development of the uniform thickness design, Eq. 24-3 is not valid for $S = 1$. At $S = 1$, $W_c/V$ should equal $1.5\ k$ as was true for the uniform stress design. Eqs. 24-3 and 24-4 are plotted in Fig. 24-5[4]. The dotted curve for the uniform thickness design should be used for $S$ between about 1.5 and 1. For values of $S$

### TABLE 24-2

### EQUATIONS FOR CONTAINER DESIGN

**Uniform Stress ($h_c = 2h_s$)**    **Uniform Thickness ($h_c - h_s$)**

$$V \frac{\pi d^3}{12}(3S - 1) \qquad\qquad V = \frac{\pi d^3}{12}(3S - 1)$$

$$h_c = \frac{P_{max}d}{2\sigma} \qquad\qquad h = \frac{P_{max}d}{2\sigma}$$

$$h_s = \frac{P_{max}d}{4\sigma}$$

$$W_c = \frac{P_{max}\rho_c \pi d^3}{4\sigma}(2S - 1) \qquad W_c = \frac{P_{max}\rho_c \pi d^3 S}{2\sigma}$$

where

$V$ = external container volume, in.$^3$

$d$ = external container diameter, in.

$S$ = $\ell/d$ ratio

$\ell$ = total length, in.

$W_c$ = container weight, lb

$P_{max}$ = maximum expected internal pressure, psi

$\rho_c$ = material density, lb/in.$^3$

$\sigma$ = maximum allowable stress in tension, psi (includes any design safety factors)

$h_c$ = cylindrical section thickness, in.

$h_s$ = hemispherical section thickness, in.

$h$ = thickness of uniform design, in.

Note: If $h_c$ or $h_s$ is < .01 in., use $h = 0.1$ for minimum design.

greater than 5, the difference in the weight to volume ratio for the two curves does not exceed 10 percent of their mean value.

## 24-2 OSCILLATORS

The base oscillator in a flueric timer is of utmost importance since it determines the precision of the overall timer. Ideally the oscillator frequency must not be sensitive to the variations in temperature or supply pressure and it must function in the following environments:

(1) −65° to +160°F (225 deg F spread)

(2) Spin up to 30,000 rpm

Oscillators have been constructed using four different sources of oscillations: (1) feedback, (2) torsion bar, (3) sonic, and (4) half-adders. Each of these oscillators is discussed. See also Ref. 6.

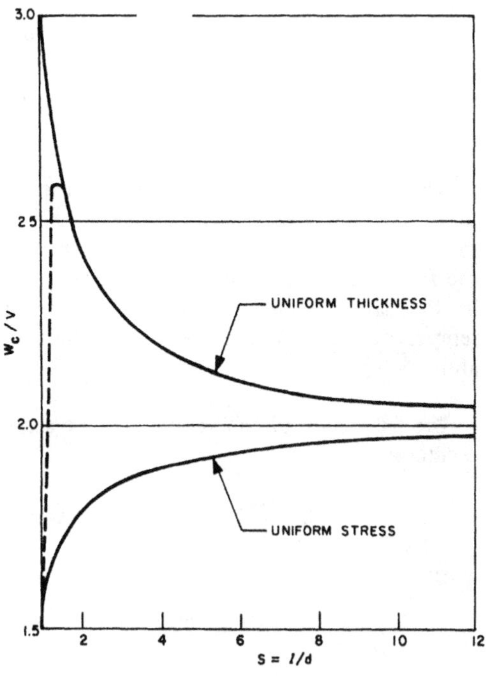

## 24-2.1 FEEDBACK OR RELAXATION OSCILLATORS

The feedback type of oscillator is advantageous since its frequency can be controlled by adjustments and modifications of the feedback path configuration in addition to design changes in the oscillator amplifier; i.e., setback, splitter location, nozzle width, etc.

Ref. 7 gives detailed results on three relaxation oscillators (designed to be insensitive to both temperature and supply pressure variations) using R-C-R feedback looks. One specially tailored oscillator achieved a ± 1 percent frequency change over a pressure range of 6 to 30 psig, and a frequency variation of less than 1 percent over a temperature range from 77° to 175°F at 10 psig. A similar oscillator at a pressure input of 10 psig did not exceed ± 1% for a temperature range from −55° to 175°F.

A typical relaxation oscillator is shown in Fig. 24-6[7]. This model used a fixed

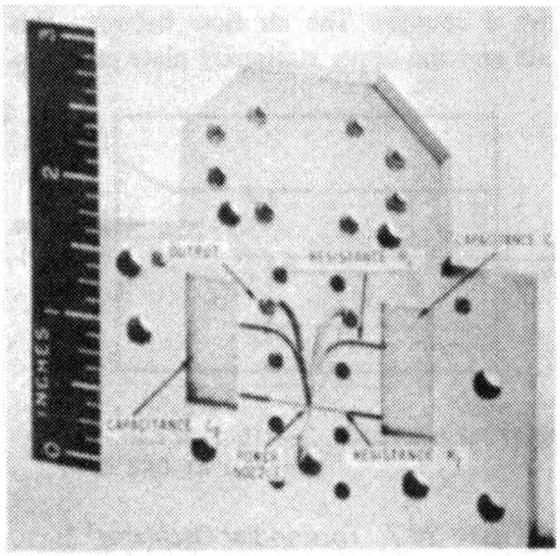

Figure 24-5. Weight-to-Volume Ratio vs Length-to-Diameter Ratio for Cylindrical Tanks With Hemispherical End Caps[16]

Figure 24-6. Relaxation Oscillator[16]

capacitance volume of 0.366 in.³ with the restriction at $R_2$ being variable. Best results were obtained when $R_2$ was 0.01 in. wide with the addition of two small adjusting screws placed in the $R_1$ line to control flow into the feedback loop. Adjusting screws were set to obtain best temperature and pressure insensitivity. Experiments illustrated that a fair degree of temperature and pressure insensitivity could be obtained by a critical choice of the $R_1$-C-$R_2$ parameters. Of the two, temperature insensitivity was considered most important, and it should be noted that above models worked over a temperature range of approximately 100 deg F. Ref. 7 gives a mathematical treatise on the conditions for pressure and temperature insensitivity.

## 24-2.2 TORSION BAR OSCILLATOR

Ref. 8 discusses an oscillator that deviates somewhat from a true flueric device in that a mechanical vibrating element is employed. A torsion bar oscillator is shown in schematic form in Fig. 24-7[7]. It consists of a circular torsion bar fixed at one end with the other end free to rotate. A flat plate attached to the free end of the bar not only provides the inertia for the system but is the element driven by the air flow across it, thus furnishing the pressure pulses necessary to drive a counter. The air flow between this plate and the upper stationary plate produces

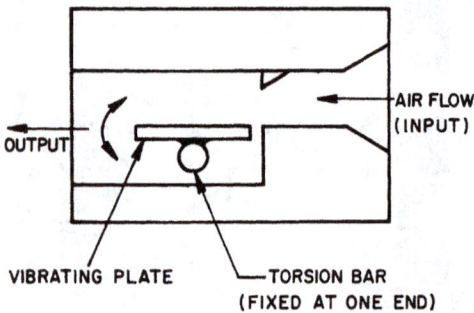

*Figure 24-7. Torsion Bar Oscillator*[16]

a lift on the plate due to Bernoulli's effect. This rotates the leading edge of the plate toward the upper plate until the lift is spoiled due to the loss of air flow. The restoring torque of the torsion bar then rotates the pressure plate back toward and beyond its neutral position. The cycle then is repeated at the natural frequency of the spring-mass system. The oscillations of the vibrating plate produce significant pressure fluctuations in the throat of the device. These are picked up via tubing and fed to a fluid counter circuit.

Since the physical dimensions of the oscillator elements are affected only slightly by temperature variations, the natural frequency of the vibrating plate remains relatively constant over a wide temperature range. It is for this reason that good frequency control can be expected using such a hybrid system.

## 24-2.3 SONIC OSCILLATORS

The sonic oscillator, as seen in Fig. 24-8[9], also depends upon feedback for its operation. However, in this case there is an unrestricted tube replacing the circuit elements. Pressure waves propagate through this tube at sonic velocity, controlling the frequency of oscillation. Since the sonic velocity varies directly as the square root of the absolute temperature, the frequency of the sonic oscillator will vary in the same manner—increasing at higher temperatures and decreasing at lower temperatures.

The frequency of oscillation $f$ of the sonic oscillator is[9]

$$f = \frac{a}{2\ell} + \frac{1}{T_s}, \text{Hz} \qquad (24-5)$$

where

$a$ = sonic velocity in the operating fluid, ft/sec

$\ell$ = length of the feedback passage, ft

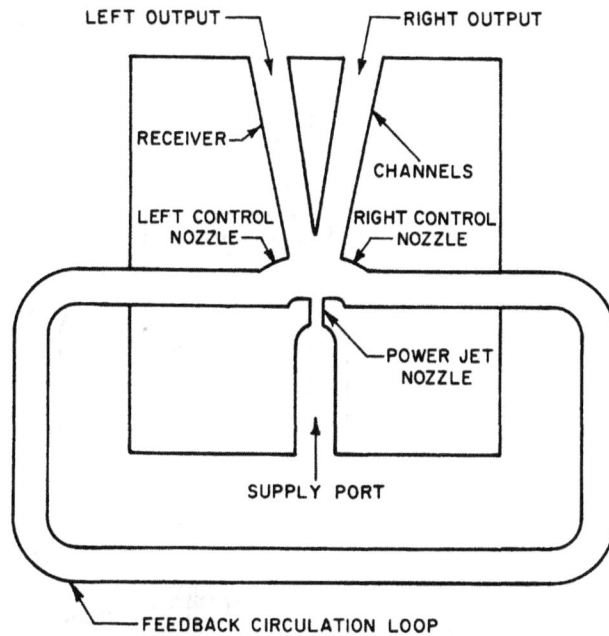

Figure 24-8. Sonic Oscillator[16]

$T_s$ = amplifier switching time, sec

One method for providing temperature compensation for the sonic oscillator is the use of a temperature sensitive element (an orifice, slide etc.) to vary the impedance of the sonic delay line as the temperature varies. A sonic oscillator with such mechanical temperature compensation is shown in Fig. 24-9. The position of the slide (the degree of obstruction) in the line is controlled by a temperature sensitive element. Changing the position of the slide in the line varies the impedance of the feedback line which changes the switching time $T_s$ (see Eq. 24-5).

An advantage of the scheme is that relatively large slide movement is required for small frequency changes. Thus large dimensional tolerances are permitted, and the scheme lends itself to miniaturization, and permits close matching of the oscillator

temperature characteristics. The disadvantage is that the temperature can be compensated only over a narrow range.

## 24-2.4 HALF-ADDER OSCILLATORS

Another type of oscillator uses the half-adder element (exclusive or, and) with the addition of a feedback loop. The element is a two input device shown in Fig. 24-10(A)[8]. If either inputs $C_1$ or $C_2$ are present (but not both), the output of the device is at port $O_2$. However, if both $C_1$ and $C_2$ are present simultaneously, the output is at $O_1$. If now, see Fig. 24-20(B), power supply fluid is applied to input $C_1$ and feedback from output port $O_2$ is applied to input $C_2$, the device will oscillate. The cyclic action continues for as long as fluid is supplied to $C_1$.

The frequency-temperature characteristics

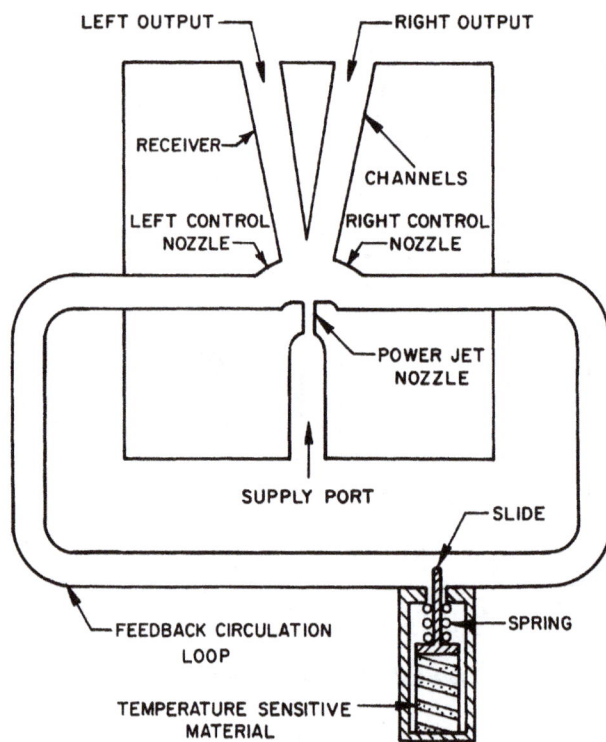

Figure 24-9. Temperature Compensating Scheme for Sonic Oscillator[16]

of the half-adder oscillator has frequency reversals as the temperature is increased. Frequency variations occur with changes in supply pressure except at the specific temperature where frequency crossover occurs. At this temperature complete insensitivity to pressure is observed. This suggests that element geometry plays an important role in establishing the frequency-temperature characteristics of flueric oscillators. In fact, theoretical studies indicate that for any given oscillator there exists a region of temperature insensitivity. By proper selection of feedback-line dimensions and operating pressure, this region can be shifted to one at which operation is desired. Unfortunately, both theory and experiment indicate that temperature insensitivity is obtained at the expense of

pressure sensitivity. Ideally for military applications, a self-compensated (i.e., no moving parts) flueric oscillator would be most desirable. This is especially true for those systems required to operate in an extreme environment such as high spin.

## 24-2.5 HIGH-PRESSURE, LOW-FLOW OSCILLATORS

A high-pressure, low-flow (HPLF) oscillator is shown in Fig. 24-11[10]. All HPLF devices are based on colinear source and receiver nozzles. A jet leaving the source nozzle impinges on the receiver port, causing increase in receiver pressure. When the receiver pressure rises above a critical value, there is a sudden change in flow direction as flow is

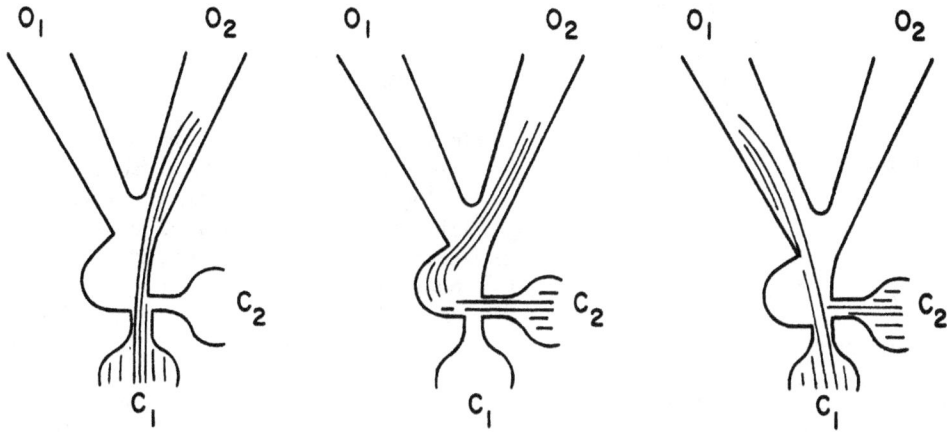

(A) HALF-ADDER (EXCLUSIVE OR, AND) ELEMENT

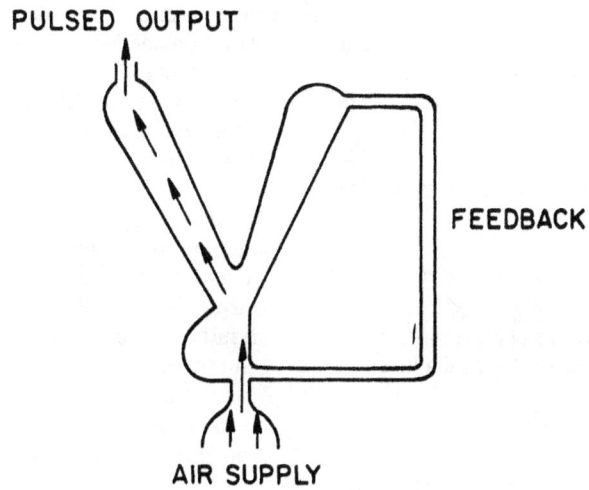

(B) HALF-ADDER OSCILLATOR ELEMENT

*Figure 24-10. Development of an Oscillator from the Half-adder Element*

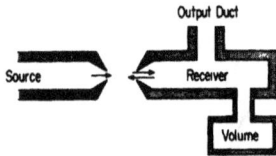

*Figure 24-11. HPLF Oscillator[16]*

expelled from the receiver. A wave front is produced by the two opposing flows. As the pressure drops in the receiver, the wave front moves toward the receiver nozzle. There is an audible snap and a change of flow direction when the wave front contacts the receiver nozzle. The receiver pressure begins to rise, and the process repeats.

## 24-2.6 DESIGN OF OSCILLATORS FROM THEORETICAL CONSIDERATIONS

Since the time base oscillator of the flueric timer must function accurately in a given environment, a great deal of attention should be given to its design. In the design, we should be cognizant of the following requirements for a time base oscillator:

(1) Temperature and pressure insensitivity within its working limitations

(2) Small size

(3) Functional with respect to its downstream components (insensitive to loading)

(4) Rigid construction.

As discussed earlier, flueric oscillators can be fabricated with a reasonable degree of temperature and pressure insensitivity, at least over the military operating range. At present, however, an oscillator cannot be designed directly from theoretical considerations, and must undergo a series of experiments to achieve the final model because of the

complexity of the equations to be solved. For example, Ref. 11 discusses a fluid oscillator employing distributed parameters as the feedback loop (ducts of constant cross section). The theoretical frequency vs temperature relationship for a physical model using a feedback line 12 in. long, 0.027 in. in diameter, and operating at 50 psig is as shown in Fig. 24-12[11]. This curve was generated using the equations presented in Ref. 11. The theoretical $T_{max}$ (point around which maximum temperature insensitivity is to be expected) is approximately $-196°F$ for this model. However, when an experimental model was fabricated and tested at 50 psig the curve of Fig. 24-13[11] was obtained with $T_{max}$ in the range 125° to 225°F. The dissimilarity was explained as follows:

"The temperature region at which the frequency peaks is shifted appreciably from the theoretical curve; however, error is to be expected because of the many variables not taken into account; these include the variations of Mach number with position, the capacitance of the amplifier itself, the switching time of the amplifier, and the nonlinear effects due to the waves being of large amplitude."

Apparently, complete use of all variables in their proper manner was too complex to be investigated at the time. The large disagreement between theoretical and experimental results indicate that the variables not considered are extremely important, and apparently additional study is required before it is possible to precisely design oscillators solely from theoretical considerations.

## 24-3 FLIP-FLOPS

### 24-3.1 GENERAL

The simplest fluid device is the binary flip-flop shown in Fig. 24-14. This is a digital device that has two stable modes; one with

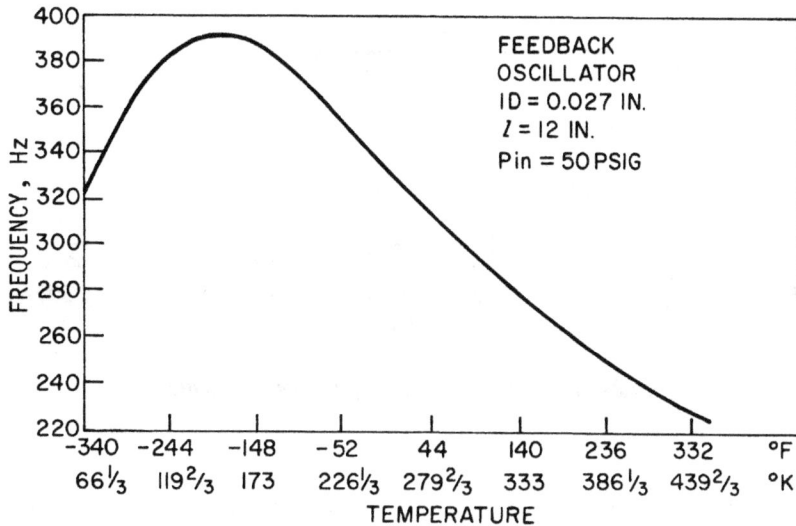

Figure 24-12. Theoretical Performance Curve for Feedback Oscillator
Employing Distributed Parameter Ducts

the stream attached to the right wall with output at $O_1$ and one with the stream attached to the left wall with output at $O_2$*. Flipping is accomplished by introducing a control pressure at $C_1$ or $C_2$ of sufficient magnitude to satisfy the entrainment characteristics of the stream. The stream then moves to the opposite wall at which time control flow can be removed and action is stable in the new position. Load vents are provided for stability and isolation. Thus, the device can be looked at as having an "on" and "off" position. Elements can be "stacked" or "staged" to produce binary counters similiar to electronic logic circuits and many of the control elements that are possible with present-day electronic circuits.

A more refined configuration of a flip-flop element is shown in Fig. 24-15. This figure portrays the fluid flow in a schematic manner while the unit is in operation. Also in this figure, relative sizes of the physical features of the device are indicated as multiples of the nozzle width. Use of the nozzle width as a scale factor permits scaling up or down of the element at will. At the present time nozzle widths of 0.010 in. are in common use. Smaller sizes to 0.004 in. or smaller may be used but the problems with particle entrapment are increased.

Fig. 24-15(A) shows the flow pattern with the outputs unobstructed, and Fig. 24-15(B) shows the flow pattern with one output partially blocked. The outputs are almost completely decoupled; blocking one output has only a small effect on the opposite output and keeps full pressure on the load. Because there is ample gain, devices designed in this manner easily may be staged and such a unit has been found capable of switching as many as 16 others with a pressure recovery as high

---

*Stream attachment is a fundamental principle sometimes referred to as the COANDA effect. It is well covered in many references [12,13] and will not be discussed here.

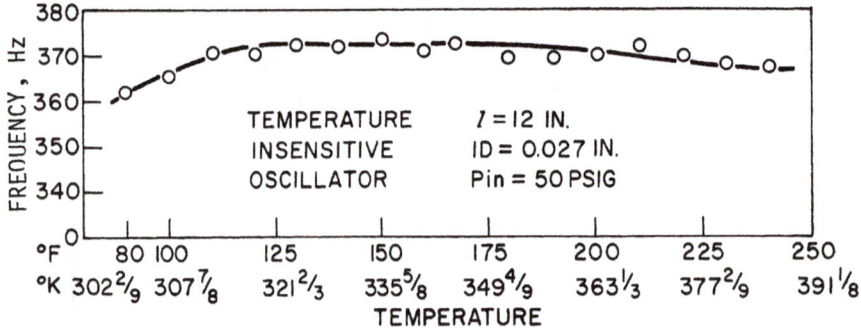

Figure 24-13. Performance of Experimental Oscillator

as 50 percent. Similar designs provide for even higher rates of pressure recovery, but with somewhat lower gain.

To maximize the effect of the side walls, it is desirable to have the nozzle of the power jet deep with relation to its width giving a high aspect ratio. It is also desirable to bring the sidewalls in close to the stream to maximize their effect, thus giving a high gain. If a wide range of operation is desired, it can be achieved at the sacrifice of some gain by moving the sidewalls back from the stream. The minimum aspect ratio for operation of a wall interaction amplifier is about 1. The optimum aspect ratio for minimum flow is about 1.75.

## 24-3.2 TYPICAL PERFORMANCE CHARACTERISTICS OF A BISTABLE FLIP-FLOP

The element of Fig. 24-16 is a schematic representation of a 5-port flip-flop, for which performance characteristics may be developed. In the following discussion assume that these conditions can be met:

(1) Supply pressure at port 1 ($P_1$) can be varied over some given finite range.

(2) Control pressures $P_2$ and $P_5$ can be opened to the atmosphere or by orifice restrictions, and auxiliary supply can also be controlled independently.

(3) Port pressures $P_1$ to $P_5$ can be measured.

(4) Flows $Q_1$ to $Q_5$ can be measured.

(5) Elements can be loaded by orifice restrictions in their output legs.

The real element will have some finite output restriction diameter (of orifice) we will call $d_o$ and the nozzle will in general be of rectangular cross section $\ell b$. We can define an equivalent circular area of diameter $d_n$ such that

$$\frac{\pi}{4}\,(d_n)^2 \;=\; \ell b; \; d_n \;=\;\left(\frac{4\ell b}{\pi}\right)^{1/2} \qquad (24\text{-}6)$$

is a constant for a given nozzle size.

We can now define a loading parameter $d_o/d_n$ where an increase in this ratio is synonymous with less restriction in the output legs. We will consider stable output to port 4, with control ports open fully. If we take measurements of flow and pressure at ports 1 to 5 for various size output orifices, we will generate typical curves as shown in

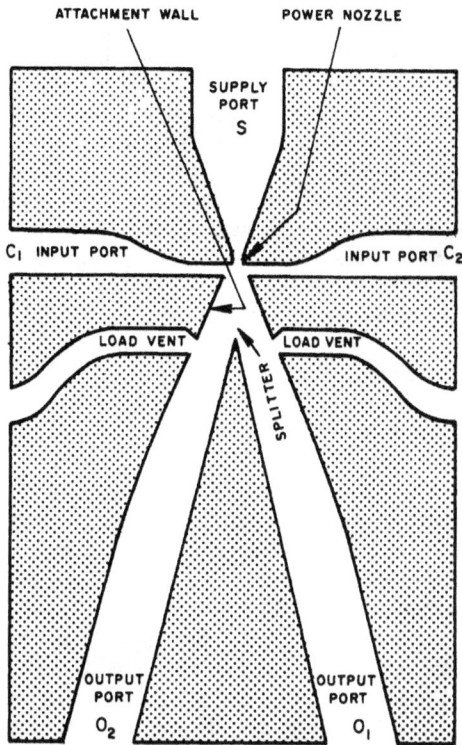

Figure 24-14. Bistable Flip-flop

Fig. 24-17[14]. These curves will be identical over a rather large input supply pressure range, thus it is valid to use normalized flow and pressure values. There will be two distinct orifice sizes (or output loads) $d_{o_1}$ and $d_{o_2}$ between which useful operation and gain will be realized. Diameter $d_{o_1}$ is found at the intersection of the $P_5/P_1$ and $P_4/P_1$ curves while $d_{o_2}$ is at the intersection of the $P_5/P_1$ and $P_3/P_1$ curves. Note that since $P_3$ is open to atmosphere or zero gage pressure, the $P_3/P_1$ ratio is zero. Due to symmetry, if we were to flip the flow (by increasing ($P_5$) so that output is at $P_3$, and repeat the above, identical curves would be obtained, and all we need do is reverse the roles of $P_3$ and $P_4$ in Fig. 24-17 and substitute $P_2$ for $P_5$. For this device the following may be defined when output is at $P_4$:

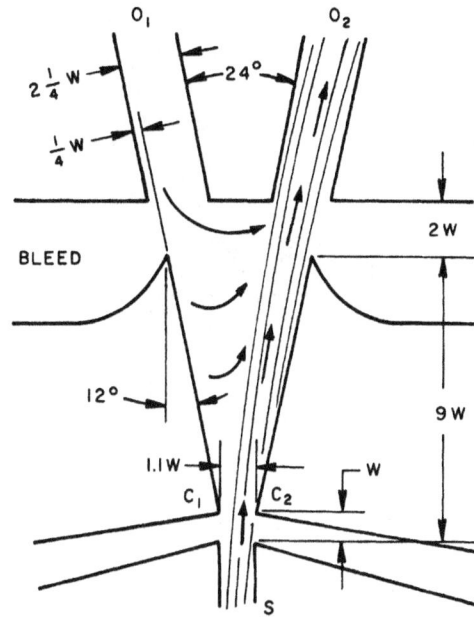

( A ) UNOBSTRUCTED OUTPUT

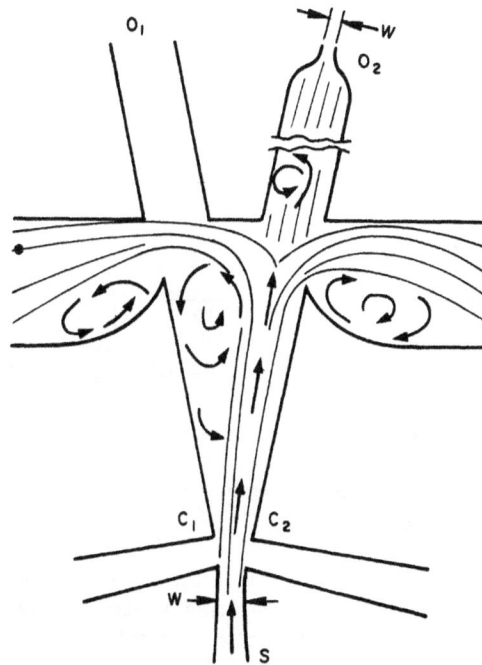

( B ) PARTIALLY RESTRICTED OUTPUT

Figure 24-15. Wall Interaction Fluid Flip-flop

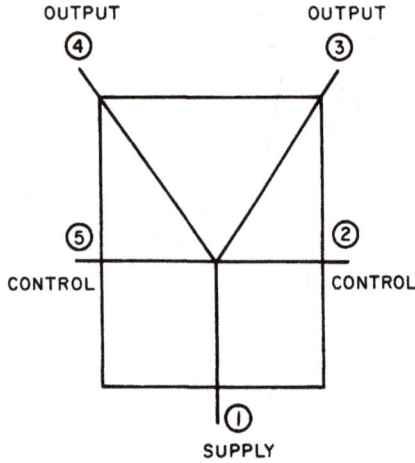

Figure 24-16. Five Port Flip-flop Schematic

Pressure Recovery $\quad R_p = P_4/P_1$

Flow Recovery $\quad R_Q = Q_4/Q_1$

Efficiency $\quad N = (R_P)(R_Q)$

Pressure gain $\quad G_P = \dfrac{P_4 - P_3}{P_5 \text{ (to switch)} - P_5}$

Flow gain $\quad G_Q = \dfrac{Q_4 - Q_3}{Q_5 \text{ (to switch)} - Q_5}$

Power gain $\quad G_T = (G_P)(G_Q)$

The gain as a function of loading will decrease in loading typically as shown in Fig. 24-18[14]. The best operating point is where $N$ is a maximum.

Ref. 14 gives detailed test results on a large bistable element constructed at US Army Harry Diamond Laboratories. In this example the nozzle dimensions were 0.12 × 0.36 in. (thus $d_n$ was 0.234 in.). The element was tested over a pressure range of 5 to 45 in. of water (0.18 to 0.62 psi) and the output channels measured 0.36 × 0.36 in. without orifices. The ratio $(d_{o_1}/d_{n_1})$, was observed at a point where $d_o = 0.3725$ in. and $(d_{o_1}/d_{n_2})$ was found at a $d_o$ of 0.338 (average values). These values were obtained with zero control port flow (open to atmosphere). At a $d_o$ of 0.348 in. best results were obtained where:

Pressure gain = 3.0

Pressure recovery = 31.5%

Efficiency = 43%

The preceding example indicates the manner in which the performance characteristics of a typical bistable flip-flop may be determined. This type of performance measurement and analysis also presents another technique that may be useful in the development of flip-flop elements. Using the empirical design approach, the designer can build a large scale model, test it to obtain the response curves as described in this paragraph, then using a linear design reduction factor scale it down to the desired miniature size.

## 24-3.3 ADDITIONAL REFINEMENTS

The addition of a concave cusp in the splitter region can result in a high-pressure, high recovery, high stability, flip-flop as shown in Fig. 24-19[10]. A vortex is established in the cusp region. When output is at $O_1$ the vortex rotates clockwise, helping to maintain flow to $O_1$. When output is at $O_2$ a vortex is established on the opposite direction, thus supporting output to $O_2$. The vortex increases the stability of attachment of the supply jet, giving high-pressure recovery. However, flow gain and switching time are degraded. With this amplifier, if one output (i.e., $O_1$) is completely blocked off, flow will commence at the opposite output. If the block is removed, output is again obtained at $O_1$ since the direction of the vortex (which has not been altered) returns the supply flow to its original path.

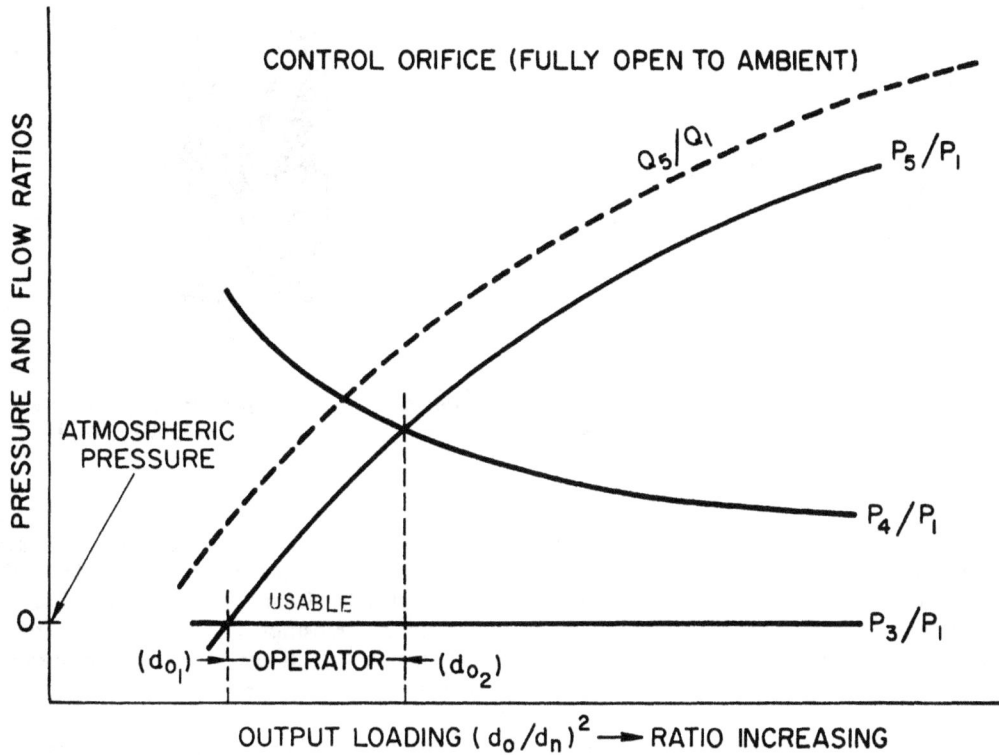

Figure 24-17. Typical Operating Curves

Switching in a high speed flip-flop capable of 0.1 msec or less is illustrated in Fig. 24-20[10]. This component uses a fluid-dynamic phenomenon known as the edgetone effect. Under proper supply conditions, oscillation will occur between the cusp of one output port and the splitter wedge. These oscillations remain in a dynamically stable condition until a small control flow shifts this action to the other side, at which time oscillation occurs between the cusp on the opposite port and the splitter wedge.

## 24-4 COUNTERS

Par. 24-2 on oscillators discusses various methods in which a stable series of pulses can be produced with some degree of temperature and pressure insensitivity. This paragraph covers the function of counters which, coupled with a stable oscillator and an output sensing device, forms the heart of the flueric timer system. If stable oscillators could be constructed such that their period between pulses was the time interval desired, there would be no need for a counter system. However, since timer application may be concerned with time durations in the interval of 0.001 to 200 sec, only the very shortest period timer could be obtained by coupling directly to an oscillator. If we consider that a useful pressure disturbance is obtained in a positive as well as negative pressure wave, then to obtain a period of 0.1 sec we would

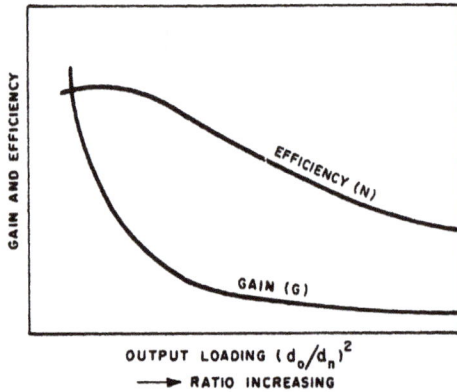

Figure 24-18. Typical Performance Curves

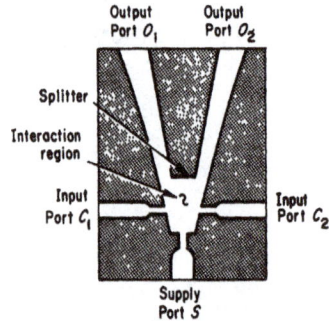

Figure 24-19. High-stability, High-pressure-recovery, Wall-attachment Amplifier[10]

need an oscillator working at 5 Hz; considered a low, unstable, and uncontrollable frequency. Also, since we would like to trigger our timer at time zero, and/or charge time intervals, an intermediate counting system is mandatory.

Therefore, any digitally sequenced and controlled system will employ some form of counter network. The complexity of these counter networks varies, depending upon the needs of the controlled system. Types of counters include binary counters, binary-coded-decimal counters, gray code counters, nonsequential counters, etc. Normally in a timer the counter design will be matched to the application; the result generally will be a unique counting and sequencing network which uses fewer components and performs the prescribed task more adequately than the pure binary form.

## 24-4.1 BINARY COUNTING

The binary counter is the most widely used counter type in both electronics and fluerics. Table 24-3[16] is a truth table for a three-stage binary counter. The table shows that the condition of each stage in the counter at time "$n$ plus one" is dependent only upon the condition of that stage at time "$n$" and

whether or not the adjacent stage of lesser significance (the stage to the right in the truth table) switches from a "one" time to a "zero" during the transition from time "$n$" to time "$n$ plus one". This logic can be mechanized in many ways.

A signal passing through a single bistable flip-flop will have its frequency halved, and time period doubled.

Example: An oscillator working at 1000 Hz coupled to a 3-stage binary counter A, B, C, provides the following time periods:

From A $\quad 2 \times \left(\dfrac{1}{1000}\right) \quad = 2$ msec

From B $\quad 4 \times \left(\dfrac{1}{1000}\right) \quad = 4$ msec

From C $\quad 8 \times \left(\dfrac{1}{1000}\right) \quad = 8$ msec

From $n$ stages $\quad 2^n \times \left(\dfrac{1}{1000}\right) \quad = 2^n$ msec

A more general example of binary counting is shown in Fig. 24-21. This figure shows how $N$ stages can be interconnected and the available outputs of each stage. Note that all

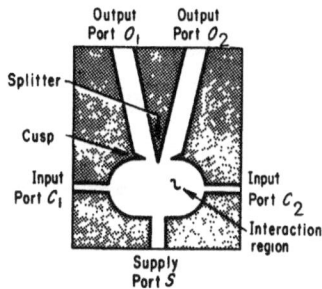

Figure 24-20. Edgetone Amplifier[10]

**TABLE 24-3**

**TRUTH TABLE FOR A THREE STAGE BINARY COUNTER**

| Count | State Time $n$ | | | State Time $n + 1$ | | |
|---|---|---|---|---|---|---|
| | C | B | A | C | B | A |
| 0 | 0 | 0 | 0 | 0 | 0 | 1 |
| 1 | 0 | 0 | 1 | 0 | 1 | 0 |
| 2 | 0 | 1 | 0 | 0 | 1 | 1 |
| 3 | 0 | 1 | 1 | 1 | 0 | 0 |
| 4 | 1 | 0 | 0 | 1 | 0 | 1 |
| 5 | 1 | 0 | 1 | 1 | 1 | 0 |
| 6 | 1 | 1 | 0 | 1 | 1 | 1 |
| 7 | 1 | 1 | 1 | 0 | 0 | 0 |

stages are coupled to a common supply S delivering manifold pressure and the $O_1$ output feeds the input of the next stage. All $O_2$ outputs are available (or simply vent) at slightly less than S due to pressure loss within the element.

The fluid counter operates due to the same pressure differential that holds the stream of a bistable amplifier against the wall[15]. The entrainment of the stream lowers the pressure between the stream and the wall. This low pressure induces a small flow from the higher ambient pressure on the $O_1$ side of the stream, around the loop to the low pressure adjacent to the wall on the $O_2$ side (see Fig. 24-22). This flow is too small to switch the stream. When a fluid pulse to be counted is inserted at the bottom of the loop, it joins the small induced flow and provides sufficient flow to satisfy the entrainment and switches the stream to the $O_1$ output. When the pulse flow ceases, the low pressure is now adjacent to the wall on the $O_1$ side of the counter. A small flow is now induced from the higher ambient pressure on the $O_2$ side of the counter to the lower pressure adjacent to the wall on the $O_1$ side of the counter. The next pulse inserted at the bottom of the loop joins the induced flow and switches the stream back to the $O_2$ side of the counter where the pattern repeats.

One also should realize that in the final timer design, a means of resetting the counter to zero, as well as a means of varying the time set over a period of perhaps 0.1 to 200 sec in 0.1-sec intervals may be desired. Some of these problems are discussed in par. 24-10.

**24-4.2 BINARY COUNTERS IN SLOW-RESPONDING SYSTEMS**

In a very slow responding system, the state of the counter may be simply decoded and used directly to generate the appropriate function. For example, if it is desired to close a contactor at count five of the counter state shown in tabular form in Table 24-3, a set of AND/NAND or OR/NOR gates (see pars. 24-5.1 and 24-5.2) may be used to decode the existence of five (101 ABC) in the counter. The decoding circuits then may be used to close the contactor. With the gating so arranged, each time the counter reaches five, the contactor would close.

**24-4.3 BINARY COUNTERS IN FAST-RESPONDING SYSTEMS**

When the counter is being used to control systems or circuits which respond almost as

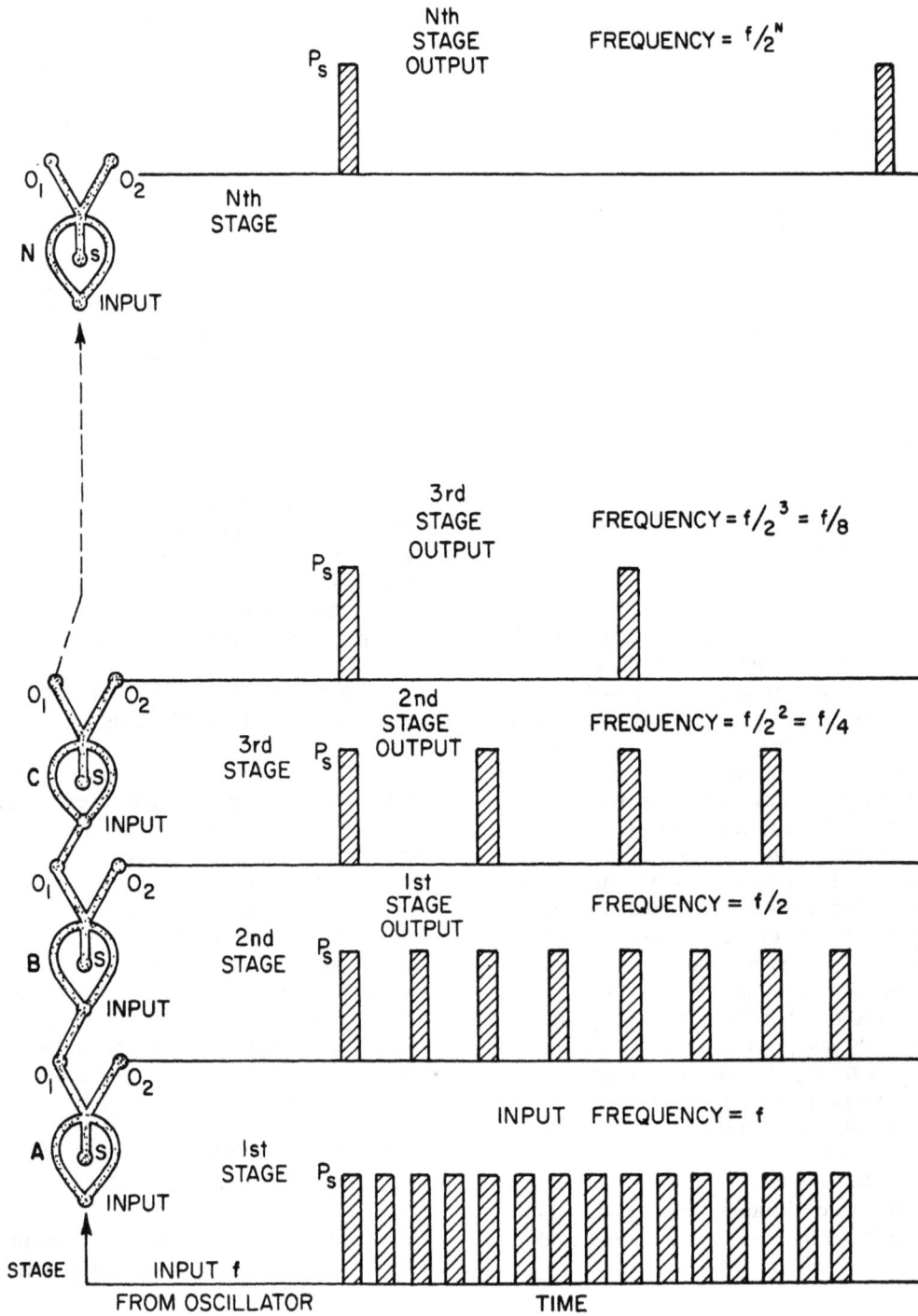

Figure 24-21. Stages of Binary Counter

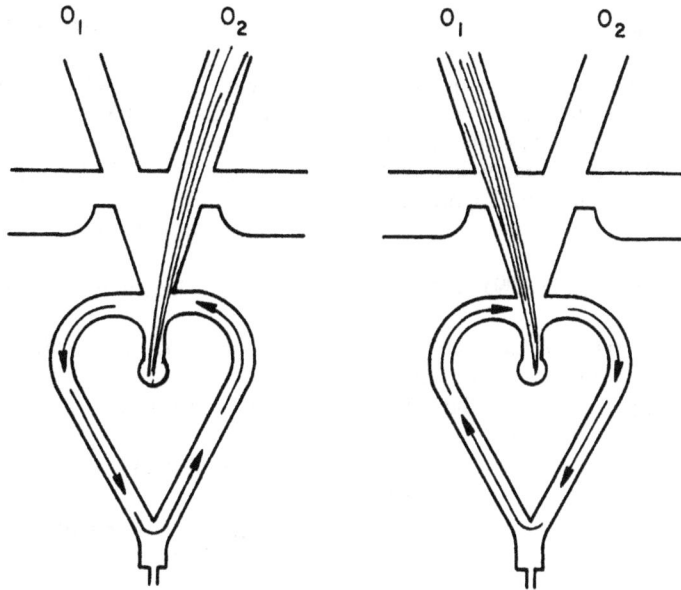

*Figure 24-22. Details of Fluid Flow in a Counter*

fast, or even faster, than the counter itself, other characteristics of the counter operation must be taken into account—such as the transient condition generated by the "flutter" of the counter. During these periods of transiency it is highly probable that false states might be decoded, thus imposing a new condition on the design of the counter. These problems can be eliminated by dividing the period between state changes into at least two phases: a counter change phase and a system control phase. Complexities added to the circuit by these new conditions include the necessity for developing two-phase clocking systems, usually accomplished by higher speed counters or the use of one-shots operating directly off the oscillator.

## 24-4.4 DECIMAL COUNTERS

A simple counter, one step more complex than the binary counter, is the counter that returns to zero prior to reaching the maximum possible count. An example of this type of counter is the binary-coded decimal counter. It is designed to count to nine, then return to zero, and is used widely to provide output control for decimal display systems. Table 24-4[16] is a truth table for the binary-coded-decimal BCD counter. Note that

**TABLE 24-4**

**BCD TRUTH TABLE**

| Count | State Time $n$ | | | | State State $n+1$ | | | |
|---|---|---|---|---|---|---|---|---|
| | D | C | B | A | D | C | B | A |
| 0 | 0 | 0 | 0 | 0 | 0 | 0 | 0 | 1 |
| 1 | 0 | 0 | 0 | 1 | 0 | 0 | 1 | 0 |
| 2 | 0 | 0 | 1 | 0 | 0 | 0 | 1 | 1 |
| 3 | 0 | 0 | 1 | 1 | 0 | 1 | 0 | 0 |
| 4 | 0 | 1 | 0 | 0 | 0 | 1 | 0 | 1 |
| 5 | 0 | 1 | 0 | 1 | 0 | 1 | 1 | 0 |
| 6 | 0 | 1 | 1 | 0 | 0 | 1 | 1 | 1 |
| 7 | 0 | 1 | 1 | 1 | 1 | 0 | 0 | 0 |
| 8 | 1 | 0 | 0 | 0 | 1 | 0 | 0 | 1 |
| 9 | 1 | 0 | 0 | 1 | 0 | 0 | 0 | 0 |

this is not the only possible design for a base ten counter. Tables 24-5[16] and 24-6[16] are other possible combinations of sequential counters which can serve the same purpose as the counter of Table 24-4 with the simple exception that different decoding matrices would be used to decode the appropriate states.

## 24-4.5 NONSEQUENTIAL COUNTERS

One interesting variety of nonsequential counters is the Gray code counter discussed in Ref. 16, which changes only a single bit when changing to an "$n$ plus one" state. Because each change of state experiences only the change of a single bit, all of the transient change problems mentioned earlier as inherent in other counter designs are eliminated. Therefore, the number of phases can be reduced to one. Table 24-7[16] shows a Gray code sequence for a counter designed to generate the control sequence for the flow diagram of Fig. 24-23[16].

The simplicity of such a Gray code design is demonstrated by the set-reset equations for the counter. They are:

Set A: (clock pulse) $\cdot$ ($\overline{C} \cdot \overline{B} \cdot$ Start + CB)

Reset A: (clock pulse)$\cdot$(delay complete)$\cdot$($\overline{C} \cdot B + C \cdot \overline{B}$)

Set B: (CA)$\cdot$(clock pulse)

Reset B: (AC)$\cdot$(clock pulse) + Malfunction

Set C: (B$\cdot \overline{A}$)$\cdot$(clock pulse)

Reset C: (Failure)$\cdot$(Monitoring) + off

Set Monitor: (C$\cdot \overline{B}$)

Reset Monitor: $\overline{C}$

For the convention of bars over the letters, see par. 24-5.1.

## 24-5 LOGIC CIRCUITS

Logic circuits are made up of a combination of elements such as flip-flops connected so that a meaningful output can be obtained. It was already stated (par. 24-3.1) that the flip-flop element can have two stable modes, either "on" or "off". These can be so chosen

### TABLE 24-5

### EXCESS THREE BCD TRUTH TABLE

| Count | State Time $n$ | | | | State Time $n + 1$ | | | |
|---|---|---|---|---|---|---|---|---|
| | D | C | B | A | D | C | B | A |
| 0 | 0 | 0 | 1 | 1 | 0 | 1 | 0 | 0 |
| 1 | 0 | 1 | 0 | 0 | 0 | 1 | 0 | 1 |
| 2 | 0 | 1 | 0 | 1 | 0 | 1 | 1 | 0 |
| 3 | 0 | 1 | 1 | 0 | 0 | 1 | 1 | 1 |
| 4 | 0 | 1 | 1 | 1 | 1 | 0 | 0 | 0 |
| 5 | 1 | 0 | 0 | 0 | 1 | 0 | 0 | 1 |
| 6 | 1 | 0 | 0 | 1 | 1 | 0 | 1 | 0 |
| 7 | 1 | 0 | 1 | 0 | 1 | 0 | 1 | 1 |
| 8 | 1 | 0 | 1 | 1 | 1 | 1 | 0 | 0 |
| 9 | 1 | 1 | 0 | 0 | 0 | 0 | 1 | 1 |

### TABLE 24-6

### EXCESS FOUR BCD TRUTH TABLE

| Count | State Time $n$ | | | | State Time $n + 1$ | | | |
|---|---|---|---|---|---|---|---|---|
| | D | C | B | A | D | C | B | A |
| 0 | 0 | 1 | 0 | 0 | 0 | 1 | 0 | 1 |
| 1 | 0 | 1 | 0 | 1 | 0 | 1 | 1 | 0 |
| 2 | 0 | 1 | 1 | 0 | 0 | 1 | 1 | 1 |
| 3 | 0 | 1 | 1 | 1 | 1 | 0 | 0 | 0 |
| 4 | 1 | 0 | 0 | 0 | 1 | 0 | 0 | 1 |
| 5 | 1 | 0 | 0 | 1 | 1 | 0 | 1 | 0 |
| 6 | 1 | 0 | 1 | 0 | 1 | 0 | 1 | 1 |
| 7 | 1 | 0 | 1 | 1 | 1 | 1 | 0 | 0 |
| 8 | 1 | 1 | 0 | 0 | 1 | 1 | 0 | 1 |
| 9 | 1 | 1 | 0 | 1 | 0 | 1 | 0 | 0 |

**TABLE 24-7**

**TRUTH TABLE FOR GRAY CODE COUNTER FOR MECHANIZING FLOW DIAGRAM FOR THE TYPICAL SEQUENTIAL CONTROL SYSTEM OF FIG. 24-23**

| Control Condition | Counter State Time $n$ | | | Counter State Time $n + 1$ | | | |
|---|---|---|---|---|---|---|---|
| | C | B | A | C | B | A | |
| State A | 0 | 0 | 0 | 0 | 0 | 1 | unconditional when start button depressed |
| State B | 0 | 0 | 1 | 0 | 1 | 1 | unconditional upon next clock pulse-energize starter |
| State C | 0 | 1 | 1 | 0 | 1 | 0 | unconditional upon next clock pulse after delay test for malfunction |
| State D | 0 | 1 | 0 | 0 | 0 | 0 | unconditional upon malfunction |
| State E | 1 | 1 | 0 | 1 | 1 | 1 | unconditional upon next clock pulse ignition on |
| State F | 1 | 1 | 1 | 1 | 0 | 1 | unconditional upon next clock pulse Fuel on |
| State G | 1 | 0 | 1 | 1 | 0 | 0 | unconditional upon next clock pulse after delay-Initiate system monitoring |
| State H | 1 | 0 | 0 | 0 | 0 | 0 | conditional upon failure or off signal |

to represent a "true" or "false" statement dependent on whether "positive" or "negative" logic is chosen, respectively. By virtue of the bistable action, an element will remain in one state until acted upon by some external control pressure, thus the concept of "memory" is inherent in each bistable element. The paragraphs that follow discuss the action of some logic elements that are refinements of the simple flip-flop element from which logic circuits are constructed. A familiarity with Boolean algebra will aid in understanding the notations used. Appendix A lists the more important identities associated with Boolean algebra.

## 24-5.1 OR/NOR ELEMENTS

The OR/NOR element is a bistable flip-flop

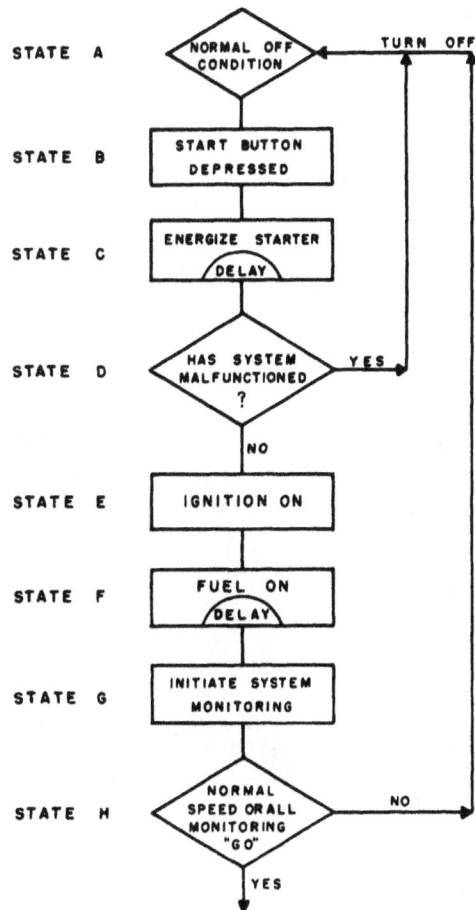

Figure 24-23. Flow Diagram for a Typical Sequential Control System[16]

that will solve the following Boolean equation, A + B = C which is to be interpreted as C equals A *or* B; i.e., C is true if either A or B is true. Obviously, if both A and B are not true, then C is false. This is portrayed most easily by means of "truth table" shown in Table 24-8[16]. Here, the 1's represent true values and the 0's false values. This relationship is called an "OR" relationship. When it is used in a system using fleuric components, it is called an "OR" element, circuit, or gate. Then $\bar{C}$, being the complement of C, functions as the NOR element.

The OR element can contain more than two inputs, thus we could write A + B + C = D for a 3 input device.

Applied to a two-input fluidic device we may interpret the preceding logic as follows (see Fig. 24-24(A))[16]:

(1) Used as an OR element. An output is obtained at C if a control input is present at either A or B, or both. No output is obtained at C if no control flow exists at A and B.

(2) Used as a NOR element. No output is obtained at C if control flow exists at A or B. Flow out of C only is present if A and B are simultaneously absent (off). OR/NOR ele-

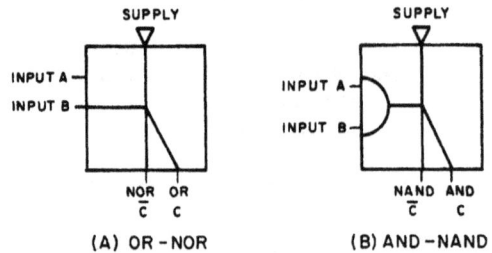

*Figure 24-24. Symbolic Logic Representation*[16]

ments may be staged to form working logic circuits, but usually are combined with other logic elements such as AND/NAND and will be discussed. (NAND means "not and".)

## 24-5.2 AND/NAND ELEMENTS

The normal AND/NAND element is likewise a bistable flip-flop with two output ports, a supply port and two input ports. It will solve the following relationship:

$$A \cdot B = C$$

which means that C equals A and B. That is, C is true only when both A and B are true. The truth table for the AND/NAND element is shown on Table 24-9[16]. $\bar{C}$ is the NAND function of the element.

### TABLE 24-8

#### TRUTH TABLE FOR A SINGLE OR/NOR ELEMENT

| A | B | C | $\bar{C}$ |
|---|---|---|---|
| 0 | 0 | 0 | 1 |
| 0 | 1 | 1 | 0 |
| 1 | 0 | 1 | 0 |
| 1 | 1 | 1 | 0 |
| | | OR | NOR* |

*NOR = not OR

### TABLE 24-9

#### TRUTH TABLE FOR A SINGLE AND/NAND ELEMENT

| A | B | C | $\bar{C}$ |
|---|---|---|---|
| 0 | 0 | 0 | 1 |
| 0 | 1 | 0 | 1 |
| 1 | 0 | 0 | 1 |
| 1 | 1 | 1 | 0 |
| | | AND | NAND* |

*NAND = not AND

Applied to a flueric device we may interpret Table 24-9 as follows (see Fig. 24-24(B)):

(1) Used as an AND element. An output is obtained at C only if control flow exists at A and B simultaneously. No output is obtained at C if only A or B are present or neither is present.

(2) Used as a NAND element. Output is obtained at C at all times, except when inputs at A and B are present simultaneously.

## 24-5.3 COMBINED LOGIC ELEMENTS

As an example of a digital system using logic elements, consider the following (assume that we have two sets of three variables with their complements):

Call them (A, B, C) and (D, E, F). The first set is varying in some manner. We would like to arrange the second set manually to a fixed combination. When the first set is exactly equal to the second set; i.e., when A and D are the same, B and E are the same and C and F are the same, there will be an output. With this given information, the following truth table is made:

| A | D | output | B | E | output | C | F | output |
|---|---|--------|---|---|--------|---|---|--------|
| 0 | 0 | 1 | 0 | 0 | 1 | 0 | 0 | 1 |
| 0 | 1 | 0 | 0 | 1 | 0 | 0 | 1 | 0 |
| 1 | 0 | 0 | 1 | 0 | 0 | 1 | 0 | 0 |
| 1 | 1 | 1 | 1 | 1 | 1 | 1 | 1 | 1 |

The overall output is true only when each combination is true. We may write (including the implements)

$$\text{Overall output} = (AD + \overline{AD}) \cdot (BE + \overline{BE})(CF + \overline{CF})$$

See Fig. 24-25[16] for a circuit arrangement to provide this operation and Fig. 24-26[16] for the flueric equivalent of this circuit.

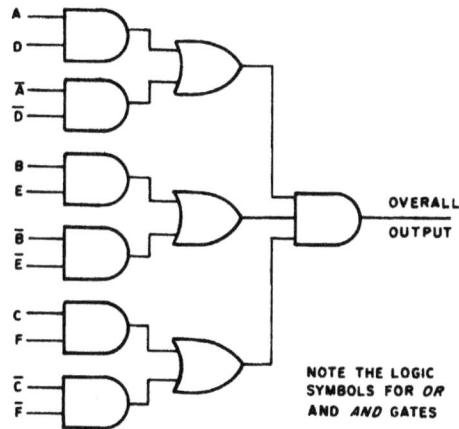

Figure 24-25. Circuit Arrangement for Comparison Circuit[16]

NOTE THE LOGIC SYMBOLS FOR *OR* AND *AND* GATES

The problem could have been solved by using only OR/NOR gates and making use of DeMorgan's laws. This solution is presented in detail in Ref. 16 (pp. 7-8).

In using combined logic elements, there is an easily defined relationship between the OR/NOR elements and the AND/NAND elements. In order to construct a system using logic components, logic levels must be defined. It makes little difference which level is chosen as the "true" level. In digital control systems, however, a name is assigned for either choice. If the higher level is chosen as the true level, the entire system is called "positive logic" system. If the higher level is chosen as the false level, then the system is called a "negative logic" system. By higher and lower levels we mean the following: for a flueric system we may choose the two outputs of a bistable switching element as A and $\overline{A}$. These outputs are, alternately, flowing at some rate, or not flowing. If we decide to use the positive logic scheme, we may call the flow condition the greater level and define it as the true condition; then, the nonflowing level is the false condition. It may be just as

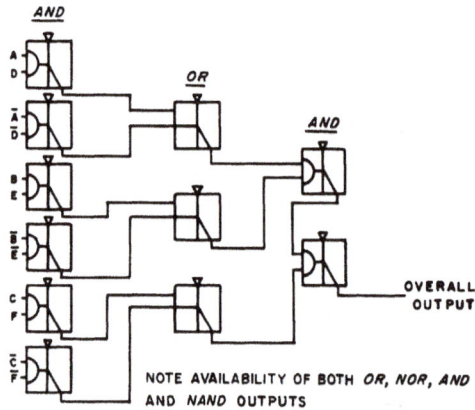

*Figure 24-26. Flueric Equivalent of Fig. 24-25[16]*

convenient, if not more so in some situations, to reverse these roles. Note from the truth tables for the AND and OR circuits that if the logic levels are reversed, the OR circuit becomes the AND circuit and the AND circuit becomes the OR circuit.

## 24-6 RESISTANCE ELEMENTS

There are several different types of resistance elements which may be used in the fluerics system:

(1) Tube resistors are small lengths of appropriate cross section having a length several times their width. These elements are sometimes called capillaries.

(2) Orifice resistors are essentially thin disks having a restricting hole placed normal to the fluid flow in the passage.

(3) Porous metal resistors are slugs of a sintered porous metal placed in the passage to resist fluid flow.

(4) Variable resistors may be a series of tapped resistors of the types described or they may be continuously variable as, for example, a needle valve. A special category of variable resistor is the settable resistor which may have its restriction varied by tools, but once having been set becomes a fixed resistor.

## 24-6.1 TUBE RESISTORS

Tube resistors can be fabricated from glass, metal, ceramics, epoxy, and like materials. Typical resistors have the characteristics indicated by the solid lines in Fig. 24-27[17], obtained with air at a fixed discharge pressure. A straightline with a slope of 45 deg on this log-log plot indicates linearity. Tube resistors have the advantage of linearity, but the disadvantage of being easily clogged.

Tube resistors can be flattened to alter their cross-sectional area (and shape) and their resistance. (Fig. 24-33 shows a crimping tool

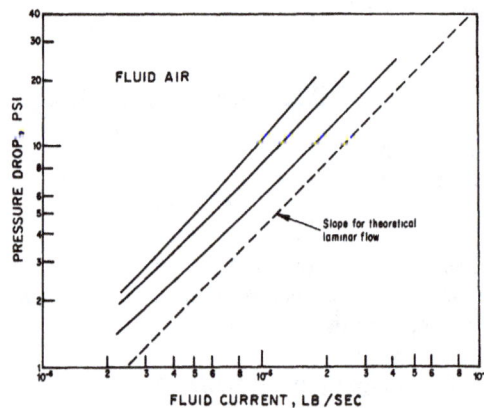

*Figure 24-27. Pressure vs Flow for Etched-glass Resistors[17]*

to be used for this purpose.) Typical linearity curves for flattened tube resistors are shown in Fig. 24-28[17] for air discharge at fixed output pressure. Flattening tends to decrease the slope below 45 deg. Tubes or capillaries are chosen for the resistors because of their more nearly linear characteristics in comparison to an orifice resistor. The resistance of a capillary tube of rectangular cross section can be approximated by the expression

$$R = \frac{\Delta P}{Q} = \frac{8\pi\mu\ell}{(bh)^2} + \frac{7\rho Q}{6(bh)^2} \qquad (24\text{-}7)$$

where

$R$     = resistance of capillary channel, lb-sec/in.[5]

$\Delta P$     = pressure differential across capillary, psi

$Q$     = flow rate, in.[3]/sec

$\mu$     = viscosity, lb-sec/m[2]

$\ell$     = length of channel, in.

$b$     = width of channel, in.

$h$     = depth of channel, in.

$\rho$     = fluid density, slug/m[3]

Each parameter in the equation will be affected by temperature changes. Thus, one expects the resistance of the capillary to vary with temperature. Theoretical analysis and experimental results indicate that the resistance of a capillary tube fabricated from high temperature epoxy resin *increases with an increase in temperature.* *

_____
*The viscosity of ideal gases increases directly with the square root of the absolute temperature. For real gases, the viscosity usually changes at a rate somewhat greater than this because of the effects of molecular attraction which are assumed zero for ideal gases.

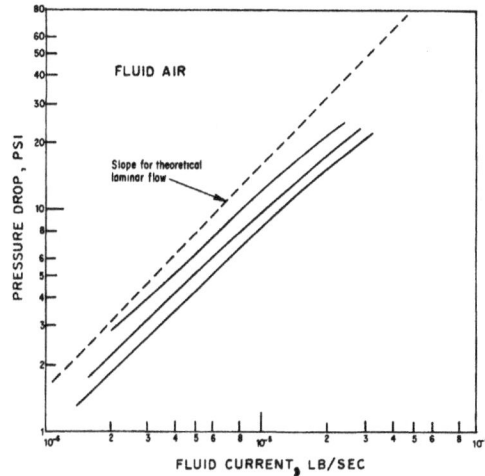

Figure 24-28. Pressure vs Flow for Flattened-tube Resistors[17]

## 24-6.2 ORIFICE RESISTORS

An orifice resistor is simply a thin disk with a restricting hole placed normal to fluid flow. Usually a coupling or adapter is provided so that different size resistors can be substituted at a given place in the flow line. Drill jig bushings available in a variety of sizes also can be used. These bushings are usually short tubes, available with a collar on one end. Since the bushings are precision drilled, reproducible resistances easily can be obtained. Fluid orifice resistors are available commercially. Dirt and contamination are cleaned more easily from orifice resistors than from tube or porous metal types. Orifice resistors are cheaper to fabricate than tube types but are nonlinear as shown in Fig. 24-29[16] where $R_{PQ}$ is the effective resistance.

## 24-6.3 POROUS-METAL RESISTORS

Although normally used as filters, sintered porous metal parts can serve as fluid resistors.

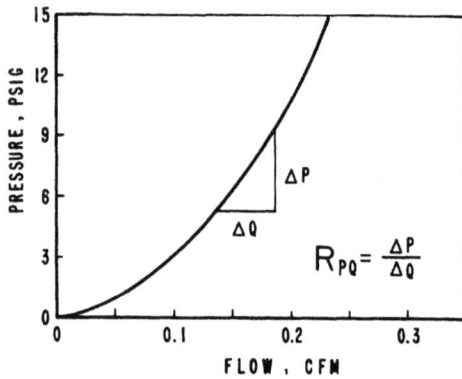

Figure 24-29. Resistance of an Orifice[16]

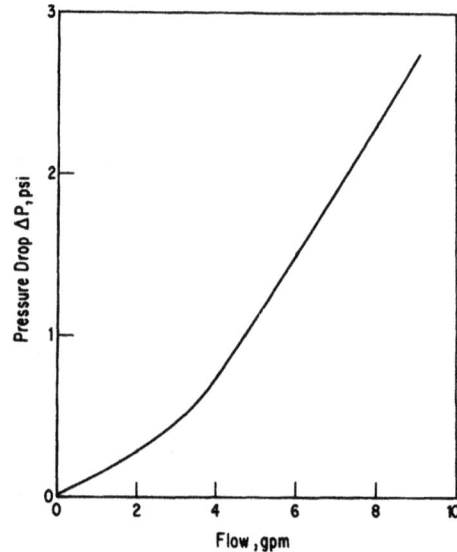

Figure 24-30. Flow of Water Through a
Porous-metal Resistor[17]

Metals that can be powdered and sintered include monel, bronze, stainless steel, gold, and silver. A wide variety of shapes—such as cones, plates, and cylinders—can be produced where required by the application.

Porous-metal resistors are nonlinear, but the pressure vs flow curves have long straight-line sections. Typical plots for water and hydraulic fluid are shown in Figs. 24-30[17] and 24-31[17].

Porous-metal resistors are not cleaned as easily as orifice or tube resistors. However, they can be cleaned by scrubbing, back-flushing, or chemical cleaning. One excellent method is to backflush while the element is immersed in an ultrasonic bath. Surfaces of porous-metal resistors must be handled with care and protected from dirt and mechanical injury. Porous-metal resistors cannot be plugged by a single dirt particle as can orifice or tube types. Instead, the dirt settles on the porous metal surface and does not change the resistance appreciably. An accumulation of dirt is required to plug a porous-metal resistor.

## 24-6.4 VARIABLE RESISTORS

All resistors discussed thus far have fixed resistances. In circuit development, variable resistors also are needed. Tapped resistors easily are fabricated by interconnecting either orifice or tube resistors, and supplying taps at the points of interconnection. Care must be exercised when using such resistors to ensure that unwanted capacitances are not introduced. One disadvantage of tapped resistors is that their resistance is not continuously variable.

The traditional variable fluid resistor is simply a valve. The needle valve is the most widely used type in flueric-circuit development. It offers the convenience of fine adjustment in flow over wide ranges. Needle valves have been constructed with only one degree of taper, producing extremely fine control of flow through the valve (see Figure 24-32[16]). If only steady-state conditions are considered, an entire circuit can be developed with needle valves as the only resistors. However, since these valves are nonlinear, it may be necessary to replace many, if not all,

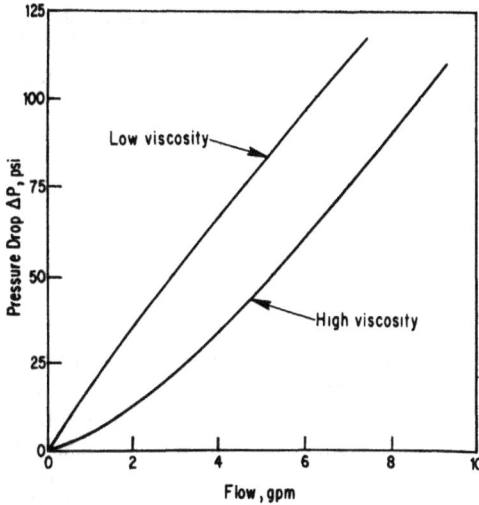

Figure 24-31.  Flow of Hydraulic Fluid
Through a Porous-metal Resistor[17]

Figure 24-32. Needle Valve With 1-deg
Taper Angle[16]

of the valves with fixed linear resistors before the small-signal behavior of the circuit can be studied.

A commonly used laboratory technique for this substitution is to adjust the needle valve until the proper values of pressure and flow are obtained. The valve is then removed and its resistance measured. Next, a piece of copper tubing is substituted for the valve in the resistance-measuring circuit and the tubing flattened with a small hand vise until its resistance equals that of the valve. The flattened-tubing fixed resistor is then installed in the circuit in place of the valve. With compressible fluids, it may first be necessary to measure the resistance of the valve and then fabricate the flattened-tube resistor while maintaining the same values of up-stream and downstream pressure as in the circuit under development.

A valuable device has been developed for solution of the circuit balancing problem: the

settable resistor. A crimpable resistor is installed in the circuit in the nonrestrictive (uncrimped) condition. A terminal-crimping tool (Fig. 24-33[16]) is used to reduce the resistor cross-sectional area until the desired downstream flow or pressure readings are attained.

## 24-7 CAPACITANCES

A fluid capacitance is simply a tank of any geometry into which the fluid may flow and be stored for a time, then discharged. Any tank or container which is safe at the maximum pressure used is suitable for a capacitor. Since the pressures are usually low, the problem can be solved easily. In American Engineering units, capacitance can be defined as volume per unit pressure, and the units are $ft^5/lb$. As in electrical circuits, fluid capacitance will effect a phase shift such that flow will lead pressure by 90 deg. In a circuit, the fluid capacitor corresponds to the electrical capacitor to ground. It cannot duplicate the action of an electrical in-line capacitor since it

24-33

Figure 24-33. Crimping Tool Placed Over
Resistor Tube[16]

favors direct current flow. Since pressure is a function of fluid density and temperature for a fixed volume, capacitance $C$ can be expressed more basically as

$$C = \frac{V}{\rho a^2} \text{ , ft}^4\text{-sec}^2/\text{lb} \qquad (24\text{-}8)$$

$V$ = volume, ft$^3$

$\rho$ = fluid density, lb/ft$^3$

$a$ = local speed of sound in fluid stream, ft/sec

Eq. 24-8 is particularly applicable in $RC$ fluid oscillators where it has been substantiated that the oscillator frequency is normally both temperature and pressure dependent. Although it is difficult to determine either analytically or experimentally the variation of capacitance with temperature, there are indications that the effective capacitance *increases with rising temperature*. In any case, it does not appear that effective capacitance decreases with rising temperature. Thus if an oscillator can be designed such that its frequency is approximated by

$$f = \frac{k}{RC} \qquad (24\text{-}9)$$

where $k$ = proportionality constant

then its frequency of oscillation should decrease as the temperature increases. This is in direct contrast to the usual assumption of increasing frequency proportional to the square root of the absolute temperature.

## 24-8 TURBULENCE AMPLIFIERS

### 24-8.1 GENERAL

Turbulence amplifiers make use of the fact that an interacting control jet can vary the degree of turbulence in the main jet, and thus vary output flow or pressure[18]. Consider a stream of fluid exiting from an input nozzle as shown in Fig. 24-34(A). At low velocities flow is laminar and, if properly designed, output pressure may be only slightly below input pressure. As the supply pressure is increased, output pressure is increased. Continuing the pressure increase (and stream velocity) a point then is reached where the flow is turbulent ahead of the output receiver (Fig. 24-34(B)) and the output pressure is reduced. Once total turbulent flow has been attained, further input pressure increases output pressure. These stages are shown on Fig. 24-35. A-B represents laminar flow; B-C is transitional region from partial turbulence to complete turbulence; and C-D represents total turbulent flow.

A control flow jet—introduced at right angles and close to the input nozzle, Fig. 24-34(C)—can increase the degree of turbulence and the shape of Fig. 24-35. Since we desire small changes in control jet pressure to effect large changes in output pressure (high gain), initial design parameter should be operational around point B, Fig. 24-35—i.e., natural turbulence should just start near output receiver with zero control pressure. The addition of a small control pressure then will increase greatly turbulence and output pressure, see Fig. 25-36[18].

(A) LAMINAR FLOW, CONTROL JET OFF

(B) TURBULENT FLOW, CONTROL JET OFF

(C) TURBULENT FLOW, CONTROL JET ON

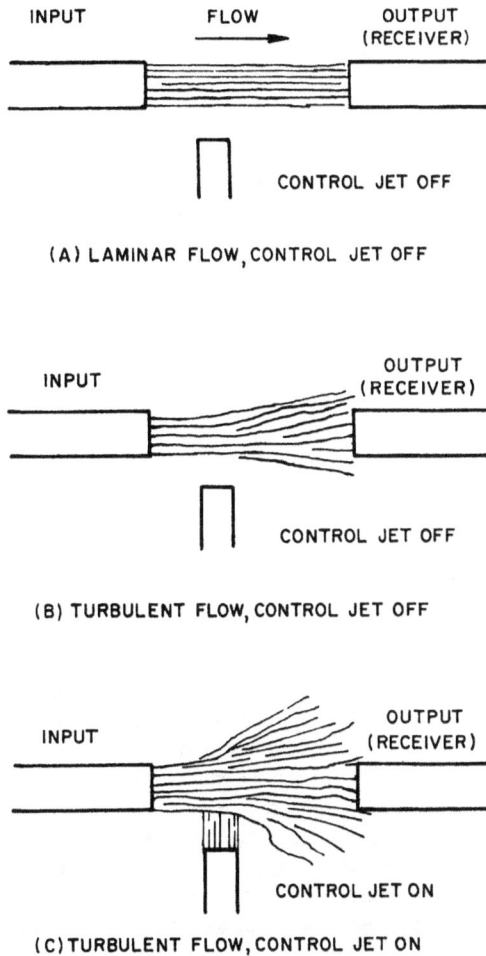

*Figure 24-34. Stages of Operation—Turbulence Amplifier*

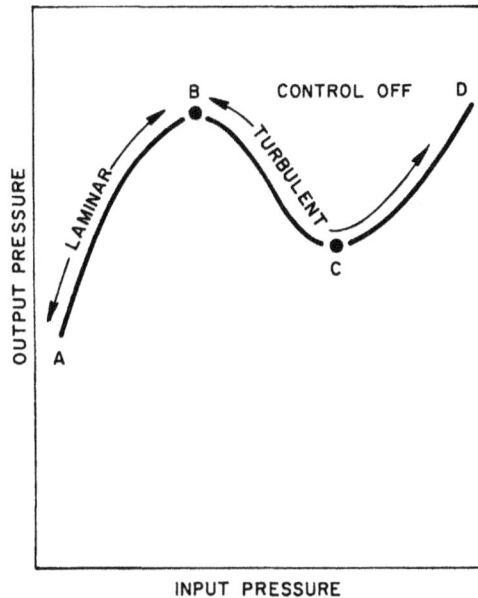

*Figure 24-35. Input-Output Pressure Relationship for Turbulent Amplifier With Control Flow Off*

## 24-8.2 DISTURBING THE MAIN JET

Small disturbances cause the point at which the stream goes turbulent to shift toward the input tube. Disturbances can be introduced into the laminar stream by other fluid streams impinging on the main jet, intrusion of a mechanical agent, or by acoustic waves.

The greater the distance between the input tube and output tube, the greater the sensitivity of the device; i.e., the greater the effect of jet disturbances. However, the maximum allowable velocity of the laminar stream decreases with increasing distance between the tubes. On the other hand, the lower the jet velocity, the lower the output pressure. The selection of a supply pressure and a distance between the tubes represents a compromise between sensitivity and useful output.

Theoretically, turbulence amplifiers can be any size, but in practice, there are upper limits. Commercially available devices are used for control rather than power purposes, and they have jet diameters of about 0.030 in. Distance between supply and output tubes ranges from 0.375 to 1.5 in. In a turbulence amplifier of this size, high sensitivity to sonic disturbances begins at about 1.3-in separation

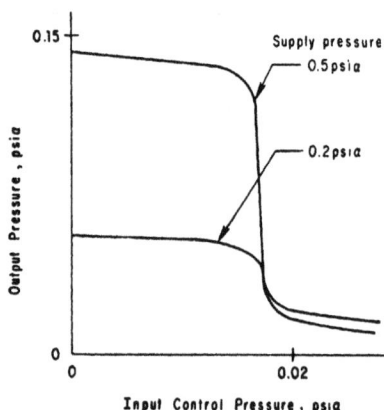

Figure 24-36. Typical Output Pressure vs
Control Input Pressure for a
Turbulence Amplifier[18]

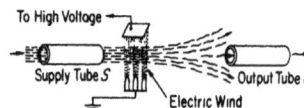

Figure 24-37. Turbulence Created by
Electrophoresis Effect[10]

between supply jet and output tubes. When the device is operated in the acoustic mode, frequency sensitivity varies with the separation. Acoustic sensitivity of these devices is so high that they can be triggered by a mouth whistle at a distance of 300 ft.

The most commonly used turbulence amplifiers are designed to be disturbed by a jet of air projected perpendicularly to the main jet. In practice, several jet inputs can be used. The number of possible inputs is limited only by available space.

An input signal also can be introduced mechanically by moving a small object, such as a wire, into the jet stream preferably near the input jet orifice. This arrangement can be used as a basis of a mechanical-flueric transducer. A small, pressure-actuated diaphragm also can be deflected into the jet stream to produce turbulence. Another manner in which the main supply can be made turbulent is through the electrophoresis effect. A pair of high voltage electrodes positioned across the main supply as shown in Fig. 24-37[10] can develop an "electric wind" that can disturb the main jet.

## 24-8.3 INHERENT ACOUSTIC RESPONSE

Acoustic waves are difficult to shield and, in fact, are generated by flueric devices themselves. If acoustic amplification is not wanted, the acoustic sensitivity of the turbulence amplifier should be low so that only flueric amplification is obtained. This can be achieved readily in any given amplifier design because the distances between input and output ports at which sound sensitivity become significant is usually much greater than the distances required when fluid jets are used as the disturbing input. Most turbulence amplifiers used for control purposes are operated in the range where they are not acoustically sensitive. These amplifiers usually are used to sense or respond to a control flow that is directed at approximately right angles to the main laminar input jet, and the distances between input and output elements are held to a minimum.

## 24-8.4 CASCADING

Since the output of a turbulence amplifier is a pressure, this output can be used to drive the control jet of a second stage. Thus, turbulence amplifiers can be cascaded. Some care is necessary, however, since the output of the amplifier never drops completely to zero over the practical operating range. Thus, if the output of one unit is directly coupled to a second, the second stage must be unresponsive to small signals from the first.

From Fig. 24-35 the required characteristic can be obtained by operating the first stage so that its output lies somewhat to the left of the peak B. Actually, the amplifier can be designed to operate considerably to the left of B, and a large range of relative unresponsive action can be obtained.

Another approach is to provide a jet stream with sufficient stability that it is undisturbed by the range of the input and can be ignored.

### 24-8.5 OVERDRIVING

When turbulence amplifiers are used in high-speed circuits, care must be taken to avoid overdriving since the resumption of laminar flow after very strong turbulence is somewhat delayed. When a control signal is turned off, the fluid stream has to become laminar again before full output pressure is regained at the output port. This effect can be observed readily in the output of oscillators. The effect on pulse circuitry is shown in Fig. 24-38[18]. Note that the recovery time is much greater than the initial reaction time.

In digital circuitry, this behavior is particularly important where a given device is turned off, and thus has a low output pressure. When a negative signal is applied to the turbulent amplifier in an attempt to turn

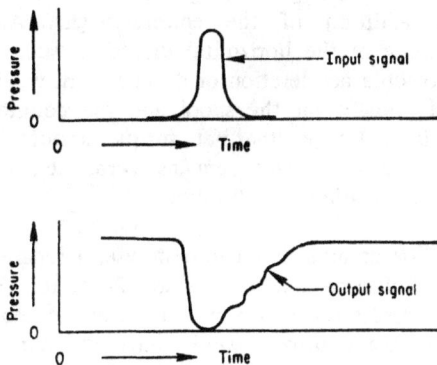

it on, it is ineffective if the input signal pulse is too short, because the amplifier does not have time to build up a laminar stream. This effect is shown in Fig. 24-39[18].

### 24-8.6 PRIMARY USES AND TYPICAL SIZES OF TURBULENCE AMPLIFIERS

Although the turbulence amplifier can be used in proportional applications, its primary use is in logic circuitry. It is particularly useful in this case because a turbulence amplifier can drive up to ten or more secondary units. Furthermore, the turbulence amplifiers are isolated completely, and a single amplifier can have a large number of inputs. Again, the limitation is one of physical room in which to place the multiple inputs rather than any theoretical limitation.

The turbulence amplifier performs a logical NOR function. A NOR function can be used as the basis for all other logic functions such as AND, OR, and counting circuits.

Amplifiers have been constructed with power gains of 40 to 80. If some acoustic sensitivity can be tolerated, pressure gains of 1000 can be obtained in a single stage. With

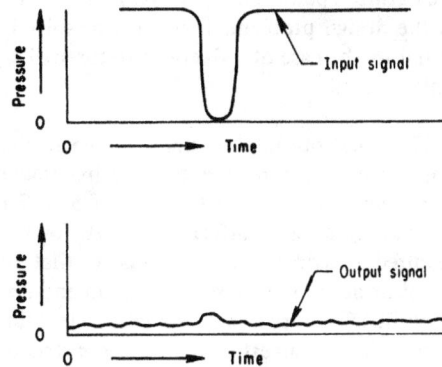

*Figure 24-38. Output Pulse for a Positive Input Pulse[18]*

*Figure 24-39. Output Pulse for a Negative Input Pulse[18]*

this sensitivity, the air currents produced by even small blowers can be detected by open amplifiers many feet from their source.

In digital applications jet sizes can vary over a wide range, with typical supply tube diameters ranging from 0.010 to 0.70 in. For diameters larger than this, the length of straight supply tube required to obtain laminar flow becomes so long as to be impractical.

By use of a typical supply tube diameter of 0.030 in. an amplifier can operate with supply pressure of approximately 0.15 psi and produce an output of 0.075 psi with a no load flow of $1.2 \times 10^{-5}$ lb/sec. If a control pressure of 0.01 psi is used, the control flow would be about $5 \times 10^{-6}$ lb/sec.

## 24-9 OUTPUTS

It is obvious that in order for the flueric timer to serve a useful purpose, some way must be found to use the energy output from the low-pressure fluid signal at the end of the set time. In designing a military timing system, the difficulty of the design problem stems from the conflicting requirements of the pneumatic signal sensitivity that must be combined with sufficient ruggedness to avoid premature operation. A convenient example of the design problems that must be solved is found in the case of a flueric fuze for artillery ammunition.

The usual output function of a mechanical time fuze is the release of a spring loaded firing pin. Since spring preloads of 5 to 7 lb are common, a significant output force is required to trip the pin. It appears that the signal or actuation pressure from an end-item sized flueric timer will be something less than 1 psi. Several investigators have constructed functional, low pressure transducers; however, in most cases these devices require more volume than that necessary for the entire flueric circuitry package.

## 24-9.1 BALANCED PISTON OUTPUT DEVICE

One type of output device uses the balanced piston principle. Applying this principle to an output mechanism results in the device shown in Fig. 24-40[2]. Since there is clearance between the spool lands and the cylinder walls, pressure supplied to port 1 enables the spool to float in a centered position. A bleed passage through the spool prevents pressure from building up behind the right-hand land thus eliminating premature operation due to pressure unbalance. Upon the elapse of the set time, an output signal of 1 psi is delivered to port 2. This signal, although small, upsets the equilibrium and causes the spool to move to the left. As soon as the trailing edge of the right-hand land passes port 1, the full supply pressure acts on the exposed piston area and results in a significant output force sufficient to perform work of fairly high magnitudes.

Compared to passive type devices, this device requires a continuous air supply throughout the timing cycle. Power consumption can be reduced by keeping the spool and cylinder clearance to an absolute minimum. Also the balanced spool is sensitive to external shock and vibration. Sensitivity to external loading along a given axis is reduced by addition of the counterweight. Any loading in the horizontal direction, such as projectile acceleration or deceleration, would act equally on the spool and the counterweight. Hence, the net torque about the balance arm pivot remains zero, i.e., the system is still in equilibrium.

For ammunition end-item use, a separate gas bottle capable of providing 20 psi for 180 sec might serve as a power source. A more desirable approach would have the ram-air driven pump supply the device. In any event, although this balanced-piston scheme may not prove to be entirely satisfactory from a power consumption standpoint, it does satisfy the

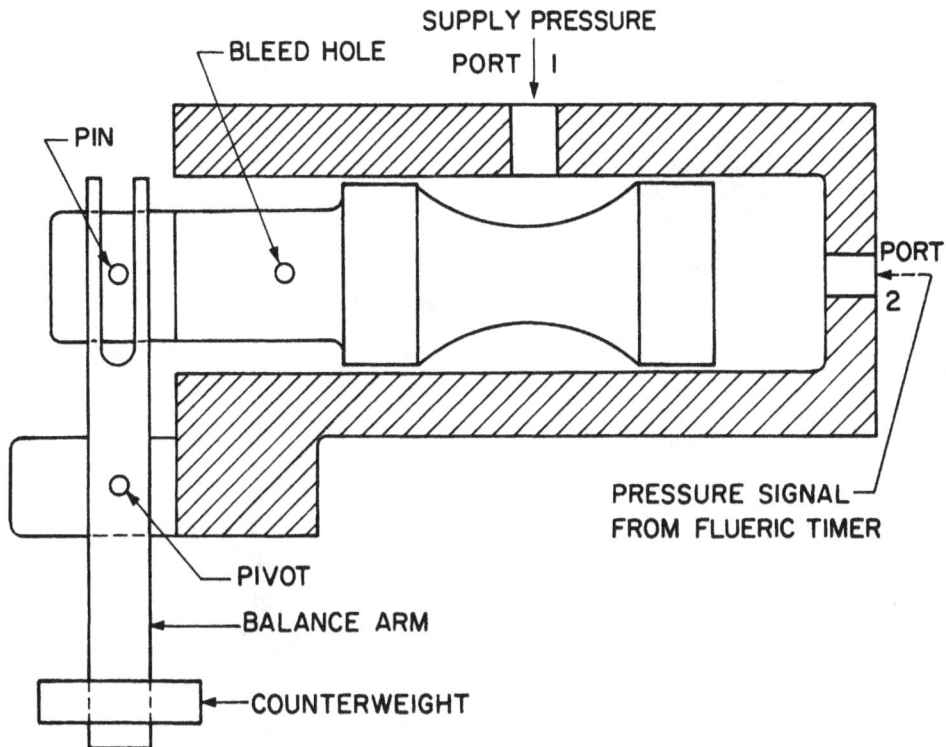

*Figure 24-40. Output Mechanism—Balanced Piston Type*

requirement for extreme sensitivity to a low level flueric counter output signal.

### 24-9.2 BELLOWS-ACTUATED OUTPUT TRANSDUCERS

An output transducer used in a study at Picatinny Arsenal is shown in Fig. 24-41[12]. It consists of a small ten-convolute bellows, an adjustable contact, and contact leads. The bellows are 0.090 in. in diameter with a wall thickness of 0.0006 in. One end of the bellows receives the output pressure of the timer and the bellows elongate to close electrical contact. The transducer functions satisfactorily up to 100 Hz. The adjustable contact is set to between 0.004 to 0.006 in.

gap to be operable at output pressures near 2 psi. The nominal linear travel of the bellows at 2 psi was determined to be about 0.012 in. The placement of these transducers is shown in Fig. 24-42[12]. No change of response was noted when the assembly was spun about an axis through the center of the timer at rotational rates up to 150 rps. The arrangement of the contacts is such that initial setback forces tend to increase the contact gap. Thus, unscheduled contact closure would occur only due to deceleration forces such as those produced by impact.

In another application, the output interface consists of a bellows-actuated rotary switch as shown in Fig. 24-43[19]. The bellows switch is

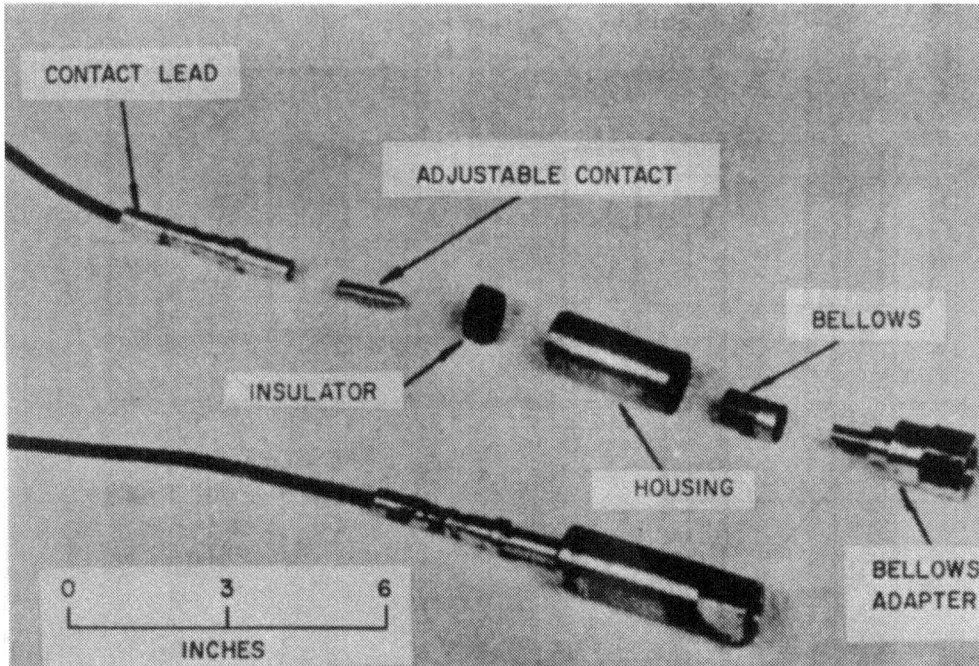

Figure 24-41. Bellows Transducer for Fluid Timer

made symmetrical to promote balancing and to lower the sensitivity to inertia forces, which would tend to operate the switch prematurely. Lateral support and damping are provided by plastic tubes surrounding the bellows. Also the bellows are designed to be stacked solid when not pressurized. This prevents any displacement in one direction and permits elastic preloading to be used to limit any inertia forces that might cause displacement in the other direction.

The present switch requires two output signals from the output logic circuit as shown in Fig. 24-44(A)[19]. Here the output circuit is built from three NOR gates; the output signal is on only when A and B are in. In order to obtain fast response of the output interface switch, with its associated volumetric capac-

ity, the third NOR gate used in this circuit has a larger nozzle (0.010 in.). Fig. 24-44(B) shows the output pressure flow characteristics of the third element that drives the bellows rotary switch.

## 24-9.3 THIN MEMBRANE

In another program, a pressure actuated switch was developed to close contacts with pressures as low as 0.5 in. of water (approximately 0.02 psi)[20]. The only moving element is a very thin stretched membrane 0.75 in.$^2$ in area on which is held one movable contact. The other contact is adjustable to actuate closing over a variation of output pressure. The switch is shown in Fig. 24-45(A)[20] along with a view of the overall timer unit in Fig. 24-45(B). After

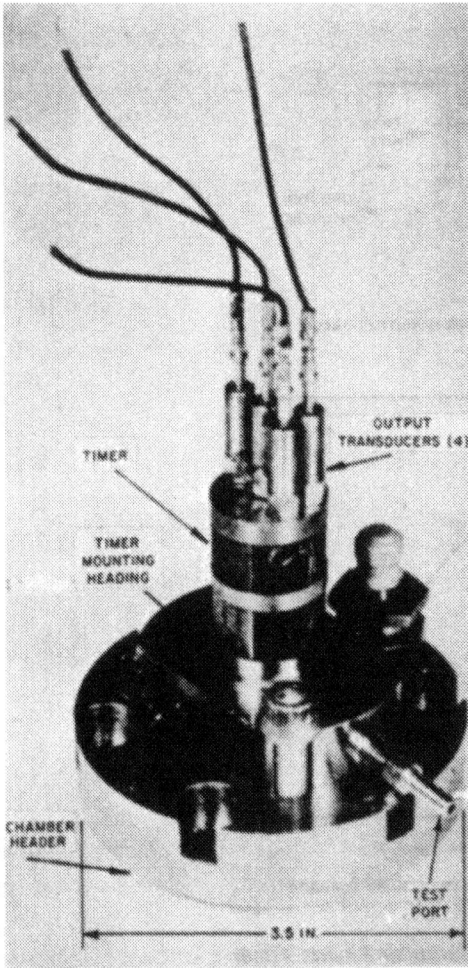

Figure 24-42. Location of Transducers on Timer Body

specification test procedure, including temperature ($-65°$ to $165°F$), shock, and vibration with a change in pressure of less than 0.5 in. of water required for closing the normally open contacts.

## 24-10 SETTING METHODS

The requirements for the setting mechanism are: (1) small size, (2) provide a decimal readout, (3) ruggedness, and (4) settable in increments to provide sufficient accuracy. The function of the setter is to program the air flow between the output channels of the binary stages comprising the flueric counter and the input (i.e., control jets) of the decoder circuit. For timers having binary stage setting ports, the setting mechanism programs the air delivered from the common setting manifold to each stage.

Some proposed setting devices have incorporated punched tape with a geared driving system or entirely geared arrangements. Printed on the tape is the decimal

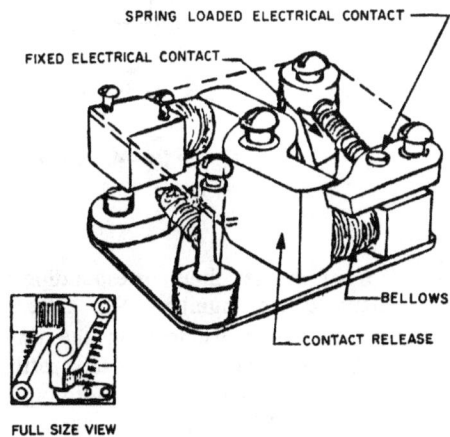

Figure 24-43. Bellows Actuated Rotary Switch

assembly and test, the switches were encapsulated in epoxy, and no further adjustment was possible or required. These switches were demonstrated to operate for billions of cycles with less than 15% change in the pressure necessary to effect closing. One package of six switches was subjected to the complete

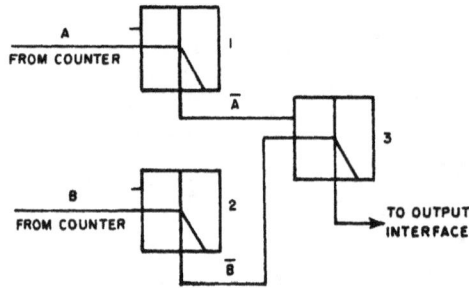

(A) APPLICATION OF THREE NOR GATES IN OUTPUT CIRCUIT

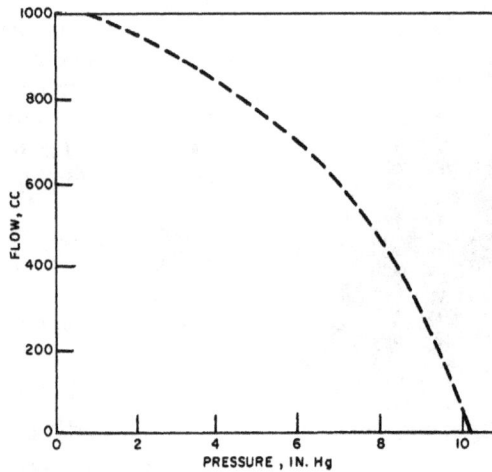

(B) PRESSURE-FLOW CHARACTERISTICS OF THIRD NOR GATE
WHICH ACTIVATES BELLOWS ROTARY SWITCH

*Figure 24-44. Output Interface Circuit for Flueric Timer*

equivalent of the set time corresponding to the binary coded number. Set time is displayed through a window on the timer housing. Although the concept of tape setting is attractive and natural for a binary to digital converter, difficulties may arise when one attempts to package such a device within the available space. For example, the tape length must be sufficient to provide for the maximum time and the tape must be capable of being driven in either direction, i.e., up

scale or down scale. Also the device must be rugged enough to withstand military operating environments. In an attempt to overcome some of these possible limitations, intermittent gearing should be considered. The apparent advantages of this system are ruggedness, positive two-way drive (no slip clutches, etc.), and compactness.

The feasibility of this scheme was demonstrated at Frankford Arsenal by the bread-

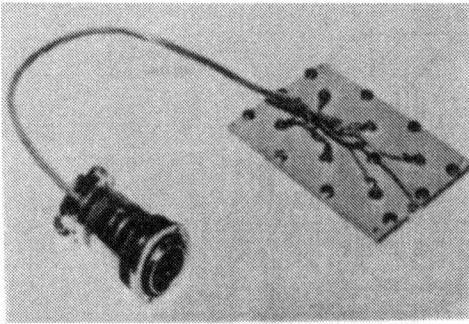

(A) SIX PORT DIAPHRAGM PRESSURE SWITCH

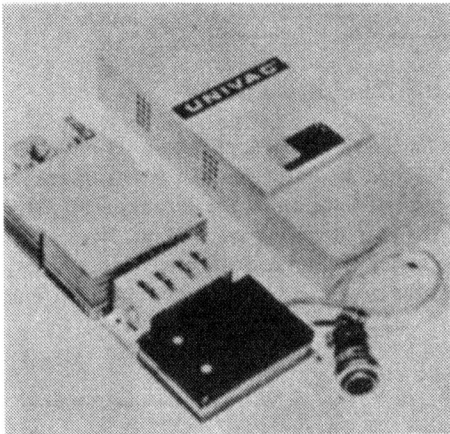

*Figure 24-46. Flueric Timer Setting Device Using Intermittent Gearing*

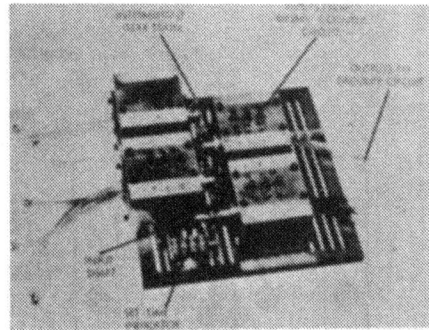

(B) SANDIA FLUID TIMER

*Figure 24-45. Pressure Sensitive Switch Used in Fluid Timer*

board device shown in Fig. 24-46[2]. The mounting plate is 8 in. square. As shown in Fig. 24-47[2] each housing contains a stationary cylinder with three equally spaced in-line tubing holes; a thin-walled sleeve with twelve holes arranged in a binary type sequence which is free to rotate about the fixed cylinder; and a spring-loaded shoe carrying three in-line hose connections. When breadboarded, the sleeves are interconnected by an intermittent gear train. A digital counter is

geared to the input shaft of the first sleeve, providing a decimal readout of the set time.

The sleeve holes are positioned in such a way that each 45 deg of sleeve rotation opens and/or closes the air passageways between the fixed cylinder and the shoe (Fig. 24-48[2]). The intermittent gear ratio between each sleeve is 8 to 1. Thus 8 revolutions of the first sleeve are required for one complete revolution of the second sleeve. Or one revolution of the first sleeve indexes the adjacent sleeve 45 deg.

To explain the operation of this device, let us take a specific example. Suppose it is desired to set 50.0 sec into the unit. Since a 4 to 5 gear ratio (reduction) exists between the digital counter and the input shaft, for a setting of 50.0 sec, 62.5 turns are required of the input shaft. Referring to Fig. 24-48, 62.5 turns, starting from the all blank position, would program 44.8 sec on the 3rd sleeve, 4.8 sec on the 2nd sleeve and 0.4 sec on the 1st sleeve. The 4th sleeve would not have moved, hence remains blank. The sum of the individual sleeve settings equals the total or desired time of 50.0 sec. If this device now were coupled to an eleven stage counter circuit and an eleven input NOR decoding circuit, upon the first coincidence of zero

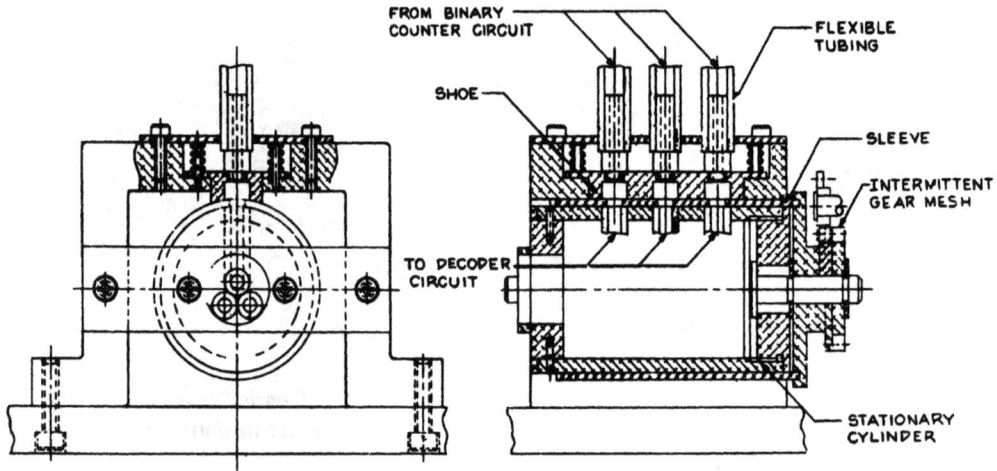

Figure 24-47. Housing Assembly—Flueric Timer Setting Device

signals from binary stages 3, 5, 6, 7, 8 and 9 an output signal from the decoder could occur 50 sec after the initiation of the timing cycle.

To make this setting arrangement suitable

for artillery fuze application, it must be redesigned to fit within the fuze contour[21]. Fig. 24-49[2] indicates the location and shape of the geared setting mechanism. Notice that the setter is positioned between the binary counter circuit and the decoder circuit. This

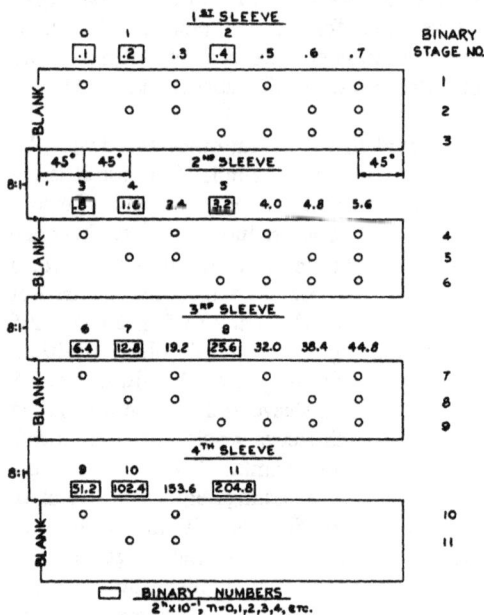

Figure 24-48. Sleeve Hole Locations

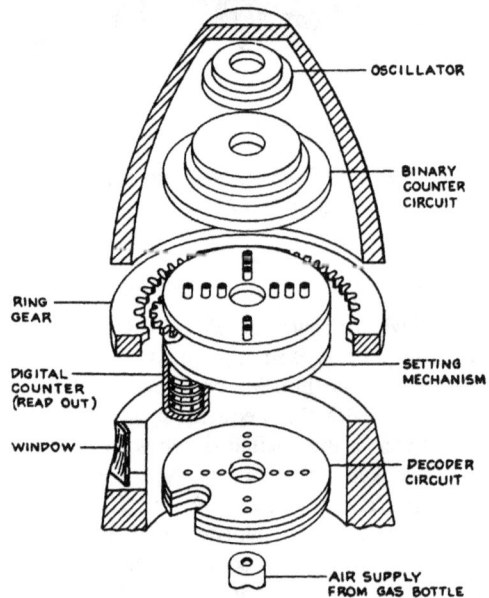

Figure 24-49. Flueric Timer Packaged for Artillery Fuze

is a natural location since the function of the setter is to interrupt the fluid signals between the binary stages and the eleven-input decoder circuit.

This miniaturized setting device consists of four intermittently driven gear wheels of approximately 0.5 in. pitch diameter sandwiched between two parallel metal plates. Each gear wheel hub contains the required twelve through holes and in function replaces the sleeves of the breadboard unit. The plates serve not only to support the gearing, but act as manifolds to permit connecting the fluid circuit to the device. Setting is accomplished by rotating an internal ring gear that drives, via a common idler gear, the first setter wheel and the digital readout counter. The large ring gear would be fixed to the upper portion of the fuze body while a readout window would be located in the lower or stationary portion of the fuze.

An example of a flueric timer developed at Picatinny Arsenal employs punched cards to set time intervals between 2 and 200 sec in increments of 0.1 sec. A sketch of the timer assembly is shown on Fig. 24-50[12]. This unit contains five dividing stages, twelve counter stages, and an output stage. With an oscillator working at 640 Hz, the 17th stage would provide a pulse to the output stage (19th) at the rate of $640/2^{17}$, or 0.0048828125 pulse per sec, or approximately 200 sec pulse separation. This is the maximum time delay if the oscillator frequency is not changed. The counter stages are presettable by inserting an encoded card into the timer which, upon receipt of a "preset pulse", initially sets a "0" or a "1" into each stage dependent on the time lapsed desired. Thus if the minimum time delay is desired, which is 1/640 sec, the first 17 stages would preset to "1's" as shown in Fig. 24-51(A)[12], the last stage to a "0". Upon initiation of the first pulse from the

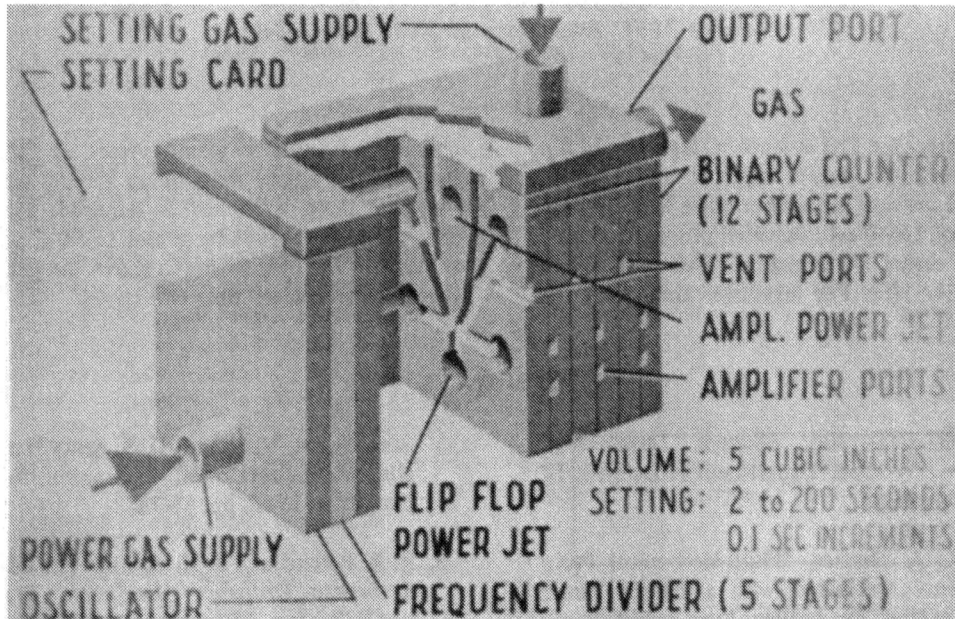

Figure 24-50. Fluid Timer With Card Preset

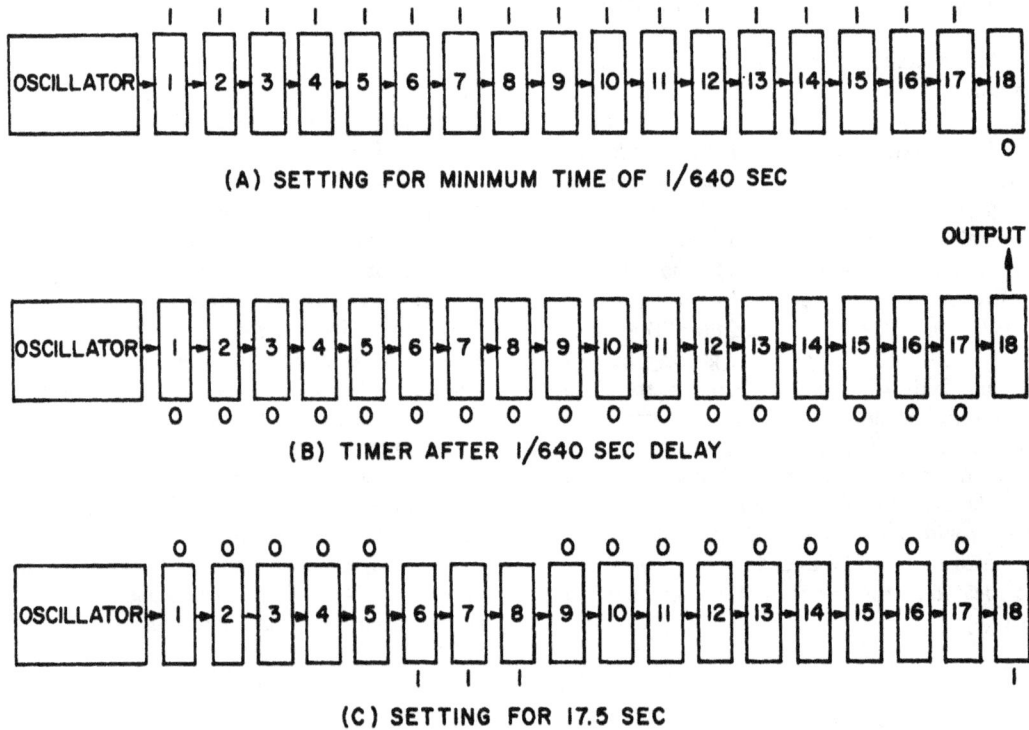

(A) SETTING FOR MINIMUM TIME OF 1/640 SEC

(B) TIMER AFTER 1/640 SEC DELAY

(C) SETTING FOR 17.5 SEC

*Figure 24-51. Preset Stage for an 18 Binary Field*

oscillator, the first seventeen stages would flip to "0's", the output stage to a "1" which would provide the output pulse in an elapsed time of 1/640 sec, provided propagation time delay errors are negligible. This is shown in Fig. 24-51(B). For maximum time delay the coded card initially would preset all "0's" into the 18 binary field. If an elapsed time between these extremes is required, some stages initially would be preset to "0's" and others to "1's". Fig 24-51(C) shows the initial preset for an elapsed time of 17.5 sec.

## REFERENCES

1. MIL-STD-1306, *Fluerics, Terminology and Symbols*, Department of Defense, 17 July 1968.

2. R. A. Shaffer, "Fluid-Mechanical Problems Associated with a Flueric Timer Designed for Artillery Fuze Application", *Proceedings of the Timers for Ordnance Symposium*, Volume I, U S Army Harry Diamond Laboratories, Washington, D.C. November 1966, pp. 58-78.

3. F. B. Pollard and J. H. Arnold, Jr., Eds., *Aerospace Ordnance Handbook*, Prentice-Hall, Inc., Englewood Cliffs, N.J. 1966.

4. D. J. Grant, et al., *Preliminary Study of Power Supplies for Pneumatic Systems,* U S Army Harry Diamond Laboratories, Washington, D.C., Report TR-847, June 30, 1960.

5. S. Katz and A. Krasnick, *Basic Considerations of Flow Parameters for Pneumatic Power Supply Systems,* U S Army Harry Diamond Laboratories, Washington, D.C., Report TM-60-7, February 16, 1960.

6. C. J. Campagnuolo and H. C. Lee, *Review of Some Fluid Oscillators,* U S Army Harry Diamond Laboratories, Washington, D.C., Report TR-1438, April 1969.

7. C. J. Campagnuolo and S. E. Gehman, *Flueric Pressure- and Temperature-Insensitive Oscillator for Timer Application,* U S Army Harry Diamond Laboratories, Washington, D.C., Report TR-1381, February 1968.

8. "The Application of Flueric Devices to Ordnance Timers", Journal Article 51.0 of the JANAF Fuze Committee, May 3, 1967 (AD-834 083).

9. C. E. Spropoulos, "A Sonic Oscillator", *Proceedings of the Fluid Amplification Symposium,* Volume III, U S Army Harry Diamond Laboratories, Washington, D.C., May 1964, pp. 27-52 (AD-601 501).

10. Fluid Power Reference Issue, Machine Design, **40**, 102-9 (September 19, 1968).

11. J. M. Kirshner and C. J. Campagnuolo, "A Temperature-Insensitive Pneumatic Oscillator and a Pressure Controlled Pneumatic Oscillator," *Proceedings of the Fluid Amplification Symposium,* Volume II, U S Army Harry Diamond Laboratories, Washington, D. C., October 1965, pp. 5-19 (AD-623 456).

12. I. B. Berg and Louis A. Parise, *Flueric Timer Evaluation for Ordnance Application,* Picatinny Arsenal, Dover, N.J., Technical Report 3613, February 1968.

13. *Fluidics,* Fluid Amplifier Associates, Inc., Boston, Mass., 1965.

14. T. A. Shook, T. F. Chen, and T. D. Reader, *Fluid Amplification, 12. Binary Counter Design,* Sperry Rand Corp., Final Report, 30 November 1964 (AD-617 699).

15. R. W. Warren, *Fluid Amplification, 3. Fluid Flip-Flops and a Counter,* U S Army Harry Diamond Laboratories, Washington, D.C., Report TR-1061, 25 August 1962 (AD-285 572).

16. *Fluidic Systems Design Guide,* Fluidonics Division of the Imperial-Eastman Corp., Chicago, Ill., 1966.

17. D. L. Letham, "Fluidic System Design, 5. Practical Fluid Resistors," Machine Design, **38**, 134-8 (June 9, 1966).

18. D. L. Letham, "Fluidic Systems Design, 7. Turbulence Amplifiers," Machine Design, **38**, 157-9 (July 7, 1966).

19. H. C. Yance, Jr. and J. Van Der Heyden, *A Fluidic Timer for Missile Applications,* Martin Company, Orlando, Florida, April 1967 (AD-812 261).

20. S. G. Martin, "Fluid Timer Development", *Proceedings of the Fluid Amplification Symposium,* Volume III, U S Army Harry Diamond Laboratories, Washington, D.C., October 1965.

21. R. A. Shaffer, *Problems Associated with a Flueric Timer Designed for Artillery Fuze Application,* Frankford Arsenal, Philadelphia, Pa., Report M68-26-1, May 1968.

# CHAPTER 25

## PRODUCTION TECHNIQUES

Techniques for fabrication of flueric devices are not difficult. In fact, the most useful techniques are well-known. However, the fact that the performance and characteristics of a flueric device are related closely to its geometric shape can lead to problems.

Although models have been constructed which have adjustable parts, in most instances, a change in a geometric shape or in a dimension requires the fabrication of a new device. Then, when a satisfactory design is attained, the problem of reproducing the design in production units arises.

Foremost among the considerations is the choice of material to use in the production units. Flueric devices are subjected to both structural and hydraulic forces. The material must have sufficient strength to withstand these forces without undue distortion.

Surface hardness of the material also must be considered, particularly if the fluid carries abrasive particles. Wear of the component is especially critical in nozzles and at the splitter in stream-interaction devices. Other factors, such as operating temperature and type of working fluid, also enter into the selection of a material.

The problems of quantity manufacture differ somewhat from those of a development laboratory. In the laboratory, ease of fabrication is desirable to permit rapid change in designs. Long-term properties of the material, such as wear, are often not as important as workability of the material.

Some of the production techniques discussed in this chapter are suitable for mass production of flueric devices, while others are useful only in development work. They are mentioned here with the thought that at some time in the future the processes may be feasible on a production basis. For additional guidance on producibility, see Ref. 1.

## 25-1 MATERIALS

### 25-1.1 GENERAL REQUIREMENTS

Materials for flueric devices should be dimensionally stable, capable of being formed with precision (nozzles with dimensions of 0.010 in. or smaller may be required), and suitable for sealing. Due to the critical dimensions of fluid elements and the trend toward miniaturization, a material must be selected that does not drastically change its physical size and shape when exposed to the intended working environment. The environmental conditions may require that the designer consider chemical, thermal, and moisture effects on the device. Also the designer should consider strength, weight, size, cost, and shelf life.

A material must be chosen so that precisely sized and reproducible elements can be fabricated. Also, since present fluid devices are built from laminates, a material must be chosen with its sealability (between layers) in mind. Leaks in a complex fluid timer are difficult to locate and will cause the unit to operate erroneously.

### 25-1.2 SUITABLE MATERIALS

Flueric devices are presently fabricated of materials of three general types—plastics, metals, and ceramics. These three provide

**TABLE 25-1**

**SUITABLE MATERIALS FOR FLUERIC DEVICES**

| PLASTICS | | METALS | CERAMICS |
|---|---|---|---|
| **Thermoplastics*** | **Thermosetting**** | **Suitable** | **Suitable** |
| a. Most suitable:<br><br>acetals, nylons, polycarbonates, ABS | a. Most Suitable:<br><br>melamines phenolics | steel<br>copper<br>aluminum<br>nickel<br>magnesium (base metals and alloys) | glasses<br>carbides<br>nitrides<br>oxides |
| b. Usable But Less Desirable:<br><br>acrylics<br>polyethers<br>polypropylenes<br>polystyrenes<br>vinyls<br>fluorocarbons | b. Slightly Less Suitable:<br><br>epoxies,<br>polyesters,<br>silicones<br>ureas | | |
| c. Not Suited:<br><br>cellulosics<br>polyethylenes<br>polyurethanes | c. Used Mostly for Coatings:<br><br>alkyds | | |

*Softened or resoftened by heat; then formed; no chemical change.

**Initial molding by heat and/or pressure—after set chemical change occurs and further shape change difficult or impossible.

properties required for each application. Also, any of these may be used in combination with each other to extend their usefulness. Table 25-1 lists materials of each type; Table 25-2[2] gives the general characteristics of plastics, metals, and ceramics, and consequent application considerations.

## 25-2 FABRICATION

Various processes have emerged in working with plastics, metals, and ceramics. These are discussed in the paragraphs that follow.

### 25-2.1 PLASTICS

The processes generally used in fabricating plastics include casting, thermoforming, photoetching, injection molding, compression molding, jet molding, and transfer molding. Typical comparisons with reference to economical quantity of production are shown in Fig. 25-1[2].

#### 25-2.1.1 Casting

This is a rapidly developing process. Starting from a two-component catalyst system in liquid or paste form, the two components are mixed and immediately poured into a mold. The process can be conducted at room temperature and with low mold pressure, and is suitable for complex precision parts. Types of thermoplastics used include methacrylates, polyesters, nylons, acrylonitrile, and epoxies.

The epoxy casting process produces a one-piece unit quickly and at low cost. With a suitable master unit, the design can be changed quickly and easily. The basic process consists of making plastic duplicates of an item, with the intermediate step of a rubber mold. The item to be duplicated is placed within a suitable form, which may merely consist of blocks of wood or metal arranged on a flat surface to contain the liquid rubber.

25-2

**TABLE 25-2**

**GENERAL CHARACTERISTICS OF MATERIALS AND
CONSEQUENT APPLICATION CONSIDERATIONS[2]**

**Characteristics:**

|  | Plastics | Metals | Ceramics |
|---|---|---|---|
| Dimensional stability | F | G | E |
| Precision formability | G | E | E |
| Seal integrity and durability | G | F | F |
| Strength | F | E | G |
| Chemical resistance | G | G | E |
| Thermal resistance | F | G | E |
| Moisture resistance | F | F | E |
| Weight | E | F | G |
| Processability | E | G | G |
| Cost per element | E | G | G |

E = Excellent    G = Good    F = Fair

**Applications:**

| Materials | Suitable for Applications Involving or Requiring: | Not Suitable for Applications Involving or Requiring: |
|---|---|---|
| Plastics | Mild environmental conditions<br>Low cost<br>Low weight | High mechanical, thermal, and moisture resistance<br>High precision<br>High dimensional stability |
| Metals | High thermal shock<br>Mechanical shock<br>High precision | Low weight<br>High moisture<br>Very high dimensional stability |
| Ceramics | High chemical, thermal, and moisture resistance<br>High precision<br>Very high dimensional stability | High thermal shock<br>Mechanical shock |

The liquid, a silicone rubber compound (RTV) that vulcanizes at room temperature, then is poured over and around the master. Use of a vacuum chamber at this time is helpful in removing bubbles of entrained air from the rubber. (Another technique for removing entrained bubbles is to work the rubber with a very fine stream of high-velocity air that punctures the bubbles.) The rubber penetrates quite well into the surface details of the master. Vulcanizing time, depending on mold temperature and the curing agent used, ranges from a few hours to 40 hr. When vulcanized, the mold is sufficiently flexible and elastic to be removed from the master without great difficulty or deformation. The mold is examined after its removal from the master, and "flash" is trimmed off. Modifications may be carved into it at this time. When the mold is fully

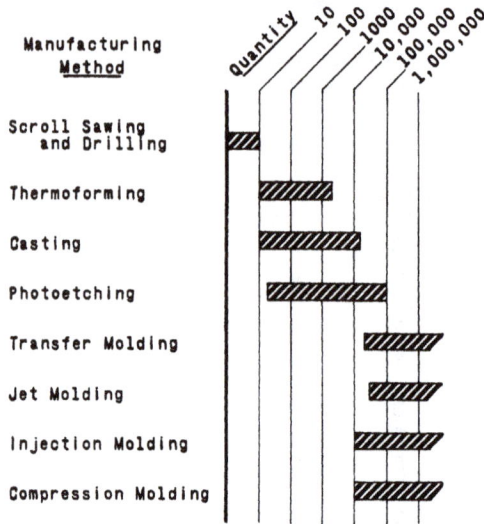

*Figure 25-1. Comparison of Plastic Forming Processes for Fabricating Economic Production Quantities[2]*

prepared, the epoxy compound is poured in. Bubbles are removed from the epoxy either mechanically, with a pick or brush, or by means of a vacuum. After the epoxy has cured, the piece is removed from the mold. Here, again, the elasticity and flexibility of the rubber allow pieces to be removed which have no draft. With little or no further work, the resulting piece accurately duplicates the desired features of the original item. Once the optimum geometry is found, the process can be used to duplicate units indefinitely and with considerable uniformity and accuracy from copy to copy. If the mold becomes damaged, second and third generation molds can be made from the epoxy pieces with little or no loss of detail.

As a material for use in flueric devices, epoxy has several desirable properties. It can be machined if necessary; it can be permanently bonded to other epoxy pieces, such as cover plates, or to itself if a unit should be broken; and it can be made with different degrees of softness and flexibility in each of several layers.

25-4

### 25-2.1.2 Thermoforming

This process uses a thermoplastic material. Heat and pressure, or vacuum are used to force the material to conform to the die contour. The process is suited to large planar flueric elements where precision and detail requirements are not too high.

### 25-2.1.3 Photosensitive Plastics

DuPont has developed a photopolymer plastic (Templex[TM]) that is being used to form flueric elements. Because of the speed and simplicity of the process, it is especially useful for breadboard experimental work, where the designer must know quickly the effect of small design changes. For large lots better methods are available. The process centers on the fact that the photosensitive plastic when exposed to ultraviolet light becomes insoluble. The unexposed part remains soluble in dilute aqueous sodium hydroxide spray. The process is shown in Fig. 25-2[2].

This process offers several advantages:

(1) Construction Ease. Three-dimensional

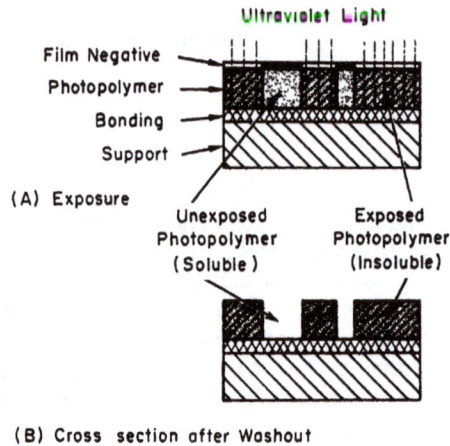

*Figure 25-2. Photoetching of Photosensitive Plastic[2]*

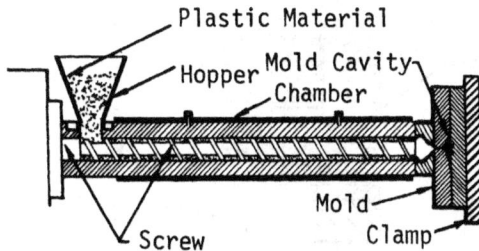

Figure 25-3. The Screw Method of Injection Molding[2]

plastic configurations can be made from two-dimensional black and white copy.

(2) Speed. Once the negative is prepared, processing requires about 20 min.

(3) Smooth Side Walls. An uneven surface can disrupt laminar flow.

(4) Simplicity. Both material and special-purpose machinery are available and, because no unusual skills are required, in-house manufacture is possible.

(5) Precision. A mean side wall taper of 2 deg can be maintained.

Several limitations also should be noted:

(1) Operating Temperature Range. Presently limited to approximately $-100°$ to $+200°F$.

(2) Aspect Ratio. Cannot exceed 3:1.

(3) Sensitivity to Moisture and Some Chemicals. The results of long-life experiments to test dimensional stability and other critical features are being determined.

## 25-2.1.4 Injection Molding

Injection molding is a process for forming a granulated thermoplastic material into a complex, precision-finished part at high speeds and with little material waste. The process is similar to the die casting of metals in some respects. Injection molding usually is accomplished by the plunger method or the screw method. The screw method, shown in Fig. 25-3[2], uses a continuous screw to transport the granulated material from the hopper, through a heating chamber, and into the mold. The movement of the screw thoroughly mixes the constituent resins, plastics, and fillers as they are being liquefied at temperatures that range between 250° and 500°F. The choice and control of temperature are important; for, while increasing temperatures facilitate the movement of the plastic into the mold cavities, too high a temperature may degrade the material. The mixture is forced at pressures of 10,000 to 20,000 psi through a system of channels into the mold cavities. Cooling channels keep the mold temperature below the fluid point of the plastic. After the part has solidified, the mold opens to eject the part and then closes as the cycle is repeated. The operation usually is entirely automatic to this stage. Little finishing work is necessary other than "degating" to remove the excess material that solidified in the channels between the nozzle and the mold cavities.

The plunger method uses a ram to force the material into the mold. This technique is not as satisfactory as the screw method because of occasionally incomplete mixing of the material charge, somewhat higher heat required, and the increased complexity of the plunger-type equipment. Roughly 80 percent of new injection molding machines purchased are of the screw type. The use of multicavity dies allows simultaneous production of many identical parts which may weigh only a fraction of an ounce. Since thermoplastics can be remelted, the excess material can be reprocessed. Manufacture of multicavity dies for injection molding is a costly and time-consuming process. Therefore, injection molding should be used for flueric device components only where the design configuration is fixed and large quantities are required.

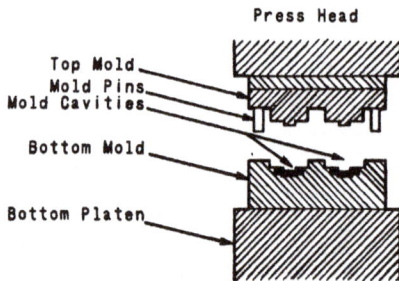

*Figure 25-4. Compression Molding[2]*

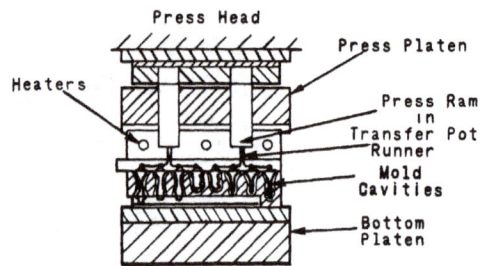

*Figure 25-5. Transfer Molding[2]*

### 25-2.1.5 Jet Molding

This process is used with thermosetting materials and is an adaptation of the plunger method of injection molding used with thermoplastics. The feature of this method is to use controlled heat so that the granules of the thermosetting material become "soft" but the material does not "set" or polymerize until some time later when the mold is filled completely. These critical controls make the process some ten or twenty times longer than injection molding.

### 25-2.1.6 Compression Molding

A die for compression molding is shown in Fig. 25-4[2]. A fairly precise "charge" of thermosetting resin is placed in the mold cavity, and the top mold is brought down to induce pressures of 2000 to 4000 psi (or greater) on the mold. The temperature—typically between 250° and 375°F—and pressure are held until polymerization occurs, then the part is removed.

Compression molding is a high-pressure process most advantageous for the forming of identical parts of relatively simple design from thermoset materials. Compression molding is more economical for high volume than small lot production. The operating cycle—a function of the material, part thickness, and size—may be a minute or longer. Tolerances can be held to ± 0.005 in. per linear inch

where necessary. Like other molding techniques, close tolerances specified but not actually required increase the cost of the finished part. Excess flash created between the top and bottom molds must be removed from the molded part. The scrap rate, which is sometimes 30%, is higher for compression molding than for injection molding mainly because thermoset rejects and flash cannot be remolded. Compression molding requires more complex machines and more highly skilled operators than the injection process.

### 25-2.1.7 Transfer Molding

In the transfer process (Fig. 25-5[2]) the molding material is placed in a chamber called a transfer pot usually located above the closed, heated molds of the machine. Mold heat and plunger pressure cause the thermosetting preform to plasticize enough so that the material flows through the runners into a mold cavity. The total time of pressure application may be as long as one minute. After the material has been cured, the mold is opened and the finished part ejected without cooling.

### 25-2.2 METALS

Fabrication processes for metals include conventional machining, powder metallurgy, die-casting, chemical machining, electroforming, electrical discharge machining, and electrochemical machining.

### 25-2.2.1 Conventional Machining

The most commonly used processes include milling and engraving (pantograph). There is a definite lower limit of size to which these processes can be used, and they are not usable for the microminiature elements that have begun to appear. However, these processes are suited for laboratory breadboard construction. By using the pantograph method, a large scale master of the element is generated; then a model is accurately milled in Bakelite, using a pantograph engraver with a lower limit tool diameter of about 0.1 in., which also becomes the channel width limit.

In conventional machining, if a separate unit is made to incorporate each design change, a great many hours of valuable shop time is used. In addition, successive units may vary slightly in those dimensions that should have remained unchanged. While not often serious, such changes add an unknown factor to the analysis of test data. To assure that only the desired dimensions are changed, sometimes only one unit is built. This unit is then tested, disassembled, modified, reassembled, and retested. This process takes less shop time than does the construction of separate units, but only a limited number of variations can be made on one unit. In addition, if the desired configuration is passed, it is usually not practical to "back up" and re-examine it. In any event, the number of combinations possible in such a study makes any process that involves machine work on each unit prohibitive in both time and money.

### 25-2.2.2 Powder Metallurgy

Powder metallurgy is a technique that produces finished parts of uniform density and high strength, starting from powdered metals. Typical metals are copper, tin, lead, zinc, nickel, and iron. The technique is suitable only for relatively high production quantities (1000 units or more). The process consists of a sequence of powder blending,

pressing, and sintering. Blending produces the desired mixture, then cold pressing between 20 and 80 tons produces a "green state" item, which is still fragile. Sintering is the next stage which is performed in an oven at high temperatures, but below the melting points of the metal powders. Sintering provides a strong chemical bond and "recrystallizes" the powdered particles. The atmosphere of the furnace usually consists of inert gases that prevent the sintered material from oxidizing. Powder metallurgy requires certain special design considerations. First, thin parts and narrow, deep sections should be avoided: the former may break and the latter may prevent complete formation of the part. These limitations may make the process unsuitable for flueric elements with high aspect ratios and for delicate splitters. The tolerances permitted in the direction of pressing should be greater than those at right angles to pressing. Dimensions perpendicular to the pressing direction can be held to about $\pm 0.0015$ in./in. Slots and channels must be tapered to two or three degrees to allow ejection of the part from the die after compacting.

### 25-2.2.3 Die Casting

Die casting is a process that forces molten metal alloys into metal dies to produce complex parts. This technique is a rapid and relatively inexpensive method of producing large quantities of precision metal parts. There are actually four principal methods of die casting: plunger, air injection, vacuum, and cold chamber methods. The first three are examined in the paragraphs that follow. Selection of the proper method is determined primarily by the kind of alloy to be cast and the production quantity desired.

For all these methods, section thicknesses are also a function of the type of alloy used. A practical minimum is 0.040 in., and maximum sections up to 0.5 in. are possible, although problems with shrinkage are likely. Tolerances attainable vary with the type of

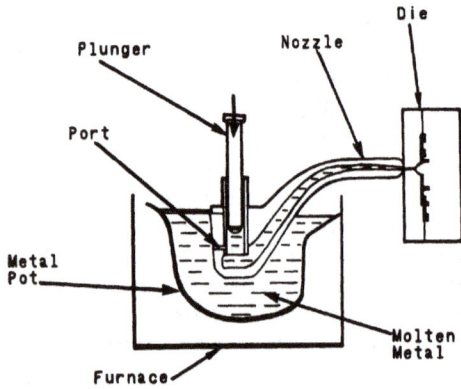

Figure 25-6. The Plunger Method of Die Casting[2]

alloy, but a figure of ± 0.004 in./in. gives a rough order of magnitude for linear dimensions.

Except for finishing, secondary operations usually are not required for die cast parts. In all instances, however, flash and cores must be separated from the finished part. Mechanical, chemical, electrolytic, or electroplated finishes may be used where they are required for proper performance of the flueric elements.

A discussion of die casting methods follows:

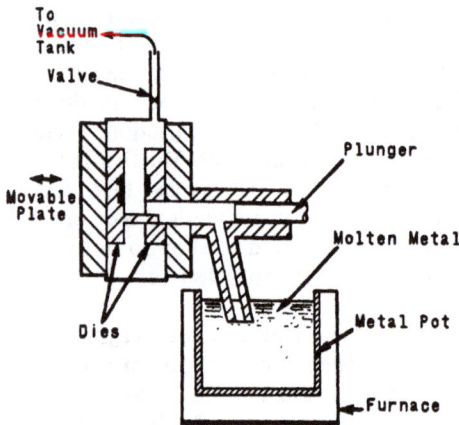

Figure 25-7. The Vacuum Method of Die Casting[2]

(1) Plunger Method. As its name implies, the plunger method (Fig. 25-6[2]) uses a plunger to force the molten metal alloy into the open die. The plunger moves vertically in a vessel immersed in a large metal pot containing the molten metal. Molten metal is allowed to flow from the pot into the vessel during the upstroke of the plunger. On the downstroke, the metal in the vessel is forced into the spout through the nozzle, and into the die cavity. When the metal in the die hardens, the plunger repeats the cycle. The die is opened and the casting is ejected during the plunger upstroke.

(2) Air Injection Method. This method differs from the plunger method in that air pressure of 300 to 600 psi, rather than a plunger, is used to force the metal into the die, and the whole vessel, rather than the plunger, moves through an upstroke and a downstroke. The air injection method is well suited to production of aluminum alloys, although low melting point alloys may also be used.

The vacuum process (Fig. 25-7[2]) offers certain improvements of particular interest to the designer of small flueric elements and circuits. Thin walls, important for miniaturization, can be achieved with the vacuum process. Wall thicknesses of 0.030 in. are common, and 0.018-in. walls are obtainable. Improved surface finishes reduce the need for additional finishing operations after casting. Lower porosity throughout the casting increases the part tensile strength and hardness.

A partial vacuum is maintained in the injection chamber and die cavity. The plunger then forces the molten metal into the partial vacuum of the die cavity. After solidification, the vacuum is released, the part ejected, and the piston withdrawn for another cycle. The vacuum process can produce parts as rapidly as conventional die casting methods, but at the higher cost of maintaining a partial vacuum.

**TABLE 25-3**

**TYPICAL SURFACE DIMENSIONAL TOLERANCES\* IN PHOTOETCH PROCESS[2]**

**Metal Thickness, in.**

| Metal | 0.001 | 0.005 | 0.010 | 0.015 | 0.020 |
|-------|-------|-------|-------|-------|-------|
| Copper & Alloys | 0.0005 | 0.002 | 0.003 | 0.005 | 0.005 |
| Steels | 0.0005 | 0.002 | 0.003 | 0.005 | 0.005 |
| Aluminum Alloys | 0.002 | 0.005 | 0.005 | 0.005 | 0.005 |
| Molybdenum | 0.0005 | 0.002 | 0.003 | 0.005 | 0.005 |
| Titanium | 0.0005 | 0.002 | 0.003 | 0.005 | 0.005 |
| Nickel | 0.001 | 0.003 | 0.005 | 0.005 | 0.0075 |

\*Tolerances ± (in.)

### 25-2.2.4 Chemical Machining

Two general types of process are included under chemical machining: photoetching and chemical milling. Each is discussed:

(1) Photoetch Process. This process is being applied to metal sheets 0.020 in. thick or less and is becoming more and more important in fabricating laminar flueric elements. It works on the principle of employing chemicals to "etch" away exposed metal areas while a photographic technique "masks" out other areas that are to remain unchanged (Fig. 25-8[2]). This process has been used with the following metals in order of increasing difficulty: copper, nickel, carbon steel, stainless steel, aluminum, titanium, and molybdenum. Etching rates vary from 0.005 to 0.003 in./min, and dimension tolerances increase with increasing metal thickness (see Table 25-3[2]).

(2) Chemical Milling. This process is similar to photoetching in that both processes use etchants to remove the metal. However, chemical milling uses tape or paint rather than the photographic method to prepare the metal and is suitable for thicker parts. It can be used for almost any metal (see Fig. 25-9[2]).

Chemical milling has several advantages

that are important to the manufacture of flueric units. A specified amount of material may be uniformly removed from the surface of a metal part to leave a complex pattern. Parts, once masked, may be etched simultaneously in numbers that are limited only by the size of the etchant bath. Chemical milling does not change the structure of the metal

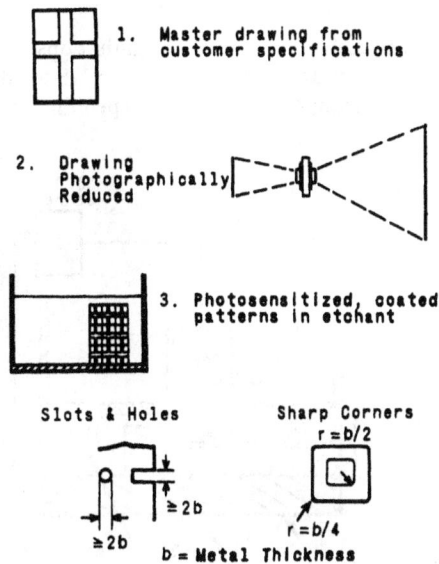

Figure 25-8. Photoetching Process and Design Considerations[2]

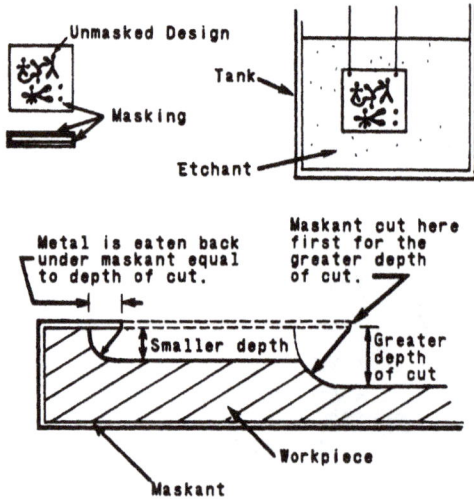

Figure 25-9. Chemical Milling, Showing the
Effect Upon the Workpiece[2]

being etched, nor does the process set up
internal stresses. Finally, operating and
overhead costs are often lower than those for
other metal-removal techniques.

Chemical milling also has limitations. The
process is relatively slow: to etch a 1/8-in. slot
requires from 40 min to 4 hr, depending upon

the etchant and material. Secondly, improper
masking may not be obvious until after the
etch has been made. Finally, the process
presents some design problems. Undercutting
is unavoidable and must be taken into
account in design. The resultant rounded
corners may have an undesirable effect for
fluid flow. There is a limit to miniaturization,
imposed by masking techniques, beyond
which chemical milling cannot be used.

### 25-2.2.5 Electrical Discharge Machining (EDM)

This method of metal removal employs a
spark discharge between the tool and the
workpiece. EDM is particularly useful for
complicated designs in hard, brittle metals. A
typical EDM system is illustrated in Fig.
25-10[2]. The tool, which is the cathode, is
shaped in the form of the desired cavity. The
workpiece is the anode and is submerged in a
dielectric fluid. Typical equipment operates
with high currents and a potential of
approximately 80 V across a gap separation of
about 0.001 in. An intense spark occurs
across the gap. The spark vaporizes a small
amount of metal from both the tool and the
workpiece. The DC source is pulse-modulated
up to 100,000 pulses per sec and the gap is
adjusted to maintain a constant voltage drop.
Metal is lost from both the tool and
workpiece but less is lost from the tool.
Ratios of workpiece penetration to tool wear
vary from 3:1 to 7:1.

### 25-2.2.6 Electrochemical Machining (ECM)

This is a deplating operation in which the
cathode tool and anode workpiece are
immersed in a flowing electrolyte. A DC
current passes through the circuit, and the
deplated metal is carried off in the electrolyte. Metal removal rates vary from 0.1
in.$^3$/sec for a 1000-A unit to 1 in.$^3$/sec for a
10,000-A unit. Typical potentials maintained
across a gap (0.001 to 0.003 in.) are 4 to 10
V. Typical electrolytes are sodium chloride,
or sodium nitrate at flow rates up to 200
ft/sec.

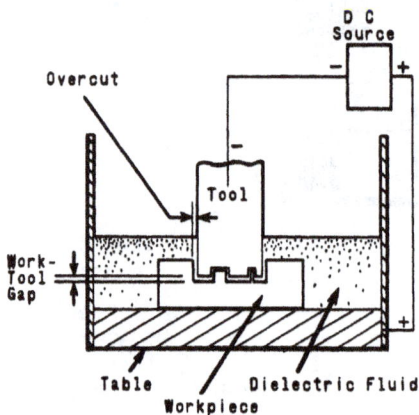

Figure 25-10. Electrical Discharge Machining
Arrangement[2]

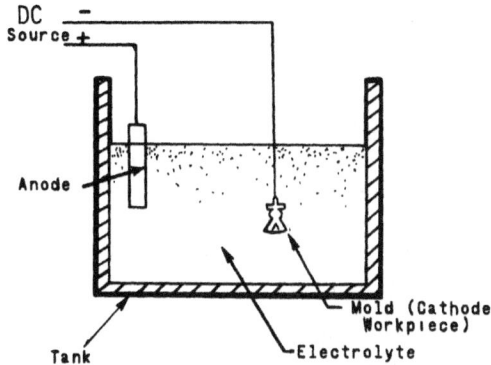

Figure 25-11. Electroforming[2]

Compared to EDM, the ECM technique is to be preferred when: (1) lot sizes are comparatively large and on a continuing schedule; (2) the volume of metal to be removed is large; (3) a good finish, relative to cutting time is required; (4) the area of engagement between workpiece and electrode is large; (5) three-dimensional contours are required; (6) the contour of a cut is critical; and (7) erosion of the electrode would disturb the form.

### 25-2.2.7 Electroforming

This is an electrolytic plating technique for manufacturing metal parts. It is similar but not identical to electroplating. Electroforming can be used to produce very complex forms with excellent tolerances, but it is expensive, limited to a few metals, and economically more suitable for small production runs. The electroforming process requires a core that duplicates the cavity or inside dimensions of the part to be produced. The core is placed in an electrolyte bath, and metal is electrolytically deposited over all surfaces of the core at a constant rate. When the required amount of metal has been deposited, the core is removed from the bath and separated from the metal deposited on it. The metal deposit then becomes the finished part having the internal dimensions of the core and a

thickness dependent upon the metal deposited from the bath (see Fig. 25-11[2]).

Cores for electroforming are made from metals or nonmetals and can either be expendable or reusable, depending upon the design of the part. Plastics, plasters, waxes, stainless steel, and aluminum are possible core materials and can be dissolved by suitable chemicals without damaging the part produced. The electroforming baths are similar to those used in electroplating. Only a few types of metal can be electroformed: copper, silver, iron, nickel, and chromium.

Electroforming techniques with expendable cores offer new design possibilities for the manufacture of complex, planar, and axisymmetric flueric subassemblies and complete units, in one piece and without interconnections or parting planes. The possibilities for making unusual and miniature configurations appear to be limited only by the ability to fabricate cores. Miniature parts from 0.030 to 0.20 in. in diameter, 1/8-in. long, and with a wall thickness of 0.001 ± 0.0001 in. have been produced. Thicknesses may vary from 0.001 in. to about 1 in.

### 25-2.3 CERAMICS

Three general processes of fabricating ceramics may be used: photoetching, ultrasonic machining, and electron beam machining.

### 25-2.3.1 Photoetched Ceramics

A process developed by Corning Glass Works in conjunction with the US Army Harry Diamond Laboratories uses a photographic technique with a thin sheet of silicate glass that contains a photosensitive ingredient such as the cesium radical, $CE^{+3}$. In the presence of ultraviolet light, the exposed glass absorbs the radiation, creating a print constant in depth. Post heating to 1200°F creates colloidal particles in the exposed areas. When dipped in hydrofluoric acid the

All dimensions in inches

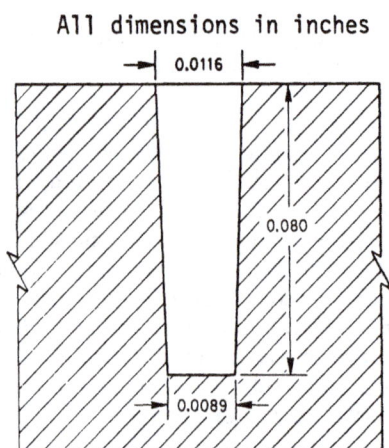

*Figure 25-12. Cross Section of Typical Channel Etched in Photosensitive Glass*[3]

exposed areas are etched away 20 to 30 times faster than the unexposed glass. With this process, the sides of a channel have a slope of less than 2.5 deg from the vertical, Fig. 25-12[3]. The channel is then, for all practical purposes, rectangular.

Typical nozzle dimensions for devices built by this technique are 0.020 by 0.080 in. Nozzles as small as 0.005 by 0.002 in. are accurately reproducible by the method. These nozzle widths are comparable to the diameter of a human hair.

Depth of the etching easily is controlled. With a 0.040-in. nominal depth, variation in depths across a plate can be kept within 0.002 in. Across a single element, tolerances can be kept within 0.001 in. Tolerances on holes and channel widths up to 0.25 in. can be kept within 0.001 in. Up to 1 in., widths can be kept within 0.002 in. Up to 5 in., center-to-center distances can be maintained within 0.003 in.

Good surface finishes are attained by etching. A typical finish for a wall is 125 $\mu$in. The bottoms are considerably smoother, 20

$\mu$in. being a typical value. The larger value of 125 $\mu$in. appears to be low enough to assure proper device performance.

Certain precautions are necessary in designing the pattern to be optically reduced. For example, nozzles should be designed at least 0.010 in. longer than the desired finished length because they etch from both ends. To produce a sharp splitter, the wedge must be designed with a radius on the end so that when the etching process is finished, the wedge has just eroded down to a sharp point. Narrow channels and small holes present problems. A fresh supply of acid cannot be maintained at the etching surface and erosion lags behind the expected rate.

When the flueric circuit is etched, a covered plate must be sealed onto it. Various gaskets have been tried but have been abandoned in favor of sealing with an epoxy cement. For a more permanent seal, a glass cover plate can be fused onto the etched base.

The glass etching method gives excellent reproducibility. The technique has been developed to the point where variations in the etched base plates have a smaller effect on the device or circuit performance than does the effect of assembly variations.

These fluidic elements offer advantages of high dimensional stability, no moisture absorption, good shock resistance, and operating capabilities to temperatures of 1000°F. The process is summarized in Fig. 25-13[2].

### 25-2.3.2 Ultrasonic Machining

Ultrasonic machining is used to work hard brittle materials, including ceramics. The method is based upon the concept of minute particle removal by fine abrasive particles traveling at high velocities. The pattern eroded is identical to the shape of the tool. The tool is held very close to the workpiece and vibrated at 15,000 to 30,000 Hz with an amplitude of 0.001-0.005 in. Between the

DESIGN

PHOTOGRAPHIC
NEGATIVE

PHOTOSENSITIVE
GLASS

ULTRAVIOLET
EXPOSURE

FURNACE

CHEMICAL
MILLING

FINISHED
GLASS

CRYSTAL
PROCESS

FINISHED
PHOTOSENSITIVE
CERAMIC

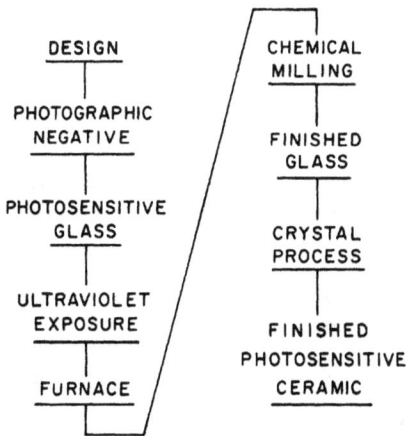

*Figure 25-13. The Photoetching Process for Ceramics and Some Examples of Etched Photosensitive Ceramic[2]*

tool and workpiece is a slurry containing the abrasive, such as boron carbide or aluminum oxide. The tool, made from cold-rolled or stainless steel, wears only slightly. No pressure is exerted on the workpiece by the tool. A typical setup is shown in Fig. 25-14[2].

### 25-2.3.3 Electron Beam Machining

Electron beam machining focuses a stream of electrons in a bundle about 0.001 in. in diameter with a cross-sectional density of $10^9$ W/in.[2] on the workpiece and vaporizes the material it impinges upon. Removal rates are small (about 0.1 mg/sec) and only relatively small cuts are economically feasible. The process makes possible the production of quite precise and fine cuts of any desired contour in any material. Fig. 25-15[2] is a typical setup. The two major limitations are that (1) a vacuum of about $10^{-4}$ mm of mercury must be maintained within the chamber, and (2) X-ray shielding must be used. However, the capability of this process to cut 0.005-in. channels separated by 0.010 in. may be very important for microminiature devices.

### 25-3 SUMMARY

As indicated, flueric devices can be formed from a variety of materials that are selected by the designer on the basis of performance properties. When production is considered, the economy of each process also must be weighed against other factors before a decision is made. Looking to the future, it appears that metals and ceramics will be used more often than plastics. Also, since the advent of microminiaturization, the requirements will be for processes that can fabricate

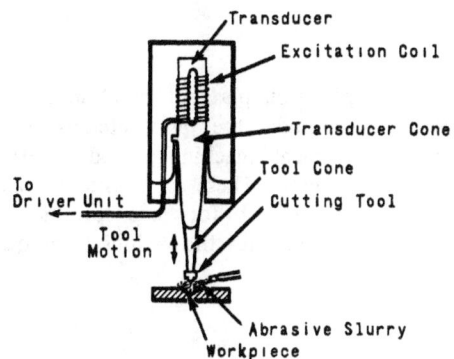

*Figure 25-14. Ultrasonic Machining[2]*

### TABLE 25-4

### COMPARISONS OF VARIOUS FABRICATION TECHNIQUES FOR FLUERIC DEVICES[2]

| | Quantity Requirements | | | Element Design | | | |
| | Low | Medium | High | Complexity | Miniaturization | Accuracy | Relative Economy |
|---|---|---|---|---|---|---|---|
| **Plastics** | | | | | | | |
| Casting | E | G | F | G | F | G | F |
| Thermoforming | E | G | F | F | P | F | F |
| Photoetching | E | G | G | E | G | G | G |
| Injection Molding | P | F | E | G | G | G | E |
| Compression Molding | P | F | E | G | G | G | E |
| Jet Molding | P | F | E | G | G | G | E |
| Transfer Molding | P | F | E | G | G | G | E |
| **Metals** | | | | | | | |
| Powder Metallurgy | P | G | G | G | F | G | F |
| Die Casting | P | F | E | G | G | G | G |
| Chemical Machining | G | E | F | E | E | E | F |
| Electroforming | G | F | P | E | G | E | F |
| Electrical Discharge Machining | P | G | G | G | F | G | F |
| Electrochemical Machining | F | G | G | G | G | G | G |
| **Ceramics** | | | | | | | |
| Photoetching | F | G | G | E | E | G | G |
| Ultrasonic Machining | P | F | G | G | G | G | P |
| Electron Beam Machining | P | P | G | E | E | E | P |

E = Excellent     G = Good     F = Fair     P = Poor

complex precision parts of small size. Such processes include the photoetching techniques, chemical machining, and electron beam machining. These processes are also economically sound for mass production. Table 25-4[2] compares the various techniques with some indication of the characteristics of each process. This table is by no means absolute but serves as a rough appraisal of the merits and limitations of each process according to present state-of-the-art fluid element fabrication techniques.

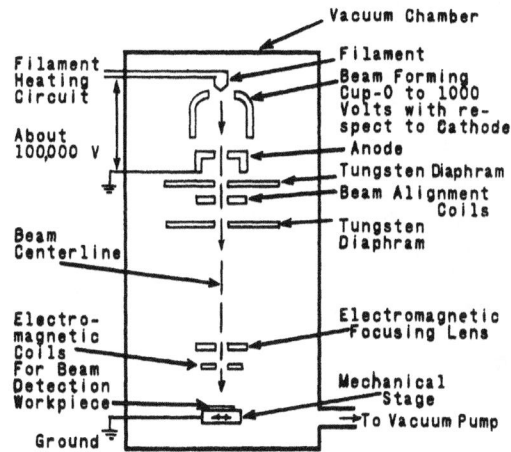

*Figure 25-15. Electron Beam Machining*[2]

## REFERENCES

1. AMCP 706-100, Engineering Design Handbook, *Design Guidance for Producibility.*

2. E.F. Humphrey and D.H. Tarumoto, Eds., *Fluidics,* Fluid Amplifier Associates, Inc., Boston, Mass., 1965.

3. D.L. Letham, "Fluidic System Design, 16. Component Fabrication", Machine Design, **39**, 215-7 (Jan. 19, 1967).

PART SIX — MISCELLANEOUS TIMING DEVICES

CHAPTER 26

ELECTROCHEMICAL TIMERS

Electrochemical timing devices are simple, small, low-cost items capable of providing delays from seconds to months.

These timers are somewhat similar to the corrosion delays used in early bomb fuzes. In the corrosion delays, the delay period is controlled by the time required for acid to dissolve a baffle and trigger an output[1]. At constant temperature, the chemical action on which the delay depends is quite accurate. The Germans reported an accuracy of ± 3% for the thrust cutoff in V-2 rockets. However, the reaction is very temperature sensitive, and there is no opportunity for temperature compensation in these devices. Therefore, the delay times obtained are not accurate in military applications where uniform operation is required over an extended temperature range.

While several electrochemical timers have been investigated for military applications, only the silver electroplating timer is in current use. This timer is convenient and reliable, and is therefore considered at length. Mention is also made of the solion and the mercury timer. For other, less common, and experimental electrochemical systems, see Ref. 2.

Because advantages and disadvantages differ for the various timer types, they are listed separately for each timer in the paragraphs that follow.

## 26-1 BASIC CONSIDERATIONS

### 26-1.1 ELECTROPLATING ACTION

The operation of electrochemical timers is based on Faraday's first two laws of electrolysis which state: (1) the amount of electrochemical decomposition is proportional to the quantity of electricity passed, and (2) the amounts of different substances deposited by a fixed quantity of electricity are proportional to their chemical equivalent weights. Mathematically, Faraday's two laws are expressed as[3]

$$W = \frac{E}{F}\int Idt, \text{g} \qquad (26-1)$$

where

$W$ = weight of deposit, g

$E$ = equivalent weight of substance deposited, g

$F$ = Faraday's constant = 96,000 C/g equivalent weight

$\int Idt$ = total number of coulombs, A-sec

Hence, when a solution is electrolyzed, the number of electrons received at the anode must equal the number delivered to the cathode. The ions arriving at the cathode are reduced (i.e., they obtain electrons) and those ions arriving at the anode are oxidized (i.e.,

26-1

ELECTROLYTE
CONTAINING
SILVER SALT

ANODE –
SILVER PLATED
NOBLE METAL
"WORKING ELECTRODE"

CATHODE –
SILVER CASE
"RESERVOIR ELECTRODE"

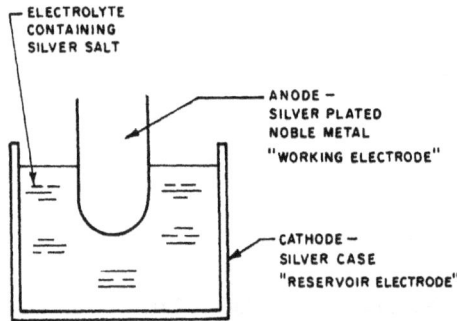

*Figure 26-1. Silver Plating Action*

they forfeit electrons). Electrochemical systems that obey these laws are called coulometric, and devices that use these principles are called coulometers.

We here are concerned specifically with the silver system, shown diagrammatically in Fig. 26-1. It consists of a silver case forming the cathode, a silver-plated anode of noble metal, and an electrolyte solution of a soluble silver salt. The anode is called the working electrode, the cathode, the reservoir electrode. These terms are needed because cell operation is reversible. As a matter of fact, the reversibility makes the electroplating system attractive as a timer. Operation is as follows: to start, the working electrode is unplated and the timer is set by passing a known time integral of current through it. During the timing cycle, the working electrode, now the anode, is deplated of its silver.

During the timing cycle, current flows through the cell to convert silver metal to silver ions at the anode ($Ag \rightarrow Ag^+ + e^-$). The ions then move to the cathode to replace the ions that have been reduced there ($Ag^+ + e^- \rightarrow Ag$). This deplating operation is preferred over plating because of its well-defined end point. During timing, while silver is being removed from the anode, cell impedance is low. The cell voltage drop will be about 0.01 to 0.1 V depending on cell temperature and other factors. When all the silver is expended, the

impedance rises rapidly and results in a sharp increase in voltage to about 1 V. This voltage step can be used as a circuit trigger. The electrolyte obeys Ohm's law like metallic conductors, except under abnormal conditions—such as very high voltage, very high frequencies, or strong concentration polarizations.

Since the chemical action is reversible, the timer can be run down, recharged, and rerun. This repeat action is ideal for testing and quality assurance purposes. However, there is a limit to the number of successful repeat cycles for some types of cell. After many cycles the timer becomes inaccurate. This is because of the exhaustion effect common to all electrochemical systems. The reversibility difficulty develops when the electrochemical oxidation-reduction process is disturbed by the "microscopic" early depletion of cations in a localized area. An overvoltage in this area then leads to the generation of hydrogen and other side effects. Although this process occurs for only a very brief time, perhaps only a microsecond in each cycle, the action is sufficient to lead to a general deterioration in the timing interval after a few cycles. On the other hand, a large number of good runs, as well as temperature independence for the timing cycle, is claimed for some of the newer cell types.

The ampere has been defined as that amount of current which, flowing for 1 sec, will deposit 1.118 mg of silver for each coulomb of charge. (This is why electroplating timers are called coulometers.) This same 1.118 mg of silver is deposited at 0.5 A for 2 sec or at 310 μA for 1 hr. The general relationship, in customary units, is

$$t = \frac{C}{I} \text{ , hr} \qquad (26\text{-}2)$$

where

$t$ = time, hr

$C$ = coulomb capacity, μA-hr

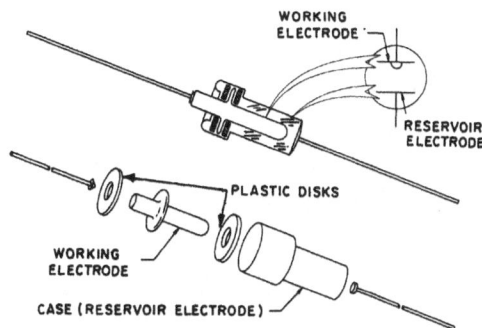

Figure 26-2. Bissett-Berman E-Cell[5]

$I$ = current, $\mu$A

## 26-1.2 THE ELECTROPLATING SYSTEM

The electroplating system consists of three parts:

(1) A battery and its circuit which deliver a constant direct current. Batteries are affected by ambient temperature but variations in their output can be controlled by temperature-compensated electronic circuits. A uniform current is a prerequisite for accurate timer operation.

(2) An electroplating delay cell in which the constant current causes a chemical (silver) to react (be deplated) at a known rate (see par. 26-1.1).

(3) A detector that senses the relative progress of the reaction, usually the depletion of one of the chemicals (silver) from one of the electrodes.

Detection can be accomplished in several ways. In electrical readout the change in output voltage is sensed by means of a suitable detection circuit; in mechanical readout a motion results that can activate a switch or plunger to perform the desired output; visual readout permits the observation of a change in appearance, such as color of electrolyte or position of components. Mili-

tary timers with electrical and mechanical readouts are described in the paragraphs that follow. Visual readout has not been used in military devices but is described in pars. 26-4 and 26-5.

Note that electroplating systems integrate the product of current and time. The current need not be continuous but may be interrupted in which instance the cell still "counts" the cumulative product. Therefore, these devices can be used both as coulometers and as integrating accelerometers in which current is a function of acceleration.

## 26-2 ELECTROPLATING TIMER WITH ELECTRICAL READOUT

### 26-2.1 PRINCIPLE OF OPERATION

When a constant current is passed through the electroplating timer (preset coulometer), it deplates silver from the working electrode onto the reservoir electrode. The increase in voltage drop across the cell at the end of deplating is used for electrical readout. The principle is described in detail in par. 26-1.1.

### 26-2.2 CELL CONSTRUCTION

Coulometers are manufactured commercially by Gibbs[4] and Bisset-Berman[5], the latter calling its device an E-Cell. The E-Cell has been used in several military applications, such as providing the delay in the Antipersonnel Mine, BLU-54/B[6,7].

Cell construction is illustrated in Fig. 26-2[5]. The cell consists of a silver case (the reservoir electrode), 1/4 in. in diameter and 5/8 in. long. The working electrode of gold over base metal is held in place by two plastic disks that serve both as seals and electric insulators. The case is filled with electrolyte containing a silver salt in a weak acid[8]. Electrical leads complete the cell. Cell weight is about 0.1 oz.

The cell illustrated is a single-anode cell,

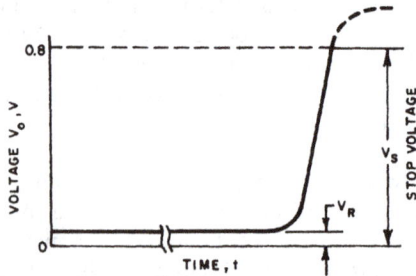

Figure 26-3. Operating Curve of Coulometer
at Constant Current[5]

permitting a single time delay. When more than one delay is desired, several anodes of different sizes may be combined in the same unit[9]. The deplating of each anode results in a sharp increase in output voltage. A dual anode cell is useful because of the common military requirement for two different time delays. For example, a mine may require an arming delay of a few minutes and a self-sterilization delay of several days.

The power requirements of E-Cells is low and accuracy of the timing interval is within ±4%. One exception to this accuracy is for short-delay setting after long storage.

## 26-2.3 TIMER DESIGN

The design of an electroplating timing system, like that of all other multicomponent systems, requires repeated consideration of the characteristics of each component until a

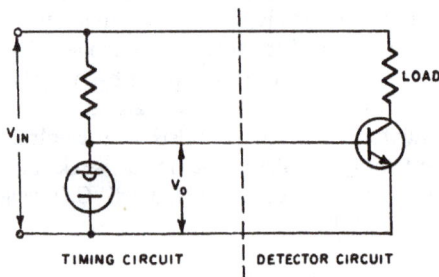

Figure 26-4. Coulometer Detector Circuit[5]

consistent combination is achieved. The first element, a source of constant DC current, is not discussed in this handbook. A special battery may be required if this power is not available from one of the circuits of the main system. Depending on the desired delay, current levels vary from about a microampere to a milliampere.

The second element is the coulometer itself. Described in par. 26-2.2, it is available in a large variety of models and timing intervals. Before a specific cell can be selected, consideration must be given to the output detection circuit.

The output signal of the coulometer is a change in voltage. During the timing interval, the voltage is low, see Fig. 26-3[5]. Upon completion of anode deplating, the voltage rises rapidly thus indicating the end on the timing interval. One way to detect this voltage rise is by use of the simple detector circuit shown in Fig. 26-4[5]. The performance of this circuit is made clear by considering its three phases of operation:

(1) While the cell deplates, the run voltage $V_R$ is below the activation voltage of the transistor. Since the cell therefore is drawing practically all the current, the equivalent circuit consists of just the cell plus its resistor.

(2) During the rapid transition to the high-voltage state, the current level through the cell is reduced as the transistor base starts to take current.

(3) When operating at the stop voltage $V_s$, the cell draws a very small residual current $I_s$, which in most cases is negligible compared with that drawn by the transistor. Thus, the equivalent circuit is essentially the original circuit without the coulometer.

Typical voltage-current characteristics at various operating temperatures are shown in Fig. 26-5[5]. Fig. 26-5(A) shows the maximum running (deplating) voltage and current while Fig. 26-5(B) shows the stop voltage and its

TEMPERATURE, °C

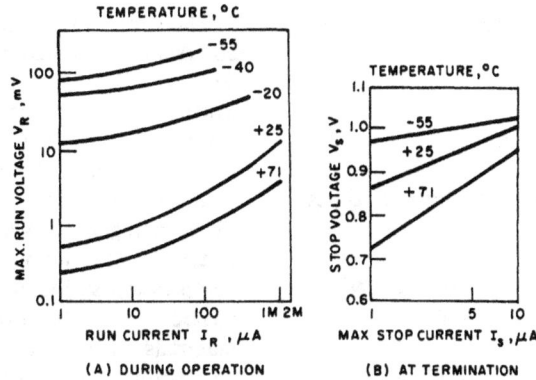

(A) DURING OPERATION

(B) AT TERMINATION

Figure 26-5. Typical Coulometer Voltage-current Characteristics[5]

associated current. The stop voltage $V_s$ is associated with the activation voltage threshold of the transistor in the detection circuit while the stop current $I_s$ is the residual current passing through the cell at that voltage. Since most circuits offer stop voltages of about 0.7 V, stop currents are about 1 $\mu A$. More sophisticated detection circuits can be considered to achieve other desired effects.

Given the required time delay and selecting a power source with a constant current in the microampere range, the designer obtains the capacity required from Eq. 26-2. The microampere-hour rating defines a particular coulometer.

Experimental solid-state types of coulometer are currently being developed[10]. Both the electrode and the electrolyte in these types are ionic pastes like those used in solid-electrolyte batteries. Operating in the same manner as the liquid-electrolyte device, they are small, rugged, and accurate within ± 5%.

## 26-2.4 ADVANTAGES AND DISADVANTAGES

The characteristics of the electroplating timer with electrical readout may be summarized as follows:

(1) Advantages:

(1) Good accuracy (within ± 4%)

(2) Good miniaturization

(3) Simple and inexpensive

(4) Wide variety of timing intervals

(5) Very low power requirements

(6) Good shock and vibration resistance

(7) Operation over military temperature range

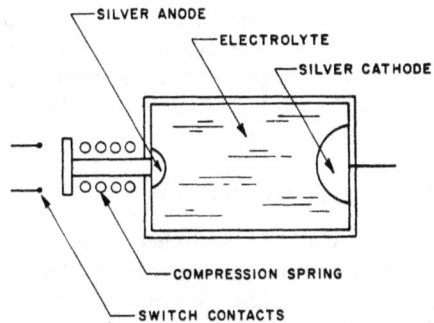

Figure 26-6. Electroplating Timer With Mechanical Readout

(8) Reusable several times (by deplating).

(2) Disadvantages:

(1) Detection circuit required

(2) Decreased accuracy for short set times after long storage.

## 26-3 ELECTROPLATING TIMER WITH MECHANICAL READOUT

### 26-3.1 PRINCIPLE OF OPERATION

Electrochemically, the mechanical-readout timer operates in the same manner as the electrical-readout timer (see par. 26-1.1). At the end of deplating, however, the action differs. A schematic arrangement of a mechanical-readout timer is shown in Fig. 26-6. The cell consists of a silver anode that may be likened to the head of a pin, a silver cathode, and an electrolytic solution containing silver ions. When a constant current is passed through the electroplating timer, it deplates silver from the anode onto the cathode. At the end of the timing interval, the anode has been reduced in size to the diameter of the pin shaft. At that instant, the spring causes mechanical motion. By pushing against a flange at the other end of the pin, the spring pulls the pin out of the hole in the cell wall.

Because of the displacement of parts, the action of a mechanical-readout timer is not reversible. Whereas the timer in Fig. 26-6 illustrates the output of closing a switch, the output motion obviously can be applied to the opening or closing of several switches or to the operation of a plunger.

### 26-3.2 CELL CONSTRUCTION

Cell construction is illustrated in Fig. 26-7. This is the Interval Timer, MARK 24 MOD 3, a component of Firing Mechanism, MARK 42. Originally designed for use in a sonobuoy, it currently is incorporated in Destructor, MARK 36[11].

The timer cell (based on a patented idea[12]) consists of a molded polychlorotrifluoroethylene (Kel-F) cup that holds the anode assembly. After filling with an electrolyte of a silver fluoroborate solution, the cup is heat sealed with an end plug that holds the silver cathode. The anode assembly consists of a silver plunger to which a contact disk is fastened. The plunger is surrounded by a compression spring and sealed with an O-ring coated with fluorosilicone lubricant. All materials were selected for their chemical compatibility with the electrolyte.

At the end of the timing interval, the anode plunger is pushed to the left. In its new position, the contact disk (1) closes an SPST (Single Pole, Single Throw) switch and (2) opens the anode switch so as to terminate the deplating action. The contact force at switch closure is 0.8 lb and contact resistance after switch closure is less than 0.3 ohm.

The timer is 5/8 in. in diameter, 1-5/8 in. long, and weighs 9 g. For some applications, the timer is sealed in Thermofit heat-shrink tubing (not shown in the figure). Timer accuracy under water (for which it was designed) at 28° to 90°F is ± 5%; over the entire military temperature range, accuracy is ± 10%. Models have withstood shocks as high as 12,000 g, low and high frequency vibrations, cold storage at −80°F, and temperature-humidity cycling.

Dual action, when required, is available from Interval Timer, MARK 25. This timer is essentially a dual MARK 24 timer. It closes two SPST switches by two separate contacts that operate from separate anodes. Construction and size are the same as for the MARK 24.

A commercial timer is shown in Fig. 26-8[4]. Its construction is similar to that of the MARK 24 timer. It also closes an SPST switch with gold-plated contacts. Body size is 5/8 in. by 7/8 in. Accuracy over the military temperature range is ± 10%.

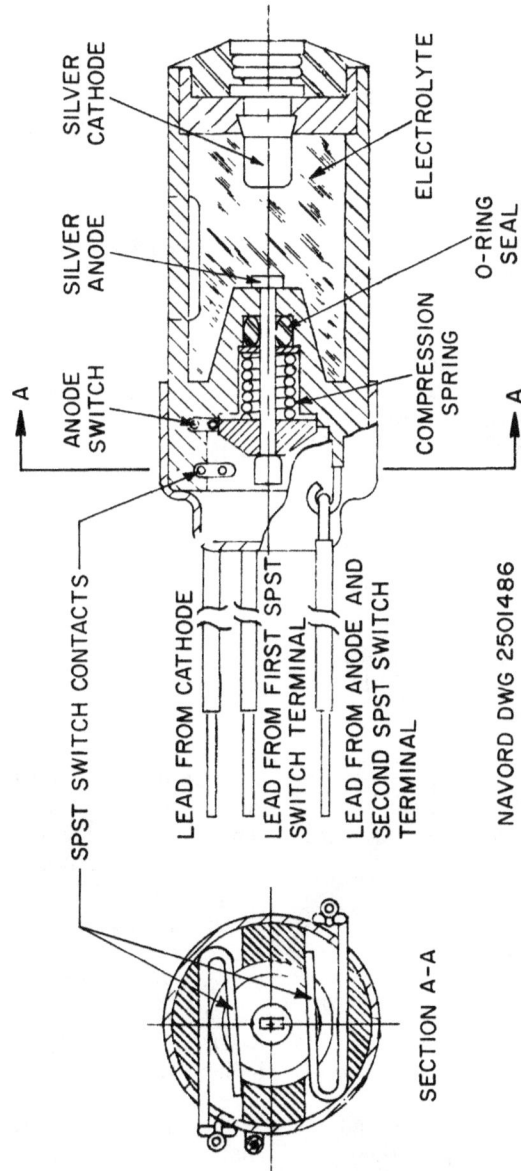

Figure 26-7. Interval Timer, MARK 24 MOD 3

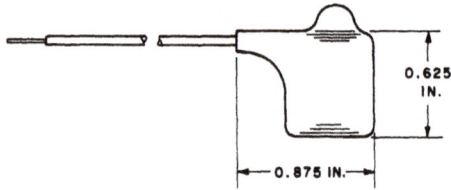

Figure 26-8. Mechanical Readout Timer by Gibbs

## 26-3.3 TIMER DESIGN

The design of a mechanical-readout timing system is simpler than that of an electrical-readout system. In effect, the mechanical-readout system has only two elements, a source of constant DC current and the coulometer itself. The output of the coulometer is the operation of a switch or plunger, and no separate output circuit is required.

Depending on the delay desired, the battery current level varies from a microampere to a milliampere. In the MARK 24 and MARK 25 timers current flow is regulated by the voltage applied and a precision resistor connected in series with the cell. Operating time is computed by

$$t = \frac{CR}{V} \text{ , hr} \tag{26-3}$$

where

$t$ = time, hr

$C$ = coulomb capacity, A-hr

$R$ = resistance, ohm

$V$ = voltage, volt

For the MARK 24 timer, $C$ = 0.005, and for the MARK 25, $C$ = 0.01. By varying voltage, resistance, or both, delays of 5 min to 30 days can be achieved.

## 26-3.4 ADVANTAGES AND DISADVANTAGES

The characteristics of the electroplating timer with mechanical read-out may be summarized as follows:

(1) Advantages:

(a) Simple and inexpensive

(b) Low power requirements

(c) Good shock and vibration resistance

(d) Operation over military temperature range

(e) No detection circuit required.

(2) Disadvantages:

(a) Not reversible

(b) Difficult to seal.

## 26-4 SOLION

### 26-4.1 PRINCIPLE OF OPERATION

The solion is an electrochemical timer using an oxidation-reduction system that uses ions in solution instead of electrons as the charge carriers. Solions have found application mainly as diodes with both electrical and visual readout. They also have been designed as tetrodes and used as electrical amplifiers and general-purpose integrators. The electrolyte contains a specified amount of iodine and a comparatively large amount of potassium iodide[2]. Hence, the solution contains triiodide $I_3^-$.

Shown schematically in Fig. 26-9[3], the solion timer consists of two identical compartments separated by a semipermeable diffusion barrier. When current is passed

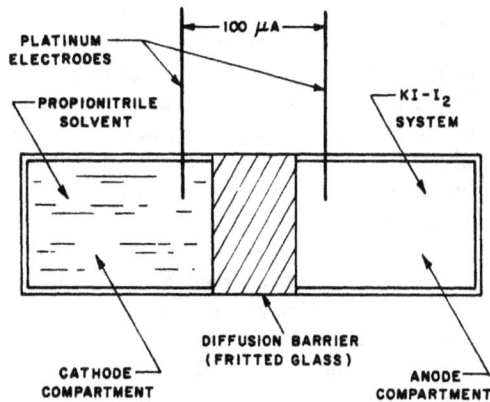

*Figure 26-9. Solion Timer*

through the cell, iodine is reduced in the cathode compartment ($I_3^- + 2e^- \rightarrow 3Ix$), and the iodide ion is oxidized in the anode compartment ($3I^- \rightarrow I_3^- + 2e^-$). The net result is that iodine is transferred to the anode compartment. The amount of iodine transferred is proportional to the time and current, and the conductivity of the electrolyte is proportional to iodine concentration, permitting an electrical readout. The color of the electrolyte is also a function of iodine concentration, so that visual readout is possible.

The polarization curves for the solion are shown in Fig. 26-10[3] for five different

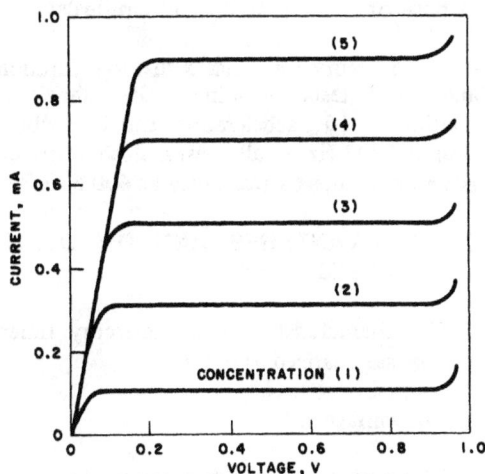

*Figure 26-10. Typical Polarization Curves for the Solion*

concentrations of the tri-iodide ion. Note that Ohm's law breaks down in the solion; the current and cell resistance maintain a constant value with increasing voltage up to approximately 1 V, where the oxidation-reduction on process terminates and Ohm's law resumes. The resistance of the solion cell varies exponentially.

For timing purposes, the electrical readout is not entirely satisfactory because the step function at the point of exhaustion of iodine ions is uncertain and inaccurate. The end points varied 162 parts per thousand at room temperature, and the device ran even slower at $-25°$ and $65°C^2$. Repeat cycles are limited by the exhaustion effect (see par. 26-1.1).

The solion has been reported to withstand vibration in all three axes, from rest to 50 g at 0 - 2000 Hz, with displacements as large as 0.5 in. Solions have been stored over a year without any detectable deviation in operational ability.

## 26-4.2 ADVANTAGES AND DISADVANTAGES

The characteristics of the solion may be summarized as follows:

(1) Advantages:

(a) Good miniaturization

(b) Good shock and vibration results

(c) Low power requirements

(d) Good storage potential

(e) Current controlled by conditions at the cathode only.

(2) Disadvantages:

(a) Poor accuracy (about ± 20%)

(b) Uncertain time end point

(c) Exponential resistance characteristics

26-9

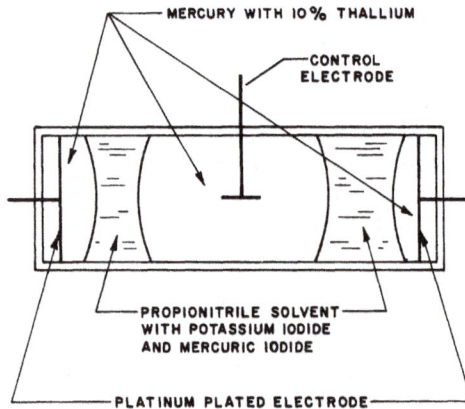

MERCURY WITH 10% THALLIUM

CONTROL ELECTRODE

PROPIONITRILE SOLVENT
WITH POTASSIUM IODIDE
AND MERCURIC IODIDE

PLATINUM PLATED ELECTRODE

*Figure 26-11. Mercury Timer*

(d) Low frequency response (lower than that of transistors by a factor of $10^6$)

(e) Degeneration of timing interval

(f) Limited temperature range.

In summary, the solion lacks precision and will not function over the military temperature range. While the solion has some commercial applications, it has not been used in a military device.

## 26-5 MERCURY TIMER

### 26-5.1 PRINCIPLE OF OPERATION

This device operates by transporting mercuric ions through a solvent that separates two bodies of mercury, eventually exhausting one of the mercury reservoirs. The chemical equations associated with the reaction are $4I^- + Hg \rightarrow Hg\,I_4^= + 2\bar{e}$ for the anode and $HgI_4^= + 2\bar{e} \rightarrow Hg + 4I^-$ for the cathode.

Shown schematically in Fig. 26-11[3], the mercury timer consists of a cell, often a glass tube, with anode and cathode at the ends and a control electrode in the center. Three pellets of mercury are shown, separated by two bodies of electrolyte consisting of an iodine solution. Upon current passage through the cell, mercury is electroplated from the anode through the electrolyte to the cathode. Buildup of mercury at the cathode causes the electrolyte gap to approach and—at the end of the delay—contact it, at which time the voltage drops to permit an electrical readout. Visual readout also is provided by observation of mercury pellet motion.

The mercury timer operates with a current of 50-100 $\mu$A. Operation is reversible except that the number of cycles is limited by the exhaustion effect (see par. 26-1.1).

Difficulties appear when an attempt is made to operate the mercury timer over the military temperature range. These include (1) the necessity for lowering of the freezing point for the cell contents because operation when frozen is destructive (mercury freezes at $-38°$F), (2) mercury expansion problems, and (3) electrolyte-mercury interface difficulties experienced with changing temperatures.

The relatively high coefficient of expansion of mercury creates the danger of a cracked housing when the environmental temperature is increased; the same coefficient of expansion will threaten the mercury-solvent interface when lower temperatures are encountered. Also, difficulties have been reported with the deposit of mercury on the end terminals[3].

The mercury timer has withstood vibration and shock tests including a 50-g shock, a continuous 16-g acceleration, and 15-g vibration at 5000 Hz on all axes. A sloshing effect is reported due to resonance at 400-600 Hz.

### 26-5.2 ADVANTAGES AND DISADVANTAGES

The characteristics of the mercury timer may be summarized as follows:

(1) Advantages:

(a) Good accuracy (over narrow temperature range)

(b) Good miniaturization

(c) Low power requirements.

(2) Disadvantages:

(1) Limited temperature range

(2) Fragile

(3) Occasional sloshing effect

(4) Setting required before operation can begin.

In summary, the mercury timer is delicate and cannot operate over the military temperature range. While the timer has some commercial applications, it has not been used in a military device.

## REFERENCES

1. AMCP 706-210, Engineering Design Handbook, *Fuzes.*

2. J.B. Lipnick, *The Use of Electrochemical Systems for Timing*, U S Army Harry Diamond Laboratories, Washington, D.C., Report TR-1156, October 30, 1963 (AD-422 899).

3. R.C. Palmer and E.C. Shults, *Literature Survey and Feasibility Study of Electrochemical Coulometers*, Edgerton, Germeshausen & Grier, Inc., Santa Barbara, Calif. Report S-239-R, Prepared for Sandia Corp. under P.O. 11-1376, November 12, 1963.

4. Gibbs Manufacturing and Research Corp., 450 N. Main St., Janesville, Wis. 53545.

5. The Bissett-Berman Corp., 3860 Centinela Ave., Los Angeles, Calif. 90066.

6. *Contractor Support Test for the BLU-54/B Antipersonnel Mine (U)*, Eglin Air Force Base, Florida, Report ADTC TR-68-23, September 1968 (Confidential Report).

7. *Engineering Evaluation of Wide Area Antipersonnel Mine, BLU-42/B and BLU-54/B (U)*, Eglin Air Force Base, Florida, Report ADTC TR-70-75, April 1970 (Confidential Report).

8. U S Patent 3,423,643, E.A. Miller, *Electrolytic Cell With Electrolyte Containing Silver Salt*, January 21, 1969.

9. U S Patent 3,423,642, E.J. Plehal, et al., *Electrolytic Cells With at Least Three Electrodes*, January 21, 1969.

10. S. Giles, *Development of a Solid State Coulister*, Atomics International Div. of American Rockwell Corp., Final Report, Prepared for U S Army Harry Diamond Laboratories under Contract DAAG39-68-C-0025, August 1968.

11. OP 3529, *Destructor MARK 36 and MARK 40, All Mods (U)*, Volumes 1-3, Department of the Navy, 2nd Revision, June 1970 (Secret Report).

12. U S Patent 3,205,321, R.J. Lyon, *Miniature Electrolytic Timer With an Erodable Anode*, September 7, 1956.

# CHAPTER 27

# NUCLEAR DECAY TIMERS

A low-frequency, high-accuracy time base may be constructed by making use of the decay of a radioactive substance in conjunction with a particle detector. In addition, electronic circuitry is required to produce a usable timer. Although nuclear decay is randomly distributed in time, accuracy can be obtained by sampling a sufficiently large number of counts. One significant advantage of the nuclear timer is that its time base rate is independent of temperature and other environmental variations. Fig. 27-1 shows the block diagram for a nuclear decay timer.

## 27-1 RADIATION SOURCE

### 27-1.1 DESIRABLE PROPERTIES

To be useful for a nuclear time base generator, the radiation source should have the following properties:

(1) Monoenergetic particles

(2) Highly ionizing particles

(3) Long half-life

(4) Decay products producing little radiation

(5) Capability for permanent mounting

(6) Ready availability.

The radiation should be nearly monoenergetic to provide for ready discrimination between output pulses produced by the detectors, and the internally generated noise in the detectors and amplifiers. All monoenergetic particles will lose approximately the same proportion of their energy in the

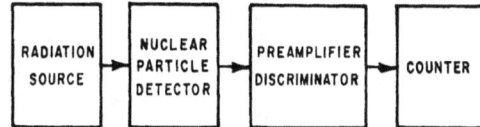

Figure 27-1. Elements of Nuclear Decay Timer

ionization process thus providing signals from the detector of approximately equal pulse heights.

Highly ionizing particles increase the signal-to-noise ratio and minimize the degree of amplification required to reach the input level of electronic scalers. There should be no significant change in the disintegration rate of the radiation source during its predicted use and storage life. Therefore, the source should have a long half-life. Fig. 27-2[1] may be used to estimate the half-life required of the radiation source, once the operation and storage periods and acceptable decrease in activity have been fixed. For example, if the radiation source is to retain 99% of its activity after a period of 10 yr, a source with a half-life greater than $6.3 \times 10^2$ or 630 yr should be chosen. A more exact estimate may be made from the following equations[1]

$$N = N_o e^{-\lambda t} \qquad (27-1)$$

where

$N$ = radioactivity at time $t$

$N_o$ = initial radioactivity

$\lambda$ = decay constant, $yr^{-1}$.

$t$ = time span, yr

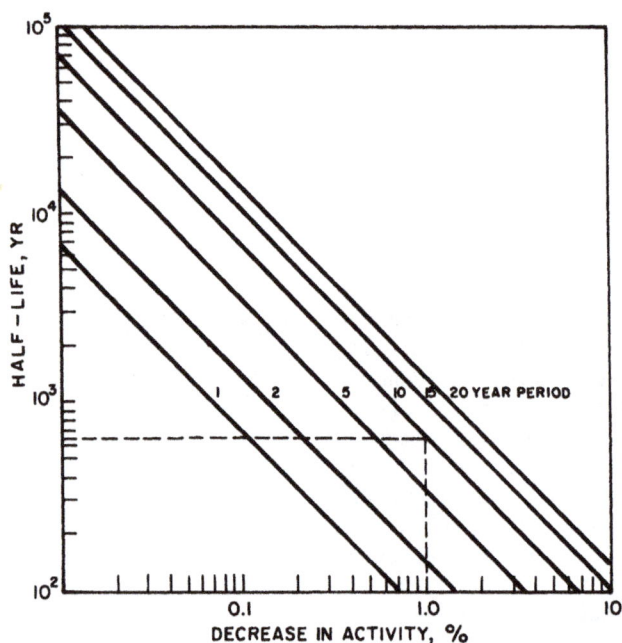

Figure 27-2. Decrease in Radioactivity as a Function of Use and Storage Period

$$T_{1/2} = \frac{0.693}{\lambda} \tag{27-2}$$

where $T_{1/2}$ = half life, yr

## 27-1.2 TYPES OF RADIATION

Four natural forms of radiation could be considered for this application: X rays, gamma rays, beta particles, and alpha particles.

Gamma rays and X rays are essentially the same radiation form and differ only in the manner in which they are produced. They are a highly penetrating radiation and therefore would be difficult to apply safely in a miniature timer because of the protective shielding required; also timer geometry could not be arranged conveniently to provide shielding.

Beta particles are high velocity electrons. The penetration characteristics of beta particles would require only a minimum of shielding, a few millimeters of aluminum. However, beta particles are not monoenergetic. The particles emitted from a particular radioisotope would have a continuous distribution of energies and would be difficult to use because of the signal-to-noise ratio problem in detectors.

Alpha particles are identical with the nucleus of a helium atom, consisting of two protons and two neutrons. These particles are emitted from the nucleus of certain of the heavy radioactive elements. Alpha particles have a discrete energy which, depending upon the producing isotope, are in the range 4 to 10 MeV. The shielding required for alpha particles is simple, several sheets of paper providing adequate shielding in nearly all instances. Alpha particles are strongly ionizing.

Considering the characteristics of each type of radiation, it is apparent that alpha particles are most suitable for use in a nuclear timer.

**TABLE 27-1**

**RADIOISOTOPES APPLICABLE TO NUCLEAR TIME BASES**

| Radioisotope | Half-Life, yr | Specific Activity, curie/g |
|---|---|---|
| Uranium 235 | $7.1 \times 10^8$ | $2.15 \times 10^{-6}$ |
| Neptunium 237 | $2.2 \times 10^6$ | $6.89 \times 10^{-4}$ |
| Uranium 238 | $4.5 \times 10^9$ | $3.34 \times 10^{-7}$ |
| Plutonium 239 | $2.4 \times 10^4$ | $6.17 \times 10^{-2}$ |

## 27-1.3 RADIOISOTOPES

Selection of the radioisotope for use in the time base can be made based on the requirements discussed:

(1) A source of alpha particles

(2) Half life sufficiently long for operational requirements

(3) Energy level (specific activity) high enough for reliable detector operation.

Table 27-1[1] lists radioisotopes that may be considered for nuclear time base applications.

## 27-2 NUCLEAR PARTICLE DETECTOR

The radiation detector used in the nuclear timer provides an electrical signal for each radiation event it detects. Radiation detectors presently in use include ionization chambers, proportional counters, Geiger-Mueller counters, scintillation counters, and semiconductor radiation detectors. Ionization chambers, proportional counters, and Geiger-Mueller counters operate on similar principles and are included in the general class of gas tube counters.

## 27-2.1 GAS TUBE COUNTERS

These counters usually consist of a cylindrical gas-filled chamber and a central electrode that is insulated from the chamber.

The chamber wall is made from a high work-function material, such as stainless steel or nickel. In a typical gas tube counter, the central electrode is usually about 0.003 in. in diameter, a compromise between physical strength and a high field strength near the electrode at a relatively low supply voltage.

With the central electrode serving as the anode and the chamber wall as the cathode, a voltage is applied to the tube. When ionizing radiation passes through the gas, the positively charged ions resulting from the ionization are attracted to the cathode where they are neutralized while the negative charges or electrons travel toward the anode.

While gas tube counters could be applied to military timers, each type presents some difficulties that must be recognized:

(1) The output pulse of the *ionization chamber* is independent of the operating voltage, provided the operating voltage is high enough so that recombination is negligible and low enough so that no secondary ionization occurs. The amplification factor is therefore unity. Thus a relatively large chamber (at least 3.5 cm long) is required.

(2) The amplification in a *proportional counter* is a strong function of gas pressure. In the fixed-volume chamber, pressure varies significantly over the military temperature range.

(3) *Geiger-Mueller* counters have an inherent recovery time that makes their response slower than desirable.

## 27-2.2 SCINTILLATION COUNTERS

The operation of a scintillation counter is based on the fact that in certain materials the ionization and excitation processes caused by nuclear radiation are accompanied by the emission of light or scintillations. The light photons are absorbed by the cathode of a photomultiplier tube that then emits photoelectrons to produce a current pulse.

(A) DIFFUSED JUNCTION DETECTOR

(B) SURFACE BARRIER DETECTOR

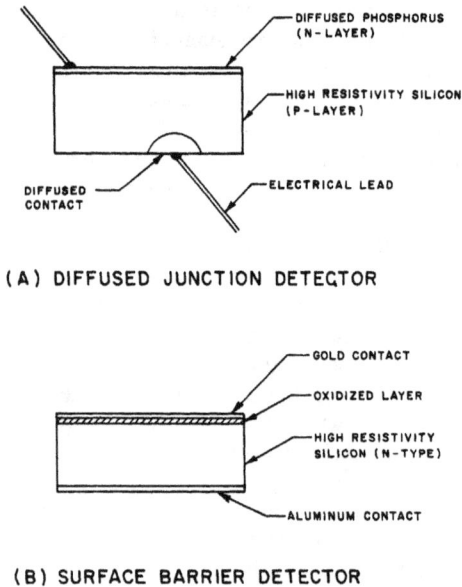

*Figure 27-3. Semiconductor Radiation Detectors*

Since photomultiplier tubes are very sensitive to shock, require a well-regulated, high-voltage power supply, and exceed the volume limits of military items, scintillation counters are not suited for the nuclear time base.

## 27-2.3 SEMICONDUCTOR RADIATION DETECTORS (SRD)

A relatively new detector, the SRD, is described in several recent references[2,3]. The SRD can be classified as a semiconductor junction detector. Its usefulness results from exploiting the phenomena occurring at the junction of two special types of material when placed together. Very pure, single-crystal silicon is used as one of the materials while the other material varies with the type of detector. Fig. 27-3[1] shows a diffused junction and a surface barrier detector. Details of construction vary with different manufacturers.

A diffused junction detector usually is prepared by diffusing phosphorus onto the surface of a pure wafer of p-type silicon. The diffusion depth is on the order of one micron and constitutes the n-layer of the diode. In the surface barrier detector, surface states are allowed to form by spontaneous oxidation on the surface of a pure wafer of n-type silicon. The formation of the oxide takes place through an evaporated gold layer that is provided for electrical contact. An inversion layer is thus formed having electrical properties similar to those of the diffused layer in a diffused junction detector. In this case the oxide layer acts as the p-type side of the diode junction.

Electrical leads to the associated electronics are attached easily to the front surfaces of both of these detectors because both the diffused and the gold surfaces are highly conductive. For the back contact to the SRD care must be taken to avoid any sharp change in resistivity that would result in the formation of a rectifying barrier. Two solutions to this problem are:

(1) Doping a portion of the silicon bulk with an impurity to lower surface resistivity.

(2) Evaporating a thick layer of aluminum onto the back surface.

The inherent energy resolution of SRD's is better by a factor of ten than that of gas tube counters. The counting rate of SRD's is limited only by the external electronics. SRD's can be made much smaller than gas tube counters, and they are more immune to physical damage due to shock. The main difficulty in the SRD approach is obtaining a detector that provides a sufficiently large signal-to-noise ratio over the military temperature range.

## 27-3 ASSOCIATED ELECTRONICS

Associated electronics must be included in

the time base system because the nuclear particle detector itself cannot provide an output usable by the counting circuits. The detector output must be amplified. The output is on the order of millivolts but a signal on the order of volts is desired by the counter. Since the capacitance of the detector varies with temperature, it is not desirable to have the voltage output of the amplifier depend on detector capacitance. Rather, a charge-sensitive preamplifier should be used that has a negative charge feedback to the input capacitance through a small capacitor (about 5 pF)[4].

In addition to amplification, a means of discriminating between the pulses to be counted and the inherent noise levels of the detector and the amplifier must be provided. Nuclear time bases use integral rather than simple discrimination[2]. The function of simple discriminators is to pass only pulses exceeding a preset threshold level, whereas differential discriminators pass pulses having amplitudes with a preset "window" above the threshold.

A serious disadvantage of a nuclear time base is the relatively long calibration time required. For typical values of 1000 Hz and 0.1% accuracy, calibration time is 4000 sec. This compares with 10 sec per reading of electronic, electromechanical, or mechanical time bases.

## REFERENCES

1. D. E. Voeller, *A Nuclear Time Base-Feasibility Study*, U S Army Harry Diamond Laboratories, Washington, D.C., Report TR-1300, 12 Auguts 1965 (AD-622 629).

2. W. J. Price, *Nuclear Radiation Detection*, Second Edition, McGraw-Hill, Inc., N.Y., 1964.

3. G. Dearnaley and D. C. Northrup, *Semiconductor Counters for Nuclear Radiations*, John Wiley and Sons, N.Y., 1963.

4. J. L. Blankenship and C. J. Borkowski, "Performance of Silicon Surface Barrier Detectors With Charge Sensitive Amplifiers", Trans. IRE PGNS, 18, 17 (January 1961).

# APPENDIX A

## BOOLEAN ALGEBRA APPLIED TO LOGIC CIRCUITS

Boolean algebra is the algebra of propositions that are permitted to be either true or false. Thus it is useful for the combination, manipulation, and simplification of control functions that can be represented by variables whose values are limited to discrete binary digits 0 and 1. These digits represent the two states of many common binary or bistable devices—e.g., nonconductive or conductive, discharged or charged, current on or off.

The basic building block used in the design of the logical elements of flueric and electronic timing systems is the *gate*. A gate is "a circuit having an output and a multiplicity of inputs so designed that the output is energized when and only when a certain definite set of input conditions are met"[1]. Gates can perform various arithmetic operations and control functions. The theory of gating networks is expressed conveniently in Boolean notation. Note that logic circuits for electronic timers are discussed in par. 11-4.2 and those for flueric timers in par. 24-5.

Fig. A-1[2] shows the symbol of an AND gate. The gate receives two input signals A and B and generates an output function C only when both A *and* B are present. Fig. A-1 also shows a switch analogy of the gate; lamp C lights only when A *and* B are closed. AND functions also are termed logical multiplications; a dot is used as the AND symbol. The following symbolizes that A *and* B produce C:

$$A \cdot B = C \qquad (A-1)$$

The basic OR gate is shown in Fig. A-2[2]. If either A *or* B is present (switch closed), there is an output C. OR functions are termed logical addition; the symbol used is a plus sign. The following symbolizes that A *or* B produces C:

$$A + B = C \qquad (A-2)$$

A third basic operation is the inversion or *not* function (Fig. A-3[2]). The symbol is a bar over the function, viz., $\overline{A}$ means that A is not present (equals 0). The *not* function inverts the state of the bistable function—i.e., it transforms a 1 into a 0 and vice versa.

The logic associated with EXCEPT, NOR, and other logical decisions can be built up from the basic logic functions of AND, OR, and their inversions.

Consider the statement: *not* A AND *not* B. Fig. A-4[2] shows a possible presentation. Note that both A and B are inverted to get *not* A AND *not* B. The mathematical expression is

$$\overline{A} \cdot \overline{B} = C \qquad (A-3)$$

The figure also shows *not* (A AND B)

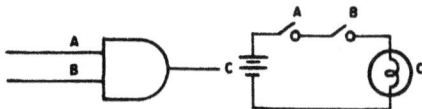

Figure A-1. Symbol and Switch Analogy for
AND Gate[2]

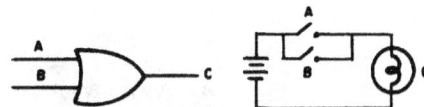

Figure A-2. Symbol and Switch Analogy for
OR Gate[2]

Figure A-3. not *Symbol is Bar Over Letter*[2]

expressed as $(\overline{A} \cdot \overline{B})$ which does not equal $\overline{A} \cdot \overline{B}$. This fact is made clear by the Venn diagram (Fig. A-5[2]) wherein all areas of true conditions are shaded while false areas are left blank.

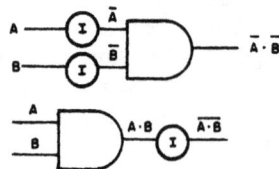

Figure A-4. *Symbolic Representation of* not *A* AND not *B, also* not *(A AND B)*[2]

A list of Boolean postulates and equations is given in Table A-1[3]. The analogy between electrical, flueric, and electromechanical systems for the basic gates in logic is contrasted in Fig. A-6. The recommended symbols for logic diagrams are contained in Ref. 4 while flueric symbols are contained in Ref. 5.

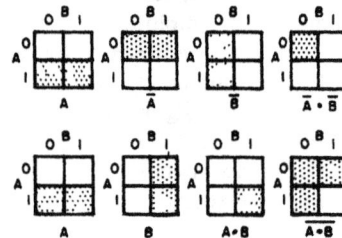

Figure A-5. *Venn Diagram for Fig. A-4*[2]

**TABLE A-1**

**BOOLEAN POSTULATES AND EQUATIONS**[3]

| | | |
|---|---|---|
| 1. | $A + A = A$ | Absorption rule for + |
| 2. | $A \cdot A = A$ | Absorption rule for · |
| 3. | $A + B = B + A$ | Commutative rule for + |
| 4. | $A \cdot B = B \cdot A$ | Commutative rule for · |
| 5. | $(A + B) + C = A + (B+C)$ | |
| | $= A + B + C$ | Associative rule for + |
| 6. | $(A \cdot B) \cdot C = A \cdot (B \cdot C)$ | |
| | $= A \cdot B \cdot C$ | Associative rule for · |
| 7. | $A \cdot (B+C) = A \cdot B + A \cdot C$ | Distributive rule for · over + |
| 8. | $A + B \cdot C = (A + B) \cdot (A + C)$ | Distributive rule for + |
| 9. | $\overline{A \cdot B} = \overline{A} + \overline{B}$ | De Morgan's rule for · |
| 10. | $\overline{A+B} = \overline{A} \cdot \overline{B}$ | De Morgan's rule for + |
| 11. | $A + I = I$ | |
| 12. | $A \cdot I = A$ | |
| 13. | $0 + A = A$ | |
| 14. | $0 \cdot A = 0$ | |
| 15. | $A + \overline{A} = I$ | |
| 16. | $A \cdot \overline{A} = 0$ | |
| 17. | $A + \overline{A} \cdot B = A$ | |
| 18. | $A + \overline{A} \cdot B = A + B$ | |
| 19. | $A \cdot \overline{B} + B \cdot C + C \cdot A = (A + \overline{B}) \cdot (B + C) \cdot (C + A)$ | |
| 20. | $\overline{A} \cdot B + A \cdot C + B \cdot C = A \cdot B + B \cdot C$ | |
| 21. | $\overline{\overline{A}} = A$ | |

GATE TYPE     ELECTRONIC     FLUERIC     ELECTROMECHANICAL

AND

OR

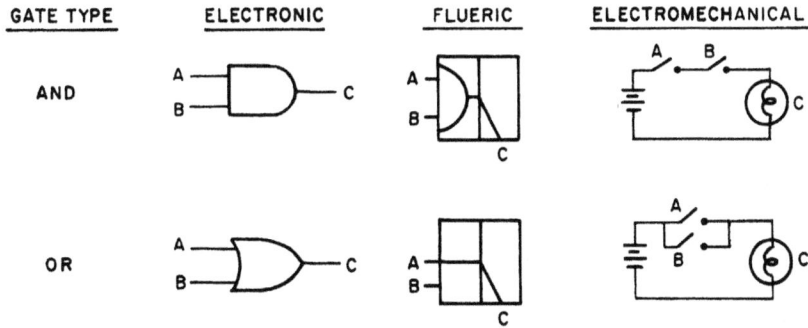

*Figure A-6. Symbols for the Basic Gates*

## REFERENCES

1. Franz Hohn, "Some Mathematical Aspects of Switching", American Mathematical Monthly, **62**, 75-90, February 1955.

2. G. A. Wires, "Boolean Algebra", Instruments and Control Systems, **33**, 270-3, February 1960.

3. *Fluidic Systems Design Guide,* Fluidonics Division of the Imperial-Eastman Corp., Chicago, Ill., 1966.

4. MIL-STD-806B, *Graphic Symbols for Logic Diagrams,* Dept. of Defense.

5. MIL-STD-1306, *Fluerics, Terminology and Symbols,* Dept. of Defense.

# GLOSSARY

This glossary is restricted to key terms for timing systems and components. Additional terms may be found in the following references:

MIL-STD-444, *Nomenclature and Definitions in the Ammunition Area*, Department of Defense, Change 2, 9 July 1964.

MIL-STD-1306, *Fluerics, Terminology and Symbols*, Department of Defense, 17 July 1968.

AMCP 706-210, Engineering Design Hand book, *Fuzes*.

*IRE Dictionary of Electronic Terms and Symbols*, Institute of Radio Engineers, Inc., N.Y. 1961.

Eric Bruton, *Dictionary of Clocks and Watches*, Arco Publications, London, 1962.

**Amplifier.** An active electric or flueric component that increases the strength of the signal applied to it.

**Arming.** The changing from a safe condition to a state of readiness for functioning. Generally a fuze is caused to arm by one or a combination of means as acceleration, rotation, mechanical drive, chemical action, electrical action, or air travel.

**Atomic Clock.** A timer that employs as its frequency regulator the vibration of the cesium atom. The atomic clock is the most accurately known frequency standard because the radio frequencies emitted by atoms and molecules at low pressures are fixed and unchanging with time and independent of temperature.

**Balance Spring.** A fine spring, usually spiral, used for oscillating the balance wheel of a clockwork. Also called hairspring.

**Balance Wheel.** A balance for clockworks to regulate their beat. It is usually shaped like a wheel.

**Bistable Device.** A device that has two stable states, such as current on or current off.

**Cesium Beam Resonator.** A device which utilizes the transition of the cesium atom designated as Fm (4,0) ↔ (3,0). The frequency of oscillation has been established as 9,192,631,770 cycles per ephemeris second.

**Counter.** A device capable of changing from one to the next of a sequence of distinguishable states upon each receipt of an input signal.

**Delay.** An explosive train component that introduces a controlled time delay in the functioning process.

**Delay, Arming.** (1) The interval expressed in time or distance between the instant an item of ammunition carrying a fuze is launched and the instant the fuze becomes armed. (2) The time interval required for the arming process to be completed in a nonlaunched item of ammunition.

**Delay, Functioning.** The interval expressed in time or distance between initiation of the fuze and detonation of the main charge.

**Ephemeris Time.** Time calculated from the orbit of the earth around the sun (as opposed to the rotation of the earth). Ephemeris time has been adopted as the fundamental unit of time.

**Escape Wheel.** The last wheel in the gear train

of a clockwork which is controlled by the escapement.

**Escapement.** The part of a mechanical clockwork which regulates the rate of transmission of energy. *See also:* specific escapements.

**Escapement, Self-starting.** An escapement that has the ability to start oscillating, without an external force, from any rest position, and to sustain the oscillations as long as energy is supplied.

**Escapement, Tuned.** An escapement consisting of a pivoted balance mass and a restoring spring, pulsed by an escape wheel.

**Escapement, Untuned.** An escapement consisting of a pivoted mass driven by an escape wheel. The mass oscillates without a spring.

**Explosive Train.** A train of combustible and explosive elements arranged in the order of decreasing sensitivity. Its function is to accomplish the controlled augmentation of a small impulse into one of suitable energy to cause the main charge of the ammunition to function. A fuze explosive train may consist of a primer, a detonator, a delay, a relay, a lead, and a booster charge, one or more of which may be either omitted or combined.

**Failure Diameter.** The minimum diameter of an explosive cylinder which will sustain propagation in the axial direction. Below this diameter, propagation is retarded or extinguished due to radial heat losses.

**Feedback.** The return to the input of a part of the output of a device, system, or process.

**Flip-flop.** A bistable device.

**Fluerics.** The area within the field of fluidics, in which fluid components and systems perform sensing, logic, amplification, and control functions without the use of moving mechanical parts.

**Fluidics.** The general field of fluid devices and systems and the associated peripheral equipment used to perform sensing, logic, amplification, and control functions.

**Fuze.** A device with explosive components which initiates a train of fire or detonation in an item of ammunition by an action such as hydrostatic pressure, electrical energy, chemical action, impact, mechanical time, or a combination of these.

**Fuze, Time.** A fuze that can be preset to function after the lapse of a specified time.

**Gear Train.** An arrangement of two or more gears meshing to transmit power from a driving shaft to a driven shaft at a different speed or in a different direction.

**Hair Spring.** *See:* **Balance Spring.**

**Igniter.** (1) Any device, chemical, electrical, or mechanical, used to ignite. (2) A specially arranged charge of a ready-burning composition, usually black powder, used to assist in the initiation of a propelling charge. (3) A device, containing such a composition, used to amplify the initiation of a primer in the functioning of a fuze.

**Initiator.** A device used as the first element of an explosive train, such as a detonator or squib, that, upon receipt of the proper mechanical or electrical impulse, produces a burning or detonating action. It generally contains a small quantity of a sensitive explosive.

**Integrator.** A device that sums discrete or variable quantities over some specified range of the selected or independent variable.

**Junghans Escapement.** A type of tuned, two-center escapement named after its original designer.

**Logic Device.** One of the general category of devices that perform logic functions; for example, AND, NAND, OR, and NOR. They can permit or inhibit signal transmission with certain combinations of control signals.

Loran. A long-range pulsed radio aid to navigation, the position lines of which are determined by the measurement of the difference in the time of arrival of initially synchronized pulses.

Monostable Device. A device with one stable and one astable state. When triggered, it goes from the stable to the astable state, and later returns to its stable state.

Oscillator, Electronic. A circuit for producing an alternating waveform, the output frequency\or repetition rate of which is determined by the internal characteistics of the device.

Oscillator, Flueric. A flueric device for producing a periodic change in fluid pressure, velocity, etc. The output frequency or repetition rate is determined by the internal characteristics of the device.

Pallet. The part of an escapement which intercepts the teeth of the escape wheel.

Primer. A relatively small and sensitive initial explosive train component that, on being actuated, initiates functioning of the explosive train but will not reliably initiate high explosive charges.

Pyrotechnic Composition. A physical mixture of finely powdered fuel and oxidant, with or without additives, to produce heat, light, smoke, or special effects when ignited. The mixture is consumed in the process of burning.

Relay. An explosive train component that provides the required explosive energy to initiate reliably the next element in the train. It is especially applied to small charges that are initiated by a delay element and, in turn, cause the functioning of a detonator.

Runaway Escapement. An untuned, two-center escapement in which the escape wheel drives an inertial mass alternately clockwise and counterclockwise. The oscillatory motion of the mass regulates the escape wheel motion. Also called verge escapement.

Safing and Arming Device. A mechanism that prevents or allows an explosive train to operate.

Setback. The relative rearward force of inertia which is created by a forward acceleration of a projectile or missile during its launching phase. Setback is directly proportional to the acceleration and mass of the parts being accelerated.

Sidereal Time. The time of the rotation of the earth as measured from a clock star, instead of from the sun.

Solion. An electrochemical timer based on an oxidation-reduction process that uses ions in solution instead of electrons as the charge carriers.

Spin. The rotation of a projectile or missile about its longitudinal axis to provide stability during flight.

Timer. A programming device that controls the time interval between an input signal and an output event or events. See also: specific timers.

Timer, Analog. A timer that obtains its programmed time delay by directly measuring the magnitude of some continuously variable parameter, often the rise in voltage from zero to some predetermined value.

Timer, Digital. A timer (1) having a numeric display of the time, or (2) measuring time by counting the output pulses of an oscillator.

Timer, Electronic. A timer using electronic digital or analog elements.

Timer, Electrochemical. A timer based on the constant rate of a chemical reaction at a constant current.

Timer, Flueric. A timer using flueric digital or analog elements.

Timer, Interval. A timer that switches electric circuits on or off for the duration of the preset time interval.

**Timer, Mechanical.** A timer based on the constant rate of a clockwork escapement.

**Timer, Pyrotechnic.** A timer based on the linear burning of a column of pyrotechnic delay material.

**Timer, Sequential.** A timer in which either each interval is initiated by the completion of the preceding interval or a sequence of timed outputs is produced. The intervals may be adjusted independently.

**Transducer.** A device that converts signals from one medium to an equivalent signal in a second medium.

**Universal Time.** The time calculated from the rotation of the earth (i.e., Greenwich Mean Time) but corrected for certain irregularities, including movement of the earth's axis and rotation.

**Verge Escapement.** *See:* **Runaway Escapement**

**Wall Attachment.** The phenomenon in flueric devices in which a fluid stream flowing close to a surface is deflected to that surface even if the surface is at a considerable angle to the original direction of flow.

# INDEX

## Overview by Chapter and Section

## INDEX (Con't.)

# INDEX (Con't.)

# INDEX (Con't.)

## INDEX (Con't.)

# INDEX (Con't.)

# INDEX (Con't.)

**INDEX (Con't.)**

# INDEX (Con't.)

## INDEX (Con't.)

# INDEX (Con't.)